Strongly Correlated Electron Systems (SCES 2020)

Journal of Physics: Conference Series Volume 2164

Campinas, Brazil
27 September – 1 October 2021

ISBN: 978-1-7138-5135-6
ISSN: 1742-6588

Printed from e-media with permission by:

Curran Associates, Inc.
57 Morehouse Lane
Red Hook, NY 12571

Some format issues inherent in the e-media version may also appear in this print version.

Institute of Physics
Dirac House, Temple Back
Bristol BS1 6BE UK

Phone: 44 1 17 929 7481
Fax: 44 1 17 920 0979

techtracking@iop.org

Additional copies of this publication are available from:

Curran Associates, Inc.
57 Morehouse Lane
Red Hook, NY 12571 USA
Phone: 845-758-0400
Fax: 845-758-2633
Email: curran@proceedings.com
Web: www.proceedings.com

TABLE OF CONTENTS

The 2020 Strongly Correlated Electron Systems Conference

Pascoal J. G. Pagliuso, Cris Adriano, Eduardo Miranda

Gleb Wataghin Physics Institute, The University of Campinas, R. Sergio Buarque de Holanda, 777, CEP 13083-859, Campinas, SP, Brazil

The International Conference on Strongly Correlated Electrons systems (SCES) is one of the most traditional conferences in Condensed Matter Physics worldwide. SCES continues to bring together, in every edition, outstanding scientists working in the frontiers of the complex and advanced phenomena of this area. The SCES 2020 Edition was planned to be an in-person event in Guarujá, SP, Brazil in September of 2020 as a continuation of the successful series of the SCES conferences: Sendai ('92), San Diego ('93), Amsterdam ('94), Goa ('95), Zurich ('96), Paris ('98), Nagano ('99), Ann Arbor ('01), Krakow ('02), Karlsruhe ('04), Vienna ('05), Houston ('07), Buzios ('08), Santa Fe ('10), Cambridge ('11), Tokyo ('13), Grenoble ('14), Hangzhou ('16), Prague ('17) and Okayama ('19). Additionally, every three years since 1997, SCES has been joining the International Conference on Magnetism (ICM) held in: Cairns ('97), Recife ('00), Rome ('03), Kyoto ('06), Karlsruhe ('09), Busan ('12), Barcelona ('15), and San Francisco ('18).

However, due to the Covid-19 pandemic, the SCES 2020 was postponed to 2021 and finally became a fully online event happening from 09/27 to 10/01/2021. The online SCES 2020/21 conference covered the following topics:

- Heavy fermion systems
- Materials Design and New/Advanced materials
- Kondo effect and valence fluctuations
- Quantum phase transitions and related phenomena
- Unconventional superconductivity
- Metal insulator transitions
- Strong spin-orbit interaction in correlated systems
- CEF effects and multipolar ordering in SCES
- Theoretical methods for strong correlations
- Correlated topological phases
- Emergent Phenomena at nanoscale
- Quantum magnetism and frustration
- Strong correlations in actinides
- Non-equilibrium phenomena in strongly correlated systems

Journal of Physics: Conference Series **2164** (2022) 011001 doi:10.1088/1742-6596/2164/1/011001

- Multiferroics and related materials
- SCES materials for applications
- Large research-facilities and Novel techniques for SCES investigation

This Proceedings volume reflects well the various areas that were represented in the conference. The conference had 462 attendees from 32 different countries as shown in Figure 1.

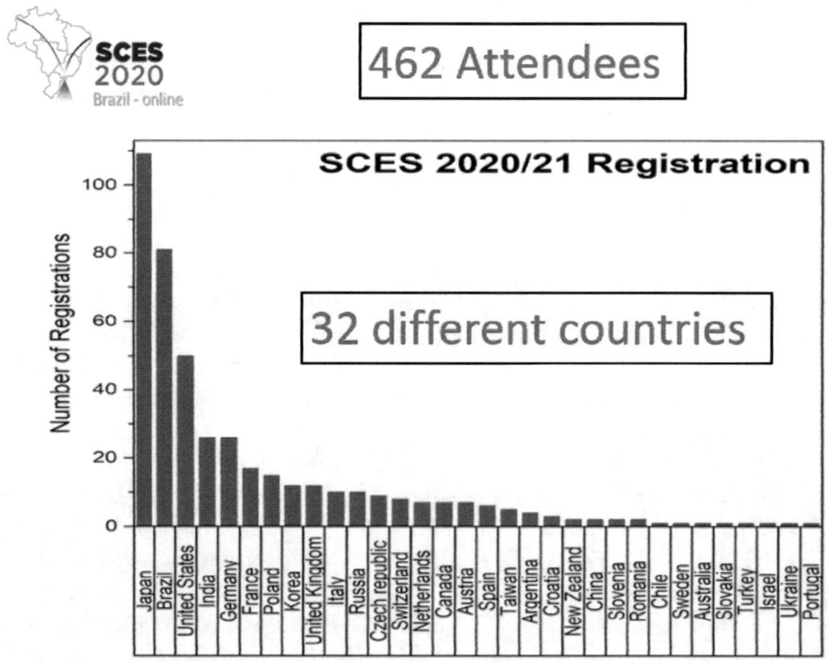

Figure 1. Number of SCES 2020/21 participants by country

Furthermore, 383 abstracts were accepted to be part of the SCES 2020/21 Program, 12% of them submitted by invited speakers, 52% accepted for oral contributed talks and 36% for poster presentation, as shown in Figure 2.

During the SCES2020/21 fully online conference, we had:

- 55 outstanding invited talks
- 171 contributed talks
- 146 poster contributions
- 15 Best Poster Award Winners
- The Prize Ceremonies for the Bryan R. Coles and the Nevill F. Mott Prize awards.

The winner of the Bryan R. Coles award was Prof. Ming Yi from Rice University while the winner for the Nevill F. Mott award was Prof. Yuxuan Wang from the University of Florida.

The 15 best poster winners, selected from the three Poster Sessions during the event, were:

- Lucas Cesar Gomes Squillante (Unesp, Brazil)
- Kazuki Yamamoto (Kyoto University, Japan)
- Virna Kisiek (Institute of Physics Zagreb, Croatia)

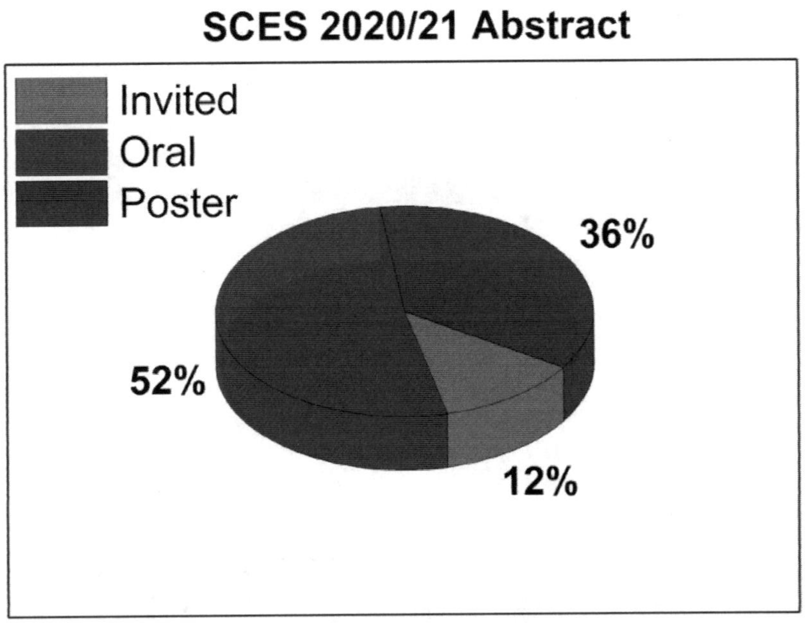

Figure 2. Distribution of SCES 2020/21 accepted abstracts

- Dana Popescu (National Institute of Materials Physics, Romania)
- Utami Widyaiswari (Meson Science Laboratory, Japan)
- Rikako Yamamoto (Hiroshima University, Japan)
- Hiroshi Tanida (Toyama Prefectural University, Japan)
- Tymoteusz Tula (University of Kent, UK)
- Lutiene Fernandes Lopes (Universidade Federal do Rio Grande do Sul, Brazil)
- Chihiro Tabata (Kyoto University, Japan)
- WooRi Ju (Chonnam National University, South Korea)
- Raquel Sánchez-Barquilla (Universidad Autónoma de Madrid, Spain)
- Bodry Tegomo Chiogo (University Of Lorraine, France)
- Hiroki Hanate (Kyushu Institute of Technology, Japan)
- Sawani Datta (Tata Institute of Fundamental Research, India)

Finally, we regret that SCES 2020 could not happen in person in Guarujá, SP, allowing for all the direct interactions that have made every previous SCES conference very memorable. However, we are certain that the SCES 2020/21 fully online conference was an exciting and inspiring conference where the enthusiasm and motivation with the great science did overcome the lack of in-person interactions. We hope we can all meet again in person sometime soon! Maybe during SCES 2022 in Amsterdam or other future SCES conferences.

Journal of Physics: Conference Series **2164** (2022) 011001 doi:10.1088/1742-6596/2164/1/011001

The international advisory committee

The prize committee

- Harima Hisatomo (Chair, Kobe University, Japan)
- Joe Thompson (Los Alamos National Lab, USA)
- Karen Hallberg (Bariloche Atomic Centre, Argentina)
- John Mydosh (Leiden University, The Netherlands)
- Laura Greene (National High Magnetic Field Lab, USA)
- Peter Riseborough (Temple University, USA)
- Hilbert v. Löhneysen (KIT Physikalisches Institut, Germany)

The publication Committee

- Eduardo Miranda (Unicamp, Brazil)
- Marcos Avila (UFABC, Brazil)
- Eduardo M. Bittar (CBPF-RJ, Brazil)
- Tobias Micklitz (CBPF-RJ, Brazil)
- Maria C. O. Aguiar (UFMG, Brazil)
- José A. Hoyos (USP-IFSC, Brazil)
- Eric C. Andrade (USP-IFSC, Brazil)
- Priscila F. S. Rosa (LANL, USA)
- José G. S. Duque (UFS, Brazil)
- Antonio Jefferson da Silva Machado (USP-EEL, Brazil)
- Adilson Jesus A. de Oliveira (UFSCar, Brazil)

The organizing committee

- Pascoal J. G. Pagliuso (Chair; Unicamp, Brazil)
- Cris Adriano (Co-chair; Unicamp, Brazil)
- Eduardo Miranda (Co-chair; Unicamp, Brazil)
- Maria Eugênia Fortes Brollo (Unicamp, Brazil)
- Mucio Amado Continentino (CBPF-RJ, Brazil)
- Elisa Maria Baggio-Saitovich (CBPF-RJ, Brazil)
- Eduardo Matzenbacher Bittar (CBPF-RJ, Brazil)
- Magda Bittencourt Fontes (CBPF-RJ, Brazil)
- Marcos de Abreu Avila (UFABC, Brazil)
- Ricardo D. Reis (LNLS, Brazil)

The local committee

- Maria Eugênia Fortes Brollo (Unicamp, Brazil)
- Ricardo Rodrigues Urbano (Unicamp, Brazil)
- Iakov Kopelevitch (Unicamp, Brazil)
- Carlos Rettori (Unicamp, Brazil)
- Eduardo Granado Monteiro da Silva (Unicamp, Brazil)
- Amir Ordacgi Caldeira (Unicamp, Brazil)
- Oscar Ferreira de Lima (Unicamp, Brazil)
- Guillermo Gerardo Cabrera Oyarzun (Unicamp, Brazil)

- Flavio César Guimarães Gandra (Unicamp, Brazil)
- Kleber R. Pirota (Unicamp, Brazil)
- Fanny Beron (Unicamp, Brazil)
- Marcelo Knobel (Unicamp, Brazil)
- Fernando Garcia (USP-SP, Brazil)
- Luis Gregorio Dias da Silva (USP-SP, Brazil)
- Julio Larrea (USP-SP, Brazil)
- Rafael Freitas (USP-SP, Brazil)
- Luiz Nunes de Oliveira (USP-IFSC, Brazil)
- José Abel Hoyos Neto (USP-IFSC, Brazil)
- Eric de Castro e Andrade (USP-IFSC, Brazil)
- José Carlos Egues de Menezes (USP-IFSC, Brazil)
- Marcos de Abreu Avila (UFABC, Brazil)
- Leticie Mendonça Ferreira (UFABC, Brazil)
- Raquel de Almeida Ribeiro (UFABC, Brazil)
- Wilson Aires Ortiz (UFSCar, Brazil)
- Aline Ramires Neves de Oliveira (IFT-Unesp, Brazil)
- Mucio Amado Continentino (CBPF-RJ, Brazil)
- Elisa Maria Baggio-Saitovich (CBPF-RJ, Brazil)
- Eduardo Matzenbacher Bittar (CBPF-RJ, Brazil)
- Magda Bittencourt Fontes (CBPF-RJ, Brazil)
- Tobias Micklitz (CBPF-RJ, Brazil)
- Raimundo Rocha dos Santos (UFRJ, Brazil)
- Thereza Cristina de Lacerda Paiva (UFRJ, Brazil)
- Eduardo Marino (UFRJ, Brazil)
- Mohammed El Massalami (UFRJ, Brazil)
- Mauro Doria (UFRJ, Brazil)
- Tatiana Rappoport (UFRJ, Brazil)
- Daniel Gustavo Barci (UERJ, Brazil)
- Caio Henrique Lewenkopf (UFF, Brazil)
- Daniel Adrian Stariolo (UFF, Brazil)
- Maria Carolina de Oliveira Aguiar (UFMG, Brazil)
- Raimundo Lora Serrano (UFU, Brazil)
- José Candido Xavier (UFU, Brazil)
- Daniele Freitas (UFU, Brazil)
- José Gerivaldo dos Santos Duque (UFS, Brazil)
- Nelson Orlando Moreno Salazar (UFS, Brazil)
- Álvaro Ferraz Filho (UFRN/IIP, Brazil)
- Rodrigo Gonçalves Pereira (UFRN/IIP, Brazil)
- Paulo Pureur Neto (UFRGS, Brazil)
- Sergio Garcia Magalhães (UFRGS, Brazil)

Journal of Physics: Conference Series **2164** (2022) 011001 doi:10.1088/1742-6596/2164/1/011001

- Narcizo Marques de Souza Neto (LNLS, Brazil)
- Ricardo D. Reis (LNLS, Brazil)
- Wendell Simões e Silva (LNLS, Brazil)
- José Albino de Oliveira Aguiar ? (UFPE, Brazil)
- Sergio Rezende (UFPE, Brazil)
- Igor Padilha (UFAM, Brazil)
- Fernanda Deus (UERJ, Brazil)
- Wei Chen (PUC-RJ, Brazil)
- Enzo Granato (INPE, Brazil)
- Daniel Reyes (IME, Brazil)

Peer Review Statement

All papers published in this volume have been reviewed through processes administered by the Editors. Reviews were conducted by expert referees to the professional and scientific standards expected of a proceedings journal published by IOP Publishing.

1. Type of peer review: single blind
2. Conference submission management system: Morressier
3. Number of submissions received: 80
4. Number of submissions sent for review: 80
5. Number of submissions accepted: 73
6. Acceptance Rate (Submissions Accepted / Submissions Received × 100): 91 %
7. Average number of reviews per paper: 1
8. Total number of reviewers involved: 25
9. Contact person for queries:
   Name: Eduardo Miranda
   Affiliation: University of Campinas - Physics
   Email: emiranda@ifi.unicamp.br

* means value has been edited

Orbital selective dynamics in Fe-based systems using time-resolved ARPES

Kalobaran Maiti

Tata Institute of Fundamental Research, Homi Bhabha Road, Mumbai - 400005, India

*E-mail : kbmaiti@tifr.res.in

Abstract. Exoticity in material properties is often linked to the complex interplay of spin, charge, orbital and lattice degrees of freedom that makes the study of the origin of such exoticity difficult. The outstanding issue is to disentangle parameter space and reveal the underlying physics. We propose a unique method to excite electrons of a selected symmetry without significant effect on other electrons using polarized pump light pulse in a pump-probe experiment. Using this technique, we show that the relaxation of itinerant electrons occurs faster than the local electrons; the first experimental identification of the orbital selective electron dynamics in a complex correlated system, $EuFe_2As_2$. In another experiment, we discover that the magnetic order in $CaFe_2As_2$ can be melted selectively without significant effect on electrons in the other energy bands. These results provide two important conclusions; (i) magnetism in Fe-based systems may not be linked to the phase space occupied by other electrons and (ii) polarized pump excitation in a pump-probe experiment is a novel method to study orbital selective dynamics.

Introduction

Discovery of superconductivity [1] in Hg in 1911 nucleated a new field of research that has been growing continuously and rapidly even after more than a century is over. Most of the materials found in the early years have superconducting transition temperature, T_c very low and the underlying physics could be captured by the condensation of cooper pairs formed due to electron-phonon coupling. The theory that captures these features is called BCS (Bardeen-Cooper-Schrieffer) theory [2]. In 1986, a new class of materials based on Cu-oxides found to show superconductivity at significantly high temperatures [3]. Extensive research in the cuprate superconductors led to finding of T_c much higher than liquid nitrogen temperature of 77 K which enabled emergence of advanced technology. The properties of these so-called high T_c superconductors cannot be captured by the BCS theory and are known as *unconventional superconductors* while the BCS superconductors are termed as *conventional superconductors*. In 2006, another new class of superconductors discovered which are Fe-based compounds [4]; there are several other classes of superconductors discovered in the meantime which are not mentioned here. The discovery of superconductivity in Fe-based systems led to significant puzzle as Fe is a highly magnetic element and superconductors are perfect diamagnets. There are several interesting common features and differences among these two classes of materials. For example, the parent compounds in cuprates are antiferromagnetic Mott insulators, where the insulating property arises due to the strong electron-electron Coulomb repulsion, U compared to the width, W of the conduction band. The charge carrier doping suppresses antiferromagnetic order and superconductivity emerges beyond a critical doping. It appears that the antiferromagnetic interaction and Coulomb repulsions are important to derive superconductivity in these materials. While origin of superconductivity in these systems remains an unresolved problem, major attention has been paid in understanding the puzzles found in the normal phase of these materials, such as, pseudogap phase, strange metallicity etc.

On the other hand, the parent Fe-based compounds are metallic and show spin density wave (SDW) phase in the ground state. These materials show structural transition from tetragonal to orthorhombic phase along with a magnetic transition to a spin density wave (SDW) ground state. Charge carrier doping in these systems suppresses both the structural and magnetic transitions and superconductivity appears at a critical doping. Interestingly, many materials show coexistence of magnetic order and superconductivity in these systems. One can achieve superconductivity via application of pressure too. The structural transition from the tetragonal phase having C_4 rotational symmetry to orthorhombic phase having C_2 symmetry often renormalizes the energy band dispersions and lifts the band degeneracies [5,6]. The lower symmetric phase is termed as nematic phase. It is observed that emergence of superconductivity in these materials also accompanies significant structural reorganizations. Thus, the superconductivity in Fe-based systems appears to be significantly more complex and involves fluctuations of spin, charge and nematic degrees of freedoms. It is a formidable task to disentangle these degrees of freedom and study their underlying physics. Here, we propose a method based on time- and angle-resolved photoemission spectroscopy (trARPES) which can be used to probe various degrees of freedom selectively.

Experimental method:
The experimental results presented here are based on the experiments carried out at CITIUS, University of Nova Gorica, Slovenia as reported elsewhere [7-9]. The time- and angle-resolved photoemission study was done using a mode-locked Ti:sapphire laser (photon energy 1.5 eV, pulse duration 40 *fs* and repetition rate 5 kHz). The laser beam was split into two parts using a beam splitter; the major part was used to generate higher-order harmonics to be used as a probe pulse. The other part was used as a pump pulse which goes through a polarizer to produce linearly polarized light and a delay stage for tuning the pump-probe time

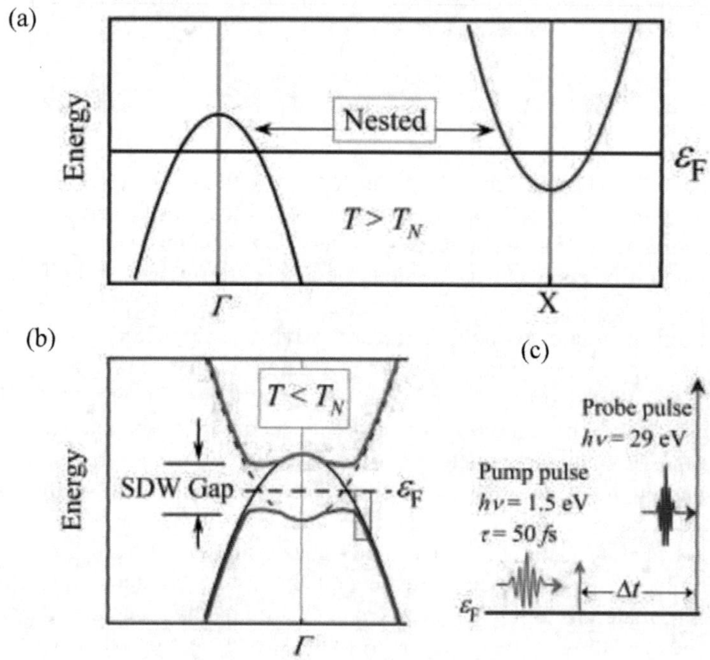

Fig. 1: (a) Schematic of the band structure of 122 materials in the paramagnetic phase. (b) SDW phase due to the nesting of the Fermi surfaces. (c) Schematic representation of trARPES.

delay, Δt. The photoemission measurements were performed using a R3000 electron analyzer and a closed cycle helium cryostat. The pressure during the measurements were maintained below 1×10^{-10} Torr. The energy resolution was set to 100 meV to get adequate signal to noise ratio. Single crystalline $CaFe_2As_2$ samples were grown by the Sn-flux method at TIFR, Mumbai, India. The samples were cleaved in-situ to generate a clean surface before every measurement. The dynamics of the photoexcited electrons is studied by varying the time delay between the pump and probe.

In Fig. 1(a), we show a schematic diagram of a hole band around Γ (0,0)-point and an electron band around X (π,π)-point of 122 class of materials which satisfy nesting condition and are responsible for the SDW ground state. The band structure in the SDW phase is shown in Fig. 1(b) exhibiting an energy gap due to the supercell of the magnetically ordered phase. The pump pulse shown in Fig. 1(c) excites electrons of selected momentum shown by a box in Fig. 1(b). The excited electrons gain an energy of 1.5 eV and occupy unoccupied states above the Fermi level. These hot electrons will relax fast by dissipating their energies via electron-electron scatterings. This process occurs at a very fast time scale and cannot be probed by the time scale of our measurements. In this process, the unoccupied bands above the Fermi level gets populated which are probed by the probe pulse coming at a time delay, Δt as shown in Fig. 1(d). Since the photoemission process occurs at an extremely fast time scale (~ attosecond), much faster than the relaxation time of various degrees of freedom in solid, the probe pulse essentially provides a snapshot of the electronic structure before the system relaxes via various dissipation processes. Thus, change in band structure and how it reconstructs can be probed by this process efficiently.

Results and discussions

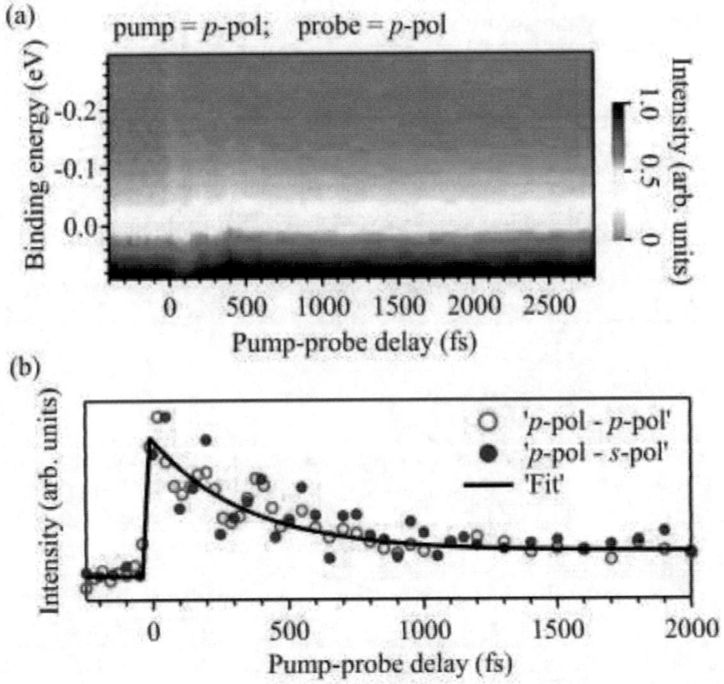

Fig. 2: (a) trARPES data of $EuFe_2As_2$ collected at 210 K using *p*-pol pump and *p*-pol probe pulse. (b) Integrated spectral intensity between -0.1 to -0.2 eV as a function of pump-probe delay.

Photoemission spectroscopy is a powerful and arguably the only tool to probe the electronic structure directly. The basic process involves the photo-electric effect, in which the sample under investigation is irradiated with a photon beam. The intensity of the excited electrons depends on the photoexcitation cross section which can be derived using Fermi's Golden rule; the photoemission cross section can be expressed as $\sigma(\epsilon) = |\langle\psi_f|H'|\psi_i\rangle|^2\delta(\epsilon_f - \epsilon_i - h\nu)$. Here, ϵ_f and ϵ_i are the energy of the final state, $|\psi_f\rangle$ and initial state, $|\psi_i\rangle$, respectively. H' $(= \vec{A}.\vec{p})$ is the perturbation due to the incident light. If we assume that the detector is placed on the *z*-axis and the *xz*-plane is the photoemission plane which contains incident light, the final state, $|\psi_f\rangle$ will be even with respect to the mirror reflection on *xz*-plane. $\sigma(\epsilon)$ will be nonzero when the integral, $\langle\psi_f|A.p|\psi_i\rangle = \int \psi_f^* A.p\psi_i d\tau \neq 0$ that requires $A.p|\psi_i\rangle$ to be even. In Fe-based systems, the hole pockets around the Γ-point are constituted by d_{xy}, d_{yz} and d_{xz} orbitals. The orbitals, d_{xy} and d_{yz} are odd with respect to the *xz*-mirror plane, and d_{xz} will be even. Therefore, if linear polarization of the incident light is perpendicular to the *xz*-plane (here defined as *s*-polarization or '*s*-pol'), it will be an odd function and will provide contributions for d_{xy} and d_{yz} electron excitations. On the other hand, the in-plane light polarization (defined as *p*-polarization or '*p*-pol') will give d_{xz} contributions. Such symmetry selection rule is often used to derive the symmetry of the initial states in various photoexcitation processes.

To verify the applicability of this selection rule to study orbital selective dynamics, we show a typical excitation spectrum in Fig. 2(a) where the sample, $EuFe_2As_2$ is excited with *p*-pol pump pulse at the sample temperature of 210 K and probed by *p*-pol probe pulse. The color scale represents the intensity of the momentum-integrated spectral functions along the ΓX line. The figure shows significant population of electrons much above the Fermi level due to pump-excitations. The modulation of intensities occurs due to lattice vibrations leading to a time-dependent change in hybridization. In order to probe the decay process, we have integrated the intensities between 0.1 and 0.2 eV above the Fermi level. This energy range will have negligible contributions due to Fermi-Dirac distribution functions. The integrated intensities for both 'p-pol' and 's-pol' probe pulse is shown in Fig. 2(b) as a function of pump-probe delay. The relaxation dynamics is found to be identical in both cases. This suggests that either the electron dynamics is not symmetry dependent or the technique is not sensitive to the detection of selective excitations.

Since the pump pulse excite electrons at the ground state, we planned to study the polarization dependence of the pump pulse while probe pulse probes the excited intermediate states. The process can be understood as follows. The intensity of the excited states due to pump pulse excitations will be nonzero if and only if

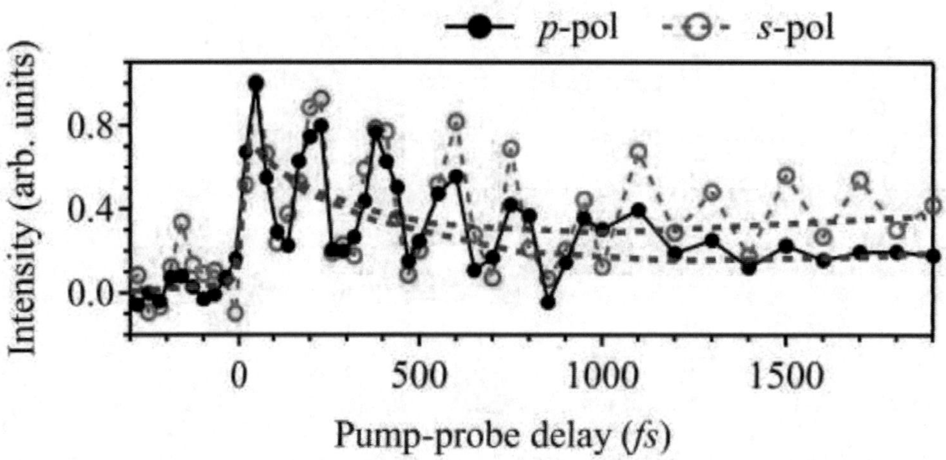

Fig. 3: Excitation dynamics using *p*-pol (solid circles) and *s*-pol (open circles) pump pulse. Sample temperature 210 K and probe energy 29 eV.

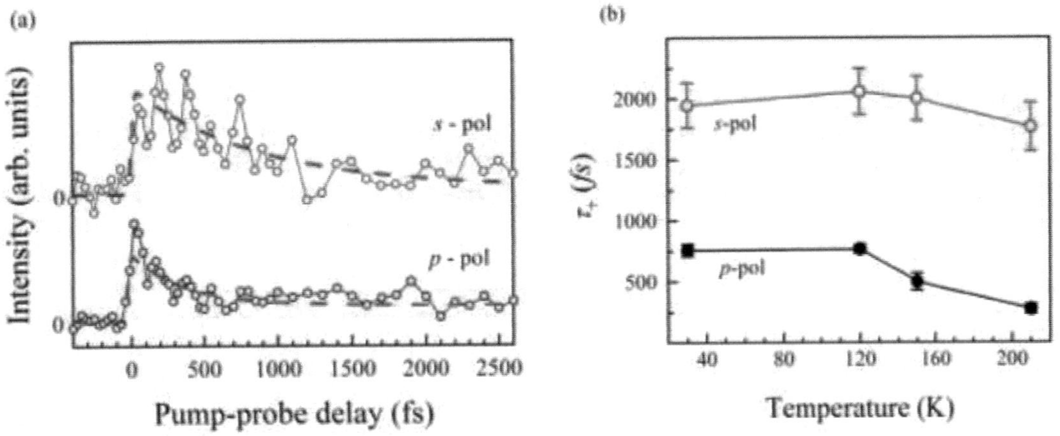

Fig. 4: (a) trARPES response of $EuFe_2As_2$ after 's-pol' and 'p-pol' pump excitation at a sample temperature of 30 K. (b) Decay time constants of hot electrons for both cases.

$\langle\psi_f|A.p|\psi_i\rangle = \int \psi_f^* A.p\psi_i d\tau \neq 0$ for pump excitations. Now, in the case of *s*-pol light, $\vec{A}$ is perpendicular to the *xz*-plane that makes $A.p|\psi_i\rangle$ zero for d_{xz} orbital excitation as $\vec{p}|\psi_i\rangle$ will be essentially lying in the *xz*-plane. Similarly, 'p-pol' light will provide major contribution from d_{xz} orbital for an incidence angle close to 45°. Such qualitative arguments, consistent with dichroic effect, indicates that polarized pump pulse is a good tool to excite electrons of selected symmetry. The experimental results are shown in Fig. 3 exhibiting significant differences in the relaxation time for 'p-pol' and 's-pol' cases.

Utilizing the above scenario, we have probed the excitation dynamics of $EuFe_2As_2$ and $CaFe_2As_2$. The results for $EuFe_2As_2$ are shown in Fig. 4. Experimental data at a sample temperature of 30 K shown in Fig. 4(a) exhibit significantly different decay time for two polarizations of the pump pulse. The temperature evolution of the decay time is shown in Fig. 4(b). While the decay time for d_{xz} electrons is about 500 *fs* at room temperature that for d_{xy}/d_{yz} electrons is close to 2000 *fs*. From the band structure calculations and ARPES studies, it is well known that the width of the d_{xy} band is significantly narrower than d_{xz}/d_{yz} bands

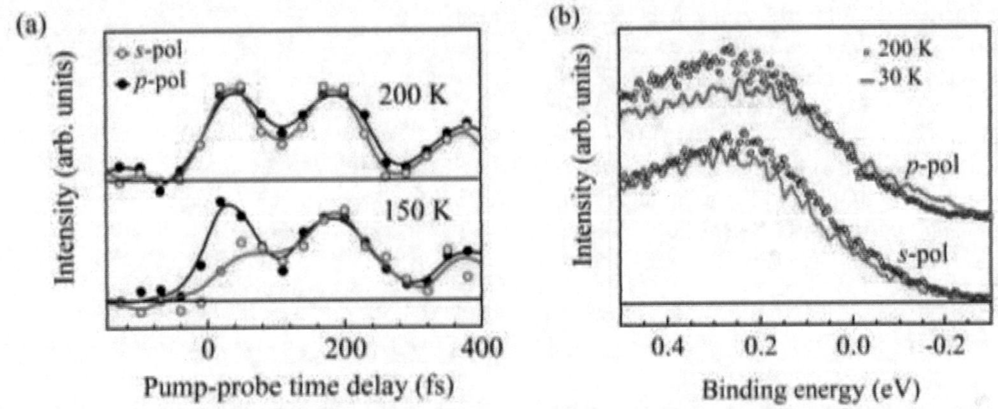

Fig. 5: (a) trARPES response of $CaFe_2As_2$ after 's-pol' and 'p-pol' pump excitation at a sample temperature of 200 K and 150 K. (b) Energy distribution curves at Γ at 50 *fs* delay after pump excitations.

[10]. Since the local electrons will be less exposed to the crystal lattice, it will be less favorable to dissipate energy by electron-phonon coupling. With the onset of antiferromagnetic order, there will be a gap in the

β-band which participate in the magnetic ordering. We observe an enhancement of decay time across the antiferromagnetic transition as expected from the behavior at higher temperatures.

We now investigate the electron dynamics close to the excitation threshold ($\Delta t \sim 0$). The results for $CaFe_2As_2$ are shown in Fig. 5. The trARPES response collected at sample temperatures of 200 K and 150 K are shown in Fig. 5(a). While the '*p*-pol' and '*s*-pol' excitations provide identical description at 200 K, the data at 150 K (below Neel temperature) are very different. The peak at 50 fs is intense in 'p-pol' case while it is almost absent in the 's-pol' case. From the ARPES studies it is well known that SDW phase leads to a gap in the d_{xz}/d_{yz} band while d_{xy} band remains unaffected [11,12]. From the above results, it appears that s-pol pump pulse response becomes different across the magnetic transition while p-pol pump excitations look very similar to the paramagnetic phase. In order to probe this scenario further, we show the energy distribution curves (EDC) at the Γ-point collected from the spectra at 50 *fs* delay time. The spectra in the 's-pol' case show signature of SDW gap while the 'p-pol' data looks very similar to the paramagnetic phase. This suggests that 'p-pol' light melted the magnetic order but 's-pol' light could not do so. This is outstanding and demonstrate that one can selectively melt the magnetically ordered phase and that the magnetic order and structural order may not be linked.

Conclusion

In summary, we present here a method to selectively excite electrons using a time- and angle-resolved photoemission spectroscopy. We demonstrate that polarized probe pulse is unable to distinguish the symmetry of the electrons in the ground state presumably due to the fact that electron-phonon coupling mixes up the symmetries at the intermediate states of the pump-probe process. On the other hand, polarized pump pulse selectively excites electrons from the ground state of the material and hence is able to manifest the orbital selective signal. We have shown the excitation dynamics of $EuFe_2As_2$ and $CaFe_2As_2$ using this technique. The results in $EuFe_2As_2$ show that local electrons have significantly longer decay time compared to the itinerant ones [7]. The order in $CaFe_2As_2$ could be melted selectively using this technique which also suggests that the magnetic and structural order in these systems may not be linked [8].

References

[1] Onnes H K, 1911 Comm. Phys. Lab. Univ. Leiden, Nos. 119, 120, 122.
[2] Bardeen J, Cooper L N, Schrieffer J R, 1957 Phys Rev **108**, 1175.
[3] Bednorz J G, Müller K A, 1986 Z Phys B **64**, 189.
[4] Kamihara Y *et al*, 2006 J. Am. Chem. Soc. **128**, 10012.
[5] Maiti K, 2015 Pramana – J Phys **84**, 947.
[6] Ali K and Maiti K, 2018 Eur Phys J B **91**, 199
[7] Adhikary G *et al.*, 2018 Phys Rev B **98**, 205142
[8] Adhikary G *et al.*, arXiv: 2005.01013
[9] Grazioli C et al., 2014, Rev Sci Instrum 85, 023104.
[10] Adhikary G *et al.*, 2013 J Phys Condens Matter **25** 225701
[11] Ali K and Maiti K, 2017 Scientific Reports **7** 6298
[12] Ali K *et al.*, 2018 Phys Rev B **97**, 054505

Journal of Physics: Conference Series **2164** (2022) 012002 doi:10.1088/1742-6596/2164/1/012002

Nodal gaps from local interactions in Sr_2RuO_4

Aline Ramires
Paul Scherrer Institut, CH-5232 Villigen PSI, Switzerland
E-mail: aline.ramires@psi.ch

Abstract. Sr_2RuO_4 has been under intensive scrutiny over the past years after new NMR measurements unveiled that the superconducting state might be spin singlet. One of the best order parameter candidates in light of these new experiments is a chiral state with E_g symmetry. This order parameter, with a horizontal nodal line, has been overlooked given the strong two-dimensional character of the normal state electronic structure. Recently, a phenomenological proposal based on local interactions showed that an even-parity orbital-antisymmetric spin-triplet (OAST) chiral state can be stable in Sr_2RuO_4 once momentum-dependent spin-orbit coupling is properly taken into account. Here we discuss the origin of the nodes and dips in this order parameter as inherited from the normal state Hamiltonian, showing that a nodal gap can emerge out of purely local interactions and connect the presence of nodes with the superconducting fitness measure.

1. Introduction

Sr_2RuO_4 was for long thought to be the best candidate material to host a chiral p-wave superconducting state [1, 2]. Recent Knight shift experiments, displaying a significant reduction of the spin susceptibility across the superconducting transition, challenged this proposal [3, 4]. Complementary experiments such as Kerr effect [5] and muon spin resonance [6, 7] indicate time-reversal symmetry breaking in the superconducting state, and ultrasound attenuation experiments [8, 9, 10] suggest a multi-component order parameter. Together, these experiments suggest an order parameter in the E_g irreducible representation of the D_{4h} point group. This order parameter has two components, one with d_{xz} and another with d_{yz} symmetry. In its chiral form, usually referred to as chiral d-wave, or $d_{xz} + id_{yz}$ state, the order parameter breaks time-reversal symmetry. This order parameter candidate was for a long time overlooked within the assumption that a superconducting state with horizontal line nodes requires strong inter-layer interactions, what is at a first sight in contradiction with the strongly two-dimensional Fermi surfaces of Sr_2RuO_4 [11].

Recently, the author and collaborators have proposed a picture with local pairing in which even-parity orbital-antisymmetric spin-triplet (OAST) chiral state is stable [12]. The main ingredients in this phenomenological theory are strong Hund's coupling and a faithful description of the normal state in three dimensions, including momentum dependent spin-orbit coupling. The projected gap at the Fermi surfaces has horizontal line nodes at the $k_z = \{0, \pm 2\pi\}$ planes, and minima along the diagonals $k_x = \pm k_y$, reminiscent of a $d_{x^2-y^2}$ gap. The aim of this proceeding article is to discuss how these qualitative features of the gap come about. The main aspect to be highlighted is the construction of the normal state and superconducting order parameters from a microscopic perspective, within the orbital basis. Once the analysis is translated to the band basis, these unexpected features emerge.

2. The normal state Hamiltonian

Following the notation introduced in [12], we can describe the normal state of Sr_2RuO_4 by a three-dimensional three-orbital model. We define $\Phi^\dagger_{\mathbf{k}} = (c^\dagger_{\mathbf{k},yz\uparrow}, c^\dagger_{\mathbf{k},yz\downarrow}, c^\dagger_{\mathbf{k},xz\uparrow}, c^\dagger_{\mathbf{k},xz\downarrow}, c^\dagger_{\mathbf{k},xy\uparrow}, c^\dagger_{\mathbf{k},xy\downarrow})$, where $c^\dagger_{\mathbf{k},\gamma s}$ creates an electron with momentum $\mathbf{k}$, orbital γ, and spin s, in terms of which we can write the single-particle Hamiltonian $\mathcal{H}_0 = \sum_{\mathbf{k}} \Phi^\dagger_{\mathbf{k}} \hat{H}_0(\mathbf{k}) \Phi_{\mathbf{k}}$, with $\hat{H}_0(\mathbf{k}) = \sum_{a=0}^{8} \sum_{b=0}^{3} h_{ab}(\mathbf{k})\, \hat{\lambda}_a \otimes \hat{\sigma}_b$. Here $\hat{\lambda}_a$ are Gell-Mann matrices encoding the orbital degrees of freedom (DOF), $\hat{\sigma}_b$ are Pauli matrices encoding the spin DOF ($\hat{\lambda}_0$ and $\hat{\sigma}_0$ are unit matrices), and $h_{ab}(\mathbf{k})$ are real even functions of momentum. In presence of spacial inversion and time-reversal symmetries, only 15 pairs of indexes (a, b) are symmetry allowed. These are described in detail in the supplementary material of Suh et al. [12]. With parameters obtained from first principle calculations, this model generates three almost cylindrical Fermi surfaces with a non-trivial orbital and spin texture, as schematically depicted in Figure 1.

For the analytical discussion below, it is useful to focus on the normal state Hamiltonian along specific planes. If we look along the XZ (YZ) plane we find only the β and γ bands, the former composed of d_{xz} (d_{yz}) and the later of d_{xy} orbitals. This allows us to construct simpler two-orbital models along these planes, $\mathcal{H}_0^{XZ(YZ)} = \sum_{\mathbf{k}} \Psi^\dagger_{\mathbf{k}} \hat{H}_0^{XZ(YZ)}(\mathbf{k}) \Psi_{\mathbf{k}}$, in the basis $\Psi^\dagger_{\mathbf{k}} = (c^\dagger_{\mathbf{k},xz(yz)\uparrow}, c^\dagger_{\mathbf{k},xz(yz)\downarrow}, c^\dagger_{\mathbf{k},xy\uparrow}, c^\dagger_{\mathbf{k},xy\downarrow})$, with $\hat{H}_0^{XZ(YZ)}(\mathbf{k}) = \sum_{a,b=0}^{3} \tilde{h}_{ab}^{XZ(YZ)}(\mathbf{k})\, \hat{\tau}_a \otimes \hat{\sigma}_b$, where the $\tilde{h}_{ab}^{XZ(YZ)}(\mathbf{k})$ are real even functions of momentum, and both $\hat{\tau}_a$ and $\hat{\sigma}_b$ are Pauli matrices encoding the orbital and the spin DOF, respectively ($\hat{\sigma}_0$ and $\hat{\tau}_0$ are identity matrices). In presence of time-reversal and inversion symmetries there are only six pairs of indexes (a, b). The symmetry-allowed terms are listed in Table 1, accompanied by the explicit form of the dominant terms for each plane, as well as the associated physical process.

Table 1. Symmetry-allowed (a, b) indexes in the two-orbital normal state Hamiltonians along the XZ- and YZ-planes and explicit form of dominant terms for each plane. Here we follow the notation introduced in [12] and define $t' = \frac{2}{3}(t_x^{11} + t_y^{11} + t_x^{33})$, $t'_x = t_x^{11} - t_x^{33}$, $t'_y = t_y^{11} - t_x^{33}$, and $\tilde{t} = 8t_{56z}^{SOC}$. The zero entries are defined by symmetry on the XZ and YZ planes.

(a, b)	$\tilde{h}_{ab}^{XZ}(\mathbf{k})$	$\tilde{h}_{ab}^{YZ}(\mathbf{k})$	Physical process
$(0, 0)$	$t'[1 + \cos(k_x)]$	$t'[1 + \cos(k_y)]$	intra-orbital hopping
$(1, 0)$	0	0	inter-orbital hopping
$(2, 1)$	η	0	atomic SOC
$(2, 2)$	0	η	atomic-SOC
$(2, 3)$	$-\tilde{t}\sin(k_x/2)\sin(k_z/2)$	$-\tilde{t}\sin(k_y/2)\sin(k_z/2)$	$\mathbf{k}$-SOC
$(3, 0)$	$t'_y + t'_x \cos(k_x)$	$t'_y + t'_x \cos(k_y)$	intra-orbital hopping

3. Local order parameters

In analogy to the normal state Hamiltonian, we parametrize the superconducting order parameters for the three orbital model as $\hat{\Delta}(\mathbf{k}) = \sum_{a,b} d_{ab}(\mathbf{k})\, \hat{\lambda}_a \otimes \hat{\sigma}_b\, (i\hat{\sigma}_2)$, with momentum-dependent functions $d_{ab}(\mathbf{k})$ carrying information about its magnitude and phase. Each order parameter component is labelled by a pair of indexes $[a, b]$ (note the different brackets used for the parametrization of the terms in the normal state Hamiltonian and for the order parameter).

In the three orbital model, the proposed chiral state is dominated by [12]:

$$\hat{\Delta}_{E_g}(\mathbf{k}) = d_0[\hat{\lambda}_5 \otimes \hat{\sigma}_3 + i\hat{\lambda}_6 \otimes \hat{\sigma}_3](i\hat{\sigma}_2). \quad (1)$$

In the microscopic basis this order parameter is momentum independent (or s-wave) with magnitude d_0, antisymmetric in the orbital DOF (given the explicit form of the Gell-Mann matrices λ_5 and λ_6), and a spin triplet state, with the d-vector along the z-axis.

Along the $XZ(YZ)$ plane we can write the local gap matrices as $\hat{\Delta}^{XZ(YZ)}(\mathbf{k}) = \sum_{ab} d_{ab}^{XZ(YZ)}(\mathbf{k})\, \hat{\tau}_a \otimes \hat{\sigma}_b\, (i\hat{\sigma}_2)$. Each order parameter component is labelled by a pair of indexes with a label concerning the plane $[a,b]_p$, with $p = \{XZ, YZ\}$. Here we focus on local, s-wave, order parameters, which should satisfy the condition imposed by fermionic anti-symmetry, $\hat{\Delta}(\mathbf{k}) = -\hat{\Delta}^T(-\mathbf{k}) \Rightarrow \hat{\Delta} = -\hat{\Delta}^T$, so we consider only order parameters associated with anti-symmetric matrices. These are listed in Table 2.

The orbitally antisymmetric order parameter, when projected into the XZ (YZ) plane takes the form $[2,3]_{XZ(YZ)}$, as $\hat{\tau}_2$ is the only anti-symmetric 2×2 matrix. More specifically, the $[6,3]$ component corresponding to an antisymmetric pair of d_{xz} and d_{xy} orbitals is associated with $[2,3]_{XZ}$ along the XZ plane. Conversely, the $[5,3]$ component corresponding to an antisymmetric pair of d_{yz} and d_{xy} orbitals is associated with $[2,3]_{YZ}$ along the YZ plane.

4. Origin of line nodes

For this discussion, the simple 2×2 models along the planes $p = \{XZ, YZ\}$ are going to be useful so we can proceed analytically. As the matrix structures are the same for both planes, we drop the p superscript from here on. In order to determine the gap in the band basis, it is useful to look directly at the eigenvalaues of the BdG Hamiltonian:

$$\hat{H}_{BdG}(\mathbf{k}) = \begin{pmatrix} \hat{H}_0(\mathbf{k}) & \hat{\Delta}(\mathbf{k}) \\ \hat{\Delta}^\dagger(\mathbf{k}) & -\hat{H}_0^*(-\mathbf{k}) \end{pmatrix}, \quad (2)$$

where $\hat{H}_0(\mathbf{k})$ is the normal-state Hamiltonian and $\hat{\Delta}(\mathbf{k}) = \Delta_0 \hat{\tau}_c \otimes \hat{\sigma}_d (i\hat{\sigma}_2)$, with different gaps associated with distinct subscripts $[c,d]$, as listed in Table 2. Here we already assume the gap to be momentum-independent with magnitude Δ_0. The eigenvalues can be explicitly written as

$$E^{\pm,\pm}(\mathbf{k}) = \pm\sqrt{\tilde{h}_{00}^2(\mathbf{k}) + \tilde{\mathbf{h}}^2(\mathbf{k}) + \Delta_0^2 \pm 2\sqrt{\tilde{h}_{00}^2(\mathbf{k})\tilde{\mathbf{h}}^2(\mathbf{k}) + \Delta_0^2 f_C^2(\mathbf{k})}}, \quad (3)$$

where $\tilde{\mathbf{h}}(\mathbf{k}) = (\tilde{h}_{10}(\mathbf{k}), \tilde{h}_{30}(\mathbf{k}), \tilde{h}_{21}(\mathbf{k}), \tilde{h}_{22}(\mathbf{k}), \tilde{h}_{23}(\mathbf{k}))$ is a five-dimensional vector. The function $f_C^2(\mathbf{k}) = \frac{1}{16}\mathrm{Tr}[\tilde{F}_C^\dagger(\mathbf{k})\tilde{F}_C(\mathbf{k})]$, is written in terms of the normalized superconducting fitness function $\tilde{F}_C(\mathbf{k}) = \hat{F}_C(\mathbf{k})/\Delta_0$, with $\hat{F}_C(\mathbf{k}) = \hat{H}_0(\mathbf{k})\hat{\Delta}(\mathbf{k}) - \hat{\Delta}(\mathbf{k})\hat{H}_0^*(-\mathbf{k})$. Note that the normal-state eigenenergies are given by $E_0^\pm(\mathbf{k}) = \tilde{h}_{00}(\mathbf{k}) \pm |\tilde{\mathbf{h}}(\mathbf{k})|$, such that the Fermi surface is defined by momenta satisfying $\tilde{h}_{00}^2(\mathbf{k}) = \tilde{\mathbf{h}}^2(\mathbf{k})$. Note also that if $f_C^2(\mathbf{k}) = 0$, the eigenenergies read $E^{\pm,\pm}(\mathbf{k}) = \pm\sqrt{(|\tilde{h}_{00}(\mathbf{k})| \pm |\tilde{\mathbf{h}}(\mathbf{k})|)^2 + \Delta_0^2}$, and the superconducting state is fully gapped.

In case $f_C^2(\mathbf{k}) \neq 0$, the energy spectra in the superconducting state at the momenta belonging to the normal-state Fermi surface can then be written as

$$E^{\pm,\pm}(\mathbf{k})\Big|_{FS} = \pm\sqrt{2\tilde{h}_{00}^2(\mathbf{k}) + \Delta_0^2 \pm 2\sqrt{\tilde{h}_{00}^4(\mathbf{k}) + \Delta_0^2 f_C^2(\mathbf{k})}}. \quad (4)$$

If we now expand for small Δ_0, we find for the low lying bands:

$$E^{\pm,-}(\mathbf{k})\Big|_{FS} \approx \pm\Delta_0\sqrt{\left(1 - \frac{f_C^2(\mathbf{k})}{\tilde{h}_{00}^2(\mathbf{k})}\right)}, \quad (5)$$

Journal of Physics: Conference Series **2164** (2022) 012002 doi:10.1088/1742-6596/2164/1/012002

and we can define the gap in the band basis as:

$$\mathcal{D}_B(\mathbf{k}) = \left[E^{+,-}(\mathbf{k})\Big|_{FS} - E^{-,-}(\mathbf{k})\Big|_{FS}\right] \approx \frac{2\Delta_0}{|\tilde{h}_{00}(\mathbf{k})|}\sqrt{\tilde{h}_{00}^2(\mathbf{k}) - f_C^2(\mathbf{k})}. \tag{6}$$

This definition is in direct analogy to the single band problem with normal state dispersion $\epsilon(\mathbf{k})$. The dispersion in the superconducting state is $E^{\pm}(\mathbf{k}) = \pm\sqrt{\epsilon^2(\mathbf{k}) + \Delta^2}$, so we can identify the gap at the Fermi surface by taking $\epsilon(\mathbf{k}) = 0$, and defining $[E^+(\mathbf{k})|_{FS} - E^-(\mathbf{k})|_{FS}] = 2|\Delta|$.

From Eq. 6, it is clear that the condition for the presence of nodes at the normal state Fermi surface is:

$$\tilde{h}_{00}^2(\mathbf{k}) - f_C^2(\mathbf{k}) = 0 \qquad \text{or} \qquad \tilde{\mathbf{h}}^2(\mathbf{k}) - f_C^2(\mathbf{k}) = 0. \tag{7}$$

These equations indicate that there can be nodes in the superconducting spectrum of Sr_2RuO_4 given the nontrivial form factors originated from the normal-state Hamiltonian $\hat{H}_0(\mathbf{k})$, even though the gap in the orbital basis is isotropic (s-wave). Note that the fitness function $\hat{F}_C(\mathbf{k})$ appears in this context as a key parameter defining the nodes, reinforcing its importance for the understanding of the phenomenology of superconducting states in complex materials [15, 16, 17, 18, 19, 12]. As a final note, keeping higher order terms in Δ_0^2 in the expansion above, we find structures associated with Bogoliubov Fermi surfaces, which can be understood as inflated nodes [20, 21]. We are not going to discuss these further here.

In Table 2 we summarise the fitness functions for each possible local order parameter labelled by $[a, b]$ in the two-orbital models (valid for both XZ and YZ planes). Based on the explicit form of the $\tilde{h}_{ab}(\mathbf{k})$ functions given in Table 1, we can directly infer the presence and position of nodes. We highlight the presence of horizontal line nodes for the order parameter labelled as $[2, 3]_p$, corresponding to the two components of the chiral d-wave order parameter $\{[5, 3], [6, 3]\}$ in the three orbital model for $p = \{YZ, XZ\}$, respectively. It can be shown that the nodes are not confined only to the XZ and YZ planes, but appear for any point with $k_z = \{0, \pm 2\pi\}$. Note that both components present horizontal line nodes, so even a complex superposition of the two components carries a horizontal line node. This is directly related to the fact that both $\tilde{h}_{23}^{XZ(YZ)}$ are proportional to $\sin(k_z/2)$.

Table 2. Normal state parameters $\tilde{h}_{ab}(\mathbf{k})$ determining the nodal structure for the s-wave gaps in the two-orbital effective model. These results are valid for both XZ and YZ planes, therefore we drop here the superscript p. The nodes as discussed assuming cylindrical Fermi surfaces. We neglect the possibility of fine-tuned accidental nodes that can appear for specific positions of the Fermi surface. Here we omit the explicit momentum dependence for conciseness, and used the identity valid at the Fermi surface $\tilde{h}_{00}^2 = \tilde{\mathbf{h}}^2$.

Gap Matrix	f_C^2	$\lvert\tilde{\mathbf{h}}^2 - f_C^2\rvert$	Line Nodes
$[0, 0]$	0	$\tilde{h}_{00}^2$	No
$[1, 0]$	$\tilde{h}_{21}^2 + \tilde{h}_{22}^2 + \tilde{h}_{23}^2 + \tilde{h}_{30}^2$	$\tilde{h}_{10}^2$	Vertical
$[2, 1]$	$\tilde{h}_{10}^2 + \tilde{h}_{22}^2 + \tilde{h}_{23}^2 + \tilde{h}_{30}^2$	$\tilde{h}_{21}^2$	No (XZ)/Vertical (YZ)
$[2, 2]$	$\tilde{h}_{10}^2 + \tilde{h}_{21}^2 + \tilde{h}_{23}^2 + \tilde{h}_{30}^2$	$\tilde{h}_{22}^2$	Vertical (XZ)/No (YZ)
$[2, 3]$	$\tilde{h}_{10}^2 + \tilde{h}_{21}^2 + \tilde{h}_{22}^2 + \tilde{h}_{30}^2$	$\tilde{h}_{23}^2$	Horizontal
$[3, 0]$	$\tilde{h}_{10}^2 + \tilde{h}_{21}^2 + \tilde{h}_{22}^2 + \tilde{h}_{23}^2$	$\tilde{h}_{30}^2$	No

5. Origin of gap minima along the diagonals

A second feature of the chiral d-wave order parameter dominated by the $\{[5,3],[6,3]\}$ components is the presence of gap minima along the diagonals, reminiscent of a $d_{x^2-y^2}$ structure. This can be understood by the fact that the two-component aspect of the order parameter in this case is not carried by the basis functions $\sin(k_x)\sin(k_z)$ and $\sin(k_y)\sin(k_z)$, but is carried by the orbital content of the gap structure. The orbital content is encoded by the Gell-Mann matrices. These matrices encode specific types of pairing: λ_5 corresponds to antisymmetric pairing between d_{yz} and d_{xy} orbitals, while λ_6 corresponds to antisymmetric pairing between d_{xz} and d_{xy} orbitals. In this case, the orbital content on the Fermi surfaces is going to limit the magnitude of the gap along specific directions. In Fig. 1 we compare the gap structure expected for a $d_{xz}+id_{yz}$ order parameter (accompanied by the $[0,0]$ matrix structure), displaying gap maxima along the diagonals, with the gap structure associated with the $[5,3]+i[6,3]$ structure, with gap minima along the diagonals.

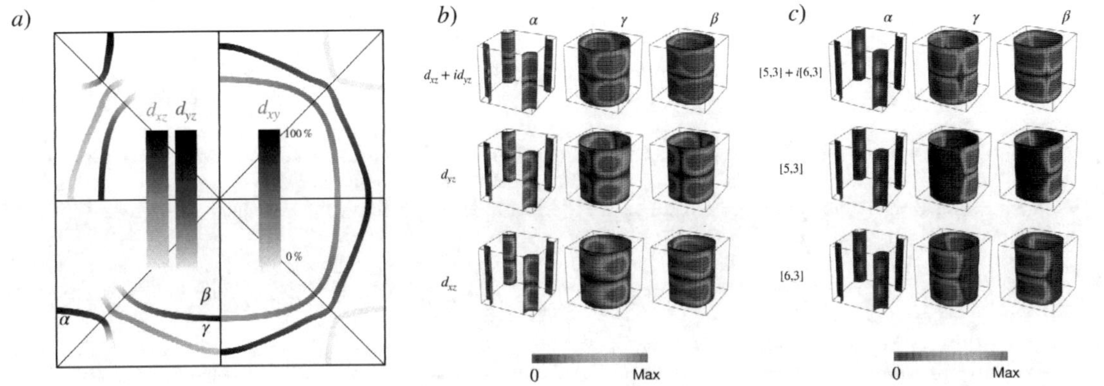

Figure 1. a) Schematic orbital content over the Fermi surfaces (FSs) of Sr_2RuO_4 in the $k_z=0$ plane. The color (red, blue, green) encode the orbital (d_{xy}, d_{yz}, d_{xz}) content as low (bright) or high (dark) in each Fermi surface sheet (labelled as α, β and γ). Adapted from [13, 14]. b) Top: Naive chiral d-wave state $\propto (\sin k_x + i\sin k_y)\sin k_z$ superimposed on the three-dimensional FSs of Sr_2RuO_4. Note the horizontal line nodes and the minima along the x- and y-directions. Middle (Bottom): separate plot of the component $\sin k_x \sin k_z$ ($\sin k_x \sin k_z$). Note the nodes along the x-(y-) direction. c) Top: Inter-orbital s-wave chiral superconducting state gap (encoded in the matrix structure $[5,3]+i[6,3]$) on the three-dimensional FSs of Sr_2RuO_4. Note the horizontal line nodes and the minima along the diagonals. Middle (Bottom): separate plot of the component $[5,3]$ ($[6,3]$). Note the extended regions with very small gap associated with the absence of certain orbitals on these regions of the Fermi surface.

6. Conclusion

In this proceeding we have discussed in detail the origin of the horizontal line nodes and vertical line dips of the superconducting gap for the recently proposed s-wave orbitally-antisymmetric spin-triplet order parameter. Even though the order parameter stem from local interactions, it develops a nodal structure inherited from the normal state electronic properties. Furthermore, we highlight that the position of the vertical dips along the diagonals is completely unexpected, and imposed by the orbital content around the Fermi surface.

Acknowledgments

The author thanks Yoshi Maeno for the question that inspired this proceeding. The author also thanks Han Gyeol Suh, Henri Menke, Philip M. R. Brydon, Carsten Timm, and Daniel F. Agterberg for discussions and previous collaborations that motivated this work. The author acknowledges the financial support of the Swiss National Science Foundation through the Ambizione Grant No. 186043.

References

[1] Rice T M and Sigrist M 1995 *J. Phys.: Condens. Matter* **7** L643–L648
[2] Mackenzie A P, Scaffidi T, Hicks C W and Maeno Y 2017 *npj Quantum Materials* **2** 40
[3] Pustogow A, Luo Y, Chronister A, Su Y S, Sokolov D A, Jerzembeck F, Mackenzie A P, Hicks C W, Kikugawa N, Raghu S, Bauer E D and Brown S E 2019 *Nature* **574** 72–75
[4] Ishida K, Manago M, Kinjo K and Maeno Y 2020 *Journal of the Physical Society of Japan* **89** 034712
[5] Xia J, Maeno Y, Beyersdorf P T, Fejer M M and Kapitulnik A 2006 *Phys. Rev. Lett.* **97**(16) 167002
[6] Luke G M, Fudamoto Y, Kojima K M, Larkin M I, Merrin J, Nachumi B, Uemura Y J, Maeno Y, Mao Z Q, Mori Y, Nakamura H and Sigrist M 1998 *Nature* **394** 558–561
[7] Grinenko V, Ghosh S, Sarkar R, Orain J C, Nikitin A, Elender M, Das D, Guguchia Z, Brückner F, Barber M E, Park J, Kikugawa N, Sokolov D A, Bobowski J S, Miyoshi T, Maeno Y, Mackenzie A P, Luetkens H, Hicks C W and Klauss H H 2021 *Nature Physics* **17** 748–754
[8] Lupien C, MacFarlane W A, Proust C, Taillefer L, Mao Z Q and Maeno Y 2001 *Phys. Rev. Lett.* **86**(26) 5986–5989
[9] Benhabib S, Lupien C, Paul I, Berges L, Dion M, Nardone M, Zitouni A, Mao Z Q, Maeno Y, Georges A, Taillefer L and Proust C 2021 *Nature Physics* **17** 194–198
[10] Ghosh S, Shekhter A, Jerzembeck F, Kikugawa N, Sokolov D A, Brando M, Mackenzie A P, Hicks C W and Ramshaw B J 2021 *Nature Physics* **17** 199–204
[11] Bergemann C, Mackenzie A P, Julian S R, Forsythe D and Ohmichi E 2003 *Advances in Physics* **52** 639–725
[12] Suh H G, Menke H, Brydon P M R, Timm C, Ramires A and Agterberg D F 2020 *Phys. Rev. Research* **2**(3) 032023
[13] Veenstra C N, Zhu Z H, Raichle M, Ludbrook B M, Nicolaou A, Slomski B, Landolt G, Kittaka S, Maeno Y, Dil J H, Elfimov I S, Haverkort M W and Damascelli A 2014 *Phys. Rev. Lett.* **112**(12) 127002
[14] Tamai A, Zingl M, Rozbicki E, Cappelli E, Riccò S, de la Torre A, McKeown Walker S, Bruno F Y, King P D C, Meevasana W, Shi M, Radović M, Plumb N C, Gibbs A S, Mackenzie A P, Berthod C, Strand H U R, Kim M, Georges A and Baumberger F 2019 *Phys. Rev. X* **9**(2) 021048
[15] Ramires A and Sigrist M 2016 *Phys. Rev. B* **94**(10) 104501
[16] Ramires A and Sigrist M 2017 *J. Phys.: Conf. Ser.* **807** 052011
[17] Ramires A, Agterberg D F and Sigrist M 2018 *Phys. Rev. B* **98**(2) 024501
[18] Ramires A and Sigrist M 2019 *Phys. Rev. B* **100**(10) 104501
[19] Andersen L, Ramires A, Wang Z, Lorenz T and Ando Y 2020 *Science Advances* **6** eaay6502
[20] Agterberg D F, Brydon P M R and Timm C 2017 *Phys. Rev. Lett.* **118**(12) 127001
[21] Brydon P M R, Agterberg D F, Menke H and Timm C 2018 *Phys. Rev. B* **98**(22) 224509

^{63,65}Cu, ^{139}La-NMR study of Electron-doped High Temperature Cuprate Superconductor T'-$La_{1.8}Eu_{0.2}CuO_{4-y}F_y$

Hideto Fukazawa[1,*], Yongsun Lee[1], Masaki Watai[1], Koma Tamura[1], Tetsuo OHAMA[1], Yoh Kohori[1], Toshiki Sunohara[2], Kota Shiosaka[2], Ryoji Nagaoka[2], Takayuki Kawamata[2], Yoji Koike[2]

[1]Department of Physics, Chiba University, Chiba 263-8522, Japan
[2]Department of Applied Physics, Tohoku University, Sendai 980-8579, Japan

E-mail: `hideto@chiba-u.jp`

Abstract. T'-$La_{1.8}Eu_{0.2}CuO_{4-y}F_y$ ($0 \leq y \leq 0.125$) was measured by ^{63,65}Cu and ^{139}La NMR. This material is an electron-doped high-T_c cuprate superconductor which has Nd_2CuO_4-type structure (so-called T'-structure). The nuclear spin-lattice relaxation rate $1/T_1$ revealed that pseudogap behavior exists in the lightly electron-doped region of $y \leq 0.075$ and electron doping suppresses the antiferromagnetic spin flutuations. These behaviors in electron-doped high-T_c cuprate superconductor are analogous with the optimum- and over-doped region in hole-doped type. ^{139}La NMR spectra shows almost no antiferromagnetic region in the samples of T'-$La_{1.8}Eu_{0.2}CuO_{4-y}F_y$.

1. Introduction

High-T_c cuprate superconductors (HTSCs) have been thought to exhibit superconductivity by doping carrier into the the parent material which is Mott insulator [1]. It has also been considered that the electron-doped HTSCs have a smaller superconducting phase and a larger antiferromagnetic phase than the hole-doped HTSCs. However, recent studies have reported that the superconducting phase becomes larger when electron-doped HTSCs are sufficiently reduced [2, 3]. This is because apical oxygens located above and below Cu are removed through the reduction annealing and the localized antiferromagnetic order is suppressed.

To investigate the phase diagram of HTSCs with the Nd_2CuO_4-type structure (so-called T'-structure), T'-$La_{1.8}Eu_{0.2}CuO_4$ (T'-LECO) is one of the most appropriate materials [4, 5, 6, 7, 8]. This compound can be both electron-doped by F with O, and hole-doped by Ca or Sr with La, respectively. Investigation of various doping clarifies the dependencies on the doping levels for both carriers in the equivalent crystal structure. In this study, we performed ^{63,65}Cu, ^{139}La NMR measurement of an electron-doped HTSC T'-$La_{1.8}Eu_{0.2}CuO_{4-y}F_y$ (T'-LECOF, $0 \leq y \leq 0.125$) in order to reveal superconducting and normal state properties of the compounds.

2. Experiments

T'-LECOF samples are polycrystals synthesized via multiple processes, according to Ref. 7. Polycrystalline samples were oxygen reduced through the annealing under a pressure of

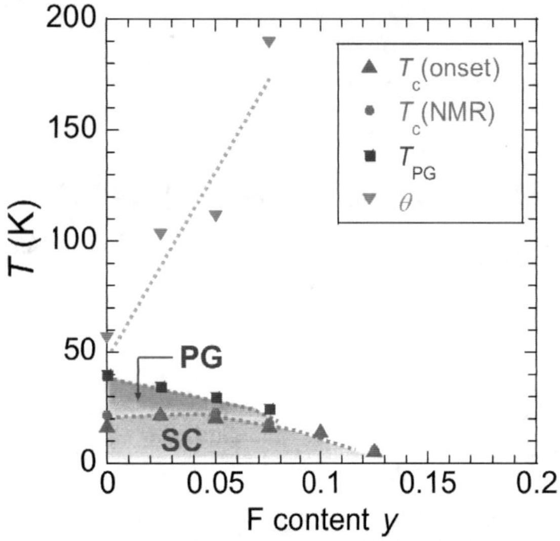

Figure 1. (Color online) Phase Diagram of T'-$La_{1.8-x}Eu_{0.2}CuO_{4-y}F_y$. (superconducting transition temperature T_c, Pseudogap (PG) temperature T_{PG}, Curie-Weiss temperature θ) The determination of each temperature is described in text.

3×10^{-5} Pa for 24 hours at 700 °C for $y = 0$ and at 650 °C for $y = 0.025$-0.125. The annealed samples were confirmed by X-ray powder diffraction. By measurement of dc magnetization with a commercial SQUID magnetometer (Quantum Design MPMS), the samples are verified in single phase with F substance dependency for T_c as Ref. 7. phase-coherent pulsed NMR spectrometers and 8 T SC magnet were used to measure the ^{63,65}Cu and ^{139}La NMR. The silver pickup coil and shield were attached to avoid the external Cu metal signals. To determine the ^{63}Cu Knight shift [9], Cu metal powder was used as a reference ($K = 0.232\%$). To evaluate the spin-lattice relaxation time T_1 for each temperature, saturation recovery method was applied for fitting the nuclear magnetization recovery curve anticipated for the nuclear spin $I = 3/2$ of the ^{63}Cu nucleus [10].

$$\frac{M(\infty) - M(t)}{M(\infty)} = A\left[0.1\exp\left(-\frac{t}{T_1}\right) + 0.9\exp\left(-\frac{6t}{T_1}\right)\right]. \quad (1)$$

3. Results and discussion

From temperature dependence of spin-lattice relaxation rate $1/T_1$ of ^{63}Cu-NMR, pseudogap (PG) behavior exists in the under electron-doped region as $y \leq 0.075$ [11]. $1/T_1$ also suggested that electron doping suppresses the antiferromagnetic spin fluctuations. In Fig. 1, we summarized substitution dependence of several characteristic temperatures (SC transition temperature T_c, PG temperature T_{PG}, Curie-Weiss temperature θ) of T'-LECOF. T_c (onset) and T_c (NMR) were determined from the superconducting onset temperatures of magnetic susceptibility χ and NMR Knight shift, respectively. T_c was defined from the local maximum of $1/T_1T$ above T_c. θ was evaluated from the fitting to the formula $1/T_1T = C/(T + \theta)$ above T_{PG}. Compared with the phase diagram of hole-doped HTSCs, there are many similarities between both electron-doped type and over optimum-doped region of hole-doped type [12, 13]. Comparing the phase diagrams of T'-LECOF and isostructural T'-$Pr_{1.3-x}La_{0.7}Ce_xCuO_4$ (T'-PLCCO) based on the nominal substitution, T'-LECOF is more effectively electron-doped than T'-PLCCO through the partial substitution [14, 15, 16, 17].

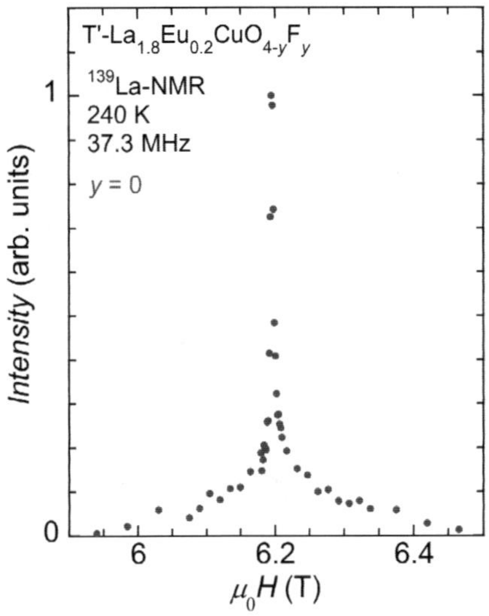

Figure 2. (Color online) ^{139}La NMR spectrum of T'-$La_{1.8}Eu_{0.2}CuO_4$ at 240 K.

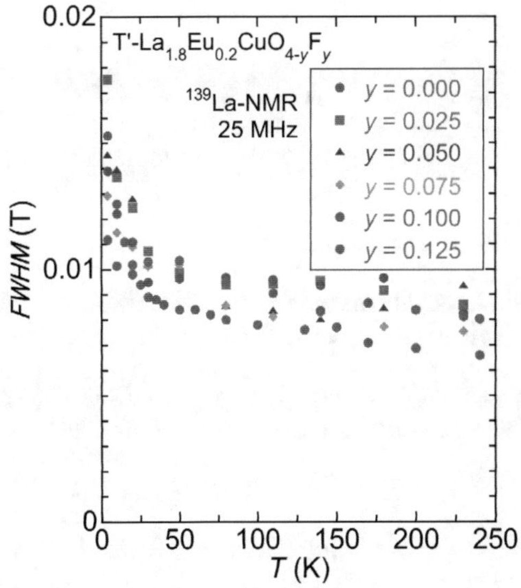

Figure 3. (Color online) Temperature dependence of the full width at half maximum of ^{139}La NMR spectra of T'-$La_{1.8}Eu_{0.2}CuO_{4-y}F_y$ ($0 \leq y \leq 0.125$).

^{139}La-NMR mainly reflects the information of the phase-separated magnetically-ordered region in the samples. In Fig. 2, the spectrum of T'-LECOF ($y = 0$) obtained at 240 K is shown. The spectrum have a sharp-edged center line originated by transition between $I_z = -1/2$ and $+1/2$. Distributed crystalline electric field occured by polycrystalline samples makes a broad structure surrounding the center line. From this structure, the electric field gradient located on La sites was estimated as $\nu_Q \sim 0.5$ MHz. As the temperature decreses, the spectrum gradually becomes broad.

A rectanglur powder originating from ^{139}La nuclei under the homogeneous internal field among the sample does not exist. In this case, the emergence of localized antiferromagnetic ordering can be observed by the sudden increase of the line width. In Fig. 3, the temperature dependence of the full width at half maximum (FWHM) of T'-LECOF is shown. Although the increase of FWHM below about 30 K was observed, this increase is five times smaller than that of T'-$Pr_{1.3-x}La_{0.7}Ce_xCuO_4$ (T'-PLCCO) in which the clear rapid increase of FWHM was observed below Néel temperature T_N [14]. Hence, the results indicate the almost no localized antiferromagnetic order in T'-LECOF.

4. Summary

In summary, T'-$La_{1.8}Eu_{0.2}CuO_{4-y}F_y$ ($0 \leq y \leq 0.125$) was measured by ^{63,65}Cu and ^{139}La NMR. This material is an electron-doped high-T_c cuprate superconductor which has Nd_2CuO_4-type structure (so-called T'-structure). The nuclear spin-lattice relaxation rate $1/T_1$ revealed that pseudogap behavior exists in the lightly electron-doped region of $y \leq 0.075$ and electron doping suppresses the antiferromagnetic spin flutuations. These behaviors in electron-doped type are analogous with the optimum- and over-doped region in hole-doped type. It is also revealed by ^{139}La NMR measurements that the localized antiferromagnetic order does not nearly exist in T'-$La_{1.8}Eu_{0.2}CuO_{4-y}F_y$.

Acknowledgments

This work was supported by JSPS KAKENHI Grant (No. 17H02915, No. 18K03505, and No. 21K03439).

References

[1] B. Keimer, S. A. Kivelson, M. R. Norman, S. Uchida, J. Zaanen, Nature **518**, 179 (2015).
[2] O. Matsumoto, A. Utsuki, A. Tsukada, H. Yamamoto, T. Manabe, and M. Naito, Physica C **469**, 924 (2009).
[3] T. Adachi, Y. Mori, A. Takahashi, M. Kato, T. Nishizaki, T. Sasaki, N. Kobayashi, and Y. Koike, J. Phys. Soc. Jpn. **82**, 063713 (2013).
[4] T. Takamatsu, M. Kato, T. Noji, and Y. Koike, Physica C **471**, 679 (2011).
[5] T. Takamatsu, M. Kato, T. Noji, and Y. Koike, Appl. Phys. Exp. **5**, 073101 (2012).
[6] T. Takamatsu, M. Kato, T. Noji, and Y. Koike, Phys. Procedia **58**, 46 (2014).
[7] T. Sunohara, T. Kawamata, K. Shiosaka, T. Takamatsu, T. Noji, M. Kato, and Y. Koike, J. Phys. Soc. Jpn. **89**, 014701 (2020).
[8] C. Lin, M. Horio, T. Kawamata, S. Saito, K. Koshiishi, S. Sakamoto, Y. Zhang, K. Yamamoto, K. Ikeda, Y. Hirata, K. Takubo, H. Wadati, A. Yasui, Y. Takagi, E. Ikenaga, T. Adachi, Y. Koike, and A. Fujimori, J. Phys. Soc. Jpn. **88**, 115004 (2019).
[9] G. C. Carter, L. H. Benett, and D. J. Kahn, *Metallic Shifts in NMR Part 1* (Pergamon, Oxford, U.K., 1977).
[10] A. Narath, Phys. Rev. **162**, 320 (1967).
[11] M. Watai, Y. Lee, H. Fukazawa, Y. Kohori, T. Sunohara, K. Shiosaka, T. Kawamata, and Y. Koike, J. Phys. Soc. Jpn. **90**, 023704 (2021).
[12] S. Ohsugi, Y. Kitaoka, K. Ishida, G. -q. Zheng, K. Asayama, J. Phys. Soc. Jpn. **63**, 700 (1994).
[13] S. Fujiyama, Y. Itoh, H. Yasuoka, and Y. Ueda, J. Phys. Soc. Jpn. **66**, 2864 (1997).
[14] M. Yamamoto, Y. Kohori, H. Fukazawa, A. Takahashi, T. Ohgi, T. Adachi, and Y. Koike, J. Phys. Soc. Jpn. **85**, 024708 (2016).
[15] Y. Lee, H. Fukazawa, S. Kanamaru, M. Yamamoto, Y. Kohori, A. Takahashi, T. Kawamata, K. Kawabata, K. Tajima, T. Adachi, and Y. Koike, J. Phys. Soc. Jpn. **89**, 073709 (2020).
[16] M. Horio, T. Adachi, Y. Mori, A. Takahashi, T. Yoshida, H. Suzuki, L. C. C. Ambolode II, K. Okazaki, K. Ono, H. Kumigashira, H. Anzai, M. Arita, H. Namatame, M. Taniguchi, D. Ootsuki, K. Sawada, M. Takahashi, T. Mizokawa, Y. Koike, and A. Fujimori, Nat. Commun. **7**, 10567 (2016).
[17] D. Song, G. Han, W. Kyung, J. Seo, S. Cho, B. S. Kim, M. Arita, K. Shimada, H. Namatame, M. Taniguchi, Y. Yoshida, H. Eisaki, S. R. Park, and C. Kim, Phys. Rev. Lett. **118**, 137001 (2017).

Journal of Physics: Conference Series **2164** (2022) 012004 doi:10.1088/1742-6596/2164/1/012004

Doping induced band renormalization in 122-type Fe-based superconductor

Ram Prakash Pandeya[1], Sawani Datta[1], Anup Pradhan Sakhya[1], Tanusree Saha[2], Giovanni De Ninno[2,3] Rajib Mondal[1], Paolo Moras[4], Matteo Jugovac[4], Carlo Carbone[4], A Thamizhavel[1], and Kalobaran Maiti[1]

[1]Department of Condensed Matter Physics and Materials Science, Tata Institute of Fundamental Research, Mumbai-400005, India
[2]Laboratory of Quantum Optics, University of Nova Gorica, 5001 Nova Gorica, Slovenia
[3]Elettra-Sincrotrone Trieste, Area Science Park, 34149 Trieste, Italy
[4]Istituto di Structtura della Materia, Consiglio Nazionale delle Ricerche, Area Science Park, I-34012 Trieste, Italy

E-mail: kbmaiti@tifr.res.in

Abstract. We study the electronic structure of a superconducting composition of 122-type Fe-based pnictide material, $CaFe_{1.9}Co_{0.1}As_2$ employing high resolution angle resolved photoemission spectroscopy technique. The experimental results exhibit three bands close to Fermi level at Γ-point of the Brillouin zone among which only one band crosses the Fermi level. In the parent compound, $CaFe_2As_2$, all the three bands cross the Fermi level and form three hole pockets. While the destruction of Fermi pockets due to electron doping (Co-substitution dopes electrons into the system) is expected, we observe significant orbital selective band renormalization with respect to the parent compound. It appears that the effect of spin-orbit coupling is stronger in the doped compound.

Introduction

The parent compounds of Fe-based superconductors (FeScs) usually forms in tetragonal paramagnetic structure at room temperature. With cooling, the system undergoes both structural and magnetic transitions; a long range anti-ferromagnetic (AFM) ordering in a two fold-symmetric orthorhombic crystal structure appears at low temperature [1]. Chemical doping in the parent FeScs at different sites suppresses the ground state orders and superconducting (SC) phase appears within a critical range of doping concentration. Although, the phase diagram obtained with doping in FeScs are similar to that observed in Cu-based superconductors, the electronic properties of FeScs are more complex due to the presence of multiple bands close to Fermi energy, complex pressure-temperature phase diagram, coexistence of mutually exclusive magnetic order and superconductivity, *etc* [1, 2]. Chemical doping is usually expected to affect the charge carrier concentration, structural distortion and electron impurity scattering due to different atomic sizes and valence electrons of the dopant atoms. Evidently, Fe-based systems are highly complex. The angle resolved photoemission spectroscopy (ARPES) studies of these materials suggest interesting doping induced effects renormalization of band dispersion, spin-orbit coupling (SOC) strength, *etc* [3, 4].

In order to probe the doping induced effects in 122-type Fe-based pnictide compounds, we studied the electronic structure of $CaFe_{1.9}Co_{0.1}As_2$ (CaCo122) employing high resolution ARPES. Similar to other FeScs parent compounds, $CaFe_2As_2$ (Ca122) undergoes both structural and magnetic transitions at 170 K from its room temperature paramagnetic-tetragonal structure to AFM-orthorhombic structure [5, 6]. Doping of transition metal elements such as Co, Ni, Rh *etc* at the Fe-site suppresses the phase transitions and at a critical doping range, SC ground state appears. The compound, CaCo122 shows SC phase with transition temperature of $T_c \sim 15$ K which is the highest in this class of materials [1]. The electronic structure studies of both the parent and doped compounds suggest weak doping induced effects in the momentum integrated spectral density of states [8, 9]. Evidentally, the role of Co-doping at the Fe-sites leading to superconducting state is complex and subtle changes close to the Fermi level needs to be studied carefully. We have investigated the electronic band structure of CaCo122 employing ARPES and find significant orbital selective band renormalization which might be a reason for the exotic ground state of this material.

Experimental Method

High quality single crystals of the CaCo122 were prepared using flux grown method [7]. No signature of magnetic or structure transition was observed in temperature dependent magnetic susceptibility and x-ray diffraction measurements. A sharp diamagnetic response was observed below 15 K indicating onset of the superconducting phase of the compound [9]. The crystallinity of the as grown samples were studied using Laue x-ray diffraction (XRD) technique. The (001)-aligned Laue XRD pattern collected on the sample is shown in the Fig. 1(b). Sharp and bright spots suggests very good quality of the single crystals.

The ARPES measurements were performed at Electron Spectroscopy Lab, TIFR, India, using GammaData Sceinta DA30L detector and a monochromatized UV lamp, VUV5k. The energy and angle resolutions were set to 10 meV and 0.3°. The ARPES measurements using 48 eV photon energy were carried out at VUV beamline of Elettra, Trieste, Italy. In the later case, the photoemission signals were detected using a Scienta R4000 WAL electron analyzer with energy and angle resolution of ∼15 meV and ∼0.3°, respectively [10]. The electronic band structure calculations were carried out following the density functional theory approach implemented in Wien2k-software using standard convergence criteria used for such systems.

Results

The experimental geometry used for the ARPES measurements is shown in Fig. 1(a). In both the ARPES measurements, the (110)-crystal axis of the sample is aligned along the slit, which allowed us to probe the electronic structure of compound along $\Gamma(0,0)$ - $X(\pi,\pi)$ direction. The incidence angle (θ shown in Fig. 1(a)), was 60° and 45° in the lab based He-I_α and synchrotron radiation source measurements, respectively. In Fig. 1(c), we show the ARPES spectra of CaCo122 collected using He-I_α source at 25 K. Due to the matrix cross section element effect, only one band is observed which crosses the Fermi energy (ϵ_F) at $k_F \sim 0.2$ Å^{-1}, in contrast to the expected no of bands close to ϵ_F [10, 11, 12]. Intense and broad feature at binding energy ∼ 300 meV have partial contribution from the surface-band; this is verified by the rigid energy shift of the band with time delay after cleaving (not shown here). To observe the dispersion of the other expected bands at the Γ-point, the ARPES spectrum was also collected using 48 eV photon energy which represents the similar k_z-value (see Fig. 1(d))[13].

The observed dispersion of the bands are shown by the dashed lines (smooth hand drawn curves) and labelled by Greek letters. While, the α_1-band crosses the Fermi level at k_F (= 0.2 Å^{-1}) close to that observed in the He-I_α ARPES spectrum, the top of the α_2-band just touches the Fermi level and does not form a finite size Fermi surface. The α_3-band appears ∼ 140 meV

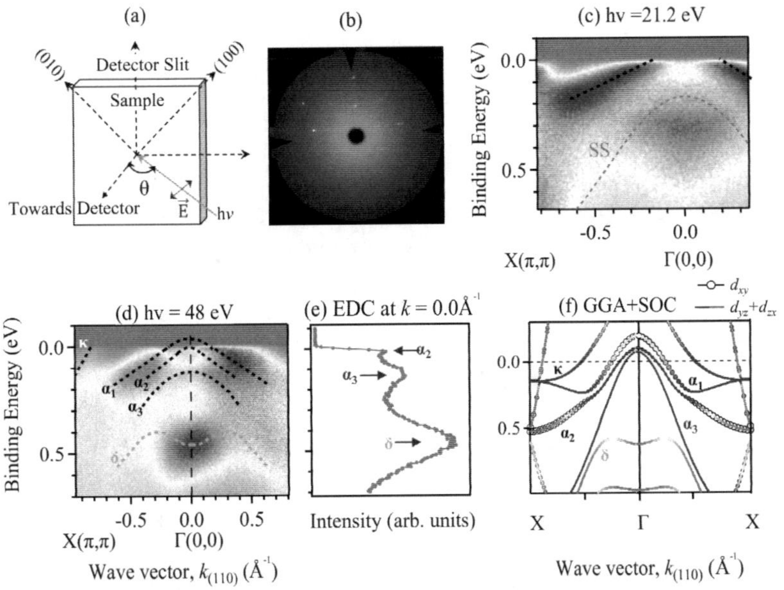

Figure 1. (a) Experimental geometry; sample surface is xy-plane. The detector is placed on z-axis with the slits along y-axis. Crystal axis (100) and (010) are shown by dashed line. (b) Laue pattern for (001)-surface. ARPES data of CaCo122 collected with photon energy (c) 21.2 eV at 25 K and (d) 48 eV at 13 K. (e) The energy distribution curve (EDC) at $k = 0.0$ Å^{-1} of 48 eV data. (f) Electronic band structure results of Ca122 calculated including spin-orbit coupling. The size of the open circles represents contributions of d_{xy}-states. The color code from blue to red represents the maximum to minimum contribution from the d_{yz}+d_{zx}-state.

from the ϵ_F which can be seen in the EDC shown in the Fig. 1(f). The κ-band crossing at k_F ~-0.79Å^{-1} is the electron-pocket centred at the X-point.

To understand the effect of Co-doping in Ca122, in Fig. 1(f), we show the theoretically calculated band structure of the Ca122 obtained with inclusion of the spin-orbit coupling (SOC) term and compare it with the experimental results discussed above. The calculated band structure is consistent with the ARPES spectra of the parent compound considering that there is some additional band renormalization effects due to electron-electron correlation [13, 10]. However, a significant deviation in the relative energy position of the bands in CaCo122 ARPES spectra is observed with respect to the band structure results shown in Fig. 1(f) and earlier ARPES studies. The calculated band structure results show a large energy difference between the α_1 and α_2 bands, and a relatively smaller gap between α_2 and α_3-bands at the Γ-point which was also observed in the experimental electronic studies carried out using ARPES technique [13, 10]. However, the ARPES spectra of CaCo122 exhibit an opposite trend; the gap between α_1 and α_2 bands is about 50 meV while the gap between α_2 and α_3 is about 140 meV. The Co-doping at Fe-site effectively adds an extra electron per atomic substitution. This will shift the energy bands; the vanishing of two hole pockets can be attributed to such effective electron doping. However, it fails to explain the change in the relative band energy difference at Γ-point with Co-doping indicating that the rigid band description is not applicable in this case.

In contrast to the shift of α-bands with the charge carrier doping, the δ-band having dominating d_{z^2}-character, appears at slightly low binding energy in comparison to the parent compound case [10]. The unequal positive energy shifts of the d_{xy}, d_{yz} and d_{zx} bands as well as the negative energy shift of the e_g-symmetric δ-band suggests orbital-dependent band renormalization in addition to a simple charge carrier doping. We observed that the energy gap between the α_2 and α_3-bands at the Γ point depends on the SOC; the gap appears only with inclusion of the SOC term in the band structure calculations. Thus, the increase in the energy difference between the bands observed above suggests significant increase in SOC strength with doping which is unusual as both Co and Fe atoms have very similar atomic weights. It is worth to note here that the changes observed in the near ϵ_F electronic structure of the parent compound with doping will enhance the total density of states close to ϵ_F, which explains the doping induced effects observed in earlier bulk electronic structure studies carried out using hard x-ray photoemission spectroscopy and x-ray absorption spectroscopy techniques [9, 8]. Clearly, the evolution of the electronic structure with Co-doping at Fe-site is complex and calls for more theoretical and experimental studies.

Conclusions

In summary, we have studied the electronic structure of $CaFe_{1.9}Co_{0.1}As_2$, a superconducting composition in the $CaFe_2As_2$ family using high resolution ARPES technique. Experimental results show signature of destruction of two hole pockets near Γ point due to electron doping via Co-substitution. In addition, we observe orbital dependent band renormalization due to Co-substitution; the band renormalization suggests role of strong spin-orbit coupling. These results reveals an unusual scenario due to electron doping in Ca122 that leads to superconductivity in this system.

Acknowledgement

The authors acknowledge financial support from the Indo-Italian Programme of Cooperation under the aegis of Department of Science and Technology and the Italian Ministry of Foreign Affairs, and Department of Atomic Energy, Government of India (Project Identification No. RTI4003, DAE OM No.1303/2/2019/R&D-II/DAE/2079 dated 11-02-2020). K.M. acknowledges BRNS, DAE, Government of India for financial support under the DAE-SRC-OI award program.

References

[1] Hosono H *et al.*, 2015 *Sci. Technol. Adv. Matter.* **16** 033503 (87pp)
[2] Stewart G R 2011 *Rev. Mod. Phys.* **83** 1589
[3] Liu Z K *et al.* 2015 *Phys. Rev.* B **92** 235138
[4] Miao H *et al.*, 2014 *Phys. Rev.* B **89** 22050 (R)
[5] Ni N, Nandi S, Kreyssig A, Goldman A I, Mun E D, Bud'ko S L and Canfield P C 2008 *Phys. Rev.* B **78** 014523
[6] Kreyssig A *et al.*, 2008 *Phys. Rev.* B**78** 184517
[7] Kumar N, Nagalakshmi R, Kulkarni R, Paulose P L, Nigam A K, Dhar S K and Thamizhavel A 2009 *Phys. Rev.* B **79** 012504
[8] Pandeya R P *et al.*, 2021 *J. Phys.: Condens. Matter* **33** 19LT01; Pandeya R P *et al.* 2020 AIP Conf. Proc. **2265** 030367;
[9] Pandeya R P, Sakhya A P, Datta S, Saha T, Nino G de, Mondal R, Schlueter C, Gloskovskii A, Moras P, Jugovac M, Carbone C, Thamizhavel A and Maiti K 2021 *Phys. Rev.* B **104** 094508
[10] Ali K *et al.* 2018 *Phys. Rev.* B **97** 054505; Maiti K, 2015 *Pramana - J. Phys.* **84** 947; Adhikary G *et al.*, 2011 *AIP Conf. Proc.* **1347** 169; Adhikary G *et al.*, 2014 *J. Appl Phys.* **115** 123901
[11] Ali K and Maiti K 2017 *Scientific Reports* **7** 6298
[12] Ali K and Maiti K 2018 *Eur. Phys. J.* B **91** 199
[13] Liu C, Kondo T, Ni N, Paleczewski A D, Bostwick A, Samolyuk G D, Khasanov R, Shi M, Rotenberg E, Bud'ko S L, Canfield P C and Kamniski A 2009 *Phys. Rev. Lett.* **102** 167004
[14] Danura M, Kudo K, Oshiro Y, Araki S, Kobayashi T C and Nohara M 2011 *J. Phys. Soc. Jpn.* **80** 103701
[15] Qi Y, Gao Z, Wang L, Zhang X, Wang D, Yao C, Wang C, Wang C and Ma Y 2011 *Europhys. Lett.* **96** 47005

Single crystal growth and de Haas-van Alphen effect in strong coupling *s*-wave superconductor $SrPt_3P$

M Horie[1], E Matsuoka[1], H Sugawara[1*], H Kotegawa[1], H Tou[1], T Terashima[2] and S Uji[2]

[1]*Graduate School of Science, Kobe University, Kobe, Japan*
[2]*National Institute for Materials Science, Tsukuba, Japan*

*E -mail: sugawara@crystal.kobe-u.ac.jp

Abstract. We have succeeded in growing a single crystal of $SrPt_3P$, which was reported as a strong coupling *s*-wave superconductor with the critical temperature of T_c = 8.4 K, and have measured the de Haas-van Alphen (dHvA) effect and the electrical resistivity under magnetic fields. Several dHvA frequency branches with the frequencies up to ~ 1 kT with the cyclotron effective masses of $m_c^* = 0.27 - 0.61\ m_0$ were observed. It was clarified that the superconducting upper critical fields $\mu_0 H_{c2}$ for fields along *H* // [001] and *H* // [100] in the tetragonal structure are almost the same in values of $\mu_0 H_{c2}$ ~4.8 T.

1. Introduction

New ternary Pt-based phosphide superconductors APt_3P (*A* = Sr, Ca, and La) with critical temperatures of T_c = 8.4, 6.6, and 1.5 K, respectively, were recently discovered [1], which crystalize in an antiperovskite-based tetragonal structure (space group No. 129, *P4/nmm*, D_{4h}^7) closely related to the heavy fermion superconductor without inversion symmetry $CePt_3Si$ [2]. Among them, $SrPt_3P$ was reported as a strong coupling *s*-wave superconductor with a large ratio between superconducting gap Δ and T_c, $2\Delta/k_BT_c \sim 5$ [1]. Temperature dependences of both the specific heat *C* and the electrical resistivity ρ at low temperatures in the normal state imply the presence of low-lying phonons, i.e., C/T-T^2 plot below 10 K shows a pronounced nonlinear behavior and ρ shows almost *T*-linear dependence below 20 K. Moreover, the presence of multiple Fermi surface pockets was inferred in $SrPt_3P$ from the nonlinear field dependence of Hall resistivity which might play a role in realizing the strong coupling of charge carriers with the low-lying phonons [1]. Thus, the investigation of the electronic state is essential to explore the superconducting mechanism in $SrPt_3P$.

The band structure calculation based on the density functional theory was performed by Kang et al. [3], and they reported that the superconductivity in $SrPt_3P$ is described well by the strong coupling phonon-mediated mechanism. In this work, we have tried to grow a single crystal of $SrPt_3P$ and measured the de Haas-van Alphen (dHvA) effect to examine the electronic state experimentally. Superconducting phase diagram; upper critical fields $\mu_0 H_{c2}$ vs. temperature *T*, and its anisotropy were also determined by the electrical resistivity measurements under magnetic fields using the high-quality single crystal.

2. Experimental details

Single crystals of $SrPt_3P$ were grown by using the Bridgman technique. Initially, we made pre-reacted products to eliminate the high vapor pressure of pure Sr and P during the sealing in a Mo crucible. The starting materials 3N (99.9% pure)-Sr, 4N-Pt, and 6N-P with the stoichiometric ratio were sealed in a quartz tube under the vacuum, heated up to 900 °C in an electric furnace, kept at this temperature for three days, and cooled rapidly down to room temperature. The pre-reacted products then sealed in a Mo crucible by using an arc furnace under 6N-ultra-pure Ar-gas atmosphere. The crucible was heated up to 1500 °C in a high-temperature MoSi-heater furnace, kept at this temperature for half-hour, then cooled to room temperature at a rate of -200 °C/h. Single crystals were grown in the small pieces and were extracted by mechanically from the ingot. The obtained single crystals are thin platelets in shape with a typical dimension of about 0.5 x 0.5 x 0.1 mm^3 as shown in the upper inset of figure 1.

The crystallographic orientations of the single crystal samples were confirmed by a back Laue X-ray diffraction method with a Mo K_α radiation. The electrical resistivity was measured by the conventional four-probe DC method down to ~0.5 K in the ^{3}He-cryostat with a 9 T superconducting magnet. The dHvA experiments were performed by a magnetic torque method using a piezoresistive microcantilever at low temperatures down to ~70 mK using a top-loading dilution refrigerator in magnetic fields up to 17.8 T using a superconducting magnet.

3. Results and discussion

Figure 1 shows the temperature dependences of the electrical resistivity ρ(T) in $SrPt_3P$ for the current along J // [100]. The overall temperature dependence is close to that of the polycrystalline sample in the previous report except the difference in absolute value [1]. $\rho(T)$ decreases with decreasing temperature exhibiting a metallic behavior, and shows almost T-linear dependence between 28 K and 15 K, followed by the saturating tendency at lower temperature and then shows the superconducting transition at T_c = 8.5 K. Below ~ 15 K under the normal state, $\rho(T)$ shows the T^2-dependence indicating a Fermi-liquid behavior as shown in the lower inset of figure 1. Assuming the relation $\rho(T) = \rho_0 + AT^2$, where ρ_0 is the residual resistivity and A is the coefficient of electron–electron scattering, we estimate ρ_0 and A to be 14 μΩcm and 1.6 × 10^{-2} μΩcm/K^2, respectively. The residual resistivity ratio ρ(290K)/ρ_0 of the present single crystal is about 7 indicating the good sample quality. Considering the specific heat coefficient γ = 12.7 mJ/K^2mol [1], the A-value is exceptionally large and does not follow the Kadowali-Woods relation; A/γ^2 ~ 1.0 × 10^{-5} μΩcm/(mol K/mJ)^2 [4], which is in marked contrast with strongly correlated systems, while the origin is not clear at this stage.

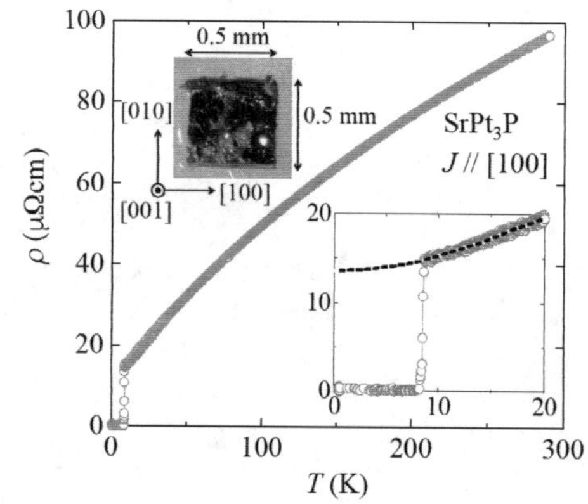

Figure 1. Temperature dependence of the electrical resistivity in $SrPt_3P$ for the current along J // [100]. The lower inset shows the expanded view at low temperature, where the dashed line represents low temperature fitting with T^2-dependence. The upper inset shows a photo of the single crystal.

Figure 2(a) shows typical dHvA oscillations in $SrPt_3P$ at the field angle θ= 76.5° from [100] direction in (010) plane, and figure 2(b) shows the corresponding fast Fourier transform (FFT) spectrum. Three fundamental dHvA signals, denoted α, β and δ, were observed as shown in figure 2(b), where 2δ is the 2nd harmonic of δ signal. The dHvA frequency F, which is expressed as a unit of magnetic field, is proportional to the extremal cross-sectional area S_F of the Femi surface (FS) with the following relation $F = c\hbar S_F/(2\pi e)$ [5]. To investigate the FS topology, we measured the angular dependence of the dHvA frequencies as shown in figure 2(c). The dHvA frequency-branches α and β are observed in the limited angle-ranges around H // [001]; these dHvA frequencies increase with increasing tilting field angles

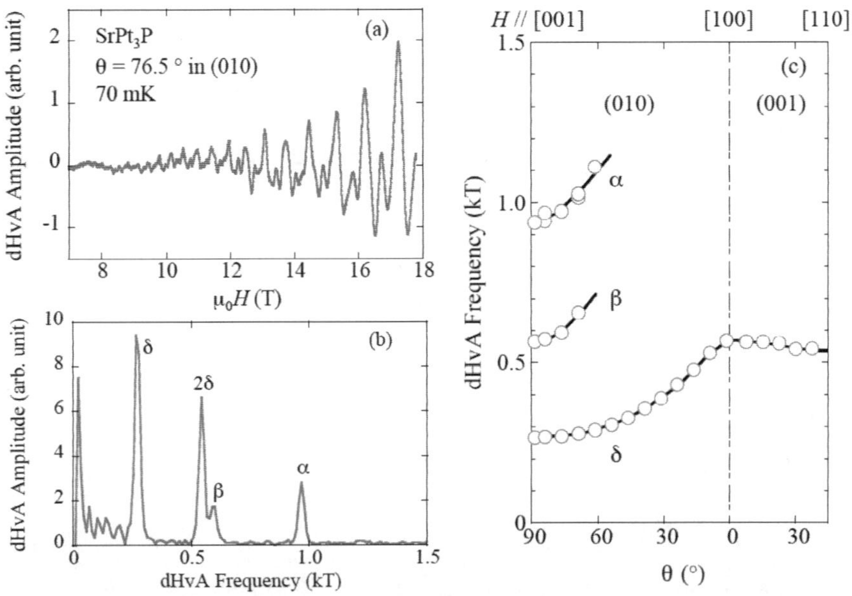

Figure 2. (a) Typical dHvA oscillations in $SrPt_3P$ at the field angle $\theta = 76.5^{\circ}$ from [100] in (010) plane, and figure 2(b) shows the corresponding fast Fourier transform (FFT) spectrum. (c) Angular dependence of the dHvA frequency. Solid lines in (c) are guides for the eyes.

from H // [001] in (010) and are no longer observable above ~ 30° from H // [001], suggesting the neck of an open FS sheet elongated along the [001] direction. The dHvA frequency-branch δ is continuously observed in the wide field-angle range, whose dHvA frequency also increases with increasing tilting field angle from H // [001] in (010), while it is almost constant between H // [100] and [110] in (001), suggesting the ellipsoidal FS which is long in the [001] direction. The size of the FS is small; S_F corresponding to the dHvA frequency of ~270 T for δ is only 2% of the area of the cross section normal to the [001] direction of the Brillouin zone. According to the results of band structure calculation [3], there are two bands crossing Fermi energy E_F, which form the multiply-connected FSs and a small ellipsoidal hole one. It seems that the α- and β-branches are corresponding to the pockets of the multiply-connected FSs and δ-branch is corresponding to the ellipsoidal one. To compare our experimental results with the band structure calculation in detail, the calculated angular dependence of the dHvA frequency is required, which will be performed in the future work.

From the temperature dependence of the dHvA amplitude, using the Lifshitz-Kosevich (LK) formula [5], we estimated the cyclotron effective mass m_c^* as shown in Table 1. The estimated m_c^*s range from 0.27 to 0.61 m_0, slightly depending on the field directions for δ-branch. These values are too small considering the experimental specific heat coefficient $\gamma = 12.7$ mJ/K^{2}mol [1]. From the observed dHvA frequencies and their cyclotron effective masses, assuming spherical Fermi surfaces, we calculated the electronic specific heat coefficient using the relation $\gamma = k_B^2 V m^* k_F/(3\hbar^2)$, where V is the molar volume, m^* is the effective mass, k_F is the caliber of the Fermi sphere; $S_F = \pi k_F^2$. The calculated γ-values for α-, β- and δ-branches are 3.0, 2.2 and 0.7 mJ/(K^{2}mol), respectively, and 5.9 mJ/(K^{2}mol) in total, which is about half of the γ-value estimated from the specific heat measurement. To compensate for this discrepancy, we expect the other dHvA signals missing in the present experiments. Farther dHvA experiments are needed at higher magnetic fields and/or in better quality single crystal samples. From the field dependence of the dHvA amplitude, using the LK formula, we estimated the Dingle temperature T_D which is inversely proportional to the scattering lifetime τ [$= \hbar/(2\pi k_B T_D)$]. The estimated values of T_D range from 6.7 to 12 K, from which we estimated the mean free path $\ell = 380$ ~ 590 Å using the relations $S_F = \pi k_F^2$, S_F $\hbar k_F = m_c^* v_F$, and $\ell = v_F \tau$, where v_F is the Fermi velocity. The

Table 1. dHvA frequency F, the corresponding cyclotron mass m_c^*, the Dingle temperature T_D and the mean free path ℓ in $SrPt_3P$.

Field angle (°)	Branch	F (T)	m_c^* / m_0	T_D (K)	ℓ (Å)
76.5 in (010)	α	973	0.61±0.02	6.7	590
	β	593	0.56±0.02	-	-
	δ	271	0.27±0.01	12	380
16.5 in (010)	δ	459	0.43±0.01	10	380

estimated value of ℓ will be compared with the superconducting coherent length ξ as shown below.

Finally, we show the superconducting phase diagram determined by the single crystal sample. Figure 3 (main panel) shows the temperature dependences of the upper critical field $\mu_0 H_{c2}$ for H // [100] and H // [001], which is evaluated form the field dependences of electrical resistivity at selected temperatures as shown in the inset; $\mu_0 H_{c2}$ is defined at the midpoint between zero resistivity and normal one. The $\mu_0 H_{c2}$–T phase diagram is basically the same as that reported in polycrystalline sample above 2 K [1]. In the present measurements, we investigated the $\mu_0 H_{c2}$ at lower temperatures down to ~0.5 K and also investigated the anisotropy between H // [100] and H // [001]. The obtained $\mu_0 H_{c2}$ is ~4.8 T at ~0 K with small anisotropy. Using the $\mu_0 H_{c2}$, the Ginzburg-Landau equation; $\mu_0 H_{c2} = \Phi_0/(2\pi\xi_0^2)$, yields the coherent length ξ_0 ~83Å, indicating the superconductivity is within the clean limit ($\xi_0 < \ell$). The small anisotropy of $\mu_0 H_{c2}$ is consistent with the s-wave superconductor.

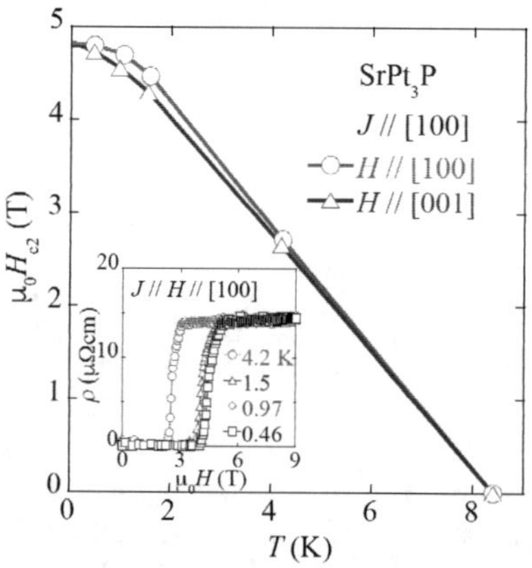

Figure 3. Temperature dependence of the upper critical field $\mu_0 H_{c2}$ in $SrPt_3P$ for H // [100] and H // [001]. Solid lines are guides for the eyes. The inset shows the field dependences of electrical resistivity at selected temperatures for H // [100].

4. Conclusions

In this study, we have succeeded in growing a single crystal of $SrPt_3P$ and have measured the dHvA effect and the electrical resistivity under magnetic fields. The dHvA frequencies up to ~ 1 kT with the cyclotron effective mass $m_c^* = 0.27 \sim 0.61\ m_0$ were observed. The superconducting upper critical fields $\mu_0 H_{c2}$ for fields along H // [001] and H // [100] in the tetragonal structure are almost the same in values of $\mu_0 H_{c2}$ ~4.8 T.

Acknowledgments
We are grateful to Professor H. Harima for the helpful discussions. This work was partly supported by the NIMS Joint Research Hub Program, JSPS KAKENHI Grant No. 18H04320 (J-Physics) and 21K03470.

References

[1] Takayama T, Kuwano K, Hirai D, Katsura Y, Yamamoto A and Takagi H 2012 *Phys. Rev. Lett.* **108** 237001
[2] Bauer E, Hilscher G, Michor H, Scheidt Ch Paul E W, Gribanov A, Seropegin Yu, Noël H, Sigrst M and Rogl P 2004 *Phys. Rev. Lett.* **92** 027003
[3] Kang C -J, Ahn K -H, Lee K -W and Min B I 2013 *J. Phys. Soc. Jpn.* **82** 053703
[4] Kadowaki K and Woods S B 1986 *Solid State Commun.* **58** 507
[5] Shoenberg D 1984 *Magnetic Oscillations in Metals* (Cambridge: Cambridge University Press)

Superconductivity in Ce_3PdIn_{11} single crystals – intrinsic phenomenon or parasitic effect?

G Chajewski, D Gnida, D Das and D Kaczorowski

Institute of Low Temperature and Structure Research, Polish Academy of Sciences, Okólna 2, 50-422 Wrocław, Poland

E-mail: g.chajewski@intibs.pl

Abstract. Single crystals of the compound Ce_3PdIn_{11} that belongs to a large family of superconducting cerium indides with the general formula $Ce_mT_nIn_{3m+2n}$ (T stands for *d*-electron transition metal) were grown from indium flux and characterized by means of low-temperature specific heat and electrical resisitivity measurements. The collected data revealed significant sample-dependent divergence in their magnetic and superconducting properties which raises the question of whether these two cooperative phenomena are intrinsic properties of Ce_3PdIn_{11}.

1. Introduction

In recent years, cerium-based indides have been intensively studied due to their very interesting physical properties, especially including heavy-fermion superconductivity and its coexistence with long-range magnetic ordering (see *e.g.* Refs.[1, 2, 3]). An example of such material is Ce_3PdIn_{11} that belongs to a group of ternaries with the general formula $Ce_mT_nIn_{3m+2n}$ (T = Co, Rh, Ir, Pd, Pt), which includes also the well-known phases $CeTIn_5$ and Ce_2TIn_8. The Ce_3PdIn_{11} indide crystallizes with a tetragonal crystal structure (space group $P4/mmm$, no. 123) that hosts two crystallographically inequivalent Ce sites in the unit cell [4]. The compound was characterized as a heavy-fermion antiferromagnet with two consecutive magnetic transitions at $T_{N1} = 1.67$ K and $T_{N2} = 1.53$ K and a large value of the Sommerfeld coefficient [5, 6]. Remarkably, Ce_3PdIn_{11} becomes superconducting below $T_c = 0.42$ K [5, 6], which suggests that long-range magnetic order and heavy-fermion superconductivity may coexists in this material at low temperatures. Thus, these unique physical properties make Ce_3PdIn_{11} an attractive platform for studying mutual relationship between antiferromagnetism and superconductivity at ambient pressure.

Earlier bulk-properties studies postulated that the antiferromagnetic ordering in Ce_3PdIn_{11} is associated with one of the Ce sublattices, while the other one is responsible for heavy-fermion behavior and superconductivity [5]. In order to verify this conjecture, more advanced measurements, such as neutron diffraction or muon spin rotation spectroscopy, must be performed. For this purpose, we made numerous attempts to synthesize large and high quality single crystals of this material. Unexpectedly, by performing basic characterization of the obtained crystals, we observed significant differences in their low-temperature properties. Here, we present a comparative study of the collected data.

2. Experimental details

Single crystals of Ce_3PdIn_{11} were grown out of indium flux, in a manner reported by Kratochvílová *et al.* [5, 7]. This technique yielded single crystals with dimensions up to 1 x 1 x 0.5 mm^3. Although most of excessive flux was centrifuged, some residuals of indium were found on the surface of crystals. Hence, individual crystals were additionally cleaned by manual dipping in isopropyl alcohol and scraping indium residues with the use of wooden toothpick (diluted hydrochloric acid, typically used to remove remaining indium flux, was found to attack also crystals of Ce_3PdIn_{11}).

The so-prepared samples were characterized by means of specific heat measurements performed in the temperature range from 0.35 K up to 20 K using a standard time-relaxation method [8] implemented in a commercial Quantum Design PPMS platform equipped with a ^{3}He cryostat. The same equipment was utilized to perform electrical resistivity measurements by means of conventional ac four-probe technique.

3. Results and discussion

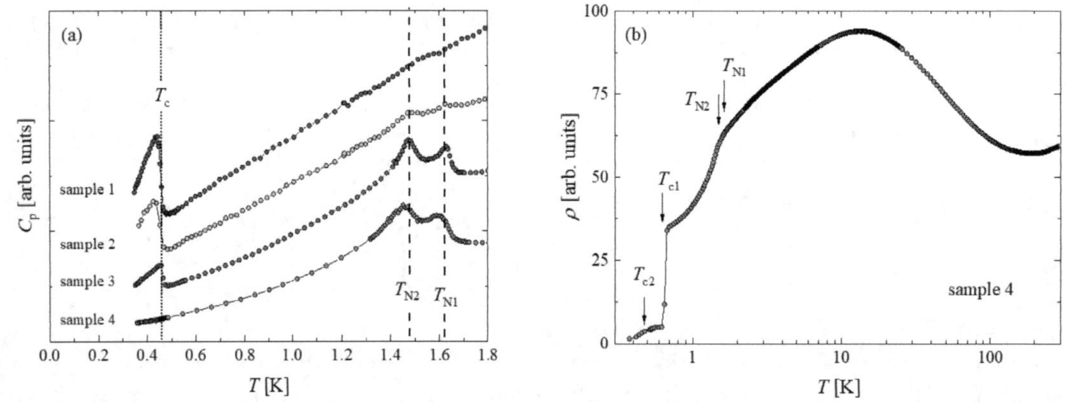

Figure 1. (a) Temperature dependencies of the specific heat measured on four different individual single crystals of Ce_3PdIn_{11} (for clarity, the data for each sample was shifted vertically). The positions of both magnetic phase transitions T_{N1} and T_{N2} are marked by dashed lines and the position of superconducting phase transition T_c is indicated by dotted line. (b) Temperature variation of the electrical resistivity of sample 4. Arrows mark the critical temperatures T_{c1} and T_{c2} of two superconducting phases and the temperatures of two magnetic transitions T_{N1} and T_{N2}.

Fig. 1(a) displays the low-temperature dependencies of the specific heat $C_p(T)$ of Ce_3PdIn_{11} measured on four different single-crystalline samples, labeled 1 to 4. Importantly, each sample was proven to be free from contamination by antiferromagnetic $CeIn_3$ (T_N = 10 K) that is a plausible impurity in the $Ce_mT_nIn_{3m+2n}$ systems [5, 6, 9, 10]. As can be inferred from the figure, notable differences were found in the $C_p(T)$ data of different crystals. While for sample 1 a distinct jump occurs at the critical temperature T_c = 0.47 K, no similar anomaly is seen for sample 4. In turn, two consecutive features related to the magnetic phase transitions at T_{N1} = 1.63 K and T_{N2} = 1.47 K can be distinguished for sample 4 and sample 3, while for sample 2 they are negligibly small and entirely absent for sample 1. It should be noted that the values of T_{N1} and T_{N2} are very close to those reported before [5, 6]. Another important observation is that the relative ratio of the specific heat magnitude at T_{N1} and T_c is different for each of the samples studied. In general, the following rule is obeyed: the larger $C_p(T)$ anomalies related to the antiferromagnetic ordering at T_{N1} and T_{N2}, the smaller specific heat jump associated with the superconducting phase transition. Since heat capacity is an additive quantity, this finding can be most straightforwardly explained by considering the presence of two independent phases in the crystals investigated. One of them plausibly represents antiferromagnetic Ce_3PdIn_{11}, while the other one, of unclear origin, would be responsible for the superconductivity. In such a scenario, sample 1 and sample 4 constitute pure (or almost pure) superconducting and antiferromagneic phases, respectively, whereas samples 2 and 3 contain a mixture of both phases in different proportion.

Although no bulk superconductivity was identified in the $C_p(T)$ data of sample 4, the electrical resisitivity $\rho(T)$ of this sample exhibits two clear drops, likely due to superconducting phase transitions (see Fig. 1(b)). Remarkably, the anomaly at $T_{c1} = 0.67$ K matches almost perfectly with the onset of superconductivity in Ce_2PdIn_8 [9, 10]. The absence of a corresponding anomaly in the heat capacity suggests that this feature may be caused by a minor contamination of sample 4 by the latter phase. The other drop in $\rho(T)$ is broader and occurs at about $T_{c2} = 0.46$ K that is the position of the specific heat jump due to the superconductivity in samples 1–3 . However, due to the lack of similar anomaly in $C_p(T)$ of sample 4 also this feature can be attributed to the presence of impurity phase rather than to intrinsic property of Ce_3PdIn_{11}. In this context it is also worth noting that down to 0.35 K, the electrical resistivity of sample 4 did not reach zero value. This finding is another indication that only small fraction of the measured specimen became superconducting below T_{c2}.

4. Conclusion

Our low-temperature measurements of the heat capacity of single-crystalline Ce_3PdIn_{11} revealed a behavior that appeared strongly sample-dependent. While some of the crystals were found to exhibit properties consistent with the data reported in the literature [5, 6], some others behaved quite differently. Quantitative analysis of the $C_p(T)$ and $\rho(T)$ curves measured for individual crystals leads us to hypothesize that the superconductivity and the antiferromagnetic order observed in the samples of Ce_3PdIn_{11} may actually originate from two distinct phases. Their differentiation in individual crystals can be very challenging because of complex phase equilibria near the ideal chemical composition Ce_3PdIn_{11}, and the formation of other ternaries with similar stoichiometry and closely related crystal structures [4]. As a result, crystals with stacking faults (or other structural defects) can be grown in a single batch that ostensibly represent the same material, but may harbor distinctly different electronic ground states.

Acknowledgment

This work was supported by the National Science Centre (Poland) under research grant no. 2015/19/B/ST3/03158.

References

[1] Knebel G, Aoki D, Braithwaite D, Salce B and Flouquet J 2006 *Phys. Rev. B* **74** 020501
[2] Knebel G, Buhot J, Aoki D, Lapertot G, Raymond S, Ressouche E and Flouquet J 2011 *J. Phys. Soc. Japan* **80** SA001
[3] Nicklas M, Stockert O, Park T, Habicht K, Kiefer K, Pham L D, Thompson J D, Fisk Z and Steglich F 2007 *Phys. Rev. B* **76** 052401
[4] Tursina A, Nesterenko S, Seropegin Y, Noël H and Kaczorowski D 2013 *Journal of Solid State Chemistry* **200** 7
[5] Kratochvílová M, Prokleška J, Uhlířová K, Tkáč V, Dušek M, Sechovský V and Custers J 2015 *Sci. Rep.* **5** 15904
[6] Das D, Gnida D and Kaczorowski D 2019 *Phys. Rev. B* **99** 054425
[7] Kratochvilova M, Dusek M, Uhlirova K, Rudajevova A, Prokleska J, Vondrackova B, Custers J and Sechovsky V 2014 *J. Cryst. Growth* **397** 47
[8] Hwang J S, Lin K J and Tien C 1997 *Rev. Sci. Instrum.* **68** 94

[9] Kaczorowski D, Pikul A P, Gnida D and Tran V H 2009 *Phys. Rev. Lett.* **103** 027003
[10] Kaczorowski D, Gnida D, Pikul A and Tran V 2010 *Solid State Commun.* **150** 411

Pressure-induced superconductivity in lightly doped $RFeAsO_{1-x}F_x$ (R=Sm and Nd)

K. Miyoshi[1,2], **A. Shiota**[1], **T. Matsuoka**[1], **K. Otsuka**[1], **S. Nishigori**[3] **and H. Kitagawa**[1,2]

[1]Department of Physics and Material Science, Shimane University, Matsue 690-8504, Japan
[2]Next Generation TATARA Co-Creation Centre, Shimane University, Matsue 690-8504, Japan
[3]Department of Materials Analysis, ICSR, Shimane University, Matsue 690-8504, Japan

E-mail: miyoshi@riko.shimane-u.ac.jp

Abstract. We performed the measurements of electrical resistivity (ρ) and dc magnetization (M) as a function of temperature (T) under pressure to investigate whether or not the superconductivity is induced by the application of pressure in lightly doped $RFeAsO_{1-x}F_x$ (R=Sm and Nd) using pulse current sintered (PCS) high density polycrystalline specimens. We have successfully observed pressure induced superconductivity with T_c of ~6 K (~8 K) above ~4 GPa (~1 GPa) for R=Sm (R=Nd). An anomaly corresponding to a magnetic phase transition into spin density wave (SDW) state was observed in the $\rho(T)$ curves in the entire pressure range, indicating that the superconducting and SDW states coexist in the T-P phase diagram. Volume fraction of the superconductivity is found to be very small. This is consistent with the coexistence of the superconducting and SDW phases.

1. Introduction

Since $LaFeAsO_{1-x}F_x$ has been found to exhibit superconductivity below T_c~26 K[1], intensive studies have been made and superconductivity has been discovered in materials which have iron-pnictide (Fe-Pn) layers consisting of edge-sharing $FePn_4$ tetrahedron, such as $RFeAsO_{1-x}F_x$ (R=lanthanide), $Ba_{1-x}K_xFe_2As_2$, AFeAs (A=Li, Na), FeSe. It is important to establish temperature (T) versus pressure (P) phase diagrams of iron-pnictide superconductors, where superconducting, antiferromagnetic (AFM) and nematic states coexist or compete, to throw further light on the superconducting mechanism. Indeed, in the $T-P$ phase diagram of FeSe, superconducting phase coexists with nematic phase for $0 \leq P \leq 1.5$ GPa and with dome-shaped AFM phase for $1.5 \leq P \leq 6$ GPa but highly enhanced above 6 GPa[2, 3]. In $RFeAsO_{1-x}F_x$ (R=lanthanide), the T-P phase diagram however remains solved, although several studies have been conducted for R=La[4, 5, 6], Ce[7, 8]. To study the relationship between the superconducting and other two phases in the phase diagram of $RFeAsO_{1-x}F_x$, we need to prepare lightly doped specimens so that the superconductivity is induced only above a critical pressure P_c, since the SDW and nematic phases do not appear in optimally doped superconducting specimens. Thus, to discover pressure induced superconductivity in $RFeAsO_{1-x}F_x$ is the first step to establish the T-P phase diagram of $RFeAsO_{1-x}F_x$. In the present work, we performed the measurements of electrical resistivity and dc magnetization under pressure for lightly doped $RFeAsO_{1-x}F_x$ with R=Sm and Nd using high density polycrystalline specimens obtained by pulse current sintering (PCS) and report the pressure induced superconductivity.

2. Experiment

High density PCS specimens of $RFeAsO_{1-x}F_x$ (R=Sm and Nd) were synthesized by the same procedure adopted in the previous study[6]. dc magnetization measurements under pressure were performed by using a miniature diamond anvil cell[9] with glycerin as pressure transmitting media (PTM), which was combined with a sample rod of commercial SQUID magnetometer. A small piece of high-purity lead (Pb) to determine the pressure from the T_c shift and $RFeAsO_{1-x}F_x$ were loaded into a sample room of 0.3-0.4 mm in diameter and 0.2 mm in height in a CuBe gasket. The magnetization data for $RFeAsO_{1-x}F_x$ and Pb were obtained by subtracting the magnetic contribution of the DAC measured in an empty run from the total magnetization[3, 11, 10]. Electrical resistivity under pressure was measured by using an indenter-type pressure cell for R=Sm and piston-cylinder cell for R=Nd with Daphne oil 7373 as PTM.

3. Results and Discussion

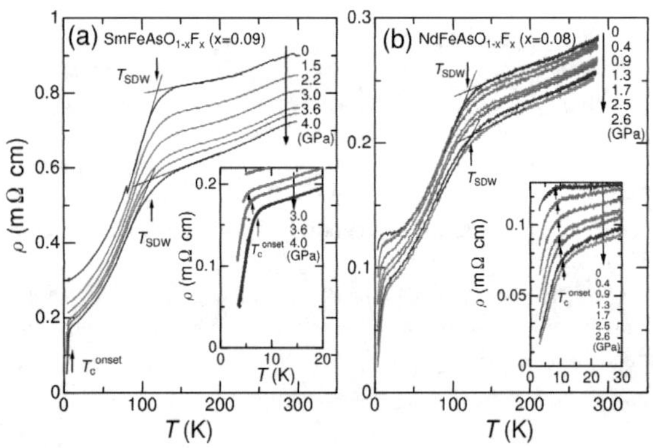

Figure 1. Temperature dependence of electrical resistivity (ρ) under various pressures for PCS specimens of $RFeAsO_{1-x}F_x$ with R=Sm (a) and Nd (b). The insets show enlarged views of $\rho(T)$ at low temperature.

To observe pressure induced superconductivity in $RFeAsO_{1-x}F_x$, we need to prepare the specimens with proper F-content x. As seen in a previous study on $LaFeAsO_{1-x}F_x$[6], if x is much lower than the critical value x_c above which the specimens exhibits superconductivity at ambient pressure, then we would fail to observe pressure induced superconductivity in the specimens, since the pressure we can generate is actually limited within ~10 GPa. According to our verification of superconductivity at ambient pressure for several specimens with various x, we assumed that x_c is around 0.10 for both R=Sm and Nd. Thus, we prepared the specimens with x=0.09 for R=Sm and x=0.08 for R=Nd and verified the pressure induced superconductivity for the specimens in the present study.

We show temperature dependence of resistivity (ρ) at various pressures for R=Sm and Nd in Figs. 1(a) and 1(b), respectively. In Fig. 1(a), the $\rho(T)$ curve for R=Sm at ambient pressure shows a rapid decrease below T_{SDW}~120 K, corresponding to the transition into the SDW phase, which is known to be ~140 K for a non-doped specimen[12]. As shown in the inset, a sharp drop begins to appear in the $\rho(T)$ curve above 3 GPa below T_c^{onset}, which is ~5 K but increases as increasing pressure, suggesting a pressure induced superconducting transition. The $\rho(T)$ curve for R=Nd at ambient pressure in Fig. 2(b) is qualitatively similar to that for R=Sm. In the inset, a downturn probably corresponding to the superconductivity is seen in the $\rho(T)$ curve even at ambient pressure below T_c^{onset}~8 K, which gradually increases as increasing pressure.

To confirm the superconductivity observed by the $\rho(T)$ measurements under pressure, zero-field cooled dc magnetization (M) measurements were performed under pressure. The results are shown in Figs. 2(a) and 2(b). In Fig. 2(a), the $M(T)$ curve for R=Sm is found to exhibit a sharp drop below ~7 K at ambient pressure, which corresponds to a superconducting transition

Journal of Physics: Conference Series **2164** (2022) 012007 doi:10.1088/1742-6596/2164/1/012007

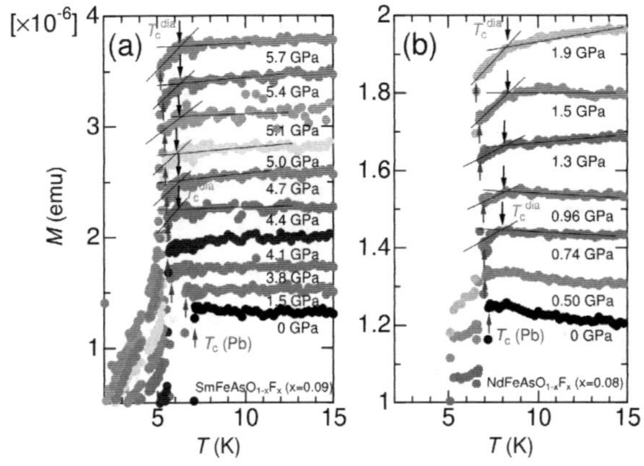

Figure 2. Temperature dependence of dc magnetization under various pressures for PCS specimens of $RFeAsO_{1-x}F_x$ with R=Sm (a) and Nd (b). The data are intentionally shifted along the longitudinal axis for clarity

of Pb used as a manometer. At 4.4 GPa, the $M(T)$ curve however appears to exhibit a small downturn below T_c^{dia}~8 K, just above the T_c of Pb, which is similarly seen in each $M(T)$ curve above 4.4 GPa, suggesting the onset of the superconductivity induced in the specimen by the application of pressure. A similar behavior is observed in the specimen with R=Nd in Fig. 2(b), where the superconductivity is found to be induced by the pressure above 0.74 GPa. As seen in the figures, it is found that the amplitude of diamagnetic response of the specimens is much smaller than that of Pb, even though the volume of the specimens is much larger than that for Pb. Thus, we roughly estimated the volume fraction of the superconductivity p from the diamagnetic amplitude obtained with extrapolating it to T=0. Assuming that the sample room is half filled with the sample, we obtained p~5 % for R=Sm and ~2.5 % for R=Nd, while p~10 % for R=La in a previous study[6]. These results suggest that the pressure induced superconductivity in $RFeAsO_{1-x}F_x$ is filamentary one.

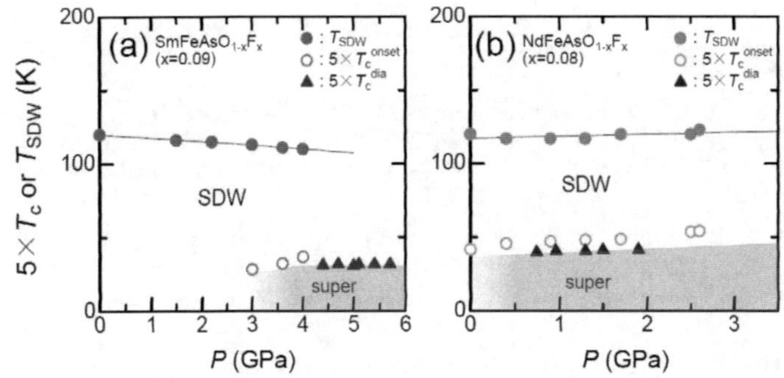

Figure 3. Plots of T_c^{onset}, T_c^{dia} and T_{SDW} versus pressure for PCS specimens of $RFeAsO_{1-x}F_x$ with R=Sm (a) and Nd (b).

Figures 3(a) and 3(b) show plots of T_c^{onset}, T_c^{dia} and T_{SDW} as a function of pressure. T_{SDW} was estimated by extrapolating the initial slope of the $\rho(T)$ curve just below T_{SDW} to that above T_{SDW}. In the figures, T_{SDW} for R=Sm decreases slightly as increasing temperature with a pressure coefficient of $dT_{SDW}/dT \sim -2.5$ K/GPa, whereas dT_{SDW}/dT is +1.4 K/GPa for R=Nd. T_c^{onset} and T_c^{dia} agree with each other, suggesting that both temperatures well correspond to the onset of the superconductivity. We note that T_c^{onset} is always higher than T_c^{dia} and a resistive drop begins to appear at lower pressure than the pressure at which a diamagnetic drop appears. Thus, we could say that electrical resistivity is more sensitive for detecting a formation of superconducting network. The superconductivity with small volume fraction is probably due to the coexistence with the SDW phase. We could expect that the SDW phase disappears by the application of much higher pressure especially for R=Sm, since dT_{SDW}/dT is negative for R=Sm. In this case, the superconducting phase would be enhanced having a maximum T_c at the pressure where the SDW phase disappears due to the enhancement of the antiferromagnetic fluctuation, if $SmFeAsO_{1-x}F_x$ hosts a quantum critical point, such as $SrFe_2As_2$[13, 14], $BaFe_2As_2$[14] and $BaFe_2(As_{1-x}P_x)$[15]. Although dT_{SDW}/dT is very small but positive for R=Nd and La[6], similar features are expected under much higher pressure. Experiments under high pressure above 10-20 GPa are necessary to establish the $T-P$ phase diagram and verify the quantum criticality.

4. Summary

In the present work, we have performed dc magnetization and electrical resistivity measurements under pressure for lightly doped $RFeAsO_{1-x}F_x$ (R=Sm and Nd) using PCS specimens. Pressure induced superconductivity is successfully observed above 3.5−4.5 GPa for R=Sm and 0.5−1.0 GPa for R=Nd, but the volume fraction of the superconductivity is very small less than 5 % probably due to the coexistence of the SDW phase. The pressure coefficient is found to be very small so that much higher pressure is necessary to establish the $T-P$ phase diagram.

5. Acknowledgment

This work was supported by JSPS KAKENHI (Grant Number JP18K03516). The authors thank T. Matsumoto, S. Tsuchiya, S. Kajikawa and K. Sugiwaki for technical assistance.

References

[1] Y. Kamihara, T. Watanabe, M. Hirano, and H. Hosono, J. Am. Chem. Soc. **130**, 3296 (2008).
[2] J. P. Sun, K. Matsuura, G. Z. Ye, Y. Mizukami, M. Shimozawa, K. Matsubayashi, M. Yamashita, T. Watashige, S. Kasahara, Y. Matsuda, J. -Q. Yan, B. C. Sales, Y. Uwatoko, J. -G. Cheng, and T. Shibauchi, Nat. Commun. **7**, 12146 (2016).
[3] K. Miyoshi, S. Yamamoto, A. Shiota, T. Matsuoka, M. Ohe, Y. Yamamoto, and S. Nishigori J. Phys. Soc. Jpn. **90**, 073706 (2021).
[4] H. Okada, K. Igawa, H. Takahashi, Y. Kamihara, M. Hirano, H. Hosono, K. Matsubayashi, and Y. Uwatoko, J. Phys. Soc. Jpn. **77**, 113712 (2008).
[5] C. A. McElroy, J. J. Hamlin, B. D. White, S. T. Weir, Y. K. Vohra, and M. B. Maple, Phys. Rev. B **90**, 125134 (2014).
[6] K. Miyoshi, K. Otsuka, A. Shiota, Y. Shimojo, G. Motoyama, K. Fujiwara, H. Kitagawa, and S. Nishigori, Physica B **536** 827 (2018).
[7] N. Takeshita, K. Miyazawa, A. Iyo, H. Kito, and H. Eisaki, J. Phys. Soc. Jpn. **78**, 065002 (2009).
[8] D. A. Zocco, R. E. Baumbach, J. J. Hamlin, M. Janoschek, I. K. Lum, M. A. McGuire, A. S. Sefat, B. C. Sales, R. Jin, D. Mandrus, J. R. Jeffries, S. T. Weir, Y. K. Vohra, and M. B. Maple Phys. Rev. B **83**, 094528 (2011).
[9] M. Mito, M. Hitaka, T. Kawae, K. Takeda, T. Kitai, and N. Toyoshima, Jpn. J. Appl. Phys. **40**, 6641 (2001).
[10] K. Miyoshi, E. Kojima, S. Ogawa, Y. Shimojo, and J. Takeuchi, Phys. Rev. B **87**, 235111 (2013).
[11] K. Miyoshi, K. Morishita, E. Mutou, M. Kondo, O. Seida, K. Fujiwara, J. Takeuchi, and S. Nishigori, J. Phys. Soc. Jpn. **83** 013702 (2014).
[12] R. H. Liu, G. Wu, T. Wu, D. F. Fang, H. Chen, S. Y. Li, K. Liu, Y. L. Xie, X. F. Wang, R. L. Yang, L. Ding, C. He, D. L. Feng and X. H. Chen, Phys. Rev. Lett. **101**, 087001 (2008).
[13] H. Kotegawa, H. Sugawara and H. Tou, J. Phys. Soc. Jpn. **78** 013709 (2009).
[14] K. Matsubayashi, N. Katayama, K. Ohgushi, A. Yamada, K. Munakata, T. Matsumoto and Y. Uwatoko, J. Phys. Soc. Jpn. **78** 073706 (2009).
[15] K. Hashimoto, K. Cho, T. Shibauchi, S. Kasahara, Y. Mizukami, R. Katsumata, Y. Tsuruhara, T. Terashima, H. Ikeda, M. A. Tanatar, H. Kitano, N. Salovich, R. W. Giannetta, P. Walmsley, A. Carrington, R. Prozorov and Y. Matsuda, Science **336** 1554 (2012).

Unusual superconductivity in the topological nodal-line semimetal candidate $Sn_xNbSe_{2-\delta}$

Riffat Munir[1], K A M Hasan Siddiquee[1], Charuni Dissanayake[1], Kapila Kumarasinghe[1], Xinzhe Hu[2], Yasumasa Takano[2], Eun Sang Choi[3], Yasuyuki Nakajima[1]

[1] Department of Physics, University of Central Florida, Orlando, Florida 32816, USA
[2] Department of Physics, University of Florida, Gainesville, Florida 32611, USA
[3] National High Magnetic Field Laboratory, Florida State University, Tallahassee, Florida 32310, USA

E-mail: yasuyuki.nakajima@ucf.edu

Abstract. We report the superconductivity of the topological nodal-line semimetal candidate $Sn_xNbSe_{2-\delta}$ with a noncentrosymmetric crystal structure. The superconducting transition temperature T_c of $Sn_xNbSe_{2-\delta}$ drastically varies with the Sn concentration x and the Se deficiency δ, and reaches 12 K, relatively higher than those of known topological superconductors. The upper critical field of this compound shows unusual temperature dependence, inconsistent with the WHH theory for conventional type-II superconductors. In a low-T_c sample, the zero-temperature limit of the upper critical field parallel to the *ab* plane exceeds the Pauli paramagnetic limit estimated from the simple BCS weak coupling model by a factor of ~ 2, suggestive of unusual superconductivity stabilized in $Sn_xNbSe_{2-\delta}$. Together with the robust superconductivity against disorder, these observations indicate that $Sn_xNbSe_{2-\delta}$ is a promising candidate to explore topological superconductivity.

1. Introduction

Topological superconductivity has attracted great interest because of the potential of Majorana fermions for technological application [1]. Promising candidates for topological superconductors are noncentrosymmetric superconductors with strong spin-orbit coupling. The lack of centrosymmetry can induce an odd-parity pairing state, which is a prerequisite of topological superconductivity [2]. Among such noncentrosymmetric systems, the topological nodal-line semimetal $PbTaSe_2$ shows superconductivity at 3.8 K [3], and its superconducting gap exhibits nematic behavior [4], similar to metal-intercalated Bi_2Se_3 [5]. In addition, a member of the $PbTaSe_2$ family, $SnNbSe_2$, is theoretically predicted to be a topological nodal-line semimetal and a superconductor [6]. The calculated superconducting transition temperature of this compound is 7 K, relatively higher than most of the known topological superconductor candidates. The higher T_c can be beneficial to probe Majorana bound states with the energy spacing Δ^2/E_F, where Δ the superconducting gap and E_F is the Fermi energy [7], allowing us to investigate topological superconductivity.

We here report unusual superconductivity of $Sn_xNbSe_{2-\delta}$ with the same noncentrosymmetric structure as stoichiometric $SnNbSe_2$. The superconducting transition temperatures are strongly dependent on Sn concentration x and Se deficiency δ, both of which are closely correlated to carrier doping. The observed upper critical fields of $Sn_xNbSe_{2-\delta}$ cannot be described by the Werthamer–Helfand–Hohenberg (WHH) theory for conventional type-II superconductors [8, 9]. Moreover, the zero-temperature value of the upper critical field of $Sn_{0.16}NbSe_{1.76}$ is beyond the BCS Pauli paramagnetic limit. The observed anomalous upper critical fields, along with the robust superconductivity against disorder, are suggestive of a possible unconventional superconducting pairing state realized in $Sn_xNbSe_{2-\delta}$.

2. Experimental methods

Single crystals of $Sn_xNbSe_{2-\delta}$ were grown by using a self-flux method [10]. The typical size of the resulting crystals is about $1 \times 1 \times 0.3$ mm^3. We observed a single crystal x-ray diffraction pattern consistent with the noncentrosymmetric crystal structure of $SnNbSe_2$ with the space group $P\bar{6}m2$ (the upper inset of fig.1). We determined the atomic ratio of the crystals with x-ray fluorescence spectroscopy.

3. Results and discussion

$Sn_xNbSe_{2-\delta}$ undergoes a superconducting transition at low temperatures [10]. Figure 1 shows the typical temperature dependence of resistivity of $Sn_xNbSe_{2-\delta}$. Overall temperature dependence of the resistivity is metallic. Because of the high normal-state resistivity right above T_c, the typical residual resistivity ratio $RRR = \rho(300\text{ K})/\rho(T_c)$ is ~ 3 and 30-50 times as small as that of $PbTaSe_2$ [11], indicating the presence of abundant disorder. The lower inset of Fig. 1 depicts the low-temperature resistivity of $Sn_xNbSe_{2-\delta}$. The superconducting transition temperatures T_c determined by the midpoints of resistive transitions vary up to 12 K, depending on Sn concentration x and Se deficiency δ, while no discernible correlation between T_c and $\rho(T_c)$ is observed. Surprisingly, despite the low RRR due to disorder, the observed highest T_c of 12 K is beyond the theoretical prediction of 7 K, as well as T_c of most known topological superconductors.

Figure 2(a) shows the Hall resistivity of $Sn_xNbSe_{2-\delta}$ as a function of magnetic fields at $T =$ 15 K. The Hall resistivity exhibits linear-in-H behavior, and its sign is positive, suggesting that a large cylindrical hole band around the Γ point [6] predominantly contributes to the charge

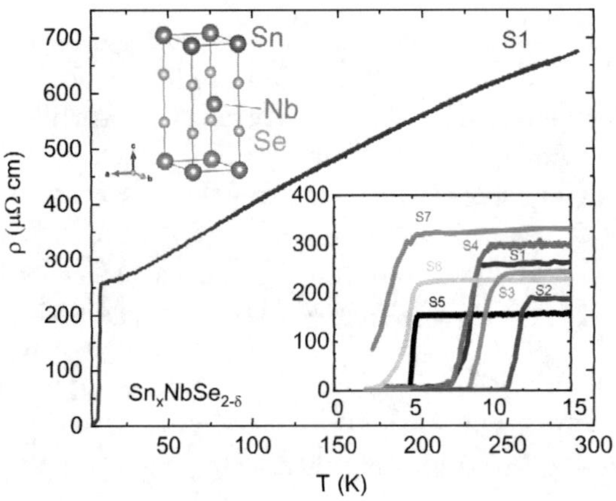

Figure 1. Main panel: Tempearture dependence of resistivity of the sample S1 ($Sn_{0.15}NbSe_{1.69}$). Upper inset: Crystal structure of $Sn_xNbSe_{2-\delta}$. Lower inset: Low-temperature resistivity of the samples S1, S2 ($Sn_{0.13}NbSe_{1.62}$), S3 ($Sn_{0.15}NbSe_{1.65}$), S4 ($Sn_{0.14}NbSe_{1.71}$), S5 ($Sn_{0.13}NbSe_{1.70}$), S6 ($Sn_{0.16}NbSe_{1.76}$), and S7 ($Sn_{0.19}NbSe_{1.73}$).

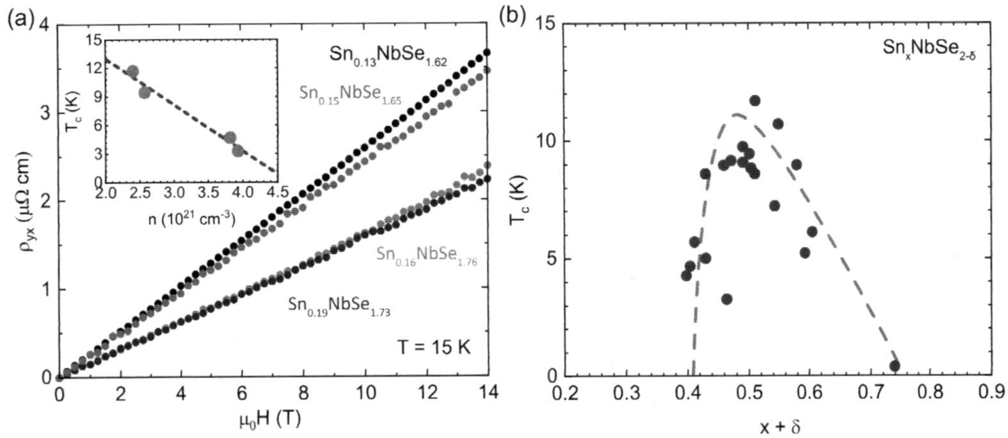

Figure 2. (a) Main panel: Hall resistivity of $Sn_xNbSe_{2-\delta}$ as a function of magnetic field at $T =$ 15 K. Inset: T_c versus the carrier concentration n obtained from the Hall coefficient $R_H = 1/ne$. A blue dashed line is a linear fit to the data. (b) Superconducting transition temperature T_c of $Sn_xNbSe_{2-\delta}$ as a function of the sum of Sn concentration x and Se deficiency δ. A red dashed line is a guide to the eye.

transport. The Hall coefficient R_H, extracted from the slope of the Hall resistivity, gives carrier densities $n \sim 10^{21}$ cm^{-3}, close to, but rather smaller than n of $PbTaSe_2$ [12]. As shown in the inset of fig.2(a), T_c linearly decreases with n in this carrier density range, suggesting that the carrier density is a key parameter to determine T_c in $Sn_xNbSe_{2-\delta}$.

In fact, T_c appears to be fine-tuned by carrier doping. Assuming that the oxidation states of Sn and Se are 2+ and 2−, respectively, the sum of the Sn concentration x and the Se deficiency δ is proportional to the number of electrons provided by intercalated Sn^{2+} and deficient Se^{2-} ions. We plot the superconducting transition temperature as a function of $x + \delta$ in Fig.2(b). We observe a superconducting dome upon varying $x + \delta$. The carrier density region in the inset of Fig.2(a) falls within the narrow range of $x + \delta = 0.40 - 0.51$ in Fig.2(b).

Figure 3 shows temperature dependence of H_{c2} of $Sn_{0.13}NbSe_{1.70}$ (T_c = 5.0 K) and $Sn_{0.14}NbSe_{1.71}$ (T_c = 8.6 K) in the in-plane magnetic field directions, $H \parallel ab$, and the out-of-plane direction, $H \parallel c$. We determined H_{c2} by the midpoints of resistive transitions, applying the electrical current I in the ab plane. We find a distinct difference in the anisotropy of H_{c2}, $\Gamma = H_{c2}^{ab}/H_{c2}^{c}$, between these samples; $\Gamma = 3.0$ in $Sn_{0.13}NbSe_{1.70}$, while $\Gamma = 1.1$ in $Sn_{0.14}NbSe_{1.71}$. Regardless of the magnetic field directions, the temperature dependence of H_{c2} of both samples increases linearly with decreasing temperatures, different from those of conventional type-II superconductors. The upper critical fields of conventional superconductors are described by the WHH theory. Its zero-temperature value $H_{c2}(0)$ is given by $H_{c2}(0) = -\alpha T_c dH_{c2}/dT \mid_{T=T_c}$, where α is 0.69 for the dirty limit and 0.73 for the clean limit [8, 9]. However, we obtained $\alpha \sim 1$ for both samples, suggesting unusual upper critical fields of $Sn_xNbSe_{2-\delta}$. Notably, the in-plane $H_{c2}(0)$ of $Sn_{0.13}NbSe_{1.70}$ is about 16 T, surpassing the simple BCS Pauli paramagnetic limit, $\mu_0 H_p = 1.84 k_B T_c = 9.3$ T, by a factor of ~ 2, suggestive of a possible odd-parity component in the pairing state allowed by the noncentrosymmetric structure. On the other hand, the out-of-plane $H_{c2}(0)$ is lower than $\mu_0 H_p$. In $Sn_{0.14}NbSe_{1.71}$, the measured zero-temperature values in both magnetic field directions are smaller than the estimated Pauli paramagnetic limit $\mu_0 H_p =$ 15.6 T, enhanced by the high T_c of 8.6 K.

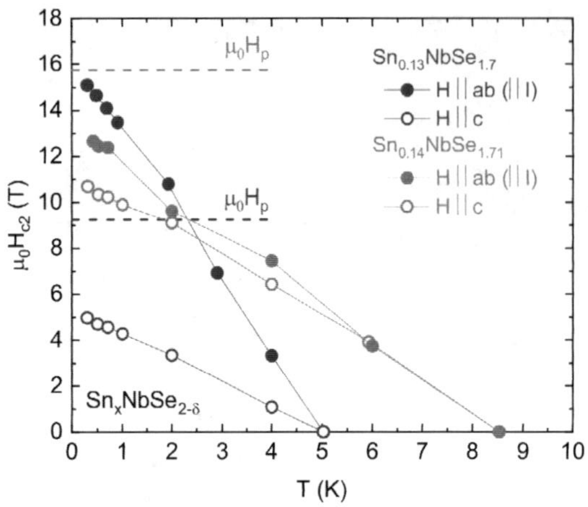

Figure 3. Temperature dependence of upper critical fields of $Sn_xNbSe_{2-\delta}$ for $H \parallel ab$ ($\parallel I$) and $H \parallel c$. A blue dashed line indicates the Pauli paramagnetic limit $\mu_0 H_p = 9.3$ T for $Sn_{0.16}NbSe_{1.76}$ and a red dashed line indicates $\mu_0 H_p = 15.6$ T for $Sn_{0.13}NbSe_{1.70}$.

4. Summary

We have investigated the resistivity, Hall effect, and upper critical fields of $Sn_xNbSe_{2-\delta}$. The superconducting transition temperatures can be fine-tuned up to 12 K by Sn concentration and Se deficiency, correlated with carrier densities. The zero-temperature value of upper critical field $\mu_0 H_{c2}(0)$ of the lower T_c sample ($Sn_{0.13}NbSe_{1.70}$) is beyond the Pauli paramagnetic limit, indicating an unusual pairing state realized in this systems. These findings suggest that $Sn_xNbSe_{2-\delta}$ can be a promising platform for exploring the relationship between superconductivity and topological properties.

Acknowledgments

This work was supported by the start-up fund from the University of Central Florida. HS and YN are supported by NSF CAREER DMR-1944975, and XH and YT by the NHMFL UCGP program. The NHMFL is supported by the National Science Foundation through NSF/DMR-1644779 and the State of Florida.

References

[1] Nayak C, Simon S H, Stern A, Freedman M and Das Sarma S 2008 *Rev. Mod. Phys.* **80** 1083–1159
[2] Sato M and Ando Y 2017 *Rep. Prog. Phys.* **80** 076501
[3] Ali M N, Gibson Q D, Klimczuk T and Cava R J 2014 *Phys. Rev. B* **89** 020505
[4] Le T, Sun Y, Jin H K, Che L, Yin L, Li J, Pang G, Xu C, Zhao L, Kittaka S, Sakakibara T, Machida K, Sankar R, Yuan H, Chen G, Xu X, Li S, Zhou Y and Lu X 2020 *Sci. Bull.* **65** 1349
[5] Pan Y, Nikitin A M, Araizi G K, Huang Y K, Matsushita Y, Naka T and de Visser A 2016 *Sci. Rep.* **6** 28632EP
[6] Chen P J, Chang T R and Jeng H T 2016 *Phys. Rev. B* **94** 165148
[7] Beenakker C W J 2015 *Rev. Mod. Phys.* **87** 1037
[8] Werthamer N, Helfand E and Hohenberg P 1966 *Phys. Rev.* **147** 295
[9] Helfand E and Werthamer N R 1966 *Phys. Rev.* **147** 288
[10] Munir R, Siddiquee K A M H, Dissanayake C, Hu X, Takano Y, Choi E S and Nakajima Y 2021 *J. Phys. Condens. Matter* **33** 23LT01
[11] Zhang C L, Yuan Z, Bian G, Xu S Y, Zhang X, Hasan M Z and Jia S 2016 *Phys. Rev. B* **93** 054520
[12] Xu C Q, Sankar R, Zhou W, Li B, Han Z D, Qian B, Dai J H, Cui H, Bangura A F, Chou F C and Xu X 2017 *Phys. Rev. B* **96** 064528

Unconventional singlet-triplet superconductivity

David Möckli
Instituto de Física, Universidade Federal do Rio Grande do Sul, 91501-970 Porto Alegre, Brazil
E-mail: mockli@ufrgs.br

Abstract. Have you been lying awake wondering what symmetries determine whether a superconductor is spin-singlet, triplet, or both? We show that if one supplies additional degrees of freedom to BCS theory, spin-singlet can coexist with spin-triplet superconductivity. We guide the reader to the most general superconducting state using symmetry arguments. If both singlet and triplet pairing channels act, a magnetic field can convert between spin-singlet and triplet states. Two possible singlet-triplet superconductors candidates are: $CeRh_2As_2$ and bilayer-$NbSe_2$.

1. Introduction

The two key symmetries in conventional Bardeen-Cooper-Schrieffer (BCS) theory are time-reversal and inversion [1, 2]. The BCS superconducting ground state $|\psi_{\mathsf{BCS}}\rangle$ has a rigid phase, which occurs through the $U(1)$ phase symmetry breaking of the normal state. Unconventional superconductors break additional symmetries such as rotation, or the key symmetries of BCS theory. Another venue to unconventional superconducting states is examining additional degrees of freedom (DOFs), such as orbital, valley, sublattice, etc., which introduce non-trivial symmetries. Here, we introduce the superconducting state-vector of a superconductor with additional DOFs. The additional DOF leads to the possibility of having a superposition of singlet and triplet components in the state-vector. This contrasts with non-centrosymmetric superconductors, where the explicit removal of inversion symmetry is required.

2. The superconducting state-vector

Consider a superconducting order parameter $\Delta^{ab}_{s_1 s_2}(\mathbf{k}) \propto \langle c_{-\mathbf{k}s_2 b} c_{\mathbf{k}s_1 a}\rangle$, where $\mathbf{k}$ is momentum, $s_1, s_2 = \{\uparrow,\downarrow\}$ is the z-spin projection, and $a, b = \{1, 2\}$ is an additional internal DOF that could refer to orbitals, sublattice or layers. We assume that corresponding superconducting state-vectors to the order parameter can be written as

$$\left|\psi^{ab}_{sm}(\mathbf{k})\right\rangle = \phi(\mathbf{k})|\chi_{sm}\rangle \otimes |\Upsilon_{ab}\rangle, \tag{1}$$

where s is the total spin and $m = s_1 + s_2$, which is $s = m = 0$ for a singlet state and $s = 1$ for the triplet states $m = \{1, 0, -1\}$ [3]. Let us write the singlet state as $|\psi^{ab}_S(\mathbf{k})\rangle = \phi(\mathbf{k})|\chi_{00}\rangle \otimes |\Upsilon_{ab}\rangle = \phi(\mathbf{k})(|\uparrow\downarrow\rangle - |\downarrow\uparrow\rangle)| \otimes |\Upsilon_{ab}\rangle$ and the triplet states $|\psi^{ab}_{Tm}(\mathbf{k})\rangle = \phi(\mathbf{k})|\chi_{1,m}\rangle \otimes |\Upsilon_{ab}\rangle$, where $|\chi_{1,0}\rangle = |\uparrow\downarrow\rangle + |\downarrow\uparrow\rangle$, $|\chi_{1,1}\rangle = |\uparrow\uparrow\rangle$ and $|\chi_{1,-1}\rangle = |\downarrow\downarrow\rangle$. For the four possibilities of $|\Upsilon_{ab}\rangle$ we choose $|\Upsilon_{11}\rangle = |1,1\rangle + |2,2\rangle$, $|\Upsilon_{22}\rangle = |1,1\rangle - |2,2\rangle$, $|\Upsilon_{12}\rangle = |1,2\rangle + |2,1\rangle$ and $|\Upsilon_{21}\rangle = |1,2\rangle - |2,1\rangle$. The four possible spin states combined with the four possible internal DOF states give 16 possible configurations, as opposed to only 4 possibilities without the additional DOF.

2.1. Permutation and inversion symmetry

The Pauli principle enforces that under the action of the permutation operator $\mathsf{P}|\psi^{ab}_{sm}(\mathbf{k})\rangle = -|\psi^{ab}_{sm}(\mathbf{k})\rangle$. The permutation operator P takes $\mathbf{k} \to -\mathbf{k}$, $s_1 \leftrightarrow s_2$, and exchanges the particle subspaces $a \leftrightarrow b$. Then we have for singlets and triplets, respectively

$$\mathsf{P}\left|\psi^{ab}_{S}(\mathbf{k})\right\rangle = \mathsf{P}\left[\phi(\mathbf{k})|\chi_{00}\rangle \otimes |\Upsilon_{ab}\rangle\right] = -\phi(-\mathbf{k})|\chi_{00}\rangle \otimes |\Upsilon_{ba}\rangle \overset{\text{Pauli}}{=} -\phi(\mathbf{k})|\chi_{00}\rangle \otimes |\Upsilon_{ab}\rangle; \quad (2)$$

$$\mathsf{P}\left|\psi^{ab}_{Tm}(\mathbf{k})\right\rangle = \mathsf{P}\left[\phi(\mathbf{k})|\chi_{1,m}\rangle \otimes |\Upsilon_{ab}\rangle\right] = \phi(-\mathbf{k})|\chi_{1,m}\rangle \otimes |\Upsilon_{ba}\rangle \overset{\text{Pauli}}{=} -\phi(\mathbf{k})|\chi_{1,m}\rangle \otimes |\Upsilon_{ab}\rangle. \quad (3)$$

If we ignore the internal DOF $|\Upsilon_{ab}\rangle$, the last equality in Eqs. (2) and (3) tell us that for a singlet (triplet) state the momentum structure $\phi(\mathbf{k})$ must be even (odd) [1, 2]. However, in the presence of $|\Upsilon_{ab}\rangle$, even-momentum-triplets and odd-momentum-singlets can now exist, because the Fermionic antisymmetry can now be carried by an additional DOF $|\Upsilon_{ab}\rangle$.

It is instructive to determine the inversion (parity) sectors of the superconducting state-vectors. The action of the inversion operator $\mathsf{I}\left|\psi^{ab}_{sm}(\mathbf{k})\right\rangle$ takes $\mathbf{k} \to -\mathbf{k}$, leaves the spins $\{s_1, s_2\}$ invariant, and we assume that $\mathsf{I}|\Upsilon_{ab}\rangle$ exchanges not the particle subspaces, but the DOFs $1 \leftrightarrow 2$, as would be the case when different sublattices are related by inversion. Then we have

$$\mathsf{I}\left|\psi^{ab}_{sm}(\mathbf{k})\right\rangle = \mathsf{I}\left[\phi(\mathbf{k})|\chi_{sm}\rangle \otimes |\Upsilon_{ab}\rangle\right] = \phi(-\mathbf{k})|\chi_{sm}\rangle \otimes \mathsf{I}|\Upsilon_{ab}\rangle = \phi(-\mathbf{k})|\chi_{sm}\rangle \otimes \begin{cases} |\Upsilon_{ab}\rangle, \ (a=1) \\ -|\Upsilon_{ab}\rangle, \ (a=2). \end{cases} \quad (4)$$

If we ignore $|\Upsilon_{ab}\rangle$, Eq. (4) tells us that singlets ($\phi(\mathbf{k}) = \phi(-\mathbf{k})$) belong to the $+1$ inversion sector, whereas triplets ($\phi(\mathbf{k}) = -\phi(-\mathbf{k})$) belong to the -1 inversion sector. Therefore, if the system is inversion symmetric, the superconducting state-vector must have a definite parity, that is, it either belongs to the parity eigenvalue sector $+1$ or -1. In this case, inversion symmetry prohibits a superposition of a singlet with a triplet state. When an inversion symmetric superconducting material lacks additional DOFs, one then knows that its superconducting state is either singlet or triplet. What if inversion symmetry lacks? Then, there is no definite parity (and no selection rule) which allows a singlet-triplet superposition in the state-vector. The main message of Eq. (4) is that in the presence of additional DOFs $|\Upsilon_{ab}\rangle$, a singlet-triplet superposition is symmetry allowed even in inversion symmetric systems.

2.2. Singlet-triplet states

As an example, consider the superposition $|\psi(\mathbf{k})\rangle = c_1|\psi^{11}_{S}(\mathbf{k})\rangle + c_2|\psi^{22}_{T0}(\mathbf{k})\rangle$, where c_1 and c_2 are complex coefficients. By applying the permutation operator, we obtain

$$\mathsf{P}|\psi(\mathbf{k})\rangle = c_1\mathsf{P}|\psi^{11}_{S}(\mathbf{k})\rangle + c_2\mathsf{P}|\psi^{22}_{T0}(\mathbf{k})\rangle = c_1\phi_1(-\mathbf{k})\left[-|\chi_{00}\rangle\right] \otimes |\Upsilon_{11}\rangle + c_2\phi_2(-\mathbf{k})|\chi_{T,0}\rangle \otimes |\Upsilon_{22}\rangle. \quad (5)$$

The Pauli principle then imposes that $\phi_1(\mathbf{k})$ is an even function and $\phi_2(\mathbf{k})$ must be odd. With this knowledge, we now apply the inversion operator to check the parity sectors. We obtain $\mathsf{P}|\psi^{11}_{S}(\mathbf{k})\rangle = (+1)|\psi^{11}_{S}(\mathbf{k})\rangle$ and $\mathsf{P}|\psi^{22}_{T0}(\mathbf{k})\rangle = (+1)|\psi^{22}_{T0}(\mathbf{k})\rangle$, where we used $\mathsf{I}|\Upsilon_{22}\rangle = -|\Upsilon_{22}\rangle$. Therefore, because of the additional DOF, $|\psi^{11}_{S}(\mathbf{k})\rangle$ singlets and $|\psi^{22}_{T0}(\mathbf{k})\rangle$ triplets belong to the same parity, which allows them to coexist in a superposition even in an inversion symmetric material. The mathematical subtlety relies on the fact that $\mathsf{P}|\Upsilon_{22}\rangle = |\Upsilon_{22}\rangle$, but $\mathsf{I}|\Upsilon_{22}\rangle = -|\Upsilon_{22}\rangle$.

The most general even and odd superpositions are

$$|\psi_{\mathsf{even}}\rangle = c_1|\psi^{11}_{S}\rangle + c_2|\psi^{12}_{S}\rangle + c_3|\psi^{21}_{S}\rangle + c_4|\psi^{22}_{T0}\rangle + c_5|\psi^{22}_{T,1}\rangle + c_6|\psi^{22}_{T,-1}\rangle; \quad (6)$$

$$|\psi_{\mathsf{odd}}\rangle = c_7|\psi^{22}_{S}\rangle + c_8|\psi^{11}_{T0}\rangle + c_9|\psi^{12}_{T0}\rangle + c_{10}|\psi^{21}_{T0}\rangle + c_{11}|\psi^{11}_{T1}\rangle + c_{12}|\psi^{11}_{T,-1}\rangle + c_{13}|\psi^{12}_{T1}\rangle + c_{14}|\psi^{12}_{T,-1}\rangle + c_{15}|\psi^{21}_{T1}\rangle + c_{16}|\psi^{21}_{T,-1}\rangle. \quad (7)$$

Here, all the 16 possible states are assigned to an even or odd sector. If the system lacks inversion, then $|\Psi\rangle = |\psi_{\text{even}}\rangle + |\psi_{\text{odd}}\rangle$. The singlet state $c_7|\psi_S^{22}\rangle$ deserves special attention since it is the only parity-odd-singlet. Suppose that the pairing mechanism of an inversion symmetric material only admits singlet instabilities. This means that because of the additional DOF, there are always mutually excluding even-singlet and odd-singlet instabilities that might *switch* depending on external parameters. This is likely to be the case in $CeRh_2As_2$ [4], and possibly also in few-layer $NbSe_2$ [5].

3. Effect of a Zeeman field

What are the interesting properties of the perhaps so far speculated singlet-tripled mixed superconductors? One interesting aspect is how the multicomponent superconducting state responds to a Zeeman magnetic field. It is intuitive to understand why. Heuristically, a Zeeman field applied to a singlet state $|\chi_{00}\rangle = |\uparrow\downarrow\rangle - |\downarrow\uparrow\rangle$ will try to polarize the spins along the magnetic field and thus breaking up the singlet configuration. In contrast, a Zeeman field applied for instance to the triplet state $|\chi_{1,\pm1}\rangle$ only favors it. Even more, if there is an attractive pairing in both singlet and triplet channels, instead of breaking up the singlets, the Zeeman field can convert the singlet configuration to a triplet. Then, the field can control the singlet-triplet volume ratio.

Let us give a simple geometric description of the singlet-triplet conversion process. Consider a 2D superconductor with the quantization axis set perpendicularly to the 2D plane. The superconducting state at zero field is taken to be the pure singlet

$$|\Psi_0(\mathbf{k})\rangle = |\psi_S^{11}(\mathbf{k})\rangle = \phi(\mathbf{k})|\chi_{00}\rangle \otimes |\Upsilon_{11}\rangle = [|\mathbf{k}\uparrow\rangle|-\mathbf{k}\downarrow\rangle - |\mathbf{k}\downarrow\rangle|-\mathbf{k}\uparrow\rangle] \otimes |\Upsilon_{11}\rangle. \tag{8}$$

We now apply a magnetic field along the y-axis. In the very large magnetic field limit, according to the heuristic conversion argument, we expect all singlets to convert to triplets. The action of the y-directed magnetic field on a $|\mathbf{k}\uparrow\rangle$ state can be implemented as a spin rotation around the x-axis $\mathsf{R}_x(\theta) = \exp(-i\sigma_x\theta/2)$. In the large magnetic field limit, and taking into account that Cooper pairing occurs at opposite momenta, the action of the magnetic field B_y on an arbitrary state is $\mathsf{B}_y = \mathsf{R}_x(\text{sgn}(\pm\mathbf{k})\pi/2)$, where the sign function determines whether the state is located at positive or negative momenta [6]. Since B_y does not act on $|\Upsilon_{11}\rangle$, we omit it in the following. We have

$$\mathsf{B}_y|\Psi_0(\mathbf{k})\rangle = R_x\left(\frac{\pi}{2}\right)|\mathbf{k}\uparrow\rangle R_x\left(-\frac{\pi}{2}\right)|-\mathbf{k}\downarrow\rangle - R_x\left(\frac{\pi}{2}\right)|\mathbf{k}\downarrow\rangle R_x\left(-\frac{\pi}{2}\right)|-\mathbf{k}\uparrow\rangle = i\left(|\psi_{T1}(\mathbf{k})\rangle - |\psi_{T,-1}(\mathbf{k})\rangle\right). \tag{9}$$

The action of B_y converted $|\psi_S(\mathbf{k})\rangle$ to $\pm i|\psi_{T,\pm1}(\mathbf{k})\rangle$. Such a Zeeman induced singlet to triplet conversion can occur in inversion symmetric superconductors with additional DOFs [7] or in non-centrosymmetric materials where no additional DOF is necessary [6, 8, 7].

4. Locally non-centrosymmetric superconductors

Materials that have an inversion center can exhibit properties of non-centrosymmetric materials. This can occur in layered crystals, where each layer alone lacks inversion, but when combined with the other layers, inversion symmetry is restored [9, 7].

4.1. $CeRh_2As_2$

We now apply the general ideas discussed above to the new two-phase superconductor $CeRh_2As_2$. The magnetic field – temperature phase diagram of $CeRh_2As_2$ displays a low and high field phase [4]. From crystal structure, one expects that the in-plane Rashba SOC components $\{\gamma_x, \gamma_y\}$ are larger than the perpendicular Ising component γ_z [9]. Therefore, as a first approximation, we can consider the magnetic field $\mathbf{B} \perp \boldsymbol{\gamma}$.

A superconducting instability analysis shows that the state-vector c_1 in Eq. (6) has the highest critical temperature at zero field [9]. Since the crystal contains the inversion element, all odd states in Eq. (7) are prohibited to coexist with the even states in Eq. (6). However, an odd state-vector might switch with the initially dominant even state at high Zeeman fields. The odd-singlet c_7 state becomes energetically favorable at high magnetic fields. This happens because the c_7 states accommodates better to the joint presence of magnetic field, spin-orbit coupling and inter-sublattice hoppings. This is a universal property of locally non-centrosymmetric superconductors, and perhaps one can speculate more discoveries of dominant c_7 states soon.

Because of the joint presence of spin-orbit coupling γ and magnetic field $\mathbf{B} \perp \gamma$, if there is (even small) attraction in the triplet channels, the field converts c_7 singlets to equal spin $\{c_{11}, c_{12}\}$ triplets. But this is not all. In the odd-phase, an additional conversion mechanism due to the joint presence of inter-sublattice hoppings and magnetic field also converts the c_7 singlets to inter-sublattice $\{c_{13}, c_{14}, c_{15}, c_{16}\}$ triplets. The zero-field $\{c_9, c_{10}\}$ triplets would require an intrinsic parity-mixed channel in the pairing interaction.

4.2. Bilayer-NbSe$_2$

Transition metal dichalcogenide bilayers with D_{3d} crystal symmetry are described by the same model as for $CeRh_2As_2$, but instead of Rashba SOC, one now has Ising SOC $\gamma(\mathbf{k}) = (0, 0, \gamma_z(\mathbf{k}), 0)$. Since γ_z locks the spins in the z-direction, these so called *Ising superconductors* are robust against in-plane fields $\mathbf{B} \perp \gamma$. The dominant superconducting component at zero field is very likely to be the c_1 state in Eq. (6) [5]. For in-plane fields, the same superconducting solutions found for $CeRh_2As_2$ for e perpendicular magnetic field apply to bilayer-TMDs. Also, because of the thin bilayer, an in-plane magnetic field couples dominantly to the spin degrees of freedom, such that orbital effects are expected to be very small. This makes bilayer-$NbSe_2$ a strong candidate for the c_7 odd-singlet state at high in-plane fields. It is possible that recent high field measurements already hint the c_7 state [5], but more high-field experimental data is needed.

5. Conclusion

If additional DOFs lack, singlet-triplet superconductivity can only occur via parity-mixing, which requires broken inversion symmetry. With additional DOFs, singlet-triplet pairing can occur within the same parity sector. The odd-parity sector contains only one singlet state. It is a universal property of locally non-centrosymmetric superconductors that the odd-parity singlet becomes favorable at high magnetic fields. If subleading pairing channels exist, the magnetic field converts the odd-singlets to both inter-sublattice triplets and intra-sublattice equal spin-triplets.

References

[1] Sigrist M 2005 *AIP Conference Proceedings* **789** 165–243
[2] Sigrist M 2009 *AIP Conference Proceedings* **1162** 55–96
[3] Sakurai J J and Napolitano J 2017 *Modern Quantum Mechanics* 2nd ed (Cambridge University Press)
[4] Khim S, Landaeta J F, Banda J, Bannor N, Brando M, Brydon P M R, Hafner D, Küchler R, Cardoso-Gil R, Stockert U, Mackenzie A P, Agterberg D F, Geibel C and Hassinger E 2021 *Science* **373** 1012–1016
[5] Kuzmanović M, Dvir T, LeBoeuf D, Ilić S, Möckli D, Haim M, Kraemer S, Khodas M, Houzet M, Meyer J S, Aprili M, Steinberg H and Quay C H L 2021 Tunneling spectroscopy of few-monolayer nbse$_2$ in high magnetic field: Ising protection and triplet superconductivity (*Preprint* 2104.00328)
[6] Möckli D and Khodas M 2019 *Physical Review B* **99**
[7] Möckli D and Ramires A 2021 Superconductivity in disordered locally noncentrosymmetric materials: an application to cerh$_2$as$_2$ (*Preprint* 2107.09723)
[8] Möckli D and Khodas M 2020 *Physical Review B* **101**
[9] Möckli D and Ramires A 2021 *Phys. Rev. Research* **3**(2) 023204

Journal of Physics: Conference Series **2164** (2022) 012010 doi:10.1088/1742-6596/2164/1/012010

Tobological feedback for superconducting states

Yutaka Tobita[1] **and Jun Goryo**[2]

[1] Department of Infromation and Computer Science, Hokkaido University of Science, 006-8585, Sapporo, Japan

[2] Department of Mathematics and Physics, Hirosaki University, 036-8561, Hirosaki, Japan

E-mail: tobita-y@hus.ac.jp

Abstract. We discuss feedback effects that stabilize the superconducting states by the induced topological term in the effective Lagrangian. The chiral feedback effect due to the Chern-Simons-like term for quasi-two-dimensional system with time-reversal symmetry breaking (TRSB) was studied in [1, 2]. We consider the extension of the chiral feedback to three-dimensional TRSB system and investigate similar feedback effects for quasi-two-dimensional or three-dimensional time-reversal symmetric systems.

1. Introduction

The p-wave superfluid ^{3}He has two distinct phases in zero fields. The pairing order parameter of A-phase has point nodes and, in the weak coupling approximation, B-phase with a fully gapped order parameter gains more condensation energy than A-phase. The reason for the overcoming stability of A-phase is the spin fluctuation feedback [3, 4]. More recently, it was found that the chiral p-wave pairing, which is a two-dimensional and superconducting analog of ^{3}He-A phase, would have another type of feedback effect [1, 2]. The chiral pairing leads to a spontaneous induction of a Chern-Simons(CS)-like term in the low energy effective action for the electromagnetic gauge field [5, 6, 7, 8]. Due to the charge-flux attachment by the CS-like term, each electron in a chiral Cooper pair feels the Lorentz force that strengthens the attractive interaction.The topological nature of pairing thus causes this chiral feedback effect and supports the pairing itself. We then present our attempt to find topological feedback mechanisms for the other types of superconducting states. We investigate the three-dimensional extension of the chiral feedback effect for TRSB Weyl superconductors with point nodes. The helical p-wave state, which is the time-reversal-symmetric (TRS) extension of the chiral p-wave state, would have a feedback effect caused by the induced BF term (TRS extension of the CS term) leading to the spin-dependent Lorentz force. The three-dimentional extension is also possible for the TRS pairing state.

2. Weyl (A-phase type) feedback

We introduce a feedback mechanism on 3D Weyl superconducting states (Weyl SC), which point nodes and break time-reversal symmetry. We consider the simplest example, i.e. ^{3}He-A-type superconducting state of

$$\vec{d}(\vec{k}) = |\Delta_0|(\hat{k}_x \pm i\hat{k}_y)\vec{e}_z,\ \hat{\Delta}(\vec{k}) = i\sigma_\nu\sigma_y d_{\nu l}\hat{k}_l,\ \hat{k}_i = \frac{k_i}{k_F}, \tag{1}$$

which has point nodes at $\vec{k} = (0, 0, \pm k_F)$. We suppose the parabolic dispersion $\epsilon(\vec{k}) = \frac{\vec{k}^2 - k_F^2}{2m_e}$, for simplicity. The CS-like term

$$\frac{\sigma_{xy\hat{z}}}{2}\epsilon_{0ij\hat{z}}(A_0\partial_i A_j + A_i\partial_j A_0),\ (i,j = x,y) \tag{2}$$

also appears in the low-energy effective action of this pairing state [5, 6, 7, 8]. We will see below that the term gives a positive feedback.

As with the chiral feedback, a charge-density–charge-current interaction induced by the CS-like term Eq. (2) contributes to the pairing interaction additionally. The density–current correlation function connected with this interaction just below T_c is

$$\pi_{0j}(\vec{q}) = k_B T_c e^2 \sum_m \int \frac{d^3k}{(2\pi)^3}$$
$$\times \mathrm{Tr}\left[\mathcal{G}(i\omega_m, \vec{k}+\vec{q})\hat{\Delta}(\vec{k}+\vec{q})\mathcal{G}(-i\omega_m, -\vec{k}-\vec{q})\frac{2k_j + q_j}{2m_e}\mathcal{G}(-i\omega_m, -\vec{k})\hat{\Delta}^\dagger(\vec{k})\mathcal{G}(i\omega_m, \vec{k})\right], \tag{3}$$

where $\mathcal{G}(i\omega_m, \vec{k}) = \frac{1}{i\omega_m - \epsilon(\vec{k})}$, $\omega_m = (2m+1)\pi k_B T_c$, and the time dependence is ignored.

Due to the symmetry, we can express

$$\pi_{0j}(\vec{q}) = i\epsilon_{ljk} q_l f_k(\vec{q}) + (\text{symmetric in } l \leftrightarrow j),\ f_k(\vec{q}) = -\frac{i}{2!}\epsilon_{ijk}\frac{\partial \pi_{0j}(\vec{q})}{\partial q_i}, \tag{4}$$

where ϵ_{ljk} is totally antisymmetric tensor. Using similar approximations used in [1, 2], we have an approximate form of $f_k(\vec{q})$ as

$$f_k(\vec{q}) \simeq -\frac{e^2 N(0)\,(\epsilon_{ijk} d^*_{\nu i} d_{\nu l})}{24 m_e\,(\pi k_B T_c)^2}\frac{1}{\sqrt{1+\gamma\xi^2 q^2}}, \tag{5}$$

where $N(0)$ is the density of state at Fermi level, $\xi = \frac{v_F}{2\pi k_B T_c}$ is a coherence length, and γ is an $O(1)$ numerical factor.

Substituting Eq. (5) into the correlation function Eq. (4), we have

$$\pi_{0\beta}(\vec{q}) = \frac{e^2 N(0)\,(\epsilon_{ijk} d^*_{\nu i} d_{\nu j})}{24 m_e (\pi k_B T_c)^2}\frac{\epsilon_{l\beta k} q_l}{\sqrt{1+\gamma\xi^2 q^2}}. \tag{6}$$

To estimate this contribution, we calculate a correction to the 4th order term in the GL free energy expressed as

$$\Delta F_{Weyl} = k_B T_c \int \frac{d^3q}{(2\pi)^3} D_{00}(\vec{q})\pi_{0\alpha}(\vec{q})D_{\alpha\beta}(\vec{q})\pi_{\beta 0}(\vec{q}), \tag{7}$$

where, $D_{00}(\vec{q}) = \frac{1}{q^2 + l_{TF}^{-2}}$, and $D_{\alpha\beta}(\vec{q}) = -\frac{\delta_{\alpha\beta}}{q^2}$. $D_{00}(\vec{q})$ and $D_{\alpha\beta}(\vec{q})$ are gauge field propagators in Coulomb gauge with the Thomas-Fermi screening length l_{TF}. Substituting Eq. (6), we have

$$\Delta F_{Weyl} \approx \frac{\alpha^2 k_F l_{TF}}{108\pi\gamma}\frac{T_c}{T_F}\frac{N(0)(\epsilon_{ijk} d^*_{\nu i} d_{\nu j})^2}{(\pi k_B T_c)^2} < 0, \tag{8}$$

where, $\epsilon_{ijk} d^*_{\nu i} d_{\nu j} = 2i\chi|\Delta_0|^2$ and $\chi = \pm 1$ denotes chirality (z-component of the relative angular momentum). The correction is negative and the effect favors the ^{3}He-A-type state. Comparing with the quasi-2D result, $1/d$ is replaced with k_F but qualitative nature is not changed.

Journal of Physics: Conference Series **2164** (2022) 012010 doi:10.1088/1742-6596/2164/1/012010

3. Extension to Time reversal symmetric system

For time reversal symmetric(TRS) systems the BF-term $\epsilon_{ij}(a_0\partial_i A_j + A_i\partial_j a_0)$, that is a time reversal symmetric extension of the CS term, is induced in the low-energy effective action. In the BF-term, the electrostatic gauge field A_0 is replaced with a "spin" gauge field a_0 which shows the spin fluctuation, and spin-density–charge-current interaction is induced.

3.1. Feedback effect for helical p-wave (quasi 2D)

In this section, we discuss the feedback effect for p-wave superconducting states in quasi-2D system with time reversal symmetry, i.e., helical p-wave states. They are categorized into two types,

$$|parallel\rangle = \frac{1}{2}\left(|s_z = +1, \chi = +1\rangle + |s_z = -1, \chi = -1\rangle\right), \tag{9}$$

$$|anti-parallel\rangle = \frac{1}{2}\left(|s_z = +1, \chi = -1\rangle + |s_z = -1, \chi = +1\rangle\right), \tag{10}$$

where $|parallel\rangle$ ($|anti-parallel\rangle$) is the superposition of two chiral p-wave states each of which has χ parallel (anti-parallel) to s_z.

The d-vectors of them are

$$\vec{d}(\vec{k}) = \frac{1}{2}\left(\vec{d}^{p(ap)}_{\uparrow\uparrow}(\vec{k}) + \vec{d}^{p(ap)}_{\downarrow\downarrow}(\vec{k})\right), \tag{11}$$

where

$$\vec{d}^{p(ap)}_{\uparrow\uparrow}(\vec{k}) = \Delta_0\left(\hat{k}_x - i\hat{k}_y\right)(\vec{e}_x + i\vec{e}_y), \quad \vec{d}^{p(ap)}_{\downarrow\downarrow}(\vec{k}) = \Delta_0\left(\hat{k}_x + i\hat{k}_y\right)(\vec{e}_x - i\vec{e}_y).$$

We analyze a feedback effect for these states. The correlation function through spin-density–charge current interaction $H_{int} = \sum_{\vec{q}} g^{\mu}_{0i}\rho^{\mu}(\vec{q})J^e_i(-\vec{q})$ has the from

$$\pi^{\mu}_{0j}(\vec{q}) = -k_BT_ce\sum_m\int\frac{d^3k}{(2\pi)^3}$$
$$\times\mathrm{Tr}\left[\sigma^{\mu}\mathcal{G}(i\omega_m, \vec{k}+\vec{q})\hat{\Delta}(\vec{k}+\vec{q})\mathcal{G}(-i\omega_m, -\vec{k}-\vec{q})\frac{2k_j+q_j}{2m_e}\mathcal{G}(-i\omega_m, -\vec{k})\hat{\Delta}^{\dagger}(\vec{k})\mathcal{G}(i\omega_m, \vec{k})\right], \tag{12}$$

where the energy band is parabolic, $\epsilon(\vec{k}) = \frac{k_x^2+k_y^2-k_F^2}{2m_e}$ and the spin-orbit interaction is ignored for simplicity. We assume a layered structure with spacing d in z-direction just like the system assumed in the chiral feedback.

Following the way used in the previous section, we obtain a spin-density–charge-current correlation function of the form

$$\pi^{\mu}_{0j}(\vec{q}) = i\epsilon_{lj}q_l f^{\mu}(\vec{q}) + (\text{symmetric in } l \leftrightarrow j), \; f^{\mu}(\vec{q}) \sim e\frac{N(0)}{(\pi k_BT_c)^2}\frac{\epsilon_{ij}\epsilon_{\mu\nu\rho}d^*_{\nu i}d_{\rho j}}{4m_e\sqrt{1+\gamma\xi^2q_\perp^2}}, \tag{13}$$

where $q_\perp^2 = q_x^2 + q_y^2$, and the circular symmetry is used for the approximation in the last line.

The spin-density–charge-current interaction contributes additionally to the pairing interaction. To estimate this contribution, we calculate a correction to the 4th order term in the GL free energy expressed as

$$\Delta F_h = k_BT_c\int\frac{d^3q}{(2\pi)^3}g^{\mu\nu}_0(\vec{q})\pi^{\nu}_{0i}(\vec{q})D_{ij}(\vec{q})\pi^{\mu}_{j0}(\vec{q}), \tag{14}$$

where the propagators for (spin) gauge fields are

$$D_{ij}(\vec{q}) = \frac{-\delta_{ij}}{q^2},\ g_0^{\mu\nu}(\vec{q}) = \frac{3k_F^2 I_0 \left(1 - \frac{q_\perp^2}{12k_F^2}\right)}{q_\perp^2 + \xi_s^{-2}} \delta^{\mu\nu}, \tag{15}$$

where $q^2 = q_x^2 + q_y^2 + q_z^2$, and a magnetic correlation length $\xi_s^2 = \frac{1}{12k_F^2}\frac{I_0 N(0)}{1-I_0 N(0)}$ shows the Stoner enhancement. Using similar approximations in Ref. [2], we have

$$\Delta F_h \approx -\frac{3\pi\alpha I_0 k_F^2}{32\gamma}\frac{\xi_s}{d}\frac{T_c}{T_F}\frac{N(0)}{(\pi k_B T_c)^2}\left(\epsilon_{ij}\epsilon_{\mu\nu\rho}d_{\nu i}^* d_{\rho j}\right)^2, \tag{16}$$

where $\left(\epsilon_{ij}\epsilon_{\mu\nu\rho}d_{\nu i}^* d_{\rho j}\right)^2 = |\Delta_0|^4\chi^2(1+s_z^2)^2$ is positive difinite for both parallel and anti-parallel states. Hence, the helical feedback effect supports both parallel and anti-parallel states equally.

3.2. ^{3}He-B phase type feedback

We study a feedback mechanism for a ^{3}He-B phase type superconductor, i.e., an extension of the helical feedback to 3-dimensional system in this section. The order parameter has the form

$$\Delta(\vec{k}) = \Delta_0(i\sigma_y\vec{\sigma})\cdot\vec{d}(\vec{k}),\ \vec{d} = \frac{1}{2}\left[(\vec{e}_x\hat{k}_x + \vec{e}_y\hat{k}_y) + (\vec{e}_y\hat{k}_y + \vec{e}_z\hat{k}_z) + (\vec{e}_z\hat{k}_z + \vec{e}_x\hat{k}_x)\right], \tag{17}$$

of the superposition of helical p-waves on each plane.

The spin-density–charge-current correlation function is almost same with the helical one except for the intermediate fermions' spatial degrees of freedom and written as,

$$\pi_{0j}^{\mu}(\vec{q}) = i\epsilon_{ljk}q_l f_k^{\mu}(\vec{q}),\ f_k^{\mu}(\vec{q}) \simeq \frac{eN(0)\left(\epsilon_{ijk}\epsilon_{\mu\nu\rho}d_{\nu i}^* d_{\rho j}\right)}{48m_e\beta}\frac{\pi}{(\pi k_B T_c)^3}\frac{1}{\sqrt{1+\gamma\xi^2 q^2}}. \tag{18}$$

We obtained Eq. (18) by similar approximations used in Eq. (5).

By using the correlation function Eq. (18), we calculate a feedback correction to the 4th order term in the GL free energy as

$$\Delta F_{B-phase} \simeq -\frac{\alpha I_0 k_F^3 \xi_s}{576\gamma\pi^2}\frac{T_c}{T_F}\frac{N(0)\left(\epsilon_{ijk}\epsilon_{\mu\nu\rho}d_{\nu i}^* d_{\rho j}\right)^2}{(\pi k_B T_c)^2}, \tag{19}$$

which is negative definite. In 3D TRS system, the feedback also arises and supports the states.

4. Summary

We study about extensions of the chiral feedback for 3D TRSB system, and quasi-2D and 3D TRS systems. It was found that those effects stabilize the states. Because of topological nature of interaction terms, all of these effects are also universal. In addition, the Stoner enhancement could strengthen the feedback significantly in TRS cases.

Acknowledgments

This work was partially supported by JSPS KAKENHI Grant No.JP15H05885 (J- Physics).

References

[1] J. Goryo and M. Sigrist, J. Phys. C, 12 L599 (2000).
[2] J. Goryo and M. Sigrist, Europhys. Lett. 57, 578 (2002).
[3] W.F. Brinkman and P. W. Anderson, Phys. Rev. A, 10 2386 (1974)
[4] Y. Kuroda, Prog. Theor. Phys., 53 349 (1975).
[5] G. E. Volovik, Sov. Phys. JETP 67 1804 (1988).
[6] J. Goryo and K. Ishikawa, Phys. Lett. A 260 294 (1999).
[7] N. Read and D. Green, Phys. Rev. B 61 10267 (2000).
[8] A. Furusaki, M. Matsumoto, and M. Sigrist, Phys. Rev. B 64 054514 (2001).

Electronic structures and effective models in BaPtSb and BaPtAs with ordered honeycomb structures

Yoshiki Imai[1], Tsuyoshi Imazu[2], Jun Goryo[2]

[1]Department of Applied Physics, Okayama University of Science, Okayama 700-0005, Japan
[2]2Department of Mathematics and Physics, Hirosaki University, Hirosaki 036-8561, Japan

E-mail: imai@dap.ous.ac.jp

Abstract. We investigate the low-energy electronic properties and construct the effective model Hamiltonians in the normal state for new materials BaPtSb and BaPtAs with ordered honeycomb structures. The preliminary experimental study suggested that the former compound may be a candidate of the time-reversal symmetry breaking superconductors. The low-energy electronic structures obtained from the first principles calculation mainly consist of Pt $5d$ and Sb/As $5p/4p$ orbitals, where there exist the strong hybridizations between both orbitals. We show that the main parts in the low-energy region including the Fermi surface are well described using renormalized 3 bands consisting of dominant Pt 2 bands and Sb/As 1 band. Our result provides a starting point for microscopic investigation in the superconducting states for the materials with the hexagonal structure.

1. Introduction

The time-reversal symmetry breaking (TRSB) superconductors have received special interest for their intriguing properties in the fields of the topological condensed matter and the strongly correlated electron systems. It is well known that the hexagonal lattice structure favors the realization of the TRSB superconducting states. The intrinsic spontaneous magnetic fields are observed in the superconducting phase in the typical material SrPtAs in the hexagonal KZnAs-type structure [1, 2]. Recently, new materials BaPtSb and BaPtAs in the SrPtSb-type structure show superconducting states [3, 4], whose lattice structures are similar to that of SrPtAs. In particular, the preliminary experimental study for BaPtSb observed the occurrence of the spontaneous magnetic field in the superconducting phase, which indicated that the superconducting state in BaPtSb may break time reversal symmetry [5].

In the present study, to identify the low-energy electronic properties microscopically, we investigate the electronic states of BaPtSb and BaPtAs in the normal phase and construct the effective model Hamiltonians by means of the first principles method and the tight-binding approach.

2. Electronic structures from first principles calculation

In order to obtain the electronic structures of BaPtSb and BaPtAs, calculations are carried out by using the Quantum Espresso Package [6, 7] based on the plane wave pseudopotential method. We employ $24 \times 24 \times 12$ $\boldsymbol{k}$-point mesh in the Brillouin zone for both materials, and the kinetic energy cutoff for wavefunctions is set to 80 Ryd.

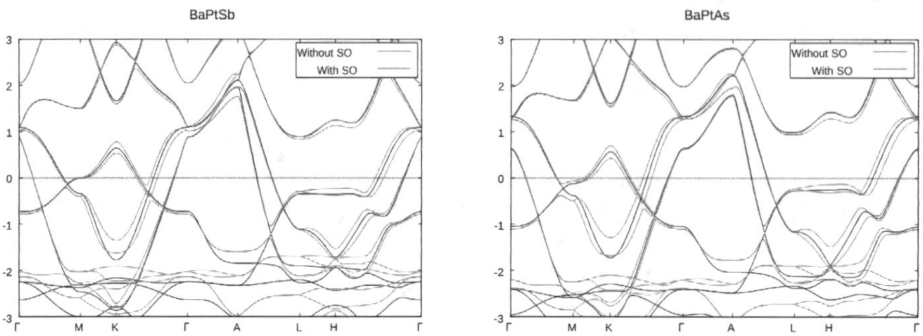

Figure 1. Electronic structures obtained from the first principles method for BaPtSb (left panel) and BaPtAs (right panel). The black (red) lines denote the corresponding bands without (with) the SO interaction.

The lattice structures of BaPtSb and BaPtAs have the SrPtSb-type one with space group $P\bar{6}m2$ (No. 187). The lattice constants obtained experimentally are $a = 4.535$Å, $c = 4.884$Å for BaPtSb [8], and $a = 4.308$Å, $c = 4.761$Å for BaPtAs [4], respectively.

We first perform the optimization for all lattice parameters. The obtained lattice constants including effects of the spin-orbit (SO) interaction are $a = 4.605$Å, $c = 4.906$Å for BaPtSb, and $a = 4.356$Å, $c = 4.800$Å for BaPtAs as the most stable states. In comparison with the experimental results, the maximum difference of lattice parameters is within 1.6%.

The band structures of BaPtSb and BaPtAs for the most stable lattice parameters are depicted in Fig. 1. Both band structures are almost identical qualitatively in the low energy region except M point. Note that while the energy eigenvalues for BaPtSb around M point touch the Fermi level, those for BaPtAs are slightly below the Fermi level.

In order to clarify the contribution of each orbital in the low-energy region, we show the Density of States (DOS) and the band structure with the projected weights of Pt 5d and Sb 5p orbitals for BaPtSb without the SO interaction in Fig. 2. The DOS around the Fermi level is mainly derived from Pt 5d and Sb 5p orbitals, where both contributions are comparable.

Note that the energy eigenvalues are divided into two groups, {Pt ($d_{x^2-y^2}$, d_{xy}), Sb (p_x, p_y)} and {Pt (d_{yz}, d_{zx}), Sb (p_z)}, shown in Fig. 2 right panel. There exist the strong hybridizations between those orbitals in the same groups. The main part of energy eigenvalues of the Pt $d_{3z^2-r^2}$ orbital is far from the Fermi level. Similar results for BaPtAs are also obtained.

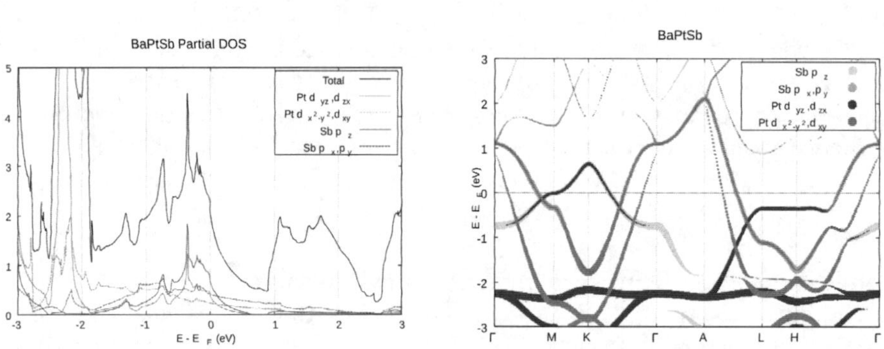

Figure 2. (Left panel) Total and partial DOSs of Pt 5d and Sb 5p orbitals for BaPtSb. (Right panel) The band structure with projected weights of Pt 5d and Sb 5p orbitals without the SO interaction.

3. Construction of effective models

In this section, we construct the effective models based on the results from the first principles calculation, which becomes a point of departure to investigate the superconducting state and other physical properties microscopically for both materials.

We focus on the low-energy electronic structure around the Fermi level which affects strongly the physical properties such as superconductivity. In the previous section, we showed the low-energy bands consist of two groups, {Pt ($d_{x^2-y^2}$, d_{xy}), Sb (p_x, p_y)} and {Pt (d_{yz}, d_{zx}), Sb/As p_z}. Although they have the strongly hybridizations in the same groups, the weights of Pt $d_{x^2-y^2}$, d_{xy}, and Sb/As p_z orbitals are comparatively larger than the other orbitals in each group. Thus, we focus on the following three orbitals Pt $d_{x^2-y^2}$, d_{xy}, and Sb/As p_z, for simplicity. Note that some of Wannier functions obtained from the Wannier90 package [9] spread over the Pt and Sb/As sites widely, which indicates that the electrons hop long distances with those basis.

We here construct the 3 band tight-binding model by means of the Slater-Koster approach [10]. The lattice structure is defined by the primitive translation vectors $\boldsymbol{a}_1 = (a, 0, 0)$, $\boldsymbol{a}_2 = \left(-\frac{a}{2}, \frac{\sqrt{3}a}{2}, 0\right)$, $\boldsymbol{a}_3 = (0, 0, c)$ where a (c) denotes the lattice constant in the a-b plane (along the c axis). Sb/As (Pt) site in the unit cell is specified with $\boldsymbol{\tau}_{\mathrm{Sb/As}} = (0, 0, 0)$ ($\boldsymbol{\tau}_{\mathrm{Pt}} = \left(\frac{a}{2}, \frac{a}{2\sqrt{3}}, 0\right)$). The diagonal SO interactions concerning the spin and orbital are added to the model in the a-b plane, for simplicity.

Then the Hamiltonian is given by

$$H = \sum_{\boldsymbol{k}} \sum_{m,m'} \sum_{\sigma,\sigma'} \varepsilon_{\boldsymbol{k}}^{(m\sigma,m'\sigma')} c^{\dagger}_{\boldsymbol{k}m\sigma} c_{\boldsymbol{k}m'\sigma'}, \tag{1}$$

$$\varepsilon_{\boldsymbol{k}}^{(m\sigma,m'\sigma')} = \sum_{i,j} t_{ij}^{mm'} e^{-i\boldsymbol{k}\cdot(\boldsymbol{r}_i - \boldsymbol{r}_j + \boldsymbol{\tau}_m - \boldsymbol{\tau}_{m'})} \delta_{\sigma\sigma'} + i\sigma\lambda_m \sum_{l=1}^{3} \sin(\boldsymbol{k}\cdot\boldsymbol{T}_l)\delta_{mm'}\delta_{\sigma\sigma'} + \epsilon_m \delta_{mm'}\delta_{\sigma\sigma'}, \tag{2}$$

where $c^{\dagger}_{im\sigma}$ ($c_{im\sigma}$) is the creation (annihilation) operator for electrons on the site i, in the orbital m (Pt $d_{x^2-y^2}$, d_{xy}, and Sb/As p_z) with the spin σ. $t_{ij}^{mm'}$ stands for the hopping obtained from the Slater-Koster approach. $\boldsymbol{r}_i$, $\boldsymbol{k}$, λ_m, and ϵ_m denote the position of the unit cell, the wave vector, the amplitude of the SO interaction, and the on-site potential. $\boldsymbol{T}_l$ stands for the nearest neighbor vector in the a-b plane between same atoms with $\boldsymbol{T}_l = \{\boldsymbol{a}_1, \boldsymbol{a}_2, \boldsymbol{a}_1 + \boldsymbol{a}_2\}$.

Diagonalizing Eqn. (2), we obtain the energy eigenvalues concerning BaPtSb, which is shown in Fig. 3. The on-site potentials are -0.80 (Pt) and -0.48 (Sb), and the amplitudes of the SO interaction are $\lambda_m = 0.02$ for all orbitals, respectively. The hopping parameters between the nearest neighbor (NN), the next nearest neighbor (NNN), and the 3rd nearest neighbor (3NN) are given in Table 1, where ($\alpha\beta m$) represents the hopping amplitude depending on the

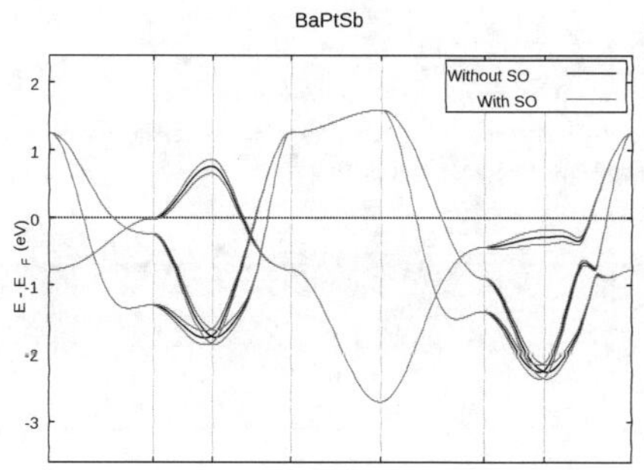

Figure 3. Energy eigenvalues for BaPtSb without/with the SO interaction.

Pt-Pt		$(dd\sigma)$	$(dd\pi)$	$(dd\delta)$		$(dd\sigma)$	$(dd\pi)$	$(dd\delta)$		$(dd\sigma)$	$(dd\pi)$	$(dd\delta)$
a-b plane	NN	0.60	0.15	0.05	NNN	−0.1	0.05	0.02	3NN	0.15	0.03	0.02
c axis	NN	0.05	0.05	0.05	NNN	−0.05	−0.05	0.00	3N	0.05	−0.02	0.00

Sb-Sb		$(pp\sigma)$	$(pp\pi)$		$(pp\sigma)$	$(pp\pi)$		$(pp\sigma)$	$(pp\pi)$
a-b plane	NN	−0.30	−0.20	NNN	0.05	0.01	3NN	0.00	−0.02
c axis	NN	0.20	0.02	NNN	0.02	0.03	3NN	0.00	0.03

Pt-Sb		$(pd\sigma)$	$(pd\pi)$		$(pd\sigma)$	$(pd\pi)$		$(pd\sigma)$	$(pd\pi)$
c axis	NN	−0.30	0.20	NNN	−0.05	0.05	3NN	0.00	−0.20

Table 1. Hopping parameters in the a-b plane and along the c axis between Pt-Pt, Sb-Sb, and Pt-Sb sites.

distance between the orbitals α and β with the component of angular momentum m around the quantization axis. The hopping between Pt-Sb sites vanishes in the a-b plane due to the lattice symmetry, and those along the c axis direction remain.

The obtained eigenvalues capture features of the result for BaPtSb from the first principles calculation around the Fermi level qualitatively. For BaPtAs, the similar procedure gives the effective model. This result indicates that the low-energy electronic structure is well described by the 3 band model, and the effect from eliminated orbitals is renormalized partially in the hopping parameters.

4. Summary and discussion

We have investigated the electronic structure of BaPtSb and BaPtAs and constructed the effective models with the first principles method and the tight-binding approach. We show that the low-energy electronic structures are dominated by the Pt $5d$ and Sb/As $5p/4p$ orbitals, and the effective model Hamiltonians are well described by Pt $d_{x^2-y^2}$, d_{xy}, and Sb/Ab p_z orbitals.

While these materials constitute the ordered honeycomb network in the a-b plane, the lattice parameters a and c are comparable. The obtained result shows that the hopping along the c axis as well as those in the a-b plane may affect physical properties. The diagonal SO interactions concerning the spin and orbital in the present study may give rise to the metallic state with the spin-Hall effect though the Pt-Sb hopping along the c axis.

[1] Nishikubo Y, *et al.* 2011 J. Phys. Soc. Jpn. **80** 055002.
[2] Biswas P K, *et al.* 2013 Phys. Rev. B **87** 180503 (R).
[3] Kudo K, *et al.* 2018 J. Phys. Soc. Jpn. **87** 063702.
[4] Kudo K, *et al.* 2018 J. Phys. Soc. Jpn. **87** 073708.
[5] Adachi T, *et al.* 2018 KEK-MSL REPORT **2017** 54.
[6] Giannozzi P, *et al.* 2009 J. Phys.:Condens. Matter **21** 95502.
[7] Giannozzi P, *et al.* 2017 J. Phys.:Condens. Matter **29** 465901.
[8] Wenski G and Mewis A 1986 Z. Anorg. Allg. Chem. **535** 110.
[9] Mostofi A A, *et al.* 2014 Computer Physics Communications **185** 2309.
[10] Slater J C and Koster G F 1954 Phys. Rev. **94** 1498.

Phase diagram of the three-orbital Hubbard-Holstein model simulating Superconducting Tungsten Bronze A_xWO_3

Takuya Sekikawa[1], **Yoshiaki Ōno**[1]
[1]Department of Physics, Niigata University, Niigata, 950-2181, Japan
E-mail: sekikawa.t.phys@gmail.com

Abstract. The superconductivity in tungsten bronze A_xWO_3 (A=alkali metal) is studied based on first-principles calculations and orbital fluctuation theory. We discuss the effects of the electron-Jahn-Teller phonon interaction and the on-site Coulomb interaction in the random phase approximation, and obtain a phase diagram on the $U-g$ plane in the low-doped regime at $x\sim0.05$, where high-temperature superconductivity has been experimentally observed.

1. Introduction

Tungsten bronze A_xWO_3 (A=alkali metal) is known to exhibit superconductivity, and the superconducting transition temperature T_c increases rapidly up to 7 K with decreasing x[1,2] and a remarkable high-T_c superconductivity with Tc=91K is reported at further low doping $x\sim0.05$ [3] and T_c=120 K for A=H [4]. The system also exhibits a rich set of structural phases, which are well explained by ab initio calculations [6]. However, relation between x and the superconducting transition temperature by first-principles calculations has not yet been clarified. In addition, the superconductivity near the structural phase transition implies strong ferro-orbital fluctuations, and the W-5d electrons have strong Coulomb interactions, which may play a crucial role in determining T_c. The aim of this work is to study the superconductivity in A_xWO_3 based on the effects of ferro-orbital fluctuations and on-site Coulomb interaction in the random phase approximation (RPA).

2. Orbital-fluctuation effect

Non-doped WO_3 is a band insulator with a band gap of 3 eV, and the conduction band consists mainly of t_{2g} three orbitals of W-5d electrons. In the case of alkali-doped A_xWO_3, angle-resolved photoemission spectroscopy (ARPES) revealed that the Fermi level shifts to the conduction band, as realized by the rigid-band shift image, while the band dispersion hardly changes[5]. Then,the tight-binding model model consisting of t_{2g} three orbitals was constructed to reproduce the conduction bands of WO_3 using WIEN2k and the maximally localized Wannier function, and this model was applied to A_xWO_3[7]. There is a significant degeneracy in the conduction band, but it is resolved by the Jahn-Teller distortion that causes the structural transition[6].

Next, we consider the electron-phonon interaction between the t_{2g} electron, the Jahn-Teller phonon, and the Coulomb interaction given by the Hamiltonian below

$$\hat{H}_{\mathrm{el-ph}} = \sum_{i,s}\sum_{m,m'}\sum_{\sigma} g_s^{mm'} \hat{c}^\dagger_{im\sigma}\hat{c}_{im'\sigma}(\hat{b}^\dagger_{is} + \hat{b}_{is}), \tag{1}$$

$$\begin{aligned}
\hat{H}_{\mathrm{el-el}} &= \frac{1}{2}U\sum_{i}\sum_{m}\sum_{\sigma\neq\bar{\sigma}} \hat{c}^\dagger_{im\sigma}\hat{c}^\dagger_{im\bar{\sigma}}\hat{c}_{im\bar{\sigma}}\hat{c}_{im\sigma} \\
&+ \frac{1}{2}U'\sum_{i}\sum_{m\neq\bar{m}}\sum_{\sigma,\sigma'} \hat{c}^\dagger_{im\sigma}\hat{c}^\dagger_{i\bar{m}\sigma'}\hat{c}_{i\bar{m}\sigma'}\hat{c}_{im\sigma} \\
&+ \frac{1}{2}J\sum_{i}\sum_{m\neq\bar{m}}\sum_{\sigma,\sigma'} \hat{c}^\dagger_{im\sigma}\hat{c}^\dagger_{i\bar{m}\sigma'}\hat{c}_{im\sigma'}\hat{c}_{i\bar{m}\sigma} \\
&+ \frac{1}{2}J'\sum_{i}\sum_{m\neq\bar{m}}\sum_{\sigma\neq\bar{\sigma}} \hat{c}^\dagger_{im\sigma}\hat{c}^\dagger_{im\bar{\sigma}}\hat{c}_{i\bar{m}\bar{\sigma}}\hat{c}_{i\bar{m}\sigma}
\end{aligned} \tag{2}$$

Here, $\hat{c}^\dagger_{im\sigma}$ is the creation operator for site i, orbit $m = d_{xy}(1),\ d_{yz}(2),\ d_{zx}(3)$, and spin σ, and $\hat{b}^\dagger_{is}$ is for site i and mode s. In addition, U and U' are intra- and inter-orbital direct terms, respectively, and$J$ and J' are Hund's coupling and pair transfer, respectively. The Hamiltonian Eq. (1) was found to reproduce the doping dependence of the structural transitions within the adiabatic approximation,, in agreement with experiment [5].

To discuss orbital fluctuations that are enhanced by electron-phonon interactions and diverge toward second-order structural transitions, the RPA method is employed for the electron-phonon interaction in Eq. (1) and the Coulomb interaction in Eq. (2). The RPA charge-orbital susceptibility $\chi^c(q)$ and spin susceptibility $\chi^s(q)$ are given by

$$\hat{\chi}^c(q) = \hat{\chi}^0(q)\left[\hat{1} + \hat{\Gamma}^c(i\omega_m)\hat{\chi}^0(q)\right]^{-1}, \tag{3}$$

$$\hat{\chi}^s(q) = \hat{\chi}^0(q)\left[\hat{1} + \hat{\Gamma}^s(i\omega_m)\hat{\chi}^0(q)\right]^{-1}, \tag{4}$$

where the bare susceptibility is defined as $\hat{\chi}^0(q) = -(T/N)\sum_k \hat{G}(k+q)\hat{G}(k)$. Here, $k = (k, i\varepsilon_n)$ and $q = (q, i\omega_m)$, and $\varepsilon_n = (2n+1)\pi T$ and $\nu_m = 2m\pi T$ are the fermionic frequency and boson Matsubara frequency, respectively. In Eq. (3), $\hat{\Gamma}^c$ represents the charge-orbital vertex and in Eq. (4), $\hat{\Gamma}^s$ represents the spin vertex, which is given as

$$\hat{\Gamma}^c(i\omega_m) = \hat{\Gamma}^{(c,0)} - 2\sum_s g_s^{m_2m_1} g_s^{m_3m_4} D_s(i\omega), \tag{5}$$

$$\hat{\Gamma}^s(i\omega_m) = \hat{\Gamma}^{(s,0)}, \tag{6}$$

$$\hat{\Gamma}^{(c,0)} = \begin{cases} U & (l_1l_2l_3l_4) = (llll) \\ -U' + 2J & (l_1l_3l_2l_4) = (l'lll') \\ 2U' - J & (l_1l_2l_3l_4) = (lll'l') \\ J' & (l_1l_4l_2l_3) = (ll'l'l) \end{cases}$$

$$\hat{\Gamma}^{(s,0)} = \begin{cases} U \\ U' \\ J \\ J' \end{cases}$$

(7)

where $D_S(i\omega_m) = 2\omega_s / (\omega_m^2 + \omega_s^2)$ is the Green's function of the local phonon for mode s. Note that the charge-orbit instability occurs when the maximum eigenvalue α_c of $\Gamma^c(i\omega_m)\hat{\chi}^0(q)$ at $i\omega = 0$ becomes 1. In this paper, we consider only the tetragonal mode with

$$\frac{2}{\sqrt{3}} g^{11}_{\text{tetra}} = -\frac{1}{\sqrt{3}} g^{22}_{\text{tetra}} = -\frac{1}{\sqrt{3}} g^{33}_{\text{tetra}} = g, \tag{8}$$

and set $\omega_s = 0.02$ eV for simplicity.

We investigate superconductivity by solving the linearized Eliashberg equation, and find the eigenvalues λ proportional to the superconducting gap function Δ and the eigenvector. The linearized Eliashberg equation is

$$\lambda \Delta_{mm'}(k) = -\frac{T}{N} \sum_{k'} \sum_{\{m_i\}} V_{mm_1,m_2m'}(k-k') G_{m_3m_1}(-k') G_{m_4m_2}(k') \Delta_{m_3m_4}(k') \tag{9}$$

where $\hat{V}(\mathbf{q},\omega)$ is expressed by the following equation,

$$\hat{V}(q) = \frac{3}{2}\hat{\Gamma}^s(i\omega_m)\hat{\chi}^s(q)\hat{\Gamma}^s(i\omega_m) - \frac{1}{2}\hat{\Gamma}^c(i\omega_m)\hat{\chi}^c(q)\hat{\Gamma}^c(i\omega_m) + \frac{1}{2}(\hat{\Gamma}^{(s,0)} - \hat{\Gamma}^{(c,0)}) \tag{10}$$

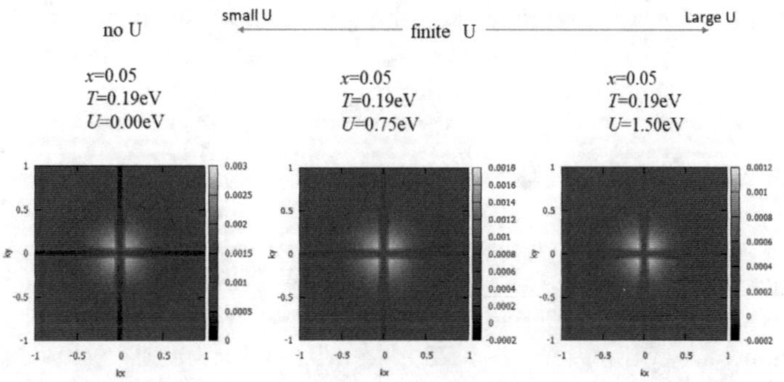

Figure 1. The d_{xy} orbital component at $k_z = 0$ has a large anisotropy due to the nature of the orbitals on the Fermi surface, and the on-site Coulomb interaction U does not change.

As T decreases, λ increases with increasing α_c and finally becomes unity at $T = T_c$. The corresponding gap function Δ_{xy} is shown in Fig. 1. Even with the gradual addition of on-site Coulomb interaction U, and the symmetry of the pair is anisotropic s-wave. Because Fig. 1 shows the d_{xy} orbital component at $k_z = 0$ has a large anisotropy, but this is due to the nature of the orbitals on the Fermi surface, and the on-site Coulomb interaction itU does not change. The d_{xy} orbital component of the susceptibility is found to be most dominant in the Jahn-Teller phonon.In this case, Δ_{xy} is the main component of the order parameter.

The value of T_c obtained for $x = 0.05$ was found to be a significant improvement over the value from standard BCS theory. And it means that it has high T_c superconductivity in the low carrier region. Effects of the on-site coulomb interaction U and the Jahn-Teller electron-phonon coupling g are taken into account within the random phase approximation (RPA). We obtain the phase diagram on the U-g plane as shown in Fig. 2, where the instability towards the magnetic order is observed for a huge value of $U{\sim}28$ eV almost independent of g while the

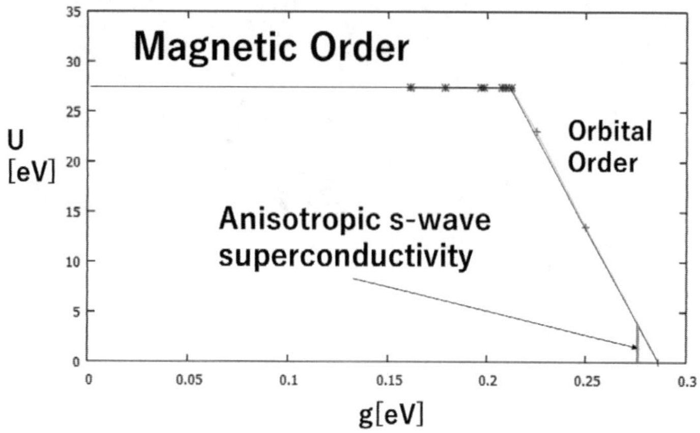

Figure 2. The phase diagram on the U-g plane, where the instability towards the ferro-orbital order accompanied by the cubic-tetragonal structural phase transition is observed for a realistic value of g=0.21∼0.28 eV depending on U.

instability towards the ferro-orbital order accompanied by the cubic-tetragonal structural phase transition is observed for a realistic value of g=0.21∼0.28 eV depending on U. Superconductivity is investigated by solving the linearized Eliashberg equation for the pairing interaction obtained by RPA, and we find that T_c for the anisotropic s-wave symmetry is largely enhanced due to the ferro-orbital fluctuation in the vicinity of the ferro-orbital order (structural phase transition) [7] as compared with that for the isotropic s-wave symmetry from the McMillan equation in which the x-dependence of the coulomb pseudopotential due to the plasmon effects is considered to reproduce the experimental x-dependence of T_c up to 7K [8].

3. Conclusion

We have investigated the three-orbital Hubbard-Holstein model simulating superconducting tungsten bronze A_xWO_3 using the RPA and the linearized Eliashberg equation, and have found that the anisotropic s-wave superconductivity is mediated by the orbital fluctuation at x =0.05 for realistic values of parameters g∼0.28 eV and U∼4 eV and is expected to be realized in the low-doped A_xWO_3 which shows a high-T_c superconductivity. To be more conclusive, we need further calculations including low temperature where the superconductivity is observed such as 90 K. Such calculations are under way and explicit results will be published in subsequent paper.

Acknowledgments

This work was partially supported by JSPS KAKENHI Grant Numbers JP 21K03399 and 21J15788.TS was supported by a JSPS Research Fellowship for Young Scientists. Numerical calculations were performed partially using OFP at the CCS, University of Tsukuba and the MASAMUNE-IMR, Tohoku University.

[1] H. R. Shanks *et al.*, Solid State Communications **15**, 753 (1974).
[2] K. Stanley *et al.*, Physical Review B **20**, 1903 (1979).
[3] A. Shengelaya *et al.*, Eur. Phys. J. B **12**, 13 (1999).
[4] S. Reich *et al.*, J. Supercond. Nov. Magn. **22**, 343-346 (2009).
[5] S. Raj *et al.*, Physical Review B **75**, 155116 (2007).
[6] A. Walkingshaw *et al.* Physical Review B **70**, 165110 (2004).
[7] T. Sekikawa *et al.* JPS Conf. Proc. 30, 011043 (2020).
[8] K. Sano, Y.Nitta and Y. Ōno, J. Phys. Soc. Jpn. 89, 113704 (2020).

Suppression of superconductivity by spin fluctuations

Hiroyuki Yamase[1,2]

[1] International Center of Materials Nanoarchitectonics, National Institute for Materials Science, Tsukuba 305-0047, Japan

[2] Department of Condensed Matter Physics, Graduate School of Science, Hokkaido University, Sapporo 060-0810, Japan

E-mail: `yamase.hiroyuki@nims.go.jp`

Abstract. We revisit the superconducting instability mediated by spin fluctuations in generic electron systems on a square lattice. We employ the standard Eliashberg theory, include the electron self-energy, keep a fine momentum resolution, and achieve numerical calculations down to very low temperatures so that the onset temperature of superconductivity (SC) can be determined. We find that spin fluctuations necessarily contain a contribution to suppress SC even though the superconducting instability can eventually occur at lower temperatures. This self-restraint effect stems from the repulsive pairing interaction induced by spin fluctuations, which leads to phase frustration of the pairing gap and consequently the suppression of SC. The self-restraint effect is a special feature of spin fluctuations and does not occur when SC is driven by the attractive pairing interaction from, for example, nematic fluctuations, orbital fluctuations, and electron-phonon coupling.

1. Introduction

It is well established that antiferromagnetic spin fluctuations can drive superconductivity (SC) [1]. This spin fluctuation mechanism was extensively discussed in a wide variety of materials such as heavy fermion systems, iron pnictides, and cuprates. In the spin fluctuation mechanism, it is presumed that the spin excitation spectrum has large spectral weight around a certain momentum $\mathbf{Q}$, which plays a major role to mediate Cooper pairs, and hence the spectral weight far away from $\mathbf{Q}$ is not relevant to the pairing mechanism. However, we find that the seemingly tiny contribution from a region far away from $\mathbf{Q}$ is also important and works to suppress the superconducting instability substantially. We report this self-restraint effect of SC by studying a minimal two-band model of iron pnictides and a canonical one-band model often used for cuprate and heavy fermion superconductors.

2. Model

We employ a minimal two-band model on a square lattice, which may be applicable to iron-based superconductors: $H = H_0 + H_I$. The kinetic energy part is given by [2, 3]

$$H_0 = \sum_{\mathbf{k},\sigma,\alpha,\beta} \epsilon_{\mathbf{k}}^{\alpha\beta} c_{\mathbf{k}\alpha\sigma}^{\dagger} c_{\mathbf{k}\beta\sigma} \tag{1}$$

where $c_{\mathbf{k}\alpha\sigma}^{\dagger}$ and $c_{\mathbf{k}\alpha\sigma}$ are the creation and annihilation operators for electrons with momentum $\mathbf{k}$ and spin orientation σ; the orbital index α takes 1 and 2, and refers to the d_{xz} and d_{yz}

orbitals, respectively. The intraorbital dispersions are given by $\epsilon_{\mathbf{k}}^{11} = -2t_1 \cos k_x - 2t_2 \cos k_y - 4t_3 \cos k_x \cos k_y - \mu$ and $\epsilon_{\mathbf{k}}^{22} = -2t_2 \cos k_x - 2t_1 \cos k_y - 4t_3 \cos k_x \cos k_y - \mu$, whereas the interorbital dispersion is $\epsilon_{\mathbf{k}}^{12} = -4t_4 \sin k_x \sin k_y$; μ is the chemical potential. The typical Fermi surfaces observed in iron pnictides are obtained by choosing the parameters as $t = -t_1$, $t_2/t = 1.5$, $t_3/t = -1.2$, $t_4/t = -0.95$, and $\mu/t = 0.6$ [3].

Instead of usual multiorbital Hubbard interactions, we introduce the SU(2) symmetric two-particle interaction in the magnetic channel so that we can study exclusively the effect of spin fluctuations on the superconducting instability:

$$H_I = \frac{1}{8N} \sum_{\mathbf{q},\mathbf{k},\mathbf{k}'} \sum_{\alpha,\beta,\sigma_j} V(\mathbf{k}, \mathbf{k}', \mathbf{q}) \boldsymbol{\sigma}_{\sigma_1\sigma_2} \cdot \boldsymbol{\sigma}_{\sigma_3\sigma_4} c^{\dagger}_{\mathbf{k}\alpha\sigma_1} c_{\mathbf{k}+\mathbf{q}\alpha\sigma_2} c^{\dagger}_{\mathbf{k}'+\mathbf{q}\beta\sigma_3} c_{\mathbf{k}'\beta\sigma_4} \tag{2}$$

where j runs from 1 to 4, $\boldsymbol{\sigma}$ are Pauli matrices, and N is the total number of the lattice sites. This interaction is the effective one close to a spin-density-wave (SDW) phase and can be obtained microscopically as a low-energy effective magnetic interaction generated by, for example, the repulsive Hubbard interaction by decreasing the energy scale in a functional renormalization group scheme [4, 5]. Since the conventional SDW order is observed in iron pnictides [6], we neglect the $\mathbf{k}$ and $\mathbf{k}'$ dependences in $V(\mathbf{k}, \mathbf{k}', \mathbf{q})$ and approximate it as $V(\mathbf{q})$. As $V(\mathbf{q})$, we may consider $V(\mathbf{q}) = 2V_1(\cos q_x + \cos q_y) + 4V_2 \cos q_x \cos q_y$ with $V_2 > V_1/2 > 0$. One may wish to consider a different form of $V(\mathbf{q})$ [7], but our conclusions do not depend on that.

We study the superconducting instability in the standard Eliashberg theory and solve the linearized gap equations numerically including the electron self-energy; see Ref. [7] for details of the Eliashberg equations. When the eigenvalue λ exceeds unity, the superconducting instability occurs. We fix $V_1 = t$ in the interaction term $V(\mathbf{q})$ and treat V_2 as a free parameter. We shall present results for $V_2 = 1.6t$ as representative ones. In the following, all quantities with the dimension of energy are measured in units of t.

3. Results

The solid line in Fig. 1(a) shows the temperature (T) dependence of the eigenvalue λ of the Eliashberg equations. λ increases rapidly at low temperatures, but does not exceed unity because the SDW order preempts the superconducting instability. We cannot obtain SC even for other choices of V_2. Thus strictly speaking, SC does not occur even at reasonably low temperatures in our model. This is consistent with results obtained in the fluctuation-exchange approximation to the multiband Hubbard model [8, 9, 10].

A new insight that we have obtained is the self-restraint effect of SC. As shown in Fig. 1(b), there are three different low-energy scattering processes, "intra", "$(\pi, 0)$", and "(π, π)". When we consider only the "$(\pi, 0)$" scattering processes, a larger λ is obtained in Fig. 1(a), whereas λ becomes very small for the "intra" and "(π, π)" scattering processes alone as one expects. However, when we take both "intra" and "$(\pi, 0)$" scattering processes into account, λ from the "$(\pi, 0)$" scattering processes is substantially suppressed to reproduce the full results almost precisely. This is our finding of the self-restraint effect. That is, a seemingly tiny effect from the intrapocket scattering processes suppresses the tendency of superconducting instability substantially.

As is well known, spin fluctuations lead to a repulsive pairing interaction [1], so that the gap function tends to have the opposite sign to each other when scattering processes occur between two different Femi momenta. Consequently, the major scattering processes of "$(\pi, 0)$" generate a gap with the positive sign on the hole pockets and the negative sign on the electron pockets, or vice versa, the so-called $s\pm$-wave state. On the other hand, the intrapocket scattering processes try to change the sign of the gap function inside each Fermi pocket, and generate phase frustration with the gap phase from the "$(\pi, 0)$" scattering processes. Figure 1(a) indicates that

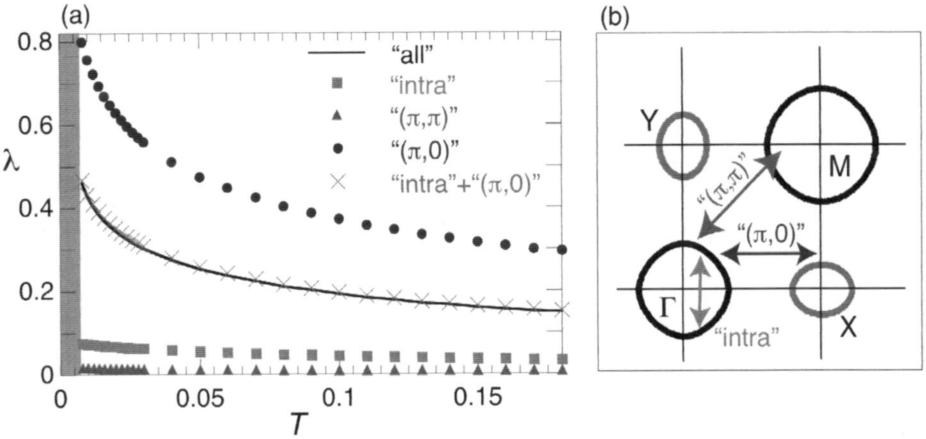

Figure 1. (a) Temperature dependence of the eigenvalue of λ. The eigenvalues are also computed by focusing on particular scattering processes denoted by "intra", "$(\pi, 0)$", and "(π, π)". Below $T = 0.008$ (shaded region), SDW order occurs before the superconducting instability. (b) Typical Ferm surfaces in iron-based superconductors: hole pockets around Γ and M points, and electron pockets around X and Y points. "intra", "$(\pi, 0)$", and "(π, π)" denote scattering processes inside each Fermi pocket, between the hole and electron pockets, and between the two hole (or electron) pockets, respectively.

this phase frustration is unexpectedly large. This self-restraint effect traces back to the fact that spin fluctuations lead to a repulsive pairing interaction. Hence we believe that the self-restraint effect is a general property of a repulsive pairing interaction.

4. Application to a one-band case

To substantiate the generality of the self-restraint effect, we also perform a similar analysis in a one-band model by discarding the orbital degrees of freedom α and β in Eqs. (1) and (2). We then take the typical tight-binding dispersion on a square lattice, $\epsilon_{\mathbf{k}} = -2t(\cos k_x + \cos k_y) - 4t' \cos k_x \cos k_y - \mu$ and the magnetic interaction $V(\mathbf{q}) = 2V(\cos q_x + \cos q_y)$ with $t'/t = -0.35$ and $V/t = 1.18$, where the system is close to the SDW phase that is realized in $V/t \gtrsim 1.28$. The chemical potential μ is tuned to reproduce the electron density per lattice site to be 0.85 so that we obtain a holelike Fermi surface around (π, π) typical of the Fermi surface in cuprates.

In the present band parameters, magnetic fluctuations occur around momentum (π, π). We thus define the scattering processes as shown in the inset in Fig. 2: "(π, π)" scattering scattering processes correspond to a region $\sqrt{(q_x - \pi)^2 + (q_y - \pi)^2} \leq 2\pi\eta_\pi$, "$(0, 0)$" a region $\sqrt{q_x^2 + q_y^2} \leq 2\pi\eta_0$, and "others" the rest. We fix $\eta_\pi = 0.25$ and compute the eigenvalue of the Eliashberg equations by changing η_0 for different scattering processes as shown in Fig. 2. Since η_π is fixed, λ for the "(π, π)" processes does not depend on η_0 and exceeds unity. On the other hand, λ for the "$(0, 0)$" processes is very small even if η_0 is increased. However, when both "(π, π)" and "$(0, 0)$" scattering processes are included, λ is suppressed with increasing η_0 and converges to the eigenvalue obtained for all scattering processes above $\eta_0 \approx 0.2$. This demonstrates the self-restraint effect of the superconducting instability, that is, the major "(π, π)" scattering processes try to drive SC, but the minor contribution from the "$(0, 0)$" scattering processes suppresses it. This effect is unexpectedly large in the sense that the contribution from the "$(0, 0)$" scattering processes alone is surely irrelevant to SC by considering its small λ.

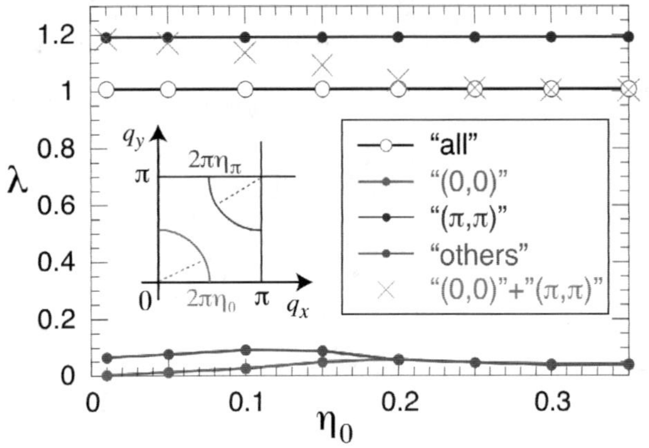

Figure 2. η_0 dependence of the eigenvalues of λ for different scattering processes at $T = 0.01$, where the superconducting instability occurs for the present parameters. The inset defines the scattering processes: "(0, 0)" and "(π,π)" correspond to a region $\sqrt{q_x^2+q_y^2} \leq 2\pi\eta_0$ and $\sqrt{(q_x-\pi)^2+(q_y-\pi)^2} \leq 2\pi\eta_\pi$ with $\eta_\pi = 0.25$, respectively, and "others" means the rest of scattering processes.

5. Conclusions and discussions

We have shown that the spin-fluctuation mechanism suffers from the self-restraint effect due to the contribution of "(0,0)" scattering processes and in this sense is not favorable to achieving high-T_c SC. This is because spin fluctuations yield a repulsive pairing interaction and thus necessarily generate phase frustration of the pairing gap. On the other hand, orbital nematic fluctuations yield an attractive pairing interaction and thus are free from the self-restraint effect. In fact, orbital nematic fluctuation mechanism leads to a phase diagram with reasonably high T_c [11] and a momentum dependence of the pairing gap [12], both of which are compatible with iron-based superconductors.

It should be noted that spin fluctuations describe a retarded pairing interaction. There is also an instantaneous interaction, which do not have an energy dependence and can escape the self-restraint effect by exploiting a sign change of the interaction owing to its momentum dependence. High-T_c cuprate superconductors might be better understood in terms of the instantaneous spin interaction, as was already emphasized by P. W. Anderson [13].

The spin fluctuation mechanism typically leads to low T_c of the order of 10^{-2} or less even in the vicinity of a SDW phase. This typical feature captures SC observed in many heavy fermion materials [14].

Acknowledgments

A study on the minimal two-band model was based on collaboration with T. Agatsuma. This work was supported by JSPS KAKENHI Grants No. JP20H01856.

References

[1] Scalapino D J 2012 *Rev. Mod. Phys.* **84**(4) 1383–1417
[2] S Raghu, X-L Qi, C-X Liu, D J Scalapino, and S-C Zhang 2008 *Phys. Rev. B* **77** 220503
[3] Z-J Yao, J-X Li, and Z D Wang 2009 *New J. Phys.* **11** 025009
[4] Husemann C and Salmhofer M 2009 *Phys. Rev. B* **79**(19) 195125
[5] Eberlein A and Metzner W 2014 *Phys. Rev. B* **89**(3) 035126
[6] Dai P, Hu J and Dagotto E 2012 *Nat. Phys.* **8** 709
[7] Yamase H and Agatsuma T 2020 *Phys. Rev. B* **102**(6) 060504
[8] Arita R and Ikeda H 2009 *J. Phys. Soc. Jpn.* **78** 113707
[9] Suzuki K, Usui H, Iimura S, Sato Y, Matsuishi S, Hosono H and Kuroki K 2014 *Phys. Rev. Lett.* **113**(2) 027002
[10] Usui H, Suzuki K and Kuroki K 2015 *Scientific Reports* **5** 11399
[11] Yamase H and Zeyher R 2013 *Phys. Rev. B* **88** 125120
[12] Agatsuma T and Yamase H 2016 *Phys. Rev. B* **94**(21) 214505
[13] Anderson P W 2007 *Science* **317** 1705
[14] Pfleiderer C 2009 *Rev. Mod. Phys.* **81**(4) 1551–1624

Journal of Physics: Conference Series **2164** (2022) 012014 doi:10.1088/1742-6596/2164/1/012014

Phase separation in high-T_c cuprates

A S Moskvin[1,2], **Yu D Panov**[1]

[1] Institute of Natural Sciences and Mathematics, Ural Federal University, 620083 Ekaterinburg, Russia

[2] Institute of Metal Physics UB RAS, 620108 Ekaterinburg, Russia

E-mail: alexander.moskvin@urfu.ru

Abstract. We develop a minimal non-BCS model for the CuO_2 planes with the on-site Hilbert space reduced to only three effective valence centers CuO_4 with different charge, conventional spin, and orbital symmetry, combined in a charge triplet, to describe the low-energy electron structure and the phase states of HTSC cuprates. Using the $S = 1$ pseudospin algebra we introduce an effective spin-pseudospin Hamiltonian which takes into account local and nonlocal correlations, one- and two-particle transport, and spin exchange. The T-n phase diagrams of the complete spin-pseudospin model for the CuO_2 planes were reproduced by means of a site-dependent variational approach within effective field approximation typical for spin-magnetic systems. Limiting ourselves to two-sublattice approximation and nn-couplings we arrived at several Néel-like phases in CuO_2 planes for parent and doped systems with a single nonzero local order parameter: antiferromagnetic insulator, charge order, glueless d-wave Bose superfluid phase, and unusual metallic phase. However, the Maxwell's construction shows the global minimum of free energy is realized for phase separated states which are bounded by the third-order phase transition line $T^{\star}(n)$, which is believed to be responsible for the onset of the pseudogap phenomenon.

1. Introduction

Many researchers argue that the main features of the cuprates superconductors phase diagram can be derived considering the disorder and intrinsic mesoscopic static/dynamic phase separation as a key property of these materials (see, e.g. Ref. [1]). However, a correct description of a phase-inhomogeneous state presupposes the use of an adequate physical model. Recent discoveries of anomalous properties of cuprates and nickelates with a T′-structure [2, 3] indicate the need to abandon the generally accepted concept of the parent composition as a Mott-Hubbard antiferromagnetic insulator. Instead, we propose to introduce a more universal definition of the "parent" system of CuO_2/NiO_2 planes with a nominal $3d^9$ configuration for Cu/Ni sites, or "half-filling", which, depending on the parameters determined by the "out-of-plane" potential and electron-lattice relaxation, can be found in various states, from an antiferromagnetic or non-magnetic insulator, Fermi metal, to a high-temperature superconductor. Furthermore, from our point of view, the explanation of the cuprate/nickelate puzzle should include some fundamentally new physics, which requires not only going beyond the Bardeen-Cooper-Schrieffer (BCS) paradigm of the superconducting state [4], but also revising a number of other well-established concepts. Following the spin-magnetic analogy proposed by Rice and Sneddon [5] to describe the three charge states (Bi^{3+}, Bi^{4+}, Bi^{5+}) of the bismuth ion in $BaBi_{1-x}Pb_xO_3$, earlier we started to develop a minimal "unparticle" model for the CuO_2 planes with the "on-site"

Hilbert space of the CuO_4 plaquettes to be a key element of crystal and electron structure of high-T_c cuprates, reduced to states formed by only three effective valence centers $[CuO_4]^{7-,6-,5-}$ (nominally $Cu^{1+,2+,3+}$, respectively), forming a "well isolated" charge triplet, or spin-pseudospin quartet ($S=1/2$ for "parent" $[CuO_4]^{6-}$ center) [6, 7].

2. Effective spin-pseudospin Hamiltonian

As for conventional spin-magnetic systems, we can integrate out the high-energy degrees of freedom, and after projecting onto the Hilbert basis of well isolated charge triplet, or spin-pseudospin quartet, we have chosen, to arrive at the effective spin-pseudospin Hamiltonian obeying the spin and pseudospin kinematic rules with a charge density constraint, $\frac{1}{N}\sum_i\langle\hat{S}_{iz}\rangle = n$, where n is the deviation from a half-filling:

$$\hat{H} = \hat{H}_{pot} + \hat{H}^{(1)}_{kin} + \hat{H}^{(2)}_{kin} + \hat{H}_{ex}\,; \tag{1}$$

$$\hat{H}_{pot} = \sum_i(\Delta_i\hat{S}^2_{iz} - \mu\hat{S}_{iz}) + \sum_{i>j}V_{ij}\hat{S}_{iz}\hat{S}_{jz}\,;$$

$$\hat{H}^{(1)}_{kin} = -\sum_{i>j}\sum_{\nu}[t^p_{ij}\hat{P}^{\nu}_{i+}\hat{P}^{\nu}_{j-} + t^n_{ij}\hat{N}^{\nu}_{i+}\hat{N}^{\nu}_{j-} + \frac{1}{2}t^{pn}_{ij}(\hat{P}^{\nu}_{i+}\hat{N}^{\nu}_{j-} + \hat{P}^{\nu}_{i-}\hat{N}^{\nu}_{j+}) + h.c.]\,;$$

$$\hat{H}^{(2)}_{kin} = -\sum_{i>j}t^b_{ij}(\hat{S}^2_{i+}\hat{S}^2_{j-} + \hat{S}^2_{i-}\hat{S}^2_{j+}),\ \hat{H}_{ex} = \sum_{i>j}J_{ij}(\hat{\mathbf{s}}_i\cdot\hat{\mathbf{s}}_j) = s^2\sum_{i>j}J_{ij}(\boldsymbol{\sigma}_i\cdot\boldsymbol{\sigma}_j)\,.$$

For the properties of the pseudospin ($\hat{S}^2_{\pm}$) and spin-pseudospin ($\hat{P}^{\nu}_{\pm}, \hat{N}^{\nu}_{\pm}$) operators, see, for example, Refs. [7, 8]. The first on-site term in $\hat{H}_{pot}$ relates with the on-site density-density interactions, $\Delta=U/2$, U being the local correlation parameter, or pair binding energy for composite boson, μ in the second term is the hole chemical potential, the third one describes the nonlocal correlations. Kinetic energies $\hat{H}^{(1)}_{kin}$ and $\hat{H}^{(2)}_{kin}$ describe a single- and two-particle transport, respectively, where t^p, t^n, t^{pn} and t^b are integrals for the correlated single-particle and the composite on-site boson hopping, respectively. Operator $\boldsymbol{\sigma} = 2\hat{\rho}^s\mathbf{s}$ in $\hat{H}_{ex}$ takes into account the on-site spin density $\hat{\rho}^s = (1-\hat{S}^2_z)$. Depending on the values of the effective Hamiltonian parameters, the model predicts the possibility of realizing even for parent cuprates/nickelates both a typical antiferromagnetic insulating state (AFMI), charge order (CO), unusual Fermi liquid (FL) phase with electron-hole interplay and a bosonic superconductor (BS) phase with effective on-site hole bosons.

The atomic limit of the spin-pseudospin model, which assumes a complete neglect for kinetic energy, was considered within the mean-field approximation (MFA), Bethe cluster approach, and classical Monte-Carlo (MC) technique with a clear manifestation of the CO-AFMI phase separation (PS) (see, e.g., Ref. [9]). In the large "negative-U" approximation, which assumes complete neglect for local correlations, single-particle kinetic energy, and spin exchange, the CuO_2 plane becomes equivalent to hard-core (hc) boson system. In addition to the conventional CO and Bose superfluid BS phases the MFA predicts the appearance of an unconventional uniform supersolid phase (SS) with the "on-site" coexistence of the insulating and superconducting properties [11]. However, detailed analysis of the hc boson model shows that the SS phase is a MFA artefact, de facto this homogeneous phase is intrinsically unstable [12]. For nn-interactions and finite temperatures the "post-MFA" Maxwell's construction shows that the phase-separated CO-BS phase has a lower energy than the uniform SS phase, though at $T=0$ the both phases have the same energy. This result is confirmed within Bethe cluster approximation [10]. Under doping away from half-filling, the CO phase undergoes PS: the superfluid BS and solid CO phases coexist but not as a single thermodynamic SS phase as predicted in the MFA [11]. Thus, the both limits of the complete spin-pseudospin system exhibit the PS effect typical of systems with competing order parameters.

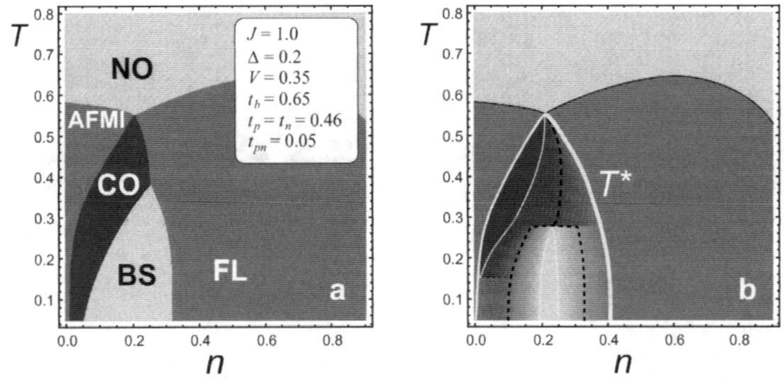

Figure 1. (Color online) Model T - n phase diagrams of the hole-doped CuO_2 plane calculated in the EF approximation ($n = p$ for the hole doping) with constant values of the Hamiltonian parameters (see inset); (a) phase diagram assuming main homogeneous phases with no allowance made for the possible coexistence of two adjacent phases; (b) phase diagram with the PS taken into account via the Maxwell's construction. Dashed curves in (b) point to fifty-fifty volume fraction for two adjacent phases, yellow curves in present the third-order phase transition lines, these limit areas with 100% volume fraction.

3. Effective-field approximation for the complete spin-pseudospin model

Making use of the effective field (EF) theory which combines the MFA with the exact accounting of local correlations, the variational approach (VA) based on the Bogolyubov inequality for the grand potential, and the Caron-Pratt model for the on-site description of the FL phase [13], we were able to numerically calculate the phase diagrams of the complete spin-pseudospin CuO_2 system within the framework of a simplified model (two sublattices, nn-coupling,...) [8]. The left hand side of the Figure 1 shows an example of the phase diagram calculated given quite arbitrarily chosen parameters of the model Hamiltonian and assuming main homogeneous single-order parameter, or "monophases" with no allowance made for the possible coexistence of two adjacent phases. However, the numerical implementation of the Maxwell's construction shows that the minimum of free energy corresponds to PS realized in the region of coexistence of phases separated by the first-order phase transition lines. This works for phases AFMI-FL, AFMI-BS, CO-BS, CO-FL, and BS-FL, but not for AFMI-CO (see Figure 1, the right hand side panel). We can immediately note that the mysterious pseudogap phase is nothing more than the AFMI-FL-CO-BS PS state with the third-order phase transition line $T^*(n)$, which separates the gapless 100% FL phase from the gapped phases, to be the highest pseudogap temperature. The PS model does predict several temperatures of the "third order" PS transitions limiting the PS phases, that is delineating areas with 100% volume fraction, and the temperatures of the percolation transitions, which can manifest itself in the peculiarities of the temperature behavior for different physical quantities. Obviously, the ground state and $T - n$ phase diagrams of the system described by the effective spin-pseudospin Hamiltonian are determined by the relationship between the values of the parameters, a relatively small change in the parameters can radically affect the phase diagram. Thus, a small increase in the local correlation parameter Δ and the transfer integral of composite bosons t_b can lead to suppression of the both AFMI and CO phases in favor of BS and FL phases. It is this situation is shown in the phase diagram in the Figure 2 a,b, calculated numerically with the parameters shown in the inset to the Figure 2a, which presents the results of conventional EF approach, while the phase diagram in the Figure 2b does illustrate the BS-FL interplay with the PS taken into account via the Maxwell's construction. Such phase diagrams with no signatures of the long-range AFMI and CO orderings turn out to be typical of cuprates with an ideal, or almost ideal, T′-structure [2],

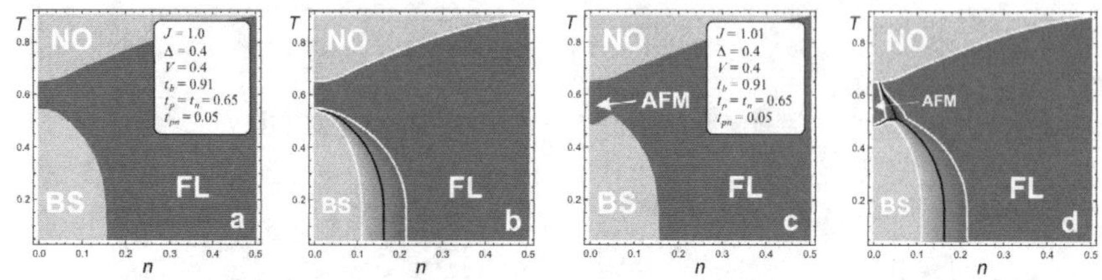

Figure 2. (Color online) Model T - n phase diagrams of the CuO_2/NiO_2 plane in the hole-doped cuprate/nickelate calculated in the EF approximation under constant values of the Hamiltonian parameters (see inset), which illustrate an AFMI-BS-FL and BS-FL interplay with (b,d) and without (a,c) PS taken into account. Solid yellow curves in (b,d) present the third-order phase transition lines, black solid curves point to fifty-fifty volume fraction for two adjacent phases.

as well as nickelates, the structure of which also lacks apex oxygen [3].

For the chosen values of the Hamiltonian parameters, the AFMI phase is energetically unfavorable, but its energy differs little from the energy of the BS and FL phases, so that the slightest 1% increase in the value of the exchange integral, or a corresponding decrease in the remaining parameters, is sufficient to restore the AFMI phase, albeit in a small region of the phase diagram (see Figures 2 c,d). Interestingly, a similar effect of suppression/restoring of the AFMI phase is observed in T′-cuprates at the slightest change in the concentration of nonstoichiometric apical oxygen [2], accompanied by corresponding change of the external potential for CuO_2-planes.

4. Conclusion

Effective field theory for the model spin-pseudospin Hamiltonian points to the phase-separated coexistence of main phases AFMI, CO, BS, and FL to be the typical one for the CuO_2/NiO_2 planes in cuprates/nickelates. Despite the many simplifications our model approach is believed to be a reliable starting point for description of the 3D phase diagrams in cuprates/nickelates with experimentally observed T_N, T_c, and T^* as the critical temperatures for the 3D phase transitions.

Acknowledgments

This research was partially supported by the Ministry of Education and Science of Russian Federation, project No FEUZ-2020-0054.

References

[1] de Mello E V L Kasal R B and Passos C A C 2009 *J. Phys.: Condens. Matter* *21* 235701
[2] Naito M Krockenberger Y Ikeda A and Yamamoto H 2016 *Physica C* *523* 28-54.
[3] Li D Lee K Wang B Y Osada M *et al.* 2019 *Nature (London)* *572* 624
[4] Hirsch J E 2009 *Phys. Scr.* *80* 035702; arXiv:0901.4099v1
[5] Rice T M and Sneddon L 1981 *Phys. Rev. Lett.* *47* 689
[6] Moskvin A S 2011 *Phys. Rev.* B *84* 075116
[7] Moskvin A S Panov Y D 2019 *J. Supercond. Nov. Magn.* *32* 61
[8] Moskvin A Panov Y 2021 *Condens. Matter* *6* 24
[9] Panov Y D Moskvin A S Chikov A A Budrin K S 2017 *J. Low Temp. Phys.* *187(5-6)* 646-653
[10] Spevak E L Panov Yu D Moskvin A S 2021 *Phys. Sol. State* *63* 1517–1523
[11] Micnas R Ranninger J and Robaszkiewicz S 1990 *Rev. Mod. Phys.* *62* 113
[12] Schmid G Todo S Troyer M and Dorneich A 2002 *Phys. Rev. Lett.* *88* 167208
[13] Caron L G Pratt G W 1968 *Rev. Mod. Phys.* *40* 802–806

Journal of Physics: Conference Series **2164** (2022) 012015 doi:10.1088/1742-6596/2164/1/012015

Possible pairing symmetries in the ordered honeycomb network superconductor BaPtSb

Masafumi Kudo[1], Shohei O. Shingu[1], Tsuyoshi Imazu[1], Jun Goryo[1], and Yoshiki Imai[2]

[1] Department of Mathematics and Physics, Hirosaki University, Japan
[2] Department of Applied Physics, Okayama University of Science, Okayama, Japan

E-mail: jungoryo@hirosaki-u.ac.jp

Abstract. We investigate the pairing symmetry of the ordered honeycomb network superconductor BaPtSb, which has a crystal structure without inversion center. There is a preliminary μSR report which implies the broken time-reversal symmetry in the superconducting state. In this paper, we classify the pairing symmetry and examine the pairing instability. Among the unconventional states with time-reversal symmetry breaking, we find that the state with a gap structure compatible with Fermi surfaces is the spin-triplet chiral p-wave state.

1. Introduction

The series of superconductors with ordered honeycomb network structure has been found, since the discovery of the superconductivity in SrPtAs [1]. In this paper we focus on the recently discovered one BaPtSb with inversion-symmetry breaking ($T_c = 1.6K$, the point group; D_{3h}) [2, 3]. The first principle calculation reveals that the anti-symmetric spin-orbit coupling works significantly and causes band splitting [4]. The preliminary result of the muon spin relaxation (μSR) experiment implies a spontaneous field below T_c and the superconducting state may break the time-reversal symmetry [5].

In this paper, we classify the pairing symmetry using group theoretical approach and examine the pairing instability. Among the states with broken time-reversal symmetry, we find that the preferable one compatible with Fermi surface structures is the spin-triplet chiral p-wave state with $S_z = 0$. This pairing is renowned as a topological state and supports the chiral surface state at the sample boundary.

2. Tight-binding model

The detail of the band structure of BaPtSb has been reported [4]. The result reveals that Pt 5d and Sb 5p orbitals are dominant in the low energy region. We have basically three Fermi surfaces with spin-orbit splitting. Two of them are cylindrical around the Γ point, and the other one is spherical around K and K′ points. We should emphasize that the anti-symmetric spin orbit coupling coming from the broken inversion symmetry plays an important role.

We construct the tight-binding model for the low energy excitations, which would be relevant for the superconductivity. For simplicity, we take three Pt 5d orbitals into account and all the hopping are diagonal with respect to the orbital degrees of freedom. The Hamiltonian is

$$
\begin{aligned}
H_0 &= \sum_{bk\sigma} \xi^{(b)}_{\boldsymbol{k}\sigma} c^{(b)\dagger}_{\boldsymbol{k}\sigma} c^{(b)}_{\boldsymbol{k}\sigma}, \qquad (1)\\
\xi^{(b)}_{\boldsymbol{k}\sigma} &= \epsilon^{(b)}_{\boldsymbol{k}} + \sigma\alpha^{(b)}\lambda_{\boldsymbol{k}},\\
\epsilon^{(b)}_{\boldsymbol{k}} &= t^{(b)}\epsilon_{\boldsymbol{k}1} + \left(t^{(b)}_{z1} + t^{(b)}_{z2}\epsilon_{\boldsymbol{k}1} + t^{(b)}_{z3}\epsilon_{\boldsymbol{k}2} + t^{(b)}_{z4}\epsilon_{\boldsymbol{k}3}\right)\cos k_z c - \mu^{(b)},\\
\lambda_{\boldsymbol{k}} &= \sum_{n=1}^{3} \sin \boldsymbol{k}\cdot\boldsymbol{T}_n,\ \epsilon_{\boldsymbol{k}1} = \sum_{n=1}^{3} \cos \boldsymbol{k}\cdot\boldsymbol{T}_n,\ \epsilon_{\boldsymbol{k}2} = \sum_{n=1}^{3} \cos \boldsymbol{k}\cdot\boldsymbol{T}'_n,\ \epsilon_{\boldsymbol{k}3} = \sum_{n=1}^{3} \cos \boldsymbol{k}\cdot 2\boldsymbol{T}_n,
\end{aligned}
$$

where b, $\boldsymbol{k}$, σ show the band, momentum, and z-component of spin, respectively. The in-plane nearest neighbor bond vectors are $\boldsymbol{T}_1 = (0, a, 0)$, $\boldsymbol{T}_2 = (-\sqrt{3}/2, -a/2, 0)$, and $\boldsymbol{T}_3 = (\sqrt{3}/2, -a/2, 0)$. We have also $\boldsymbol{T}'_1 = \boldsymbol{T}_1 - \boldsymbol{T}_2$, $\boldsymbol{T}'_2 = \boldsymbol{T}_2 - \boldsymbol{T}_3$, and $\boldsymbol{T}'_3 = \boldsymbol{T}_3 - \boldsymbol{T}_1$. This model consists of the in-plane nearest neighbor hopping $t^{(b)}$, and out-of-plane l-th nearest hoppings $t^{(b)}_{zl}$ $(l = 1, 2, 3, 4)$. We also have the anti-symmetric spin-orbit coupling $\alpha^{(b)}$ with $\lambda_{\boldsymbol{k}} = -\lambda_{-\boldsymbol{k}}$ coming from the broken inversion symmetry. In this system, the symmetry breaking occurs in the in-plane direction and the spin-orbit coupling is diagonal with respect to σ. The spectrum satisfies $\xi_{\boldsymbol{k}\uparrow} = \xi_{-\boldsymbol{k}\downarrow}$ and the spin-singlet pairing or the spin-triplet pairing with $S_z = 0$ (i.e. S_z conserving states) are stabilized as we will see later.

We fit this model to the result of *ab initio* calculation [4]. The fitting parameters and Fermi surfaces are shown in Table 1 and Figure 1.

Table 1. The fitting parameters in the tight-binding model Eq. (1).

	$t^{(b)}$	$t^{(b)}_{z1}$	$t^{(b)}_{z2}$	$t^{(b)}_{z3}$	$t^{(b)}_{z4}$	$\mu^{(b)}$	$\alpha^{(b)}$
Band 1 ($b = 1$)	0.85	0.07	-0.09	-0.01	0.01	1.7	0.3
Band 2 ($b = 2$)	1	0.1	-0.25	-0.25	0.2	0.4	0.25
Band 3 ($b = 3$)	-0.65	0.9	-0.3	0.05	0.01	1.9	0.07

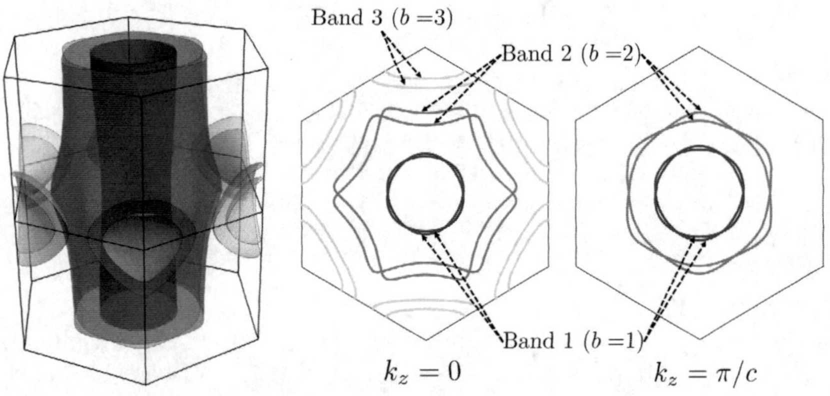

Figure 1. Fermi surfaces obtained from the tight-binding model Eq. (1) with parameters listed in Table 1. See also the result of *ab initio* calculation [4].

Journal of Physics: Conference Series **2164** (2022) 012015 doi:10.1088/1742-6596/2164/1/012015

3. Pairing instability

Using the point group D_{3h} we find a basis of gap matrices in an irreducible representation Γ

$$\hat{\Delta}_{\boldsymbol{k}}^{\Gamma,m} = \left[\Delta_s \psi_{\boldsymbol{k}}^{\Gamma,m} + \Delta_t \boldsymbol{d}_{\boldsymbol{k}}^{\Gamma,m} \cdot \hat{\boldsymbol{\sigma}}\right] i\hat{\sigma}_y, \tag{2}$$

where $\psi_{\boldsymbol{k}}^{\Gamma,m}$ and $\boldsymbol{d}_{\boldsymbol{k}}^{\Gamma,m}$ are the basis functions for spin-singlet and triplet channels normalized in the Brillouin zone, and $m = 1, ..., N$ in N-dimensional irreducible representation. The details of this classification would be discussed elsewhere.

We then argue the pairing instability. Although it may occur due to the inversion-symmetry breaking in principle, the mixing between spin-singlet and triplet channels is not expected to be large and neglected hereafter. The linearized gap equations are

$$\frac{1}{g_s^{\Gamma,m(b)}} = \sum_{\boldsymbol{k}\sigma} \frac{|\psi_{\boldsymbol{k}}^{\Gamma,m}|^2}{2\xi_{\boldsymbol{k}\sigma}^{(b)}} \tanh \frac{\xi_{\boldsymbol{k}\sigma}^{(b)}}{2k_B T_c}, \tag{3}$$

$$\frac{1}{g_t^{\Gamma,m(b)}} = \sum_{\boldsymbol{k}\sigma} \frac{|\boldsymbol{d}_{\boldsymbol{k}}^{\Gamma,m}|^2}{2\xi_{\boldsymbol{k}\sigma}^{(b)}} \tanh \frac{\xi_{\boldsymbol{k}\sigma}^{(b)}}{2k_B T_c}. \tag{4}$$

We fix T_c and estimate the integral in the *r.h.s.*, and compare the obtained coupling constants in the *l.h.s.* of the equations. A basis would be more favorable if the obtained coupling constant is smaller. The most stable basis for each pairing range for each band is listed in Table 2. We note that spin-singlet states or triplet states with $S_z = 0$ ($\boldsymbol{d} \parallel \hat{\boldsymbol{z}}$) are stabilized strongly by the spin-orbit coupling $\sigma\alpha^{(b)}\lambda_{\boldsymbol{k}}$, which causes the band splitting with keeping the relation $\xi_{\boldsymbol{k}\uparrow} = \xi_{-\boldsymbol{k}\downarrow}$. We also see that the E_2 representation is favored in several cases. The locations of the peaks of these basis functions are well overlapped with that of Fermi surfaces, and thus the condensation energies are enhanced (see Figure 2).

Table 2. The list of the most stable basis for each pairing range and for each band without normalization constants. Definitions of the functions of $\boldsymbol{k}$ are; $e_{\boldsymbol{k}} = \sum_n \cos \boldsymbol{k} \cdot \boldsymbol{T}_n$, $o_{\boldsymbol{k}} = \sum_n \sin \boldsymbol{k}\cdot\boldsymbol{T}_n$, $o_{\boldsymbol{k}}^+ = \sum_n \omega^{n-1} \sin \boldsymbol{k}\cdot\boldsymbol{T}_n = (o_{\boldsymbol{k}}^-)^*$, $\bar{o}_{\boldsymbol{k}} = \sum_n \sin \boldsymbol{k}\cdot\boldsymbol{T}'_n$, $\bar{o}_{\boldsymbol{k}}^+ = \sum_n \omega^{n-1} \sin \boldsymbol{k}\cdot\boldsymbol{T}'_n = (\bar{o}_{\boldsymbol{k}}^-)^*$, $\tilde{e}_{\boldsymbol{k}} = \sum_n \cos \boldsymbol{k} \cdot 2\boldsymbol{T}_n$, $\tilde{o}_{\boldsymbol{k}}^+ = \sum_n \omega^{n-1} \sin \boldsymbol{k} \cdot 2\boldsymbol{T}_n = (\tilde{o}_{\boldsymbol{k}}^-)^*$, and $\omega = e^{i2\pi/3}$. We omit the index m for the one-dimensional representations.

pairing range	Band 1	Band 2	Band 3
on-site	$\psi^{A_1} = 1$	$\psi^{A_1} = 1$	$\psi^{A_1} = 1$
in-plane nearest	$\psi^{A_1} = e_{\boldsymbol{k}}$	$\boldsymbol{d}_{\boldsymbol{k}}^{E_2,1} = o_{\boldsymbol{k}}^+ \hat{\boldsymbol{z}}$ $\boldsymbol{d}_{\boldsymbol{k}}^{E_2,2} = o_{\boldsymbol{k}}^- \hat{\boldsymbol{z}}$	$\boldsymbol{d}_{\boldsymbol{k}}^{A_1} = o_{\boldsymbol{k}} \hat{\boldsymbol{z}}$
in-plane 2nd nearest	$\boldsymbol{d}_{\boldsymbol{k}}^{E_2,1} = \bar{o}_{\boldsymbol{k}}^+ \hat{\boldsymbol{z}}$ $\boldsymbol{d}_{\boldsymbol{k}}^{E_2,2} = \bar{o}_{\boldsymbol{k}}^- \hat{\boldsymbol{z}}$	$\boldsymbol{d}_{\boldsymbol{k}}^{A_2} = \bar{o}_{\boldsymbol{k}} \hat{\boldsymbol{z}}$	$\boldsymbol{d}_{\boldsymbol{k}}^{E_2,1} = \bar{o}_{\boldsymbol{k}}^+ \hat{\boldsymbol{z}}$ $\boldsymbol{d}_{\boldsymbol{k}}^{E_2,2} = \bar{o}_{\boldsymbol{k}}^- \hat{\boldsymbol{z}}$
out-of-plane nearest	$\psi_{\boldsymbol{k}}^{A_1} = \cos k_z c$	$\psi_{\boldsymbol{k}}^{A_1} = \cos k_z c$	$\psi_{\boldsymbol{k}}^{A_1} = \cos k_z c$
out-of-plane 2nd nearest	$\psi_{\boldsymbol{k}}^{A_1} = \tilde{e}_{\boldsymbol{k}}$	$\boldsymbol{d}_{\boldsymbol{k}}^{E_2,1} = \tilde{o}_{\boldsymbol{k}}^+ \hat{\boldsymbol{z}}$ $\boldsymbol{d}_{\boldsymbol{k}}^{E_2,2} = \tilde{o}_{\boldsymbol{k}}^- \hat{\boldsymbol{z}}$	$\psi_{\boldsymbol{k}}^{A_1} = \tilde{e}_{\boldsymbol{k}}$

The pairing states in the E_2 representation is

$$\boldsymbol{d}_{\boldsymbol{k}} = \eta_1 \boldsymbol{d}_{\boldsymbol{k}}^{E_2,1} + \eta_2 \boldsymbol{d}_{\boldsymbol{k}}^{E_2,2}, \tag{5}$$

where $\vec{\eta} = (\eta_1, \eta_2)$ is the two-dimensional order parameter, and we mainly have two possible pairing states; the chiral p-wave state and the nematic one [6] (see Table 3). Both states are topological but only the chiral p-wave state breaks the time-reversal symmetry. The chiral state is fully gapped, whereas the nematic one has line nodes. Thus, the chiral state would be more favorable at least in the weak coupling regime.

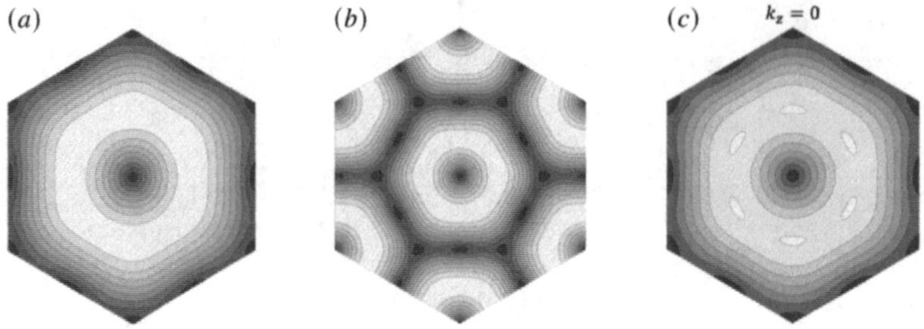

Figure 2. The contour plots of the amplitudes of stable E_2 bases listed in Table 2; (a) in-plane nearest basis $|o_{\boldsymbol{k}}^{+}\hat{\boldsymbol{z}}|$, (b) in-plane 2nd nearest basis $|\bar{o}_{\boldsymbol{k}}^{+}\hat{\boldsymbol{z}}|$,and (c) out-of-plane 2nd nearest basis $|\tilde{o}_{\boldsymbol{k}}^{+}\hat{\boldsymbol{z}}|$ at $k_z = 0$. We see that the peaks in (a) and (c) ((b)) are on the Fermi surfaces of Band 2 (Band 1 and 3).

Table 3. Possible pairing symmetries in the E_2 representation.

pairing	(η_1, η_2)	small k exp. of Eq. (5)	TRS	topological surface state
chiral p-wave	$(1,0),(0,1)$	$k_x \pm ik_y$	broken	chiral
nematic	$(1,1),\ (1,-1)$	$k_x,\ k_y$	unbroken	flat band

4. Summary

We discuss the pairing states favored by the Fermi surface structures in BaPtSb using the linearized gap equations. The candidate with broken time-reversal symmetry is the chiral p-wave state with non-trivial topology. In other words, the Fermi surface structures favor the chiral p-wave state more than any other states with broken time-reversal symmetry, such as the chiral d-wave ($d_{x^2-y^2} \pm id_{xy}$-wave) state included in E_1 and E_2 representations. The chiral d-wave state could be stabilized if there were a cylindrical Fermi surface in the proximity of the saddle points [7, 8, 9, 10].

We need further discussion for the comprehensive understanding of the superconducting state in this material. The polar Kerr effect measurement would be a relevant probe to examine the time-reversal symmetry breaking [11]. More information about the normal state properties and electronic correlation effects would be necessary to reveal the pairing mechanism that works in this system.

Acknowledgment

This work was supported by JSPS KAKENHI Grant Number JP20K03826.

References

[1] Nishikubo Y, *et al.* 2011 J. Phys. Soc. Jpn. **80** 055002.
[2] Kudo K, *et al.* 2018 J. Phys. Soc. Jpn. **87** 063702.
[3] Kudo K, *et al.* 2018 J. Phys. Soc. Jpn. **87** 073708.
[4] Uzunok H Y, Tütüncü H M, Ertuğrul Karaca, Srivastava G P 2019 Intermetallic **108** 109.
[5] Adachi T, *et al.* 2018 KEK-MSL REPORT 2017 54.
[6] Yonezawa S, *et al.* 2017 Nature Phys. **13** 123.
[7] Nandkishore R, Levitov L S, and Chubukov A V, 2012 Nature Physics **8** 158.
[8] Black-Schaffer A M and Honerkamp C, 2014 J. Phys.: Condens. Matter **26** 423201.
[9] Goryo J, Fischer M H, and Sigrist M, 2012 Phys. Rev. B **86** 100507 (R).
[10] Fischer M H, *et al.* 2014 Phys. Rev. B **89** 020509 (R).
[11] Xia J, Maeno Y, Beyersdorf P T, Fejer M M, and Kapitulnik A, 2006 Phys. Rev. Lett. **97** 167002.

Spontaneous spin polarization at the edge of the chiral superconducting states in a spin Hall metal

Tsuyoshi Imazu and Jun Goryo

Department of Mathematics and Physics, Hirosaki University, Hirosaki 036-8561, Japan

E-mail: jungoryo@hirosaki-u.ac.jp

Abstract. We discuss the chiral superconducting state in a square lattice system with an intrinsic spin-orbit coupling causing the spin Hall effect. We estimate the spontaneous spin polarization (local moment) $\langle m_i \rangle$ at the i-th site and its sum $M = \sum_i \langle m_i \rangle$ in ribbons. In the zigzag ribbon, we see $\langle m_i \rangle \neq 0$ near the edge. The energy spectra of the chiral edge states show the spin splitting and then we have non-zero M. Moreover, we find that M is enhanced significantly at a topological phase transition point, where the energy gap is closing. We also have non-zero $\langle m_i \rangle$ near the edge of the straight ribbon, but in contrast, there is no spin splitting in the spectra and M vanishes.

1. Introduction

The chiral superconductor is the topological superconducting state with broken time-reversal symmetry supporting the chiral edge state at the sample boundary and has been keeping a lot of attention. The role of the spin-orbit coupling in a superconductor has also been studied intensively, and its relation to the chiral edge state is the central issue of this paper. In this point of view, Imai, Wakabayashi and Sigrist argued the spontaneous spin polarization due to the chiral edge states in the model of the chiral superconductor Sr_2RuO_4 [1]. The same mechanism was also examined in the ordered honeycomb network superconductor SrPtAs, which is a potential candidate for the chiral d-wave superconductor[2, 3, 4, 5]. Both systems possess the spin-orbit coupling which gives in the normal state the spin-dependent Aharonov-Bohm phase to an electron and causes the spin Hall effect.

To gain the insight into this intriguing phenomena, it would be important to examine a minimal model intensively. We then utilize the simple model on the square lattice proposed by Bernevig, Hughes and Zhang (BHZ) to discuss the quantum spin Hall insurator in the HgTe quantum wells [6, 7].

2. The chiral p-wave pairing in the BHZ metal

In the BHZ model we have four electron fields on a cite $\boldsymbol{r}$; $\psi_{s\uparrow,\downarrow}(\boldsymbol{r})$, $\psi_{p_+\uparrow}(\boldsymbol{r})$, and $\psi_{p_-\downarrow}(\boldsymbol{r})$, where $p_\pm = p_x \pm ip_y$. We consider the metallic state by shifting the Fermi level and introduce the chiral p-wave order parameter. The mean field Hamiltonian is

$$
\begin{aligned}
H &= H_{\mathrm{BHZ}} + H_\Delta, \qquad (1)\\
H_{\mathrm{BHZ}} &= \sum_{\boldsymbol{r}\sigma=\pm}\sum_{\boldsymbol{\delta}=\pm\boldsymbol{e}_x,\pm\boldsymbol{e}_y} \Big\{ -t_{ss}\psi^\dagger_{s\sigma}(\boldsymbol{r}+\boldsymbol{\delta})\psi_{s\sigma}(\boldsymbol{r}) + t_{pp}\psi^\dagger_{p_\sigma\sigma}(\boldsymbol{r}+\boldsymbol{\delta})\psi_{p_\sigma\sigma}(\boldsymbol{r})
\end{aligned}
$$

$$
\begin{aligned}
& -t_{sp}e^{i\theta_\delta\sigma}\psi^\dagger_{s\sigma}(\boldsymbol{r}+\boldsymbol{\delta})\psi_{p_\sigma\sigma}(\boldsymbol{r}) - t_{sp}e^{-i\theta_\delta\sigma}\psi^\dagger_{p_\sigma\sigma}(\boldsymbol{r}+\boldsymbol{\delta})\psi_{s\sigma}(\boldsymbol{r})\Big\} \\
& - \sum_{r\sigma=\pm}\Big\{(\mu+\delta\mu)\psi^\dagger_{s\sigma}(\boldsymbol{r})\psi_{s\sigma}(\boldsymbol{r}) + (\mu-\delta\mu)\psi^\dagger_{p_\sigma\sigma}(\boldsymbol{r})\psi_{p_\sigma\sigma}(\boldsymbol{r})\Big\}, \\
H_\Delta \;=\; & \frac{1}{2}\sum_{r\sigma=\pm}\sum_{\boldsymbol{\delta}=\pm\boldsymbol{e}_x,\pm\boldsymbol{e}_y}\Big\{\Delta e^{i\theta_\delta}\psi^\dagger_{s\sigma}(\boldsymbol{r}+\boldsymbol{\delta})\psi^\dagger_{s\bar{\sigma}}(\boldsymbol{r}) + \Delta e^{i\theta_\delta}\psi^\dagger_{p_\sigma\sigma}(\boldsymbol{r}+\boldsymbol{\delta})\psi^\dagger_{p_{\bar{\sigma}}\bar{\sigma}}(\boldsymbol{r}) + h.c.\Big\},
\end{aligned}
$$

where the up (down) spin is denoted by + (−), $\bar{\sigma} = -\sigma$, and $\theta_{\boldsymbol{\delta}} = \theta_{-\boldsymbol{\delta}} + \pi = 0, \pi/2$ for $\boldsymbol{\delta} = \boldsymbol{e}_x, \boldsymbol{e}_y$. The term t_{sp} is the spin-orbit coupling, which gives the spin-dependent Aharonov-Bohm phase to an electron hopping along a certain closed path and causes the spin Hall effect.

We may diagonalize H_{BHZ} by the Fourier transformation in the bulk, and 3D plots of the energy bands and Fermi surfaces are given in Figures 1(a) and 1(b). The topology of the chiral p-wave state is characterized by the Chern integer $\mathcal{N}_{Ch}$ [8, 9], which counts the phase winding of the k-space gap functions around the Fermi surface and gives the number of the topologically protected chiral edge states in a finite-size system.

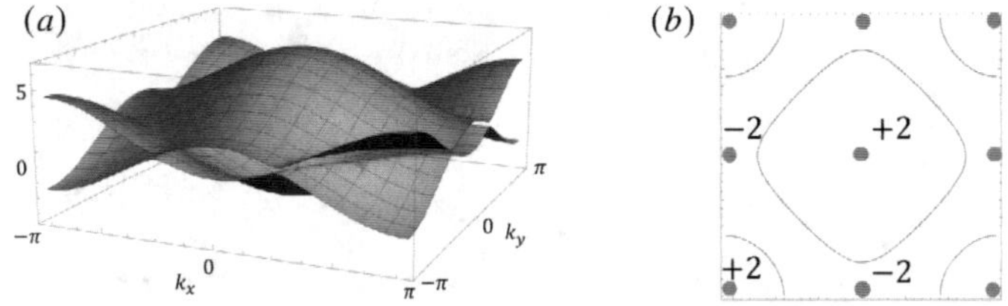

Figure 1. (a): 3D plots of the energy bands of H_{BHZ} with parameters $t_{ss} = t_{pp} = 1.0$, $t_{sp} = 0.1$, $\mu = -1.5$, $\delta\mu = 1.0$. Each sheet has two-fold degeneracy with respect to spin. (b): Fermi surfaces of H_{BHZ} with parameters $t_{ss} = t_{pp} = 1.0$, $t_{sp} = 0.1$, $\mu = -1.5$, $\delta\mu = 1.0$. Purple dots and integers show the zeros of the k-space chiral p-wave gap function $\Delta(\sin k_x + i\sin k_y)$ and their contributions to the Chern integer $\mathcal{N}_{Ch}$ including the spin degrees of freedom. In this case, $\mathcal{N}_{Ch} = 4$.

3. Spontaneous spin polarization at the edge

Besides the spontaneous charge current, we would have the spin current and therefore spin at the edge due to the spin orbit coupling [1, 5]. In this paper, we focus on the spin polarization in our model and discuss its enhancement.

We examine straight and zigzag ribbons (see Figure 2), and obtain the energy spectra and spin polarization (local moment) at the i-th site in the unit cell ($i = 1, ..., N$)

$$
\langle m_i\rangle = \sum_{k\sigma}\sigma\left(\langle\psi^\dagger_{iks\sigma}\psi_{iks\sigma}\rangle + \langle\psi^\dagger_{ikp_\sigma\sigma}\psi_{ikp_\sigma\sigma}\rangle\right), \tag{2}
$$

where $\langle ...\rangle$ denotes the expectation value in the superconducting ground state and k wave number. We verify that $\langle m_i\rangle$ would be zero in the absence of the spin-orbit coupling t_{sp}.

We then use the parameters $t_{ss} = t_{pp} = 1.0$, $t_{sp} = 0.1$, $\mu = -1.5$, $\delta\mu = 1.0$, which give the bulk Fermi surfaces shown in Figure 1(b) , and $\Delta = 1.0$ and $N = 100$. The results are summarized in Figure 3. We have $\mathcal{N}_{Ch} = 4$ and four chiral states at each edge. In the straight ribbon, all edge spectra are degenerate with respect to spin degrees of freedom. The spin polarization $\langle m_i\rangle$ arises near the edge, whereas its sum

$$
M = \sum_{i=1}^{N}\langle m_i\rangle \tag{3}
$$

vanishes completely. On the other hand, in the zigzag case the degeneracy is lifted for the edge spectra with respect to spin. We thus have a difference between the total occupation numbers for up and down spin and $M \neq 0$ besides the spin polarization near the edge.

We then examine the μ dependence of M. The result is summarized in Figure 4. We see a sharp enhancement slightly below $\mu = -1$, where the topological phase transition occurs and $\mathcal{N}_{Ch}$ shows a jump from $+4$ to 0 with increasing μ. This enhancement comes from the fact that the split edge spectra become almost flat when the gaps at $k = \pm\pi$ become tiny, and we have a large difference between the occupation numbers of up and down spin.

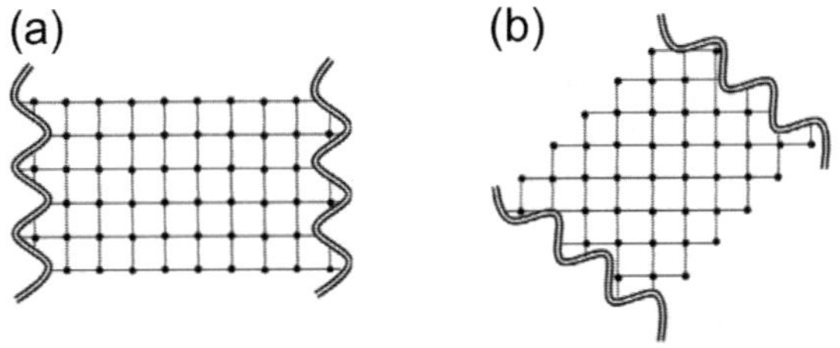

Figure 2. (a) The straight and (b) zigzag ribbons we have examined.

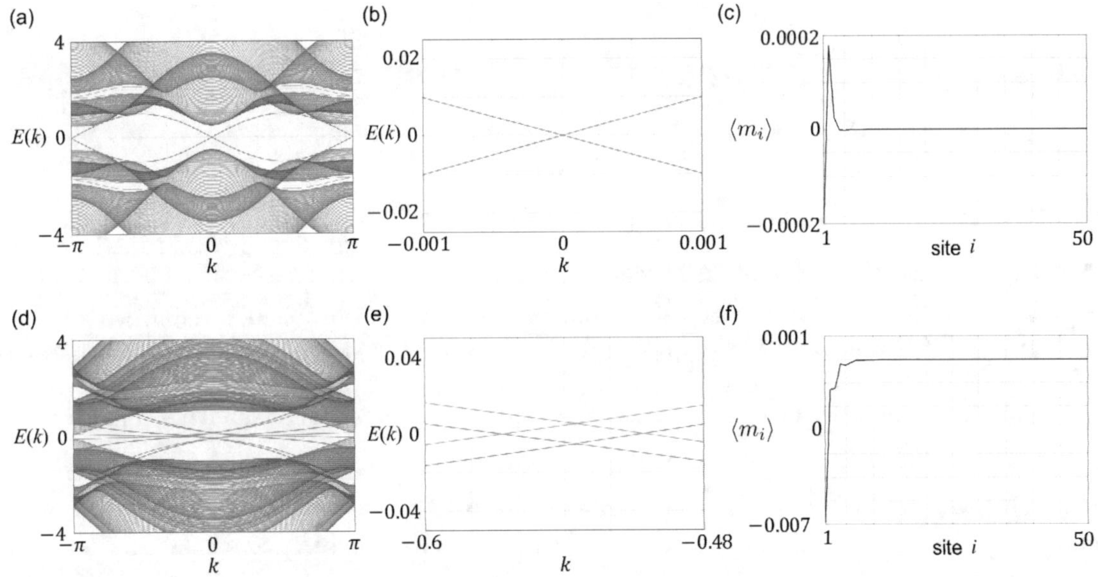

Figure 3. The energy spectra in (a) the straight and (d) zigzag ribbons ((b) and (e) are magnifications for (a) around $k = 0$ and (e) around $k = -0.5$, respectively), and the spin polarization $\langle m_i \rangle$ in the left half of (c) the straight and (f) the zigzag ones. Each result is symmetric in the right half. Note that the spectra in (b) do not but that in (e) show the spin splitting; Orange and green lines in (e) show the spectra of chiral edge states with up and down spin. The spectra in (b) are two-fold degenerate.

4. Summary and discussion

We discuss the chiral p-wave pairing in the BHZ spin Hall metal and estimate the spin polarization $\langle m_i \rangle$ in Eq. (2) and its sum M in Eq. (3) in the ribbon geometries. The result depends on the shape of the ribbon significantly. We have $\langle m_i \rangle$ around the edge but no M in the straight ribbon. In the zigzag ribbon, on the other hand, we have both since the energy

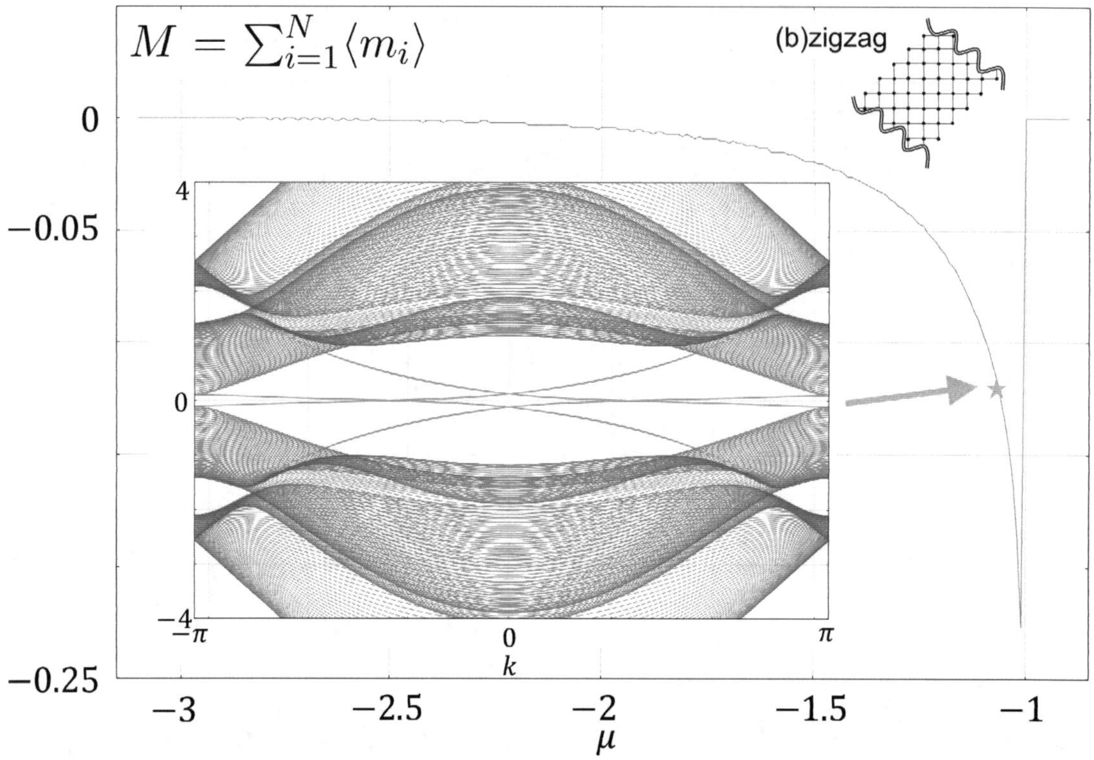

Figure 4. The μ dependence of M. We see a sharp peak slightly below $\mu = -1$, where we have the topological phase transition (gap closing) at $k = \pm\pi$. M vanishes suddenly when $\mu > -1$, since $N_{Ch} = 0$ and the edge state is absent.

spectra of the chiral edge states show spin splitting. M is significantly enhanced at the vicinity of a topological phase transition (gap closing) point, since the spin-split edge spectra become almost flat.

We would discuss elsewhere the relation between the ribbon geometry and the spin splitting in the edge spectra [10].

Acknowledgment
This work was supported by JSPS KAKENHI Grant Number JP20K03826.

References
[1] Imai Y, Wakabayashi K, and Sigrist M,
[2] Nishikubo Y, *et al.* 2011 J. Phys. Soc. Jpn. **80** 055002.
[3] Bisbas P K, it et al. 2013 Phys. Rev. B **87** 180503(R).
[4] Fischer M H, *et al.* 2014 Phys. Rev. B **89** 020509(R).
[5] Goryo J, Imai Y, Rui W B, Schnyder A P, and Sigrist M, 2017 Phys. Rev. B **96** 140502(R).
[6] Bernevig B A, Hughes H L, and Zhang S -C, 2006 Science **314** 1757.
[7] König M, *et al.* 2007 Science **317** 766.
[8] Thouless D J, Kohmoto M, Nightingale M P, and den Nijs M, 1982 Phys. Rev. Lett. **49** 405.
[9] Kohmoto M, 1985 Ann. Phys. (N. Y.) **160** 355.
[10] The same issue has been pointed out also in Ref. [5].

Temperature and doping dependence of the singlet and triplet pair susceptibilities in the one-band Hubbard model based on the dynamical mean-field theory

Yusuke Inokuma[1] **and Yoshiaki Ōno**[1]

[1]Department of Physics, Niigata University, Niigata 950-2181, Japan

E-mail: `f21j002k@mail.cc.niigata-u.ac.jp`

Abstract. The superconductivity in the one-band Hubbard model on a Bethe lattice is investigated on the basis of the dynamical mean-field theory (DMFT) in which the irreducible vertex function has no $\boldsymbol{k}$-dependence and then only the s-wave superconductivities with the spin-singlet even-frequency and the spin-triplet odd-frequency pairings are possibly realized. We calculate the pair susceptibility by solving the Bethe-Salpeter equation with the use of the linearized DMFT and find that the spin-singlet pair susceptibility is suppressed by the on-site Coulomb interaction U especially for the region near the Mott transition. On the other hand, the spin-triplet pair susceptibility is apparently enhanced by U in the strong correlation regime at very low temperatures where the effective interaction for the triplet pairing is considered to be attractive and then the triplet superconductivity is expected to be realized.

1. Introduction

The Hubbard model has a long history of research as the fundamental model for describing electron correlation effects in magnetism and Mott metal-insulator transition. Especially after the discovery of high-T_c cuprate superconductors, it is actively used for the studies as a fundamental model which describes superconductivity by electron correlation effects. The Dynamical mean-field theory (DMFT) [1], which can take into account local electron correlation effects non-perturbatively, has been developed for many studies in the Hubbard model including superconductivity, because it becomes exact in infinite spacial dimensions and is expected to be a good approximation in three dimensions. In the previous DMFT studies, both the singlet and triplet superconductivities have not been observed in the one-band Hubbard model [2,3] in contrast to the case with the two-band Hubbard model where the singlet and/or triplet superconductivities are found to be realized [3,4,5]. However, the detailed results of the pair susceptibilities for superconductivity over the whole parameter regime including quite low temperatures where the superconductivity is expected to be realized were not been explicitly shown there [2,3].

In the present paper, we calculate the singlet and triplet pair susceptibilities in the one-band Hubbard model for a wide range of parameters: the electron number per site n, the temperature T and the on-site Coulomb interaction U. The singlet (triplet) pair susceptibility is constructed by ladder diagrams of the particle-particle channel with respect to the singlet (triplet) irreducible vertex function Γ_s (Γ_t) which corresponds to an effective interaction leading singlet (triplet) pairing and is obtained by solving the Bethe-Salpeter equation with the use of the linearized DMFT [6] developed for describing the electronic state close to the Mott transition.

2. Model and formulation

We consider the one-band Hubbard model on a Bethe lattice with infnite connectivity $Z = \infty$. The Hamiltonian is given by

$$H = \sum_{ij\sigma} t_{ij} c^{\dagger}_{i\sigma} c_{j\sigma} + U \sum_{i} n_{i\uparrow} n_{i\downarrow}, \tag{1}$$

where $c_{i\sigma}$ is the annihilation operator of an electron with spin $\sigma(=\uparrow,\downarrow)$ at site i and $n_{i\sigma} = c^{\dagger}_{i\sigma} c_{i\sigma}$, U is the on-site Coulomb interaction and $t_{ij} = t/\sqrt{Z}$ is the transfer integral between the nearest-neighbor sites and we set $t = 1$eV resulting in the bare band width of 4eV. To solve the model eq. (1), we use the DMFT in which the lattice model is mapped onto an impurity Anderson model embedded in an effective medium which is determined so as to satisfy the self-consistency condition for the single-particle Green's functinos [1].

The s-wave pair susceptibilities with the spin-singlet even-frequency and the spin-triplet odd-frequency pairings χ_s and χ_t are respectively given by

$$\chi_s = \sum_{m,l} \chi_s(\nu_m, \nu_l) \tag{2}$$

$$\chi_t = \sum_{m,l}^{\nu_m>0,\nu_l>0} \chi_t(\nu_m, \nu_l) + \sum_{m,l}^{\nu_m<0,\nu_l<0} \chi_t(\nu_m, \nu_l) - \sum_{m,l}^{\nu_m<0,\nu_l>0} \chi_t(\nu_m, \nu_l) - \sum_{m,l}^{\nu_m>0,\nu_l<0} \chi_t(\nu_m, \nu_l) \tag{3}$$

with the frequency-dependent pair susceptibilities defeined by

$$\begin{aligned} \chi_{s(t)}(i\nu_m, i\nu_l) &= \int_0^{\beta} d\tau_1 \int_0^{\beta} d\tau_2 \int_0^{\beta} d\tau_3 e^{-i\nu_m\tau_1} e^{i\nu_m\tau_2} e^{-i\nu_l\tau_3} \\ &\times \sum_{\mathbf{k},\mathbf{k}'} \left[\left\langle T_\tau [c_{\mathbf{k}\uparrow}(\tau_1) c_{-\mathbf{k}\downarrow}(\tau_2) c^{\dagger}_{\mathbf{k}'\uparrow}(\tau_3) c^{\dagger}_{-\mathbf{k}'\downarrow}] \right\rangle - (+) \left\langle T_\tau [c_{\mathbf{k}\uparrow}(\tau_1) c_{-\mathbf{k}\downarrow}(\tau_2) c^{\dagger}_{\mathbf{k}'\downarrow}(\tau_3) c^{\dagger}_{-\mathbf{k}'\uparrow}] \right\rangle \right] \end{aligned} \tag{4}$$

where $\nu_m = (2m+1)\pi/\beta$ is an odd Matsubara frequency with the inverse temperature $\beta = 1/k_B T$. As shown in Figure 1, the singlet (triplet) pair susceptibility $\chi_s(\chi_t)$ is constructed by ladder diagrams of the particle-particle channel with respect to the singlet (triplet) irreducible vertex function $\Gamma_s(\Gamma_t)$ which corresponds to an effective interaction leading singlet (triplet) pairing and is obtained by solving the Bethe-Salpeter equation. Within the DMFT, $\Gamma_s(\Gamma_t)$ has no $\boldsymbol{k}$-dependence and can be obtained by solving the Bethe-Salpeter equation in the effective impurity Anderson model. For numerical simplicity, we calculate $\Gamma_s(\Gamma_t)$ by employing the linearized DMFT developed for describing the electronic state close to the Mott transition [6].

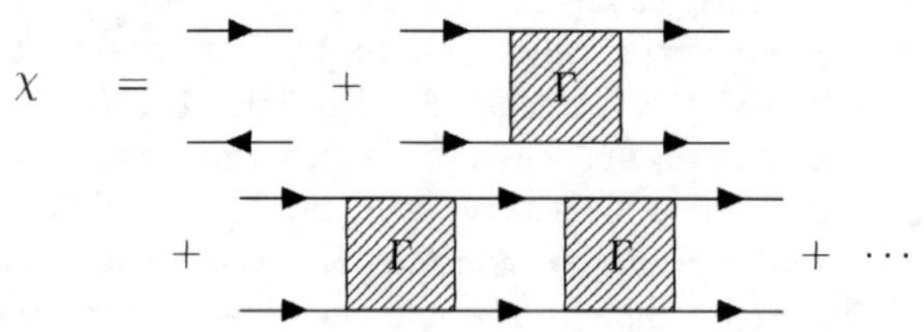

Figure 1. Diagrammatic representation for the pair susceptibility χ constructed by ladder diagrams with respect to irreducible vertex function Γ, where solid lines represent the single-particle Green's functions.

3. Results

Figure 2 (a) shows the T dependence of the spin-singlet pair susceptibility χ_s which is negative and then the absolute value $|\chi_s|$ is plotted there. We see that χ_s is suppressed by the repulsion U especially at half-filling $n = 1$, where n is the number of electrons per site. We plot the U dependence of χ_s at $T = 10$K in Fig. 2 (b) and find that χ_s is suppressed by U especially for the region near $U = 6$eV where the Mott transition takes place at $n = 1$ within the linearized DMFT [6]. From these results, it is considered that the s-wave superconductibity with the spin-singlet even-frequency pairing is not realized for $U > 0$ as consistent with the previous DMFT studies [2,3].

The T dependence of the spin-triplet pair susceptibility χ_t is shown in Fig. 3 (a). We see that the divergent property of χ_t down to $T = 1$K for $U = 5.3$eV is enhanced as compared to that for $U = 0$

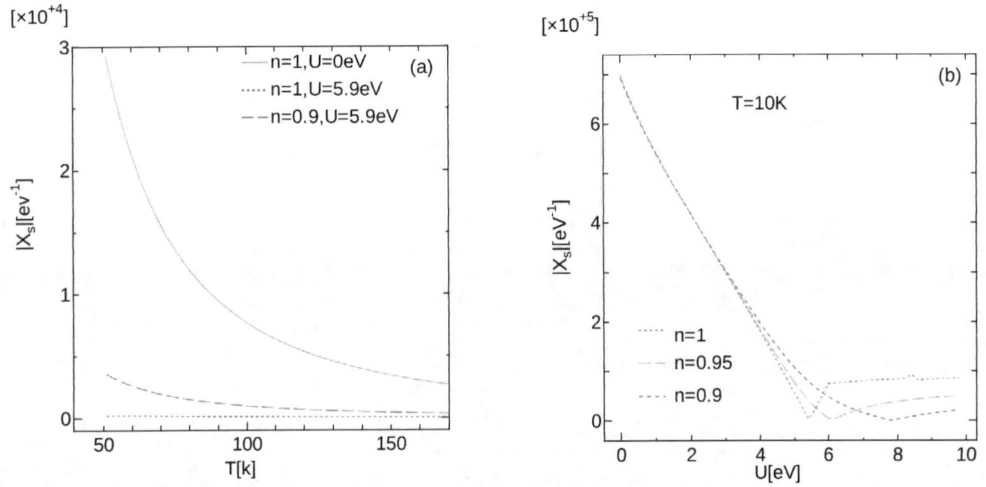

Figure 2. Absolute values of the spin-singlet pair susceptibilities $|\chi_s|$ as functions of T for several values of n and U (a), and those as functions of U for several values of n at $T = 10$K.

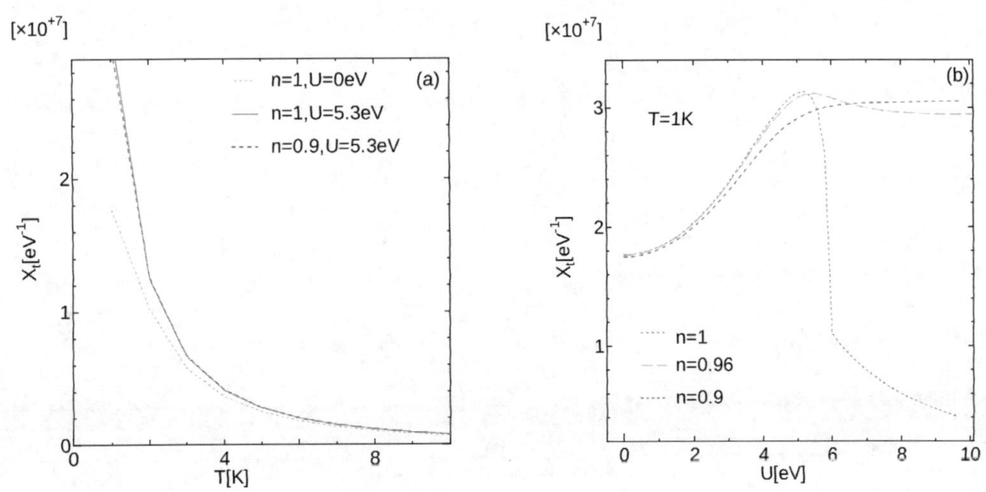

Figure 3. The spin-triplet pair susceptibilities χ_t as functions of T for several values of n and U (a), and those as functions of U for several values of n at $T = 1$K.

where χ_t diverges with $T \to 0$. We plot the U dependence of χ_t at $T = 1$K in Fig. 3 (b) and find that χ_t is enhanced by U especially for the region just below the Mott transition $U_c = 6$eV in contrast to the case with χ_s which is suppressed by U as shown in Fig. 2 (b). These results suggest that, in the region just below the Mott transition, χ_t diverges at a finite temperature $T_c(< 1\text{K})$ below which the s-wave superconductibity with the spin-triplet odd-frequency pairing is expected to be realized. To be more conclusive, we need further calculations of χ_t at very low temperatures by using the full DMFT.

References

[1] A. Georges, G. Kotliar, W. Krauth and M.Rozenberg, Rev. Mod. Phys. **68**, 13(1996).
[2] M. Jarrell and T. Pruschke, Z. Phys. B **90**, 187 (1993).
[3] M. Caffarel and W. Krauth, Phys. Rev. Lett. **72**, 1545 (1994).
[4] A. Georges, G. Kotliar and W. Krauth. Z. Phys. B **92**, 313 (1993).
[5] A. Koga and P. Werner, Phys. Rev. B **91**, 085108 (2015).
[6] R. Bulla and M. Potthoff, Eur. Phys. J. B **13**, 257 (2000).

Journal of Physics: Conference Series **2164** (2022) 012018 doi:10.1088/1742-6596/2164/1/012018

Joint machine learning analysis of muon spectroscopy data from different materials

T Tula[1], G Möller[1], J Quintanilla[1], S R Giblin[2], A D Hillier[3], E E McCabe[1,4], S Ramos[1], D S Barker[1,5] and S Gibson[1]

[1] School of Physical Sciences, University of Kent, Park Wood Rd, Canterbury CT2 7NH, United Kingdom
[2] School of Physics and Astronomy, Cardiff University, Cardiff CF24 3AA, United Kingdom
[3] ISIS Facility, STFC Rutherford Appleton Laboratory, Chilton, Didcot Oxon, OX11 0QX, United Kingdom
[4] Department of Physics, Durham University, Durham, DH1 3LE, United Kingdom
[5] School of Physics and Astronomy, University of Leeds, Leeds, LS2 9JT, United Kingdom

E-mail: tt343@kent.ac.uk

Abstract. Machine learning (ML) methods have proved to be a very successful tool in physical sciences, especially when applied to experimental data analysis. Artificial intelligence is particularly good at recognizing patterns in high dimensional data, where it usually outperforms humans. Here we applied a simple ML tool called principal component analysis (PCA) to study data from muon spectroscopy. The measured quantity from this experiment is an asymmetry function, which holds the information about the average intrinsic magnetic field of the sample. A change in the asymmetry function might indicate a phase transition; however, these changes can be very subtle, and existing methods of analyzing the data require knowledge about the specific physics of the material. PCA is an unsupervised ML tool, which means that no assumption about the input data is required, yet we found that it still can be successfully applied to asymmetry curves, and the indications of phase transitions can be recovered. The method was applied to a range of magnetic materials with different underlying physics. We discovered that performing PCA on all those materials simultaneously can have a positive effect on the clarity of phase transition indicators and can also improve the detection of the most important variations of asymmetry functions. For this joint PCA we introduce a simple way to track the contributions from different materials for a more meaningful analysis.

1. Introduction

In recent years, machine learning tools have become increasingly familiar to the scientific community, but they are not yet regarded as standard practice for analysing small-scale experimental measurements in condensed matter physics. The strength of ML techniques lies in their ability to recognise patterns and correlations in data [1, 2, 3], which is often the goal of scientific enquiries (such as the search for phase transitions). Usually, for the ML methods to work, a large data set is required and since data acquisition can be costly for many experiments (in both time and resources), it is understandable that ML tools are not always thought to be a suitable tool. In this work, we want to test whether experimental muon spectroscopy measurements can be analysed with ML even when using smaller sets of data. Specifically we apply a linear principal component analysis algorithm to muon spectroscopy measurements [4] in order to find signs of phase transitions for a range of different materials.

In this article we report on a new development from our previous research [5], in which we demonstrated that PCA can be successfully applied to a range of different materials simultaneously, thus increasing the effectiveness of this ML method, even when the underlying physics of the materials are vastly different. To have more insight into the PCA results in this case, we present a cumulative principal component score metric, which tells us how important each PC output is for a given material. That allows us to look at phase transition indicators in correct principal component scores, and this improved our previous analysis of magnetic materials.

2. Methods

2.1. Muon spectroscopy experiments

The main quantity to analyse from muon spectroscopy experiments is the asymmetry function, which tells us the distribution of spin directions of muons implanted into the sample. This can be directly linked to the internal magnetic field of the material studied [6, 7, 8, 9]. We are interested in studying changes in the asymmetry functions at different temperatures, which can indicate phase transitions. A more comprehensive introduction to muon spectroscopy can be found in [4, 5].

2.2. PCA

Principal component analysis is an unsupervised machine learning technique [3, 10], which means that we are not required to label the input data, and given that there are some correlations between asymmetry functions, PCA will be able to find those. Applying PCA is equal to performing singular value decomposition on a matrix containing all asymmetry functions $\mathbf{A}$. Principal components (PC) are given by left-singular vectors of $\mathbf{A}$, with singular values signifying the importance of each PC. By looking at projections of asymmetry functions into the most important PCs, we can see changes in the behaviour that could indicate phase transitions. More on the PCA method as applied to μSR data can be found in our previous article [5]

3. Results and Discussion

We applied PCA to a range of different materials. For each of them we had around 10-30 asymmetry functions, measured at different temperatures and at vanishing external magnetic field throughout. Among the systems studied are time-reversal symmetry breaking superconductors ($LaNiGa_2$, $LaNiC_2$, $LaNi_{0.9}Cu_{0.1}C_2$, $LaNi_{0.75}Cu_{0.25}C_2$), a proposed spin liquid ($LuCuGaO_4$) and a layered antiferromagnet ($BaFe_2Se_2O$). We initially tried performing the analysis for each material individually, but found out that for some materials the small number of measurements resulted in most PCs having a significant contribution to covariance. To overcome that, we performed PCA on the joint dataset of measurements from all materials simultaneously. This improved the visibility of phase transitions for those measurements that were previously hard to analyse, and did not significantly alter the other results. This is remarkable given that the materials in the data set have very different underlying physics. We can interpret this result as follows: by feeding all the data to the PCA simultaneously, the algorithm is not learning anything new about any particular material, but it is learning more about the experimental technique being employed in all cases (μSR). A comparison of the shapes of the ensuing principal components – with PCA performed on one material at a time or simultaneously on all of them – can be found in our previous work [5].

We now demonstrate that more insight can be gained from the results of a PCA analysis performed on multiple samples by calculating the total score of a given PC for each individual material. Unlike analysis for a single material, where PC scores are naturally ordered in

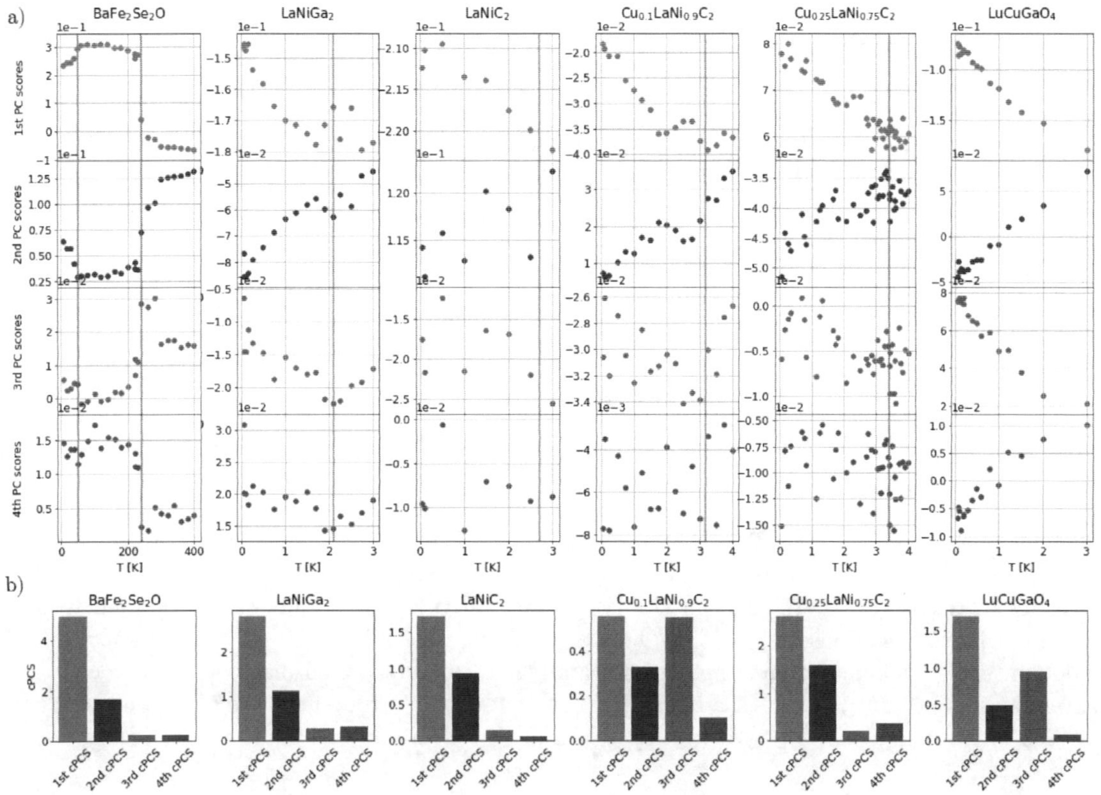

Figure 1. a) Temperature dependence for principal component scores when applying PCA to multiple materials simultaneously. Red vertical lines correspond to expected phase transitions, taken from the literature; b) Cumulative principal component scores for all materials.

terms of decreasing significance, in a sample of several materials considered jointly, the order of significance can depend on the material: Some PC shapes might be dominant in most of materials, meaning that they will have large contribution to the covariance, but for other materials another PC might be more prominent. To interpret the analysis, we should always examine the most important PC(s) for each sample. To identify the relevant scores, we define a cumulative principal component score (cPCS) as

$$\mathrm{cPCS}(i;n) = \sum_{j=1}^{N_i} |\mathrm{PCS}_n^i(j)|, \tag{1}$$

where i iterates over different materials, n stands for n-th principal component, $\mathrm{PCS}_n^i(j)$ corresponds to n-th principal component score for a j-th measurement of a given material and N_i stands for number of measurements for i-th material. We then use this measure to identify the principal component scores with the largest cPCS for each material.

Our findings are presented in figure 1, which is an extension of figure 5 from [5]. We present the temperature dependence of the first four principal component scores for all the materials in a), and their cPCS in b). From the cPCS we can see that although the 1st PC has the largest contribution to all of measurements, the 2nd PC is less important than the 3rd one for $Cu_{0.1}LaNi_{0.9}C_2$ and $LuCuGaO_4$. If we compare the principal component scores for $Cu_{0.1}LaNi_{0.9}C_2$, we can clearly see that the 3rd PC has a much more visible indication of the expected phase transition (red vertical line) than the 2nd or even 1st one. For $LuCuGaO_4$ material, we expected improvement from our previous joint analysis, and although we find that 3rd PC is more important than 2nd, there seems to be no feature indication at 0.6 K, which we have found in principal component analysis of just this material ([5], figure 4, last column, third row). This tells us that joint analysis is most helpful whenever the separate PCA is not successful, which we can always tell by looking at scree plots. In the case of $LuCuGaO_4$, the cPCS metric looks worse than scree plot of its separate analysis ([5], figure 4, last column, last row), and we should take that into consideration when analysing the material. One important note here is that adding $LuCuGaO_4$ did not affect negatively the results for other materials, which we have showed previously [5].

4. Conclusions

We have suggested a new analysis method to extend our recently propsed PCA of μSR data [5]. We are more confident that performing a joint PCA of measurements taken for different materials is a viable method of analysis, given that for each set of measurements we are looking at the right PC scores, which can be ensured by calculating a simple metric – cumulative principal component scores. We should also compare the cPCS to scree plots from separate analyses conducted for each material, in case the simultaneous PCA might not improve the clarity of each individual study. The ability to thus enhance our understanding gained from a new experiment based on existing data goes beyond the possibilities of preexisting approaches, where data for each material is necessarily analyzed and fitted in isolation, and overarching commonalities are anticipated in advance by the formulation of a suitable fitting function. We anticipate this could offer great advantage when deployed in large-throughput user facilities.

References

[1] Zdeborová L 2017 *Nat Phys* **13** 420–421
[2] Carleo G, Cirac I, Cranmer K, Daudet L, Schuld M, Tishby N, Vogt-Maranto L and Zdeborová L 2019 *Rev. Mod. Phys.* **91** 2773
[3] Géron A 2019 *Hands-on machine learning with Scikit-Learn, Keras, and TensorFlow: Concepts, tools, and techniques to build intelligent systems* (O'Reilly Media)
[4] Blundell S J 1999 Muon-spin rotation : a brief introduction [Online; accessed 11. Sep. 2020] URL `http://users.ox.ac.uk/ sjb/musr/musr.html`
[5] Tula T, Möller G, Quintanilla J, Giblin S R, Hillier A D, McCabe E E, Ramos S, Barker D S and Gibson S 2021 *Journal of Physics: Condensed Matter* **33** 194002
[6] Schenck A G E 1985 *Muon Spin Rotation Spectroscopy Principles and Applications in Solid State Physics* (CRC Press) ISBN 978-085274551-9
[7] S F J Cox 1987 *Journal of Physics C: Solid State Physics* **20** 3187
[8] Blundell S J 1999 *Contemporary Physics* **40** 175–192
[9] Lee S L, Cywinski R and Kilcoyne S H (eds) 1999 *Muon science: Muons in physics, chemistry and materials* (CRC press)
[10] Jolliffe I T and Cadima J 2016 *Philosophical Transactions of the Royal Society A: Mathematical, Physical and Engineering Sciences* **374** 20150202

Journal of Physics: Conference Series **2164** (2022) 012019 doi:10.1088/1742-6596/2164/1/012019

Shear modulus anomaly of multi-conponent superconductor: the case of M_xBi$_2$Se$_3$

Pye Ton How[1]

[1] Institute of Physics, Academia Sinica, 128 Academia Road Section 2, Taipei 11520, Taiwan

E-mail: p.t.how@outlook.com

Abstract. A pair of split superconducting transitions in the presence of a symmetry breaking field is a very definitive signature of multi-component superconductivity[1, 2]. We theoretically study the shear modulus anomaly across such pair of split transitions[3]. The talk will be focused on MxBi2Se3, a nematic superconductor candidate, for which no experimental confirmation of the split transition has been made so far. We propose that the shear modulus C_{66} must vanish at the lower transition: a very clear signature detectable by experiments. The observation of shear modulus anomalies would be a conclusive test for the nematic superconductivity hosted by the material.

1. Introduction

Bi_2Se_3 is an archetypical topological insulator[4, 5]. By intercalating with electron donors (M = Cu, Sr or Nb), the material become superconducting at low temperature[6, 7, 8, 9]. Numerous experiments with an applied magnetic field parallel to the basal plane indicate a two-fold symmetric response emerging in the superconducting phase[10, 11, 12, 13, 14], inconsistent with the trigonal crystal structure. This prompts the suggestion that the superconductivity may be *nematic*: along with the usual $U(1)$ symmetry-breaking, it also breaks the lattice rotationally symmetry spontaneously.

Spontaneous symmetry breaking necessitates that the two-fold direction be *randomly chosen* from three equivalent high-symmetry directions every time a sample turns superconducting in zero field. In contrary, the two-fold direction for a given sample is *persistent*[15, 16, 17]. A pre-existing, symmetry-breaking pinning field must be added to the theory by hand to account for this effect[18]. This pinning field has been linked to small lattice distortions[15].

With the lattice rotational symmetry already broken in the normal phase by the pinning field, the original symmetry argument for the nematic superconductivity is lost. One is faced with a dilemma: either there is some exotic pairing mechanism[19, 20, 21] that explicitly disfavors the s-wave pairing, or the superconductivity is single-component (i.e. not nematic) yet unusually susceptible to directional perturbations. To settle the issue, we believe a directly test for the multi-component nature of the order parameter in some other way is needed. To this end, the present author and collaborator have examined the extended phase diagram of the material under applied stress[18], and the physics of half vortices in such a nematic superconductor[22]. As an s-wave order parameter is forbidden by symmetry to couple linearly couple with shear strain, the shear modulus anomaly may prove to be the most promising test yet.

Journal of Physics: Conference Series **2164** (2022) 012019 doi:10.1088/1742-6596/2164/1/012019

2. Ginzburg-Landau theory for nematic superconductor

The order parameter $\vec{\eta}$ for a nematic superconductor can be thought of as a two-dimensional vector in the basal plane of the crystal. To be exact, the two-component quantity transforms in a two-dimensional irreducible representation of the lattice group D_{3d} for $M_x\mathrm{Bi_2Se_3}$. The Ginzburg-Landau (GL) free energy compatible with the symmetry can be written down:

$$\begin{aligned} \vec{\eta} &= (\eta_x, \eta_y) \equiv e^{i\chi}\,|\eta|\,(\cos\theta, \sin\theta) \\ \mathcal{F}_0 &= a|\eta|^2 + \lambda|\eta|^4 + (\Gamma_1 + \Gamma_2\cos 6\theta)|\eta|^6 \end{aligned} \tag{1}$$

In fact $\vec{\eta}$ behaves as a double-headed vector, since an overall negative sign is the same as a global phase shift of π, and two opposite directions are equivalent. For concreteness we assume $\Gamma_2 > 0$, leading to the "natural preference" of $\theta = \pm\pi/6$ or $\pi/2$ in equilibrium in the absence of the pinning field.

The pinning field is experimentally reported to always favors either the a or a^* crystallographic direction. It can be modelled by a term in the GL theory:

$$\mathcal{F}_\Delta = -\Delta\,|\eta|^2\,\cos 2\theta. \tag{2}$$

The sign of Δ determines the favored direction and is sample-dependent. A positive (negative) Δ favors $\theta = 0$ ($\pi/2$).

When Γ_2 and Δ are of the same sign, the mismatch in their preferences splits the superconducting transition into two[1, 2]. We stress that both signs of Δ are naturally found in the samples[16, 17]: without any need of further tuning, *some* of the crystals must exhibit the physics discussed here. Consider cooling down from the normal phase. At the so-called *upper transition*, superconductivity sets in, and the order parameter is pinned at $\theta = 0$: see Figure 1. It should be noted that there is no *spatial* symmetry breaking here: the horizontal direction has already been singled out by the pinning field, and both the normal and the (upper) superconducting phases have only a two-fold spatial symmetry. The physics is controlled by the $O(\eta^2)$ term in the free energy. At a lower temperature, when the $O(\eta^6)$ terms become important, the competition between Δ and Γ_2 drives the so-called *lower transition*. The remaining two-fold symmetry is spontaneously broken, and the order parameter drifts away from $\theta = 0$ toward $\theta = \pm\pi/6$: see Figure 2 . The double transitions are unique to a multi-component superconductor, but the observation of the lower transition can be challenging[18]. But we will presently show that the shear modulus anomaly signature is quite dramatic and, we believe, hard to miss in an experiment.

3. The interplay between shear strain and nematic superconductivity

The in-plane shear strain is coupled to the superconducting sector in the usual fashion:

$$\begin{aligned} &\epsilon_1 = \epsilon_{xx} - \epsilon_{yy},\ \epsilon_2 = 2\epsilon_{xy}\,; \\ &\mathcal{F}_\epsilon = -g|\eta|^2\epsilon_1\cos 2\theta - g|\eta|^2\epsilon_2\sin 2\theta + \frac{c}{2}(\epsilon_1^2 + \epsilon_2^2). \end{aligned} \tag{3}$$

The geometry of ϵ_1 and ϵ_2 are sketched in Figure 3 and 4. We stress that a multi-component order parameter is *required* to even write down (3): the linear coupling of ϵ_1 and ϵ_2 to an s-wave order parameter is forbidden by symmetry. The 2×2 shear modulus tensor $c_{i,j}$ can be computed with the standard method; we refer the reader to ref [3] for the full treatment that leads to the plot Figure 5. However, the most important qualitative features can be understood using symmetry arguments alone, and we will attempt to do so in this paper.

Below the upper transition, $|\eta| \neq 0$ nucleated along the $\theta = 0$ direction, as shown in Figure 1. Either by comparing with the geometry of ϵ_1 as depicted in Figure 3, or by consulting the

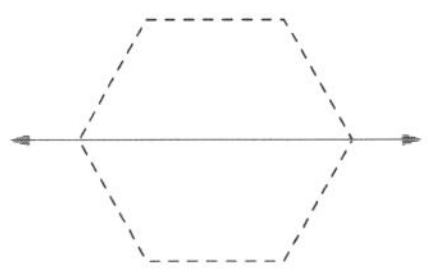

Figure 1. Upper superconducting phase. The order parameter is pinned at the horizontal by the symmetry-breaking pinning field.

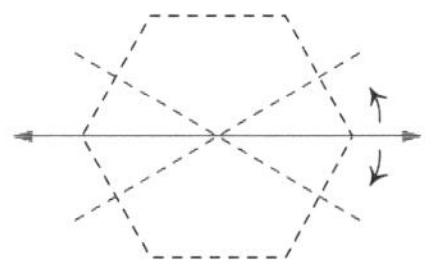

Figure 2. Order parameter in the lower superconducting phase. The two-fold spatial symmetry is broken by the order parameter tilting away from the initial orientation.

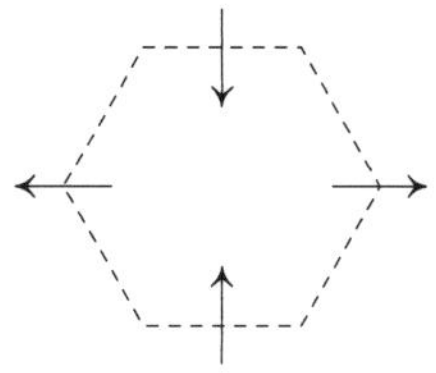

Figure 3. In-plane shear strain component ϵ_1.

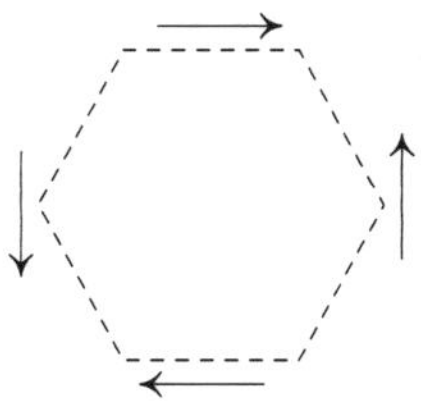

Figure 4. In-plane shear strain component ϵ_2.

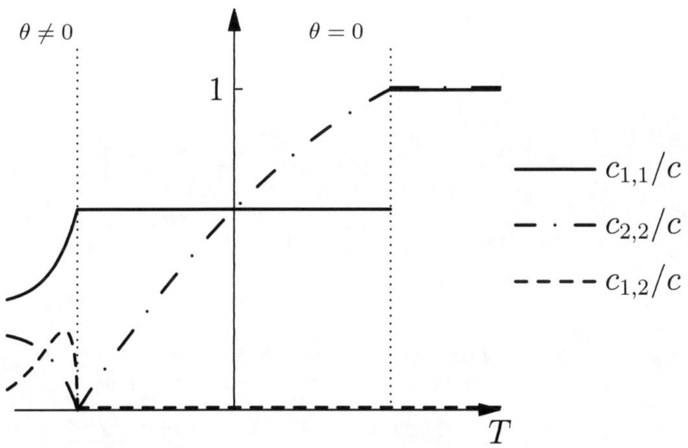

Figure 5. The qualitative behavior of the shear modulus tensor $c_{i,j}$ in the (ϵ_1, ϵ_2) basis. In Voigt's notation, $c_{1,1} = (C_{11} + C_{22} - 2C_{12})/4$, $c_{2,2} = C_{66}$, and $c_{1,2} = (C_{16} - C_{26})/2$.

GL equation $\partial \mathcal{F}_\epsilon / \partial \epsilon_1 = 0$, it is easy to convince oneself that ϵ_1 is spontaneously induced by the superconductivity below the upper transition, with a definite sign determined by $\mathrm{sgn}(g)$. A *step discontinuity* in $c_{1,1}$ is the result.

In contrast, $c_{2,2}$ merely shows a *kink*: ϵ_2's coupling to $\vec{\eta}$ implies that the response must change in the superconducting phase, but the two-fold symmetry forbids spontaneous generation of ϵ_2 in equilibrium, resulting in a weaker anomaly. The off-diagonal $c_{1,2}$ is inconsistent with the two-fold symmetry and remains zero in the upper phase.

Below the lower transition, the equilibrium θ becomes non-zero, spontaneously breaking the remaining two-fold symmetry, as shown in Figure 2. As a result, ϵ_2 (see Figure 4) is induced spontaneously below the transition. Furthermore, at the critical point, a deformation mode must become completely soft, as the crystal is ready to go one way or the opposite at any infinitesimal push, and it is easily seen that ϵ_2 is such mode. This is in contrast with the upper transition, where no spatial symmetry is broken (the broken symmetry is $U(1)$ only), and no soft mode is resulted.

The component $c_{1,2}$ is no longer symmetry-forbidden, and becomes non-zero below the lower transition. Its asymptotic behavior (the square root of reduced temperature) just below the transition point simply follows from the standard meanfield critical exponent.

4. Conclusion

In summary, we argue that more test for the nematic superconductivity hosted by $M_x\mathrm{Bi_2Se_3}$ is needed, beyond the various two-fold anisotropic responses probed in existing experiments, and we believe that the shear modulus anomalies associated with the double superconducting transitions may be the best opportunity yet. Any observation of shear modulus anomaly constitutes strong support for multi-component superconductivity, but the vanishing of the shear modulus component $c_{2,2}$ at the lower transition is an especially dramatic signal that, we believe, is hard to miss in an experiment. We hope our findings help to shed new light on the nature of superconductivity hosted by $M_x\mathrm{Bi_2Se_3}$.

Acknowledgement. This work was supported by the Ministry of Science and Technology (MOST), Taiwan through grants MOST-107-2112-M-001-035-MY3 and MOST-110-2112-M-001-051-MY3.

References

[1] Sigrist M, Joynt R and Rice T M 1987 *Phys. Rev.* B **36** 5186–5198
[2] Volovik G E 1988 *Pis'ma Zh. Eksp. Teor. Fiz.* **48** 39–42
[3] How P T and Yip S K 2021 *Phys. Rev.* B **104** L020506
[4] Zhang H, Liu C X, Qi X L, Dai X, Fang Z and Zhang S C 2009 *Nat. Phys.* **5** 438–442
[5] Liu C X, Qi X L, Zhang H, Dai X, Fang Z and Zhang S C 2010 *Phys. Rev.* B **82** 045122
[6] Hor Y S, Williams A J, Checkelsky J G, Roushan P, Seo J, Xu Q, Zandbergen H W, Yazdani A, Ong N P and Cava R J 2010 *Phys. Rev. Lett.* **104** 057001
[7] Kriener M, Segawa K, Ren Z, Sasaki S and Ando Y 2011 *Phys. Rev. Lett.* **106** 127004
[8] Shruti, Maurya V K, Neha P, Srivastava P and Patnaik S 2015 *Phys. Rev.* B **92** 020506
[9] Asaba T, Lawson B J, Tinsman C, Chen L, Corbae P, Li G, Qiu Y, Hor Y S, Fu L and Li L 2017 *Phys. Rev.* X **7** 011009
[10] Matano K, Kriener M, Segawa K, Ando Y and Zheng G Q 2016 *Nat. Phys.* **12** 852–854
[11] Pan Y, Nikitin A M, Araizi G K, Huang Y K, Matsushita Y, Naka T and de Visser A 2016 *Sci. Rep.* **6** 28632
[12] Yonezawa S, Tajiri K, Nakata S, Nagai Y, Wang Z, Segawa K, Ando Y and Maeno Y 2017 *Nat. Phys.* **13** 123–126
[13] Smylie M P, Willa K, Claus H, Koshelev A E, Song K W, Kwok W K, Islam Z, Gu G D, Schneeloch J A, Zhong R D and Welp U 2018 *Sci. Rep.* **8** 7666
[14] Sun Y, Kittaka S, Sakakibara T, Machida K, Wang J, Wen J, Xing X, Shi Z and Tamegai T 2019 *Phys. Rev. Lett.* **123** 027002
[15] Kuntsevich A Y, Bryzgalov M A, Prudkoglyad V A, Martovitskii V P, Selivanov Y G and Chizhevskii E G 2018 *New J. Phys.* **20** 103022
[16] Kuntsevich A Y, Bryzgalov M A, Akzyanov R S, Martovitskii V P, Rakhmanov A L and Selivanov Y G 2019 *Phys. Rev.* B **100** 224509
[17] Kawai T, Wang C G, Kandori Y, Honoki Y, Matano K, Kambe T and Zheng G Q 2020 *Nat. Commun.* **11** 235
[18] How P T and Yip S K 2019 *Phys. Rev.* B **100** 134508
[19] Brydon P M R, Das Sarma S, Hui H Y and Sau J D 2014 *Phys. Rev.* B **90** 184512
[20] Wan X and Savrasov S Y 2014 *Nat. Commun.* **5** 4144
[21] Wang J, Ran K, Li S, Ma Z, Bao S, Cai Z, Zhang Y, Nakajima K, Ohira-Kawamura S, Čermák P, Schneidewind A, Savrasov S Y, Wan X and Wen J 2019 *Nat. Commun.* **10** 2802
[22] How P T and Yip S K 2020 *Phys. Rev. Research* **2** 043192

Journal of Physics: Conference Series **2164** (2022) 012020 doi:10.1088/1742-6596/2164/1/012020

Revisiting Anderson-Higgs mechanism: application of Lieb-Schultz-Mattis theorem

Yasuhiro Tada

Quantum Matter Program, Graduate School of Advanced Science and Engineering, Hiroshima University, Higashihiroshima, Hiroshima 739-8530, Japan

E-mail: `ytada@hiroshima-u.ac.jp`

Abstract. We consider an electron model of superconductivity on a three-dimensional lattice where there are on-site attractive Hubbard interaction and long-range repulsive Coulomb interaction. It is claimed that fully gapped s-wave superconductivity within this model, if present, exhibits spontaneous translation symmetry breaking possibly related to a charge order. Our discussions are based on an application of the Lieb-Schultz-Mattis theorem under some physical assumptions. The inconsistency between the proposed supersolid and experiments can impose some constraints on a reasonable choice of a theoretical model.

1. Introduction

Superconductivity and superfluidity are well understood based on the spontaneous broken U(1) symmetry of the particle-number conservation. One of the essential differences between a charged superconductor and neutral superfluid is that there is no Nambu-Goldstone (NG) boson associated with the U(1) symmetry in the former, while the linearly dispersive sound mode exists as an NG boson in the latter. The absence of an NG boson in a charged superconductor is often explained by the screening effect of the long-range Coulomb interaction, which could be regarded as a variant of Anderson-Higgs mechanism [1, 2, 3, 4, 5, 6]. On the other hand, in presence of dynamical (compact) electromagnetic fields, superconductivity is regarded as an intrinsic $\mathbb{Z}_2$ topological order and the gapped spectrum is a defining property of this state [7, 8, 9, 10, 11]. It has been naively considered that these two approaches are consistent and the Coulomb interaction could stabilize the topological order.

However, the physics of these two explanations seem to be slightly different; the gapped NG boson is a consequence of an electric field coupled to a charge density in the former approach, while it is driven by a fluctuating magnetic field coupled to a charge current in the latter approach. This may be a physically relevant difference in a non-relativistic condensed matter system where the electric field and magnetic field can be distinguished [12]. Furthermore, in the former, the Meissner effect can be described with use of the Maxwell equation for a magnetic field even without the long-range Coulomb interaction, which would suggest that the Meissner effect and the gapped NG boson are independent phenomena. In contrast, these two phenomena take place simultaneously in a gauged model, which is common for general gauge theories. Besides, there could be experimentally measurable differences between these two approaches in some superconductors such as a superconductor whose normal state is a semimetal or an insulator with a vanishing Fermi surface and a purely two-dimensional superconductor at zero temperature. In such a superconductor, screening effects may be too weak to gap out the NG boson in the charge screening approach, while a gapped spectrum will be by definition robust in a gauged model.

In this study, we revisit this subtle problem based on the Lieb-Schultz-Mattis theorem for low energy spectra and ground state degeneracy (GSD) for general quantum many-body systems [13, 14]. We argue that a conventional s-wave superconductor described by an attractive Hubbard model with a long-range Coulomb interaction has GSD $\geq q$ when the particle filling per unit cell is fractional $\rho = p/q$ with coprime integers p, q. The ground states exhibit translation symmetry breaking possibly corresponding to a charge-density-wave formation with period q, and consequently, the superconducting state will be a supersolid. The inconsistency between the proposed supersolid and experiments can impose some constraints on a reasonable choice of a theoretical model.

2. Possible supersolid in electron model of superconductivity

The aim of this study is to investigate possible outcomes of a reasonable model of superconductivity based on the well established experimental fact that an s-wave superconductor is fully gapped. This in turn can impose some constraints on a model choice. For this purpose, we consider the following Hamiltonian defined on a cubic lattice of a linear size L with the periodic boundary condition,

$$H = -\sum_{\langle ij\rangle\sigma} t c^\dagger_{i\sigma} c_{j\sigma} - U\sum_i n_{i\uparrow}n_{i\downarrow} + \frac{1}{2}\sum_{i\neq j} V_{ij}(n_i-\rho)(n_j-\rho), \quad (1)$$

where $\langle ij\rangle$ is a pair of nearest neighbor sites and $\sigma = \uparrow, \downarrow$ is the spin index. $c_{i\sigma}$ is an annihilation operator of an electron at site i and spin σ, and $n_{i\sigma} = c^\dagger_{i\sigma}c_{i\sigma}, n_i = \sum_\sigma n_{i\sigma}$. The second term is an attractive interaction arising from e.g. phonons, and the third term is the repulsive Coulomb interaction on a lattice, $V_{ij} = V/|i-j|$ for $i \neq j$. (More precisely, the attractive term should be regarded as a sum of a phonon mediated interaction and on-site repulsive Coulomb interaction.) The charge density is measured from the average density $\rho = p/q$ per unit cell with coprime integers p, q. The total number of electrons is $N = \rho L^d$ with $d = 3$ in the present model. Equation (1) is considered to be one of the simplest models for charged superconductivity and one can generalize the discussions to more complex systems with local orbitals on other lattices. Note that although our argument is applicable also to a two-dimensional system, we focus on three-dimensional cases which are more non-trivial since a plasmon is not gapped for a normal state in pure two dimensions and screening effects are expected to be similarly weak in a purely two-dimensional superconductor at zero temperature [15]. We simply assume that eigenvalues of the Hamiltonian are extensive and the system has a well-defined thermodynamic limit. Throughout our discussion, we focus on zero temperature, and strictly keep the U(1) particle-number conservation since we consider an isolated superconductor. In this framework, the U(1) symmetry is not broken although there can be a corresponding long-range order due to the attractive interaction U, and the ground state is a Schrödinger's cat state in terms of the global U(1) phase of the Cooper pair order parameter [16, 17, 18]. Any gauge invariant quantity can be correctly evaluated within this fixed particle-number description, and differences are seen only for gauge dependent quantities such as the order parameter which is not an observable.

In absence of the Coulomb interaction ($U \neq 0, V = 0$), Eq. (1) is reduced to the well-known attractive Hubbard model, and the corresponding ground state is expected to be a spatially uniform s-wave superfluid [19, 20, 21, 22, 23, 24]. On the other hand, when $U = 0, V \neq 0$, the system would be either metallic or charge ordered as seen in the previous studies of extended Hubbard models [20, 25, 26, 27]. Naively, one would expect that superconductivity and a charge order can coexist to form a supersolid state at some parameter region with both non-zero U, V. For example, in the strong coupling limit ($t \ll U$ and $V \neq 0$), ground states for $\rho = 2/3$ would be

$$\begin{aligned} |\tilde{\Psi}_0\rangle &= |\cdots, \uparrow\downarrow, 0, 0, \uparrow\downarrow, 0, 0, \cdots\rangle + (\text{corrections}), \\ |\tilde{\Psi}_1\rangle &= |\cdots, 0, \uparrow\downarrow, 0, 0, \uparrow\downarrow, 0, \cdots\rangle + (\text{corrections}), \\ |\tilde{\Psi}_2\rangle &= |\cdots, 0, 0, \uparrow\downarrow, 0, 0, \uparrow\downarrow, \cdots\rangle + (\text{corrections}), \end{aligned} \tag{2}$$

where they are related by the translation operator as $\mathcal{T}_x |\tilde{\Psi}_0\rangle = |\tilde{\Psi}_1\rangle, \mathcal{T}_x |\tilde{\Psi}_1\rangle = |\tilde{\Psi}_2\rangle$, and $\mathcal{T}_x |\tilde{\Psi}_2\rangle = |\tilde{\Psi}_0\rangle$. In this case, the on-site $\uparrow\downarrow$ Cooper pair is regarded as a boson which has an effective hopping $\sim t^2/U$ in the second order perturbation when $U \gg t$, and a supersolid of the effective bosons could be stabilized by the interaction V [28, 29, 30]. This is a strong coupling picture for $U \gg t$ corresponding to a BEC regime. Although basic structures of these candidate ground states are reasonable, it is highly non-trivial whether or not the true ground states can indeed exhibit both superconductivity and a charge order simultaneously in some parameter region especially for small U corresponding to a BCS regime. For example, it would be possible that a superfluidity driven by U and a charge order stabilized by V are separated by a phase transition and there is no coexisting region.

Here, we do not directly examine existence or absence of a superconducting state in our model with both non-zero U, V, but instead we introduce some physical assumptions which are necessary to describe desired properties of superconductivity. We first focus on low energy states in the attractive Hubbard model at $V = 0$ and later take the Coulomb interaction into account. Now we simply assume that there is a long-range order of the spin-singlet Cooper pairing in the ground state of the attractive Hubbard model, $\langle\Psi_0| \mathcal{O}^\dagger \mathcal{O} |\Psi_0\rangle \sim (L^d)^2$. The bulk order parameter is defined as $\mathcal{O} = \sum_i c_{i\uparrow} c_{i\downarrow}$. In this case, single-particle fermion excitations and spin excitations will be gapped and thus it is natural to assume that there are only two kinds of low energy states in the N-particle Hilbert space $\mathcal{H}_N$, namely, the ground state $|\Psi_0\rangle$ and states with NG excitations. It is known that the NG boson is related with collective particle density excitations, where the variational states (up to normalization)

$$|\Phi_k\rangle = N_k |\Psi_0\rangle, \qquad N_k = \sum_j n_j e^{ikr_j} \tag{3}$$

are good approximations of the exact excited states [5]. According to the NG theorem, the variational energy of these states are $\simeq$ Const $\cdot k \sim L^{-1}$ measured from the ground state energy E_0. It is important to remark that there are other low energy states associated with the U(1) symmetry breaking if one extends the Hilbert space to include different particle-number sectors [16, 17, 18]. In this case, the entire Hilbert space is $\mathcal{H} = \bigoplus_{N=0}^{2L^d} \mathcal{H}_N$ and the tower of states has energies $\sim L^{-d}$ measured from E_0. The tower of states is well approximated by the variational states $\mathcal{O}^m |\Psi_0\rangle$ and $(\mathcal{O}^\dagger)^m |\Psi_0\rangle$ up to normalization. However, these states belong to $\mathcal{H}_{N-2m}$ and $\mathcal{H}_{N+2m}$ respectively, but not to $\mathcal{H}_N$ which contains $|\Psi_0\rangle$. As mentioned before, these states are not relevant in the present study for an isolated superconductor with the particle-number conservation law.

Next, we consider the full Hamiltonian with both non-zero U, V under the assumption that the system is superconducting [31]. To have a constraint on possible low energy spectrum and GSD, we use the Lieb-Schultz-Mattis theorem for a long-range interacting system recently proved by the author [13, 14]. For the proof of the theorem, variational low energy states $|\Psi_1\rangle, \cdots, |\Psi_{q-1}\rangle$ are constructed with use of translation symmetry and U(1) symmetry, and these states including $|\Psi_0\rangle$ are distinct each other when the filling per unit cell $\rho = p/q$ is fractional. The existence of such states implies that only the following two possibilities are allowed in the present model.

(i) There are gapless excitations above the ground state(s).

(ii) There is a gap between the ground state sector and excited states, and the ground state sector has GSD $\geq q$.

This is a non-perturbative statement which is valid for any parameters, but we cannot fully identify which one will be the case in the present system. However, as was discussed in the introduction, the long-range Coulomb interaction can lead to the screening effect and the NG boson will be gapped out. Therefore, there are no low energy excited states in $\mathcal{H}_N$ above the ground states, and the first possibility (existence of gapless excitations) is excluded. The remaining possibility is that there are at least q-hold degenerate ground states with a gap above them. We stress that the system cannot be a simple uniform $\mathbb{Z}_2$ topologically ordered state, because GSD $\geq q$ is filling dependent and is clearly distinguished from the $\mathbb{Z}_2$ topological degeneracy which depends only on the global topology of the underlying manifold on which the Hamiltonian is defined. Besides, these degenerate ground states are different from the tower of states since the former belong to $\mathcal{H}_N$. The remaining resolution is that they are related to translation symmetry breaking [32, 33]. Indeed, each ground state is essentially an eigenstate of the translation operator, $\mathcal{T}_x |\Psi_n\rangle = e^{iP_n} |\Psi_n\rangle$ with $P_n = 2\pi n/q$, and one can construct physical ground states (up to normalization) from these cat states [33],

$$|\tilde{\Psi}_n\rangle = \sum_{n'=0}^{q-1} e^{iP_n n'} |\Psi_{n'}\rangle . \tag{4}$$

Then, the translation symmetry is spontaneously broken in the physical ground states and they are related each other as $\mathcal{T}_x |\tilde{\Psi}_n\rangle = |\tilde{\Psi}_{n+1}\rangle$ (mod q) just like the naive guess Eq. (2). From the previous studies for extended Hubbard models [25, 26, 27], it is reasonable to suppose that the translation symmetry breaking in the present Hamiltonian is related to a charge-density-wave order, although other non-uniform states such as a Fulde-Ferrell-Larkin-Ovchinnikov state are not completely excluded. To summarize, the resulting states are superconducting and break translation symmetry simultaneously possibly corresponding to a charge order. This means that if the Hamiltonian Eq. (1) can describe the desired properties of superconductivity as we have assumed, the outcome is that the ground state is generally a supersolid. We stress that our argument is based on the non-perturbative Lieb-Schultz-Mattis theorem, and thus the statement holds for any parameters as long as the assumptions are met. Note that if the two approaches for the gapped NG boson based on the screening effect and gauge fluctuations respectively are consistent as mentioned in the introduction, the supersolid state will also be simultaneously a $\mathbb{Z}_2$ topologically ordered state.

3. Discussion and summary

In the previous section, we have shown that a system described by the electron Hamiltonian with both a local attractive interaction and long-range Coulomb interaction will exhibit a supersolid state. Our argument is non-perturbative and general, although it is just qualitative and magnitude of a charge order could be tiny in a realistic parameter region. Although we have introduced several assumptions (U(1) long-range order, no gapless excitations other than the NG boson at $V = 0$, and gapping NG boson by the Coulomb interaction) so that the argument goes step by step, they are combined into simpler assumptions that the Hamiltonian Eq. (1) can describe superconductivity and has a gapped spectrum [34]. Of course, from a purely theoretical point of view, these assumptions are non-trivial and should be examined, but a breakdown of the assumptions implies that the chosen model is not appropriate for superconductivity. Our proposal of a supersolid state sharply contradict many existing experiments, although there are some superconductors with charge orders. In this view point, our result implies that the simple model of superconductivity Eq. (1) may not be a fully correct choice. One can add some other density-density type interactions for which the Lieb-Schultz-Mattis theorem is still applicable [14], and also our statement would be perturbatively robust to more general short-range interactions since the system is assumed to be gapped. In this way, the inconsistency between the proposed supersolid and experiments can impose some constraints on a reasonable choice of a theoretical model.

As mentioned in the introduction, superconductivity can be alternatively described by a gauged model such as the compact U(1) Higgs model corresponding to the Ginzburg-Landau theory. In this phenomenological model, a spatially uniform $\mathbb{Z}_2$ topological ordered state is realized. Similarly, superconductivity in a gauged electron model coupled to a dynamical electromagnetic field is not characterized by a local order parameter [35, 36]. However, to our best knowledge, there is no theoretical study on a possible charge order in such a gauged electron system. One may naively expect that a uniform state is stable since the interactions are short ranged in the $\mathbb{Z}_2$ ordered phase [11], but a charge ordered state could be stable as well since it is gapped. Besides, there is another subtle issue for gauged models about compactness of the U(1) electromagnetic field. It is known that a non-compact U(1) Higgs model with a gauge fixing can show a long-range order of a bosonic condensation and the correlation function decays algebraically implying existence of an NG boson [37]. Although one would naively expect that this candidate NG boson is decoupled from the rest of the system and thus cannot be observed through local gauge invariant quantities, such excitations could contribute to thermodynamic quantities such as specific heat, which is inconsistent with experiments. Nevertheless, we believe that gauged models are promising models, since they necessarily include magnetic fields and coupling to charge currents. Especially in compact gauge theories, gapping photons (Meissner effect) and gapping NG bosons occur at the same time once a coupling between gauge and matter fields is introduced, in contrast to a pure electron model with Coulomb interaction where these two phenomena take place separately. This is consistent with the conventional Ginzburg-Landau description of superconductivity which well explains experiments [38]. Therefore, we expect that superconductivity can be better described by a gauged model.

The problem discussed in the present study is a highly subtle issue, and indeed the two theoretical approaches for the absence of the NG boson have been thought basically equivalent. The key is a coupling to the U(1) gauge field, which is to some extent common for both approaches. However, as discussed above, there is no gauge degrees of freedom and no magnetic field in a Coulomb interacting model which takes only an electric field into account (in this sense the model is non-gauged), while dynamical electromagnetic fields are fully coupled to electrons in a gauged model. This may bring a clearer difference in their outcomes than the charge order proposed in the present study. Our crude discussion is just a single step, and further investigation will be necessarily for this subtle but fundamental problem.

Acknowledgments

We are grateful to Y. Yao, M. Oshikawa, and A. Koga for valuable discussions. This work was supported by JSPS KAKENHI Grant No. JP17K14333.

References

[1] Anderson P W 1958 *Phys. Rev.* **110**(4) 827–835
[2] Anderson P W 1958 *Phys. Rev.* **112**(6) 1900–1916
[3] Rickayzen G 1959 *Phys. Rev.* **115**(4) 795–808
[4] Nambu Y 1960 *Phys. Rev.* **117**(3) 648–663
[5] Wagner H 1966 *Z. Physik* **195** 273–299
[6] Schrieffer J R 1971 *Theory of Superconductivity* (Westview Press)
[7] Osterwalder K and Seiler E 1978 *Annals of Physics* **110** 440–471 ISSN 0003-4916
[8] Fradkin E and Shenker S H 1979 *Phys. Rev. D* **19**(12) 3682–3697

[9] Hansson T, Oganesyan V and Sondhi S 2004 *Annals of Physics* **313** 497–538 ISSN 0003-4916
[10] Oshikawa M, Kim Y B, Shtengel K, Nayak C and Tewari S 2007 *Annals of Physics* **322** 1477–1498 ISSN 0003-4916
[11] Wen X G 2004 *Quantum Field Theory of Many-body Systems: From the Origin of Sound to an Origin of Light and Electrons* (Oxford University Press)
[12] For example in a three-dimensional metal, the electric field is short ranged due to a screening effect, while the magnetic field is long ranged.
[13] Lieb E, Schultz T and Mattis D 1961 *Ann. Phys.* **16** 407–466 ISSN 0003-4916
[14] Tada Y, arXiv:2106.12222. To be published in *Phys. Rev. Lett.*
[15] Ando T, Fowler A B and Stern F 1982 *Rev. Mod. Phys.* **54**(2) 437–672
[16] Koma T and Tasaki H 1993 *J. Stat. Phys.* **76** 745–803
[17] Tasaki H 2019 *J. Stat. Phys.* **174** 735–761
[18] Tasaki H 2020 *Physics and Mathematics of Quantum Many-Body Systems* (Springer)
[19] Lieb E H 1989 *Phys. Rev. Lett.* **62**(10) 1201–1204
[20] Auerbach A 1994 *Interacting Electrons and Quantum Magnetism* (Springer-Verlag)
[21] The η-pairing with a charge modulation on different sublattices can be stable in the ground state of an attractive Hubbard model on some bipartite lattices with particle-hole symmetry [22, 23, 24]. In this case, the modulation period is fixed by the lattice structure and the translation symmetry is not broken. These states are beyond the scope of our study.
[22] Shen S Q and Qiu Z M 1993 *Phys. Rev. Lett.* **71**(25) 4238–4240
[23] Tian G S 1994 *Physics Letters A* **192** 278–284 ISSN 0375-9601
[24] Shen S Q and Qiu Z M 1998 *Int. J. Mod. Phys. B* **12**(25) 709–779
[25] Giamarchi T 2003 *Quantum Physics in One Dimension* (Oxford University Press)
[26] Zhang Y and Callaway J 1989 *Phys. Rev. B* **39**(13) 9397–9404
[27] van Loon E G C P and Katsnelson M I 2018 *Journal of Physics: Conference Series* **1136** 012006
[28] Wessel S and Troyer M 2005 *Phys. Rev. Lett.* **95**(12) 127205
[29] Heidarian D and Damle K 2005 *Phys. Rev. Lett.* **95**(12) 127206
[30] Melko R G, Paramekanti A, Burkov A A, Vishwanath A, Sheng D N and Balents L 2005 *Phys. Rev. Lett.* **95**(12) 127207
[31] We do not explicitly assume the U(1) long-range order at $V \neq 0$, because the system might be a gapped $\mathbb{Z}_2$ topologically ordered state where any correlation functions are short ranged. Instead, we say that an electron system is superconducting when it shows Meissner effect.
[32] Koma T 2000 *J. Stat. Phys.* **99** 313–381
[33] Oshikawa M 2000 *Phys. Rev. Lett.* **84**(7) 1535–1538
[34] Our argument can be greatly simplified if we simply assume that the spectrum of the Hamiltonian Eq. (1) is gapped so that it is consistent with the experimental results of the gapped superconducting nature. By applying the Lieb-Schultz-Mattis theorem to the Hamiltonian, we immediately conclude that the ground states are at least q-hold degenerate under this assumption. This implies a supersolid state when the system is superconducting, as discussed in the main text. Although this is just a straightfoward application of the Lieb-Schultz-Mattis theorem, it has not been discussed so far to the best of our knowledge.
[35] Tada Y and Koma T 2016 *J. Stat. Phys.* **165** 455–470
[36] Tada Y and Koma T 2018 *J. Math. Phys.* **59** 031905
[37] Kennedy T and King C 1985 *Phys. Rev. Lett.* **55**(8) 776–778
[38] Note that a static Ginzburg-Landau theory with no time dependence should provide a sufficient description of the Meissner effect and gapped NG boson, since they are equilibrium properties of superconductivity. In this description, no electric field is necessary.

Effect of the strong electron correlations on the formation of Wigner crystal in n-GaAs/AlGaAs structure: Surface Acoustic Waves studies

I Yu SMIRNOV[1], I L DRICHKO[1], A V SUSLOV[2], L N PFEIFFER[3], K W WEST[3], Y M GALPERIN[1,4]

[1] Ioffe Institute, St. Petersburg 194021, Russia
[2] National High Magnetic Field Laboratory, Tallahassee, FL 32310, USA
[3] Department of Electrical Engineering, Princeton University, Princeton, NJ 08544, USA
[4] Department of Physics, University of Oslo, 0316 Oslo, Norway

ivan.smirnov@mail.ioffe.ru

Abstract. The complex high-frequency conductance of high-mobility n-GaAs/AlGaAs heterostructures was determined in magnetic fields 12–18 T using the Surface Acoustic Waves technique. Analysis of frequency and temperature dependences of conductance showed that in the investigated magnetic field range and at low temperatures, T<200 mK, the electron system forms a Wigner crystal deformed due to pinning by disorder. We have also defined an effective melting temperature of the Wigner crystal.

1. Introduction

When the correlation energy of a two-dimensional electron system (2DES) $E_c=e^2/\varepsilon l$ exceeds the kinetic (Fermi) one $E_F=\pi n\hbar^2/m$, formation of a Wigner crystal (WC) [1] could be expected. This to be happen when the interaction parameter $r_s=E_c/E_F$ achieves the critical value of 37 as calculated in [2]. This implies a very low density of 2DES of $n=10^9$ cm^{-2}, at which electrons in a semiconductor are actually strongly localized on chaotic potential fluctuations at low temperatures, and a WC could be hardly realized. In a real 2DES, due to disorder, a transition to an electron solid state can happen at much smaller r_s down to 10 [3]. Moreover, the electron crystallization can be stimulated by applying a strong magnetic field perpendicular to a 2DES [4] via the vanishing magnetic length $l_B=(\hbar/eB)^{1/2}$ in the increased correlation energy $E_c^{MF}=e^2/\varepsilon l_B$. Presence of a disorder makes the WC state pinned. This insulating phase is not quite experimentally distinguishable from the single-particle localization, especially in current transport measurements. In the present work we employ the surface acoustic wave (SAW) technique to identify the nature of the states observed in high magnetic fields at the filling factors $0.125<\nu<0.18$, where r_s come up to the value of 12. This contactless method allows one finding the complex AC conductance, $\sigma^{AC}(\omega)=\sigma_1(\omega)-i\sigma_2(\omega)$, as a function of magnetic field, temperature and frequency. Knowing of both the real and imaginary parts is important for determination of the conductance mechanisms.

2. Experiment and Discussion

The system under study was a single 65 nm quantum well AlGaAs/GaAs/AlGaAs structure with the carriers concentration of n=5×10^{10} cm^{-2}, with a high mobility of μ=8.5×10^{6} cm^{2}/Vs, and with the zero-field r_s≈2.7. In our experiments a surface acoustic wave propagates along the surface of the piezodielectric lithium niobate ($LiNbO_3$), and the studied structure is pressed onto the same surface of the $LiNbO_3$. An electric field produced by the SAW interacts with the carriers in the 2D channel. Via this interaction both the attenuation Γ and velocity v of the acoustic wave are governed by the conductance of the 2DES. From simultaneously measured attenuation and velocity of SAW both tht real $\sigma_1(\omega)$ and imaginary $\sigma_2(\omega)$ components of the complex AC conductance $\sigma^{AC}(\omega)$ were calculated as in [5]. The measurements have been carried out in the frequency region of 28-300 MHz at temperatures 40-380 mK in the magnetic field of up to 18 T.

Magnetic field induces a rich oscillation pattern in AC conductance of our system containing features of both integer and fractional quantum Hall effect (QHE) thus evidencing high quality of the sample. We however focus here on high fields B>12 T (ν<1/5), i.e., away from the QHE oscillations. Frequency dependences of σ_1 and σ_2 for ν = 0.18 (B=12.2 T) and T=40 mK are shown in Fig. 1, where a maximum is observed in σ_1 (f) at the frequency $f \approx$ 86 MHz, and σ_2 changes sign at the same frequency. It corresponds to a resonant character of AC conductance $\sigma^{AC}(f)$.

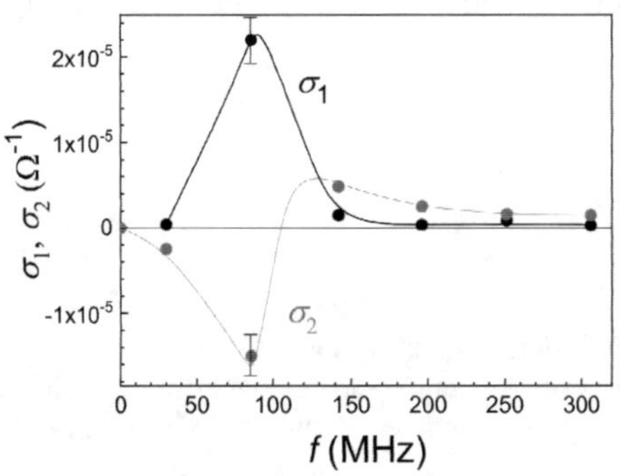

Figure 1. Frequency dependences of $\sigma_{1,2}$ for ν=0.18 (B=12.2 T), T=40 mK.

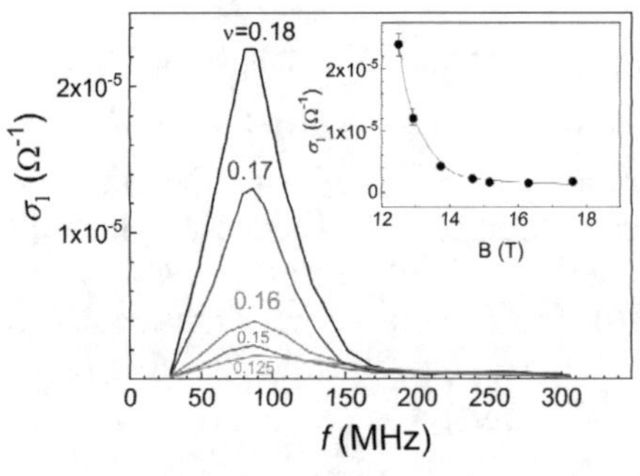

Figure 2. Frequency dependences of σ_1 for different filling factors (shown near the curves). Inset: Magnetic field dependence of σ_1 (86 MHz). T=40 mK.

The presence of a resonance in $\sigma^{AC}(f)$ agrees with the results of works [6, 7, 8], where the features in the dependences of the conductivity and resistance on the magnetic field, temperature and frequency are associated with the formation of a pinned WC. $f \approx$ 86 MHz is the pinning frequency.

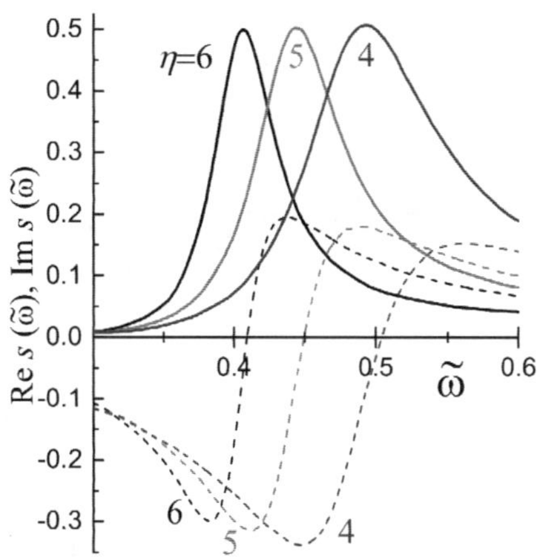

Figure 3. Real and imaginary parts of normalized conductance $s(\varpi)$ for $\eta = 4$, 5, and 6 modelled with help of [9].

Shown in Fig. 2 are the frequency dependences of σ_1 for different filling factors. Each curve has a maximum at $f \approx 86$ MHz, and the maximum value decreases when the magnetic field increases (see the inset to the figure Fig. 2). In accordance with the theoretical results of [9] AC conductance of a pinned Wigner crystal can be written as

$$\sigma(\omega) \equiv \sigma_0 s(\varpi);\ \sigma_0 \equiv \frac{e^2 n\eta^2}{2m^* \omega_c},\ (1)$$

where $\varpi=\omega/\Omega$, $\Omega \sim \omega_{p0}^2 \eta/\omega_c$, ω_{p0} is the pinning frequency at B=0, ω_c is the cyclotron frequency, $\eta=(\lambda/\beta)^{1/2}$, λ and β are the longitudinal and transverse elastic moduli of the WC. We calculated the frequency dependences of the real and imaginary components of $s(\varpi)$ near the resonance frequency of the pinned mode of a Wigner crystal for different η. Graphs of real and imaginary parts of $s(\varpi)$ for η = 4, 5, and 6 are shown in Fig. 3. The best qualitative agreement between theoretical calculations and experimental data is achieved at η=5. Dependence of $\sigma_1(\omega)$ on magnetic field presented on Fig. 2 is also well explained by Eq. (1), namely via magnetic field dependence of the term σ_0.

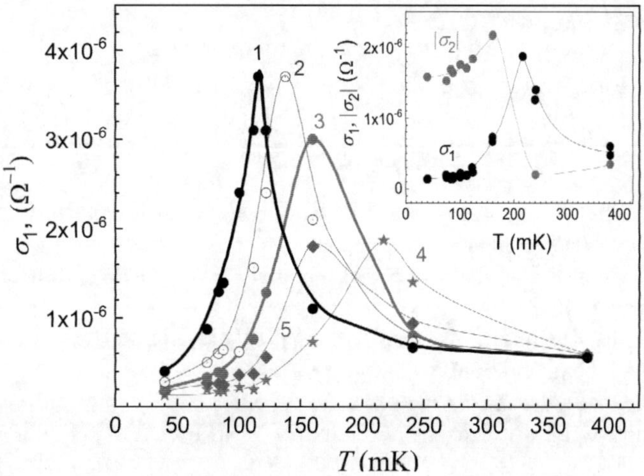

Figure 4. Temperature dependences of σ_1 for different ν: 1: 0.18; 2: 0.17; 3: 0.16; 4: 0.145; 5: 0.125. f=28.5MHz. Inset: Temperature dependences of σ_1 and $|\sigma_2|$ for ν=0.13, f=28.5 MHz.

Now, knowing η it is possible to determine the pinning frequency in the absence of a magnetic field and, accordingly, the average correlation length (size) of WC domains, which turned out to be 3×10^{-4} cm. It is much larger than both the distance between the electrons, $a=2.5\times10^{-6}$ cm, and the magnetic length at $B=12.2$ T ($\nu = 0.18$) $l_B = 7.3\times10^{-7}$ cm.

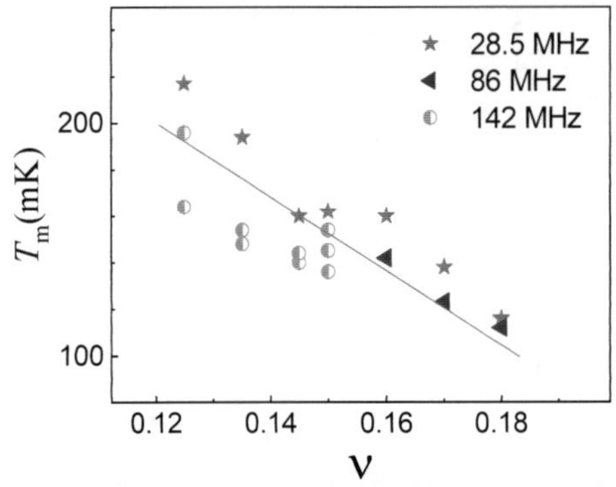

Figure 5. Dependence of the "melting temperature" T_m (see text) on the filling factor ν for different frequencies.

The temperature dependences of σ_1 for $0.18 > \nu > 0.125$ are presented in Fig. 4. $\sigma_1(T)$ has maxima. To the left of the maxima, the $\sigma_1(T)$ exhibits a dielectric behavior. In this case, as shown in the inset of Fig. 4, $|\sigma_2|>\sigma_1$. This fact, along with the described above frequency dependences of σ_1, can be attributed to the manifestation of a pinned Wigner crystal. To the right of the maximum, $|\sigma_2|$ decreases rapidly with increasing temperature. σ_1 also decreases with increasing temperature but much more slowly than $|\sigma_2|$. Therefore, to the right of the maxima, $|\sigma_2|<\sigma_1$. Thus, at the maxima there is a change of the conduction mechanisms, which can be interpreted as a WC melting point T_m. As the magnetic field increases, T_m increases. Using this approach, a new method for determining the WC melting temperature and its dependence on the magnetic field was constructed and applied at various SAW frequencies (Fig. 5) [10].

References

[1] Wigner E 1934 *Phys. Rev.* **46** 1002
[2] Tanatar B and Ceperley D M 1989 *Phys. Rev. B* **39** 5005
[3] Pudalov V M, D'Iorio M, Kravchenko S V and Campbell J W 1993 *Phys. Rev. Lett.* **70** 1866
[4] Lozovik Y E and Yudson V I 1975 *JETP Lett.* **22** 11
[5] Drichko I L, Diakonov A M, Smirnov I Yu, Galperin Y M and Toropov A I 2000 *Phys. Rev. B* **62** 7470.
[6] Paalanen M A, Willett R L, Littlewood P B, Ruel R R, West K W, Pfeiffer L N and Bishop D J 1992 *Phys. Rev. B* **45** 11342
[7] Paalanen M A, Willett R L, Littlewood P B, Ruel R R, West K W, Pfeiffer L N and Bishop D J 1992 *Phys. Rev. B* **45** 13784
[8] Chen Y P, Lewis R M, Engel L W, Tsui D C, Ye P D, Wang Z H, Pfeiffer L N and West K W 2004 *Phys. Rev. Lett.* **93** 206805
[9] Fogler M M and Huse D A 2000 *Phys. Rev. B* **62** 7553
[10] Drichko I L, Smirnov I Yu, Suslov A V, Galperin Y M, Pfeiffer L N and West K W 2016 *Phys. Rev. B* **94** 075420

Journal of Physics: Conference Series **2164** (2022) 012022 doi:10.1088/1742-6596/2164/1/012022

Superconducting and chiral-glass to insulator transition in phase-glass models in a magnetic field

Enzo Granato

Laboratório Associado de Sensores e Materiais, Instituto Nacional de Pesquisas Espaciais, 12227-010 São José dos Campos, SP, Brazil

E-mail: enzograna@gmail.com

Abstract. We investigate the quantum phase transitions of a phase-glass model in a magnetic field with frustration parameter $f = 1/2$, describing the effects random negative Josephson-junction couplings in two-dimensional superconductors. The critical behavior is obtained by a scaling analysis of path-integral Monte Carlo simulations at zero temperature, including corrections to finite-size scaling. A single superconducting and chiral-glass to insulator transition occurs above a disorder threshold with divergent nonlinear magnetic susceptibility, unaffected by the additional magnetic-field frustration. The relevance of this transition for nanohole superconducting thin films doped with magnetic impurities is discussed.

1. Introduction

Negative Josephson junctions distributed randomly lead to striking effects in the transport and magnetic properties of superconducting systems. In high-T_c superconductors, these effects have been studied by XY spin-glass or chiral-glass models, where they represent the π junctions arising from grain misorientations [1, 2]. A quantum version of these models, the phase-glass model, which includes the charging energy of local superconducting regions, has been employed [3, 4] to explain a Bose metal phase separating the superconductor and insulator phases in thin films, appearing experimentally as an anomalous metallic phase [5, 6]. Although recent numerical simulations [7, 8] of this model in two dimensions found no evidence of such metallic phase, the model finds applications for nanohole superconducting thin films doped with magnetic impurities in an applied magnetic field [9].

Here, we investigate in further details the quantum phase transitions of the phase-glass model in a magnetic field with frustration parameter $f = 1/2$. From a scaling analysis of the Monte Carlo (MC) data including finite-size scaling corrections, we find a single superconducting and chiral-glass to insulator transition above a disorder threshold, unaffected by the additional magnetic-field frustration. It is also argued that the nonlinear magnetic susceptibility diverges at transition for decreasing temperatures. Recent measurements on nanohole superconducting thin films are interpreted in terms of this chiral-glass regime.

2. Phase-glass model and MC simulation

We consider a two-dimensional phase-glass model given by the Hamiltonian [3, 7]

$$\mathcal{H} = \frac{E_C}{2} \sum_i n_i^2 - \sum_{<ij>} E_{ij} \cos(\theta_i - \theta_j - A_{ij}), \tag{1}$$

where θ_j is the phase operator at site j of a square lattice with lattice spacing a and $n_j = -i\partial/\partial\theta_j$ is the corresponding canonically conjugate operator. $E_C = 4e^2/C$ is charging energy, where e is the electron charge and C is an effective capacitance to the ground. E_{ij} is the Josephson-junction coupling of nearest-neighbor grains, with an asymmetric probability distribution $P(E_{ij}) = x\delta(E_{ij} + E_J) + (1 - x)\delta(E_{ij} - E_J)$, where x is the fraction of negative junctions, a measure of the degree of disorder. The gauge variables A_{ij} are constrained by $\sum'_{ij} A_{ij} = 2\pi f$, where the summation is a direct sum around a lattice plaquette. The frustration parameter $f = Ba^2/\Phi_o$ corresponds to the average number of flux quantum, $\Phi_o = hc/2e$, per plaquette, due to the external magnetic field **B**. The presence of negative junctions can introduce quenched vortices [1] even when $f = 0$, due to frustration effects. This can be described by a chiral order parameter [2], $\kappa_p = \frac{1}{\kappa_0} \sum'_{<ij>} E_{ij} \sin(\theta_i - \theta_j - A_{ij})$, where p labels a plaquette and κ_0 is a normalization factor.

The MC simulations were implemented in the equivalent (2+1) dimensional classical model [10, 7], with reduced Hamiltonian

$$H = -\frac{1}{g}[\sum_{\tau,i} \cos(\theta_{\tau,i} - \theta_{\tau+1,i}) + \sum_{<ij>,\tau} e_{ij} \cos(\theta_{\tau,i} - \theta_{\tau,j} - A_{ij})], \qquad (2)$$

where $g = (E_C/E_J)^{1/2}$, $e_{ij} = E_{ij}/E_J$ and τ labels the sites in the time direction. MC simulations employed the parallel tempering method with periodic boundary conditions [7] in systems of linear size L.

3. Results and discussion

For increasing disorder, a transition from a superconducting phase to a chiral-glass phase occurs at a critical disorder value x_C. This threshold value can be obtained from the behavior of the domain-wall energy [7] E_W in the classical model of Eq. (2). Figure 1a shows the disorder averaged $[E_W]_d$ for $f = 1/2$. At small disorder, it increases with L, corresponding to long-range phase coherence while at large disorder it decreases with L in the disordered glass phase. From the change of behavior, we estimate a disorder threshold $x_C \sim 0.9$.

We now consider the quantum phase transitions in the chiral-glass regime ($x > x_C$) at $f = 1/2$, tuned by the coupling parameter g. The critical properties can be obtained from the scaling behavior of the size-dependent correlation length ξ in the imaginary time direction, defined as

$$\xi(L, g) = \frac{1}{2\sin(k_0/2)}[S(0)/S(k_0) - 1]^{1/2}. \qquad (3)$$

$S(k)$ and k_0 are the Fourier transform of the correlation function and the smallest nonzero wave vector, respectively [11]. The phase correlation length ξ_G is obtained from the correlation function of the overlap order parameter [12, 7] $q_{\tau,j} = \exp(i(\theta^1_{\tau,j} - \theta^2_{\tau,j}))$, where 1 and 2 are two different replicas with the same coupling parameters. Likewise, the chiral correlation length ξ^c_G is obtained as $q^c_{\tau,p} = \kappa^1_{\tau,p}\, \kappa^2_{\tau,p}$. For a continuous phase transition, $\xi(L, g)$ is expected to satisfy the scaling form $\xi/L = A(L^{1/\nu}\delta g)$, where $\delta g = g - g_c$ and ν is the critical exponent. The scaled correlation length $\xi(L, g)/L$ for different L, plotted as a function of g, should then cross at the critical coupling g_c.

Figures 1b and 1c show the correlation lengths ξ_G and ξ^c_G at $x = 0.2$, with the dynamic exponent $z = 1.2$, the same value used [7, 8] when $f = 0$. For large system sizes, the curves for ξ_G/L and ξ^c_G/L as a function of g cross at a common point g_c. However, there are significant corrections to finite-size scaling since the curves for smaller system sizes cross at different points. From data for large system sizes ($L = 16$ to 30), estimates of $g_c = 1.50(2)$ and $\nu = 1.4(2)$ for phase-coherence and $g_c = 1.42(2)$ and $\nu = 1.1(2)$ for chiral-glass transitions have already been obtained [8] in absence of scaling corrections. There is a small difference in the estimates of g_c

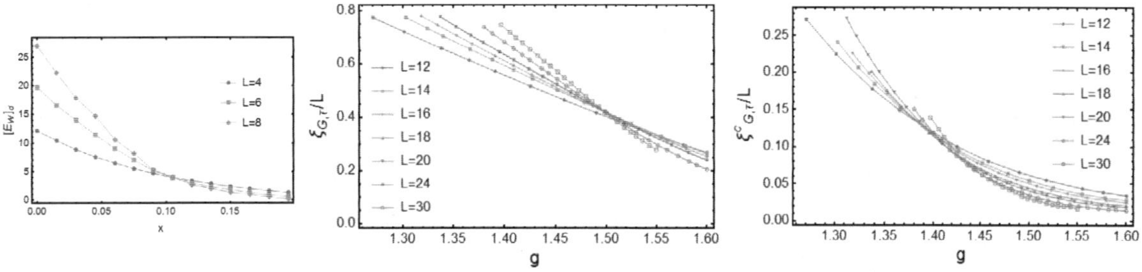

Figure 1. (a) Domain-wall energy $[E_W]_d$ at $f = 1/2$ for increasing disorder x and system sizes L. (b) Scaled phase correlation length $\xi_{G,\tau}/L$ near the chiral-glass to insulator transition at $x = 0.2$. (c) Same as (b) but for the chiral correlation length $\xi^c_{G,\tau}/L$.

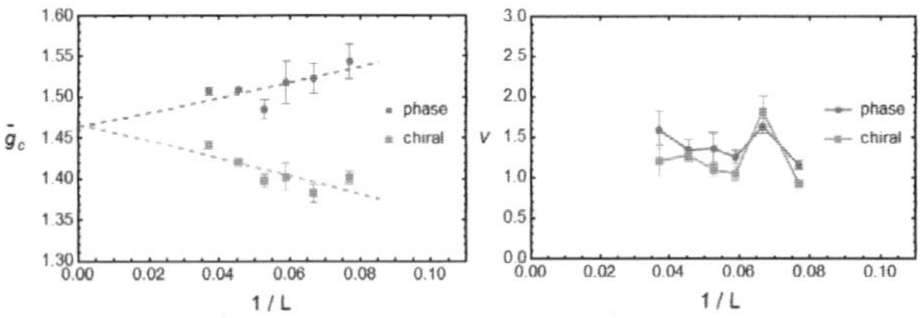

Figure 2. Finite-size behavior of critical couplings $\bar{g}_c(L)$ and exponent ν. Dotted lines are fittings according to Eq. (4).

for the two transitions. As the chiral variables are obtained from phase correlations of nearest-neighbors sites, chiral-glass disorder leads to phase incoherence. It is then expected that the critical couplings for chiral-glass order and phase coherence should be equal or the chiral-glass transition occurs at a larger coupling [13]. This discrepancy of the estimates of g_c is probably the effect of corrections to scaling and the two transitions occur as a single transition.

To check this possible scenario, we performed an approximate extrapolation to the infinite system of finite-size estimates of the critical coupling $\bar{g}_c(L)$ including the smaller system sizes of Figs 1b and 1c. $\bar{g}_c(L)$ was obtained from the crossing points of $\xi(L, g)/L$ for successive system sizes, averaged over six independent calculations, and required to follow the finite-size scaling [11]

$$\bar{g}_c(L) = g_c + a/L^{\omega+1/\nu}, \tag{4}$$

where a is constant and ω is a correction-to-scaling exponent. As shown in Fig. 3a, an extrapolation is possible for both phase and chiral variables with a common critical coupling $g_c = 1.46$ and critical exponents $\omega + 1/\nu = 1.07$. The corresponding estimates of the critical exponent ν are shown in Fig 3b. Although the estimates of ν fluctuate with L, there is no particular trend and for the largest systems sizes they are consistent with previous estimates [8, 7] for $f = 1/2$ and $f = 0$, considering the errorbars, indicating that the superconducting chiral-glass to insulator transition is unaffected by the additional magnetic-field frustration. Qualitatively, this effect results from the disorder of the quenched vortices, which dominate in the chiral-glass regime for $x > x_C$. On the other hand, for $x < x_C$, magnetic-field frustration reduces the critical coupling of the superconductor-insulator transition. In particular, at $x = 0$, it changes its universality class [14, 15].

The superconducting chiral-glass transition should also affect the magnetic properties, like the linear and nonlinear magnetic susceptibility [2] χ and $\chi_{nl} = \delta\chi/\delta H^2$. Since the chirality κ is an Ising-like variable, the singular contributions to χ and χ_{nl} due to the random circulating currents in the phase-glass model, should have the same scaling behavior of the quantum Ising-spin glass. Using the results for the quantum Ising-spin glass; [12], we then expect that $\chi \sim 1/L_\tau^{\beta/z\nu-1}$ and $\chi_{nl} \sim 1/L_\tau^{(2\beta-2\nu)/z\nu-3}$, at the transition. From the estimates of β, ν and z for the phase-glass model [7, 8], and the scaling of $L_\tau \propto 1/T$ with the temperature T, we find that these quantities have a power-law dependence with the temperature at the transition as $\chi \sim T^{0.19}$ and $\chi_{nl} \sim T^{-2.28}$. Thus, as for the thermal transition in absence of quantum fluctuations [2], the zero-temperature chiral-glass transition shows a divergence of the nonlinear magnetic susceptibility with decreasing temperatures.

The phase-glass model can explain some magnetic-field effects in superconducting thin films perforated with nanoholes, containing magnetic impurities [9]. Since magnetic impurities can introduce negative junctions (π junctions) [16], the transition to the insulating phase can be described by the phase-glass model. The activation energy for the resistance observed experimentally in the insulating phase corresponds to the energy gap E_G, which vanishes near the transition as $E_{G,f} \propto (g - g_{c,f})^{z\nu}$. For small disorder, the magnetic-field frustration reduces significantly the critical coupling for the superconductor-insulator transition, $g_{c,1/2} << g_{c,0}$, with the corresponding reverse behavior for the energy gap, $E_{G,1/2} >> E_{G,0}$. However, above the chiral-glass threshold x_C, a single superconducting and chiral-glass to insulator transition occurs, unaffected by the additional magnetic-field frustration, leading to $g_{c,1/2} \sim g_{c,0}$, $E_{G,1/2} \sim E_{G,0}$, with essentially the same critical behavior. The absence of magnetic-field frustration effects at large doping in the experimental system is therefore a signature of such chiral-glass regime [8].

4. Conclusions

We have studied the quantum phase transitions of a phase-glass model with frustration parameter $f = 1/2$. From a finite-size scaling analysis of MC data at zero temperature including corrections to scaling, we find evidence of a single superconducting and chiral-glass to insulator transition above a disorder threshold. Therefore, an intervening Bose metal phase between the superconductor and insulator phases is absent, in sharp contrast with predictions from mean-field theory. This single transition also affects the magnetic properties, leading to a nonlinear magnetic susceptibility diverging with decreasing temperatures, which could be tested experimentally. The absence of magnetic-field frustration effects at large magnetic-impurity doping in superconducting nanohole thin films [9] is consistent with such chiral-glass regime.

Work supported by FAPESP (Grant No. 18/19586-9), CNPq and CENAPAD-SP.

References

[1] Sigrist M and Rice T M 1995 *Rev. Mod. Phys.* **67**(2) 503–513
[2] Kawamura H 1995 *Journal of the Physical Society of Japan* **64** 711
[3] Dalidovich D and Phillips P 2002 *Physical Review Letters* **89** 027001
[4] Phillips P W 2016 *Nature Physics* **12** 206–207
[5] Kapitulnik A, Kivelson S A and Spivak B 2019 *Reviews of Modern Physics* **91** 011002
[6] Yang C, Liu Y, Wang Y, Feng L, He Q, Sun J, Tang Y, Wu C, Xiong J, Zhang W, Lin X, Yao H, Liu H, Fernandes G, Xu J, Valles J M, Wang J and Li Y 2019 *Science* **366** 1505–1509
[7] Granato E 2020 *Phys. Rev. B* **102** 184503
[8] Granato E 2021 *The European Physical Journal B* **94** 1–8
[9] Zhang X, Joy J C, Wu C, Kim J H, Xu J and Valles Jr J M 2019 *Physical Review Letters* **122** 157002
[10] Wallin M, Sorensen E S, Girvin S M and Young A P 1994 *Phys. Rev. B* **49**(17) 12115–12139
[11] Ballesteros H G, Cruz A, Fenández L A, Martin-Mayor V, Pech J, Ruiz-Lorenzo J J, Tarancón A, Téllez P, Ullod C L and Ungil C 2000 *Phys. Rev. B* **62** 14237
[12] Rieger H and Young A P 1994 *Physical Review Letters* **72** 4141
[13] Lee L W and Young A P 2003 *Phys. Rev. Lett.* **90**(22) 227203
[14] Granato E and Kosterlitz J M 1990 *Phys. Rev. Lett.* **65**(10) 1267–1270
[15] Cha M C and Girvin S M 1994 *Phys. Rev. B* **49** 9794
[16] Bulaevskii L, Kuzii V and Sobyanin A 1977 *JETP lett* **25** 290

Journal of Physics: Conference Series **2164** (2022) 012023 doi:10.1088/1742-6596/2164/1/012023

Variation of the elliptical Fermi surface for a two-dimensional electron gas with anisotropic mass

Orion Ciftja
Department of Physics, Prairie View A&M University, Prairie View, Texas 77446, USA
E-mail: ogciftja@pvamu.edu

Abstract. We consider a two-dimensional electron gas in the thermodynamic (bulk) limit. It is assumed that the system consists of fully spin-polarized (spinless) electrons with anisotropic mass. We study the variation of the shape of the expected elliptical Fermi surface as a function of the density of the system in presence of such form of internal anisotropy. To this effect, we calculate the energy of the system as well as the optimum ellipticity of the Fermi surface for two possible liquid states. One corresponds to the standard system with circular Fermi surface while the second one represents a liquid anisotropic phase with a tunable elliptical deformation of the Fermi surface that includes the state that minimizes the kinetic energy. The results obtained shed light on several possible scenarios that may arise in such a system. The competition between opposing tendencies of the kinetic energy and potential energy may lead to the stabilization of liquid phases where the optimal elliptical deformation of the Fermi surface is non-obvious and depends on the density as well as an array of other factors related to the specific values of various parameters that characterize the system.

1. Introduction

The model of a two-dimensional electron gas (2DEG) of electrons with constant isotropic (effective) mass is one of the most studied models in condensed matter physics [1, 2, 3, 4, 5, 6, 7]. At absolute zero temperature and in absence of interactions, the electrons occupy all the quantum states inside a circular Fermi disk in reciprocal space [8]. These quantum states are typically associated with plane wave states that arise when we solve the quantum problem for a two-dimensional (2D) box after imposing periodic boundary conditions. In the common jellium approximation one guarantees the overall charge neutrality of the system by filling the 2D box with uniformly distributed positive charge that represents the neutralizing background.

On the other hand, the 2DEG model for electrons possessing an anisotropic (effective) mass [9] is much less studied. For such a case, based on naive considerations, one expects that an (effective) anisotropic mass will lead to a specific elliptical deformation of the Fermi surface in order to minimize the kinetic energy. However, the role of the potential energy term is overlooked in this discussion. It turns out that kinetic energy and potential energy terms have opposing tendencies under an elliptical deformation of the Fermi surface. This fact calls for a more detailed study of the interplay between these two terms in presence of such form of internal anisotropy of the system as represented by the (effective) mass anisotropy of the electrons.

In our work we study a 2DEG in the thermodynamic (bulk) limit. For simplicity, we assume that the system consists of fully spin-polarized (spinless) electrons with anisotropic mass. The jellium approximation is used and the quantum state of the electrons is described by an anti-symmetrized Slater determinant wave function of plane waves [10] as required by Pauli's exclusion principle. Two distinct liquid states of electrons are considered: (i) A standard isotropic liquid phase with circular Fermi surface; and (ii) A liquid anisotropic phase with a tunable elliptical deformation of the Fermi surface that includes the state that minimizes the kinetic energy. It is shown that the competition between opposing tendencies of the kinetic energy and potential energy may lead to the stabilization of liquid phases where the optimal elliptical deformation of the Fermi surface is non-obvious and depends on several factors.

2. Theory and model

Let us consider a 2DEG system of N electrons with anisotropic mass, m_x and $m_y(\neq m_x)$ immersed in a 2D region, $\Omega : \left\{ -\frac{L}{2} \leq x, y \leq +\frac{L}{2} \right\}$. We assume a spin-polarized (spinless) system of electrons. The electron number density of the system reads:

$$\rho_0 = \frac{N}{L^2} \ . \tag{1}$$

Density remains constant in the $N \to \infty$, $L \to \infty$ limit. We must remember that, differently from other conventional 2DEG systems treated in the literature [11, 12], the current model assumes an anisotropic (effective) mass for the electrons. The Hamiltonian is:

$$\hat{H} = \hat{T} + \hat{U} \ , \tag{2}$$

where

$$\hat{T} = \sum_{i=1}^{N} \left[\frac{\hat{p}_{ix}^2}{2\, m_x} + \frac{\hat{p}_{iy}^2}{2\, m_y} \right] \ , \tag{3}$$

is the kinetic energy operator and

$$\hat{U} = \hat{U}_{ee} + \hat{U}_{eb} + \hat{U}_{bb} \ , \tag{4}$$

is the potential energy operator. The potential energy operator in Eq.(4) includes the electron-electron (ee), electron-background (eb) and background-background (bb) interaction energy terms written as:

$$\hat{U}_{ee} = \frac{1}{2} \sum_{i=1}^{N} \sum_{j \neq i}^{N} v(|\vec{r}_i - \vec{r}_j|) \ , \tag{5}$$

$$\hat{U}_{eb} = -\rho_0 \sum_{i=1}^{N} \int_{\Omega} d^2 r' \, v(|\vec{r}_i - \vec{r}\,'|) \ , \tag{6}$$

and

$$\hat{U}_{bb} = \frac{\rho_0^2}{2} \int_{\Omega} d^2 r \int_{\Omega} d^2 r' \ v(|\vec{r} - \vec{r}\,'|) \ , \tag{7}$$

where $v(|\vec{r}_i - \vec{r}_j|) = k_e\, e^2 / |\vec{r}_i - \vec{r}_j|$ represent the Coulomb interaction energy between two electrons with charge, $-e$ $(e > 0)$ at a distance $r_{ij} = |\vec{r}_i - \vec{r}_j|$ and k_e is Coulomb's electric constant.

If electrons fill a circular Fermi surface, all vectors, $\vec{k}$ would be in the region $|\vec{k}| \leq k_F$ where:

$$k_F^2 = 4\, \pi\, \rho_0 \ . \tag{8}$$

where k_F represents the value of the Fermi wave number for a circular Fermi surface. However, if the electrons fill an elliptically deformed Fermi surface, all the vectors, $\vec{k}$ would be in the reciprocal space domain, $D_{\vec{k}}$ written as:

$$\vec{k} \in D_{\vec{k}} : \left\{ \frac{k_x^2}{k_a^2} + \frac{k_y^2}{k_b^2} \leq 1 \right\} \quad ; \quad k_a = \alpha\, k_F \quad ; \quad k_b = \frac{k_F}{\alpha} \ , \tag{9}$$

where we introduced a real positive parameter, $\alpha > 0$ in Eq.(9) consistent with the constraint: $k_a\, k_b = k_F^2$. We describe the ground state of the 2DEG by a normalized Slater determinant wave function of ortho-normalized plane waves

$$\Psi(\alpha) = \frac{1}{\sqrt{N!}}\, Det\left\{\phi_{\vec{k}_1}(\vec{r}_1), \ldots, \phi_{\vec{k}_N}(\vec{r}_N)\right\} \quad ; \quad \vec{k} \in D_{\vec{k}} \ , \tag{10}$$

where the parameter α is shown as argument of the wave function in order to draw asttention to the fact that we are dealing, in general, with an elliptically deformed Fermi surface. The state $\Psi(\alpha = 1)$ would represent the case of an isotropic liquid phase with circular Fermi surface. However, any $\Psi(\alpha \neq 1)$ state represents an anisotropic one with elliptical Fermi surface. The plane wave states are written as: $\phi_{\vec{k}_j}(\vec{r}_i) = \frac{1}{\sqrt{A}}\, e^{i\,\vec{k}_j\,\vec{r}_i}$ where $A = L^2$ is the area of the 2D box.

The total energy of the system, which is the expectation value of the Hamiltonian with respect to the Slater determinant wave function for a system with an elliptical Fermi surface can be calculated through well-established procedures. The total kinetic energy is:

$$T(\alpha) = \frac{\langle\Psi(\alpha)|\hat{T}|\Psi(\alpha)\rangle}{\langle\Psi(\alpha)|\Psi(\alpha)\rangle} = \sum_{\{\vec{k}\in D_{\vec{k}}\}} \left[\frac{\hbar^2 k_x^2}{2\, m_x} + \frac{\hbar^2 k_y^2}{2\, m_y}\right] . \tag{11}$$

The total potential energy is written as:

$$U(\alpha) = \frac{\langle\Psi(\alpha)|\hat{U}|\Psi(\alpha)\rangle}{\langle\Psi(\alpha)|\Psi(\alpha)\rangle} = -\frac{1}{2}\int_{\Omega} d^2r_1 \int_{\Omega} d^2r_2\, |\rho(\alpha, \vec{r}_1, \vec{r}_2)|^2\, v(r_{21}) \ , \tag{12}$$

where Ω is the 2D square box region with area, $A = L^2$ in the thermodynamic limit of $L \to \infty$, $v(r_{21})$ is the pair Coulomb interaction potential and $\rho(\alpha, \vec{r}_1, \vec{r}_2) = \sum_{j=1}^{N} \phi_{\vec{k}_j}(\vec{r}_1)^*\, \phi_{\vec{k}_j}(\vec{r}_2) = \frac{1}{A}\sum_{\{\vec{k}\in D_{\vec{k}}\}} e^{i\,\vec{k}\,(\vec{r}_2 - \vec{r}_1)}$ is the so-called one-particle density matrix. Note the negative sign of the (exchange) potential energy in Eq.(12).

3. Results and conclusions

The kinetic energy per particle denoted as $t(\alpha, \beta)$ is calculated to be:

$$t(\alpha, \beta) = \frac{T(\alpha)}{N} = \frac{1}{4}\frac{\hbar^2 k_F^2}{2\, m_0}\left(\frac{\alpha^2}{\beta} + \frac{\beta}{\alpha^2}\right) \quad ; \quad \beta = \sqrt{\frac{m_x}{m_y}} \quad ; \quad m_0 = \sqrt{m_x\, m_y} \ , \tag{13}$$

where $\beta > 0$ is a dimensionless parameter that measures the mass anisotropy and has a given value. Calculation of the potential energy per particle gives:

$$u(\alpha) = \frac{U(\alpha)}{N} = -\frac{8}{3\pi^2}\, k_F\, k_e\, e^2\, F(\alpha) \ , \tag{14}$$

where

$$F(\alpha) = \frac{1}{\alpha} K\left(m = 1 - \frac{1}{\alpha^4}\right) \ , \tag{15}$$

is an auxiliary function that is expressed in terms of $K(m)$, a complete elliptic integral of the first kind with parameter, m [13]. Note that $F(\alpha = 1) = \pi/2$ represents the largest possible value for this function. This means that the minimum potential energy per particle is always obtained for $\alpha = 1$ and has the value:

$$u_{min} = u(\alpha = 1) = -\frac{4}{3\pi}\, k_F\, k_e\, e^2 \ . \tag{16}$$

The key point that we make here is that, for a given mass anisotropy parameter β, the value of kinetic energy depends on it, but the potential energy does not. This means the value of ellipticity that optimizes the kinetic energy depends on β but that value may not be the optimal one for the potential energy. In fact, it can be proved that kinetic energy has its minimum for the specific value: $\alpha_0 = \sqrt{\beta}$, resulting in:

$$t_{min} = t(\alpha = \sqrt{\beta}, \beta) = \frac{1}{2}\frac{\hbar^2 k_F^2}{2\,m_0} \ . \qquad (17)$$

On the flip side, the potential energy has its minimum for $\alpha = 1$ (irrespective of the value β) and, thus, favors a circular Fermi surface. In short, an elliptical deformation of the Fermi surface for $\beta \neq 1$ decreases the kinetic energy, but increases the potential energy. This means that a sizeable elliptical deformation is not at all obvious or expected to happen in those instances where the dominant energy term is the potential energy. This is precisely the situation that happens when the density of the 2DEG system is relatively low. Density enters in the expressions for both kinetic and potential energy via the Fermi wave number, k_F. After writing the electron number density as $\rho_0 = 1/[\pi\,(r_s\,a_B)^2]$ where r_s is the Wigner-Seitz radius parameter and $a_B = \hbar^2/(k_e\,m_0\,e^2)$ is some "effective" Bohr radius (related to mass, $m_0 = \sqrt{m_x\,m_y}$) one can prove that $(k_F\,r_s\,a_B)^2 = 4$. Therefore, one concludes that kinetic energy ($\propto 1/r_s^2$) and potential energy ($\propto -1/r_s$) dominate, respectively, at high density (small r_s) and low density (large r_s).

Thus, for $\beta \neq 1$, at relatively low density, the isotropic liquid state, $\Psi(\alpha = 1)$ may have a total energy per particle, $\epsilon(\alpha = 1, \beta) = t(\alpha = 1, \beta) + u(\alpha = 1)$ lower than the total energy per particle, $\epsilon(\alpha = \sqrt{\beta}, \beta) = t(\alpha = \sqrt{\beta}, \beta) + u(\alpha = \sqrt{\beta})$ of its anisotropic counterpart, $\Psi(\alpha_0)$. At all other higher densities, a 2DEG with elliptical Fermi surface might be favored although it is very likely that $\Psi(\alpha_0)$ is not necessarily the optimum state with the lowest global energy.

In conclusion, the results obtained shed light on interesting scenarios that may arise in a 2DEG system with mass anisotropy. The competition between opposing tendencies of the kinetic energy and potential energy may lead to the stabilization of liquid phases where the optimal elliptical deformation of the Fermi surface is non-obvious and may depend on the density of the system. It is also possible that very little deformation of the Fermi surface may happen at small density where the potential energy term dominates over the kinetic energy.

Acknowledgments

This research was supported in part by National Science Foundation (NSF) grant no. DMR-2001980.

References

[1] Tanatar B and Ceperley D M 1989 *Phys. Rev.* B **39** 5005
[2] Kwon Y, Ceperley D M and Martin R M 1993 *Phys. Rev.* B **48** 12037
[3] Attaccalite C, Moroni S, Gori-Giorgi P and Bachelet G B 2002 *Phys. Rev. Lett.* **88** 256601
[4] Gori-Giorgi P, Moroni S and Bachelet G B 2004 *Phys. Rev.* B **70** 115102
[5] Ceperley D 1978 *Phys. Rev.* B **18** 3126
[6] Zong F H, Lin C and Ceperley D M 2002 *Phys. Rev.* E **66**, 036703
[7] Gill P M W and Loos P-F 2012 *Theor Chem Acc.* **131** 1069
[8] Dugdale S B 2016 *Phys. Scr.* **91**, 053009
[9] Ciftja O, Livingston V and Thomas E 2017 *Am. J. Phys.* **85** 359
[10] Slater J C 1929 *Phys. Rev.* **34** 1293
[11] Ciftja O, Sutton B and Way A 2013 *AIP Adv.* **3** 052110
[12] Ciftja O 2015 *Physica* B **458** 92
[13] Ciftja O 2017 *Results Phys.* **7** 1674

Journal of Physics: Conference Series **2164** (2022) 012024 doi:10.1088/1742-6596/2164/1/012024

Stability of ordered and disordered phases in the Heisenberg-Kitaev model in a magnetic field

Pedro M Cônsoli[1] **and Eric C Andrade**[2]

[1] Institut für Theoretische Physik and Würzburg-Dresden Cluster of Excellence ct.qmat, Technische Universität Dresden, 01062 Dresden, Germany

[2] Instituto de Física de São Carlos, Universidade de São Paulo, C.P. 369, São Carlos, SP, 13560-970, Brazil

Abstract. The $S = 1/2$ Kitaev honeycomb model has attracted significant attention as an exactly solvable example with a quantum spin liquid ground state. In an properly oriented external magnetic field, the system exhibits chiral Majorana edge modes with an associated quantized thermal Hall conductance, and a distinct spin-disordered phase emerges at intermediate field strengths, below the polarized phase. However, since material realizations of Kitaev magnetism invariably display competing exchange interactions, the stability of these exotic phases with respect to additional couplings is a key issue. Here, we report a 24-site exact diagonalization study of the Heisenberg-Kitaev model in a magnetic field applied in the [001] and [111] directions. By mapping the full phase diagram of the model and contrasting the results to recent nonlinear spin-wave calculations, we show that both methods agree well, thus establishing that quantum corrections substantially modify the classical phase diagram. Furthermore, we find that, in a [111] field, the intermediate-field spin-disordered phase is remarkably stable to Heisenberg interactions and may potentially end in a novel quantum tricritical point.

1. Introduction

Over the past two decades, strong spin-orbit coupling has been recognized as a key ingredient to generate novel states of matter. In the context of magnetic insulators, the interplay of strong spin-orbit coupling and strong electronic correlations provides a route to realize bond-directional interactions between magnetic degrees of freedom. Several instances of this connection arise in Mott insulators constituted of magnetic ions with partially filled $4d$ or $5d$ shells [1–5]. Among such systems, the so-called Kitaev materials have attracted enormous interest as hosts for the Ising-like bond-dependent interactions that are characteristic of Kitaev's honeycomb model [6–13]. The Kitaev model describes a system of spins $S = 1/2$ located at the vertices of a honeycomb lattice and, remarkably, admits an exact solution over its entire parameter space. In the vicinity of its isotropic point, the ground state is a quantum spin liquid (QSL) [14] in which gapless Majorana fermions hop in a static $\mathbb{Z}_2$ background. Interestingly, a weak magnetic field opens a gap in the Majorana spectrum and induces a transition to a nonabelian QSL that exhibits chiral Majorana edge modes [6].

By now, it is well established that Kitaev interactions are strong in the honeycomb iridates [15,16] and α-$RuCl_3$ [17–19], wherein Ir^{4+} and Ru^{3+} ions form effective $j_{\text{eff}} = 1/2$ local magnetic moments distributed in stacked honeycomb planes. However, the observation of long-range magnetic order in these materials at low temperatures indicates that other exchange interactions come into play and prevent the stabilization of the Kitaev QSL. In α-$RuCl_3$, the magnetic order can be suppressed by an in-plane magnetic field [20–22]. For certain field directions, a new phase, appearing between the low-field ordered state and the high-field polarized state, exhibits a half-quantized thermal Hall conductance [23,24] – a unique signature for a gapped topological QSL [25–27]. Evidence supporting this conclusion was also obtained from magnetocaloric effect experiments [28], as well as from thermal expansion and magnetostriction measurements [29].

Numerical studies have uncovered yet another striking feature of the Kitaev model in a magnetic field: For an antiferromagnetic (AF) Kitaev coupling, an intermediate-field spin-disordered phase (IFDP) appears sandwiched between the low-field Kitaev QSL and the polarized phase for a wide range of field orientations [30–36]. An appealing scenario is that this phase is a U(1) QSL with a spinon Fermi surface [30, 34], which could explain quantum oscillations observed in the longitudinal thermal conductivity in α-$RuCl_3$ at low temperatures [37, 38]. However, recent studies suggest that such a U(1) QSL may be unstable at low temperatures [39,40], and further investigations are needed to elucidate the nature of the IFDP.

2. Model and details of the ED calculation

As a minimal model to describe the physics of Kitaev materials, we consider the nearest-neighbor Heisenberg-Kitaev (HK) Hamiltonian [8] in the presence of an external magnetic field

$$\mathcal{H} = J \sum_{\langle ij \rangle} \mathbf{S}_i \cdot \mathbf{S}_j + K \sum_{\langle ij \rangle_\gamma} S_i^\gamma S_j^\gamma - \mathbf{h} \cdot \sum_i \mathbf{S}_i, \qquad \gamma \in \{x, y, z\}, \tag{1}$$

where γ labels the different links on the honeycomb lattice. For convenience, we use the parametrization $J = A\cos\varphi$ and $K = 2A\sin\varphi$, with $A > 0$ being an overall energy scale. The field directions are given in the cubic spin basis $\{\hat{\mathbf{x}}, \hat{\mathbf{y}}, \hat{\mathbf{z}}\}$, so that $\mathbf{h} \parallel [xyz]$ reads $\mathbf{h} \propto x\hat{\mathbf{x}} + y\hat{\mathbf{y}} + z\hat{\mathbf{z}}$. At zero field, the Hamiltonian (1) has four collinear ordered ground states as a function of φ [9]. While one of these is ferromagnetic (FM), the other three present AF spin correlations and are thus amenable to spin canting once h is switched on. However, because the Kitaev term breaks spin-rotational symmetry, the response of the system strongly depends on the direction of the field. In particular, spin canting can only occur uniformly if $\mathbf{h}$ is perpendicular to one of the cubic axes [22,41]. For all other field orientations, the inhomogeneously canted extensions of the zero-field states compete not only with the polarized state, but also with new symmetry-broken phases that allow more favorable canting mechanisms. This gives rise to complex magnetic orderings and metamagnetic transitions at large S [41–44].

Here, we study the Hamiltonian (1) for two representative field directions, [001] and [111], by means of two complementary approaches: exact diagonalization (ED) and nonlinear spin-wave theory (NLSWT), as originally presented in Ref. [44]. Since the latter method does not suffer from finite-size effects, it provides valuable means to investigate the fate of all semiclassical phases. Our ED calculations were performed on a 24-site cluster that preserves the threefold rotational symmetry of the honeycomb lattice and was subjected to periodic boundary conditions. We exploited the translational invariance of Eq. (1) to operate in a subspace of momentum $\mathbf{k}$ of our choice and determined ground-state properties via the Lanczos method. To capture phase transitions, we monitored two quantities: The second derivative of the ground-state energy with respect to a control parameter r and the ground-state fidelity $f(r) = |\langle \psi(r + dr) | \psi(r) \rangle|$, which measures the overlap between the ground states at r and $r + dr$. Sharp and smooth peaks in $f(r)$ are typically associated with first-order and continuous phase transitions, respectively. For the most part, our results were derived by taking $r = h$, in which case we also computed the magnetization per site, m_h.

3. ED results and phase diagrams

We begin by analyzing the results for the AF Heisenberg point, $\varphi = 0$. In Figs. 1(a-c), we see that both the second derivative of the ground-state energy and the fidelity exhibit a

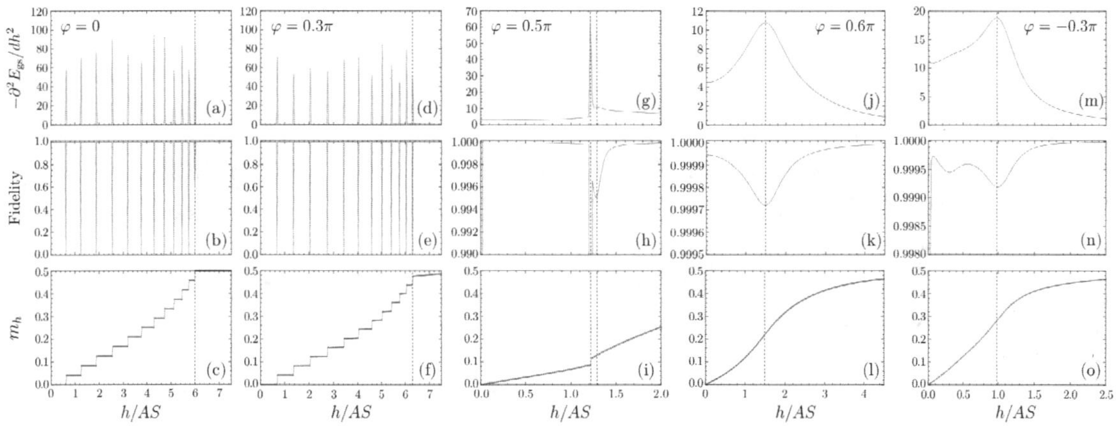

Figure 1. 24-site ED results for the HK model in a magnetic field $\mathbf{h} \parallel [001]$ with (a-c) $\varphi = 0$, (d-f) $\varphi = 0.3\pi$, (g-i) $\varphi = 0.5\pi$, (j-l) $\varphi = 0.6\pi$, and (m-o) $\varphi = -0.3\pi$. A is a global energy scale related to the coupling constants by $J = A\cos\varphi$ and $K = 2A\sin\varphi$ and $S = 1/2$. From top to bottom, we show the second derivative of the ground state energy with respect to the magnetic field, the fidelity, and the magnetization per site in the direction of the magnetic field.

series of sharp peaks as a function of h, whereas the magnetization increases in steps. These features simply reflect the finite size of the system and the fact that the total magnetization is a conserved quantity of Heisenberg Hamiltonians. The peak at $h/AS = 6$ marks the transition to the polarized phase and survives the thermodynamic limit because the positive curvature of $m_h(h)$ necessarily leads to a cusp at the critical field. In Figs. 1(d-f), we observe a similar behavior for $\varphi = 0.3\pi$, but with two important distinctions: (i) The magnetization does not saturate at the last plateau and (ii) the slope of the plateaus now increases with h. Both of these observations are consistent with the fact that the total magnetization ceases to be a conserved quantity of the Hamiltonian when $K \neq 0$. Still, the conclusion that all peaks in Figs. 1(d,e) but the last one are washed out in the thermodynamic limit continues to apply, and we are left with a single continuous phase transition at $h/AS \approx 6.49$.

As one moves further away from $\varphi = 0$, the magnetization plateaus become steeper and less separated from each other, until they begin to merge near the AF Kitaev point. Fig. 1(i) shows that the $\varphi = 0.5\pi$ magnetization curve has a single discontinuity at $h/AS \approx 1.22$, in agreement with Ref. [31]. Yet, as shown in Figs. 1(g,h), the second derivative of ground-state energy and the fidelity both present a smaller peak at $h/AS \approx 1.29$ (the feature close to $h = 0$ is likely to be spurious). If one interprets this as a transition to the polarized phase, then the results suggest the existence of an intermediate-field phase. This is precisely the conclusion reached in a previous ED study [35], wherein the intermediate phase was identified as the IFDP.

Past the AF Kitaev point, we find continuous magnetization curves. This is what happens, for instance, at $\varphi = 0.6\pi$, see Fig. 1(l). Since we know that the system is in the zigzag phase at zero field [9], the single smooth peak in Figs. 1(j,k) must indicate a direct continuous transition to the polarized phase. The situation is not as simple when φ belongs to the domain of the stripy phase. As illustrated in Figs. 1(m,n), besides a well-defined peak at $h/AS \approx 0.98$, two features appear at lower fields. The sharp peak in the fidelity around $h = 0$, in particular, is related to an abrupt growth in the magnitude of the Bragg peaks characterizing the two stripy domains selected by the [001] field. We did not encounter, on the other hand, signs in the static structure factor capable of explaining the intermediate-field peak in Fig. 1(n).

Similar analyses for different φ lead to Fig. 2(a), which compares the ED results with

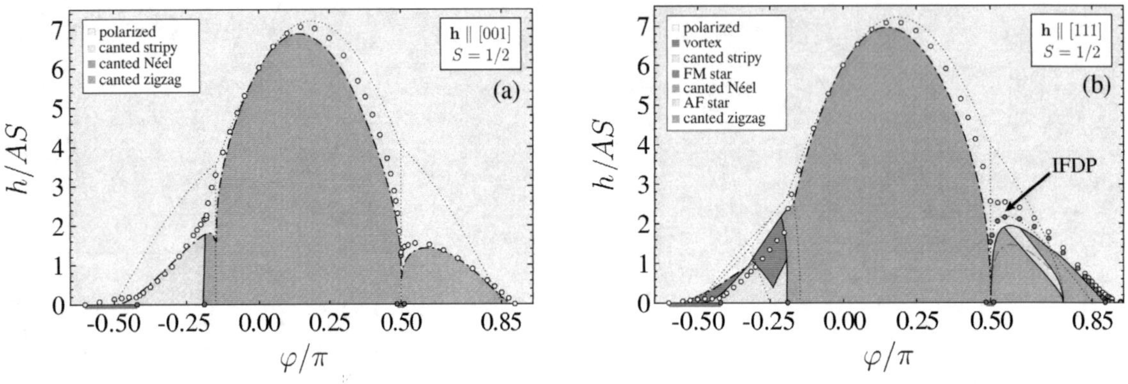

Figure 2. 24-site ED results (dots) superimposed with the $S = 1/2$ phase diagrams of the HK model derived within NLSWT in Ref. [44] for (a) $\mathbf{h} \parallel [001]$ and (b) $\mathbf{h} \parallel [111]$. Yellow, green, and red dots indicate the last field-induced transition at a given φ, the lower boundary of the IFDP, and zero-field results, respectively. Red stripes below the horizontal axes highlight the domain of spin-liquid phases at $h = 0$, and the dotted lines mark the $S \to \infty$ boundaries.

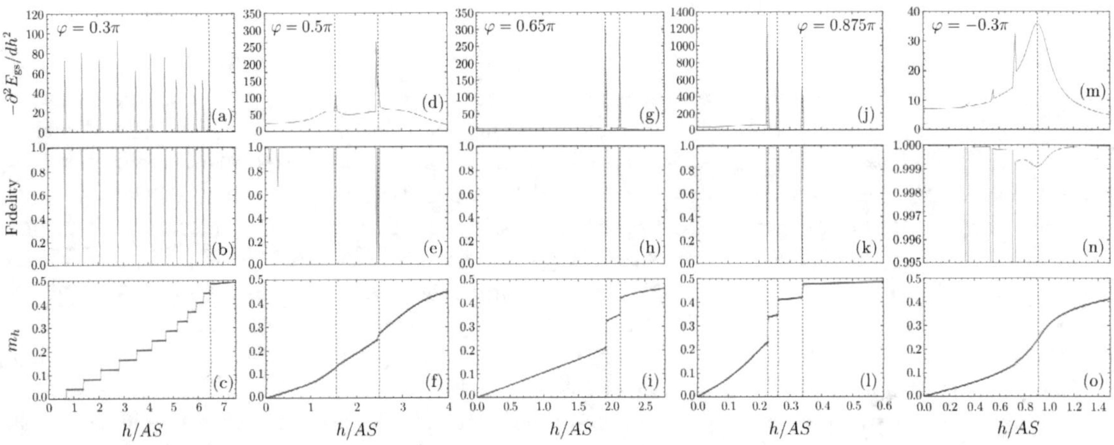

Figure 3. Same as Fig. 1, but for a magnetic field $\mathbf{h} \parallel [111]$. (a-c) $\varphi = 0.3$, (d-f) $\varphi = 0.5\pi$, (g-i) $\varphi = 0.65\pi$, (j-l) $\varphi = 0.875\pi$, and (m-o) $\varphi = -0.3\pi$.

the NLSWT of Ref. [44]. We do not show data points below the critical fields, however, because secondary features such as those in Figs. 1(j-k) intermingle with the signatures of phase transitions and render the precise identification of intermediate-field phase boundaries difficult. Overall, we find that both methods are in good agreement for the ordered phases.

We now discuss the more involved case of $\mathbf{h} \parallel [111]$. For a more in-depth discussion of all semiclassical phases, we refer the reader to Refs. [41, 44]. In Figs. 3(a-c), we see that the data for $\varphi = 0.3\pi$ show no qualitative difference with respect to the case of $\mathbf{h} \parallel [001]$. Accordingly, we consider that a single continuous transition occurs between the canted Néel and the polarized phases at the final peak of the second derivative of the ground state and the fidelity.

On the other hand, striking differences appear close to the AF Kitaev point. Figs. 3(d-f) show two distinct features at $h/AS \approx 1.55$ and 2.48 delimiting a considerably larger intermediate-field regime than that of Figs. 1(g-i). By now, several numerical studies have converged to the conclusion that these peaks represent the boundaries of the IFDP [32–36, 45, 46]. A remarkable property of this phase is its stability to the inclusion of Heisenberg interactions in a [111] field. While Figs. 3(g-i) confirm that the IFDP is present at $\varphi = 0.65\pi$, we find that it also extends past $\varphi = 0.75\pi$, where the $|J| > K$. In fact, Figs. 3(j-l) suggest that it might even be present at $\varphi = 0.875\pi$, near the right end of the zigzag phase. The meaning of the third peak appearing in the second derivative of the ground-state energy and in the fidelity remains unclear to us.

Figs. 3(m-o) display the results for $\varphi = -0.3\pi$. Once again, we observe a series of features below the smooth peak that signals the transition to the polarized phase. Yet, this time, none of the sharp peaks are associated with the selection of a subset of the three stripy domains, since the C_3^* symmetry of the Hamiltonian is preserved. This might imply the existence of intermediate-field ordered phases, such as those presented in Refs. [41, 44]. Our preliminary static structure factor results suggest that the vortex phases are absent, as we do not find their corresponding Bragg peaks, but we cannot rule out a transition to the FM star phase.

After assembling results for several other values of φ, we obtained the phase diagram shown in Fig. 2(b). As in the [001] case, we refrained from determining the locations of phase boundaries below the upper critical fields. However, we emphasize that our data do not show any evidence for the existence of an ordered phase besides the canted zigzag in the interval $\varphi \in [0.5\pi, 0.87\pi]$. This supports the conclusion from Ref. [44] that the AF vortex is completely suppressed for $S = 1/2$, while suggesting that higher-order terms in $1/S$ should account for the disappearance of the remaining slither covered by the AF star. Apart from that, Fig. 2(b) reveals again a good agreement between our ED and the NLSWT results. We also find consistency with other numerical studies [34, 46, 47] that focused on reduced portions of the phase diagram.

One novel feature coming from our ED results is the behavior of the upper critical field near the right end of the zigzag phase. The ED results indicate that the two-peak structure from Figs. 3(g,h) does not give way to a single peak ending at the zero-field transition point $\varphi \approx 0.900\pi$. Instead, Fig. 2(b) suggests that, for $\varphi > 0.900\pi$, a [111] magnetic field could stabilize the IFDP between the zero-field FM state and the high-field polarized phase. While this unlikely scenario is probably linked to finite-size effects, our results raise the question of whether the IFDP actually reaches down to zero field. If so, does it end at a quantum tricritical point shared with the canted zigzag and polarized phases? This intriguing question invites further investigation on the stability of the IFDP and its extent in the phase diagram.

4. Conclusions and outlook

In this paper, we have studied the stability of various ordered phases of the $S = 1/2$ HK model in an external magnetic field. In contrast to spin models with spin-rotational symmetry, where collinear zero-field states typically become simple canted states, the HK model shows, in addition to QSLs, various complex large-unit-cell states in which nonuniform canting may occur.

Our primary goal has been to characterize the effects of quantum fluctuations on the stability of magnetically ordered phases. To this end, we discussed the NLSWT results from Ref. [44], which capture the reduction and/or destruction of ordered phases with larger unit cells and the expansion of the polarized phase. We combined this semiclassical approach with and an ED study on a 24-site cluster. This last step confirmed the main features observed in Ref. [44], while also allowing us to investigate the fate of the spin-liquid phases. The ED results confirm that the Kitaev spin liquid is more stable for $K > 0$ in the presence of a uniform magnetic field along the [111] direction. We also encounter the IFDP in between the low-field QSL and the high-field polarized phase. Our ground-state results do not allow us to discuss its nature, but we find it is stable with respect to the Heisenberg coupling, even for $|J| \gtrsim K$. Interestingly, our results hint at the existence of a new quantum tricritical point, involving the canted zigzag, the IFDP and the [111] polarized phase. If confirmed, this can shed more light on the intriguing nature and stability of this phase and further highlight the richness of the physics contained in the HK model.

Acknowledgments
We acknowledge discussions with, L. Janssen, H.-H. Tu, and M. Vojta. P.M.C. was supported by FAPESP (Brazil), Grants No. 2017/22133-3 and 2019/02099-0, and from the Deutsche Forschungsgemeinschaft through SFB 1143 (Project-id No. 247310070). E.C.A. was supported by CNPq (Brazil) and FAPESP (Brazil).

References
[1] Khaliullin G 2005 *Prog. Theor. Phys. Suppl.* **160** 155
[2] Witczak-Krempa W, Chen G, Kim Y B and Balents L 2014 *Annu. Rev. Condens. Matter Phys.* **5** 57
[3] Rau J G, Lee E K H and Kee H Y 2016 *Annu. Rev. Condens. Matter Phys.* **7** 195–221
[4] Natori W M H, Andrade E C, Miranda E and Pereira R G 2016 *Phys. Rev. Lett.* **117**(1) 017204
[5] Natori W M H, Andrade E C and Pereira R G 2018 *Phys. Rev. B* **98**(19) 195113
[6] Kitaev A 2006 *Ann. Phys. (N.Y.)* **321** 2
[7] Jackeli G and Khaliullin G 2009 *Phys. Rev. Lett.* **102**(1) 017205
[8] Chaloupka J, Jackeli G and Khaliullin G 2010 *Phys. Rev. Lett.* **105**(2) 027204
[9] Chaloupka J, Jackeli G and Khaliullin G 2013 *Phys. Rev. Lett.* **110**(9) 097204
[10] Winter S M *et al. J. Phys.: Condens. Matter* **29** 493002
[11] Trebst S 2017 Kitaev materials (*Preprint* cond-mat.str-el/1701.07056)
[12] Hermanns M, Kimchi I and Knolle J 2018 *Annu. Rev. Condens. Matter Phys.* **9** 17 – 33
[13] Takagi H, Takayama T, Jackeli G, Khaliullin G and Nagler S E 2019 *Nat. Rev. Phys.* **1** 264–280
[14] Savary L and Balents L 2017 *Rep. Prog. Phys.* **80** 016502
[15] Choi S K *et al.* 2012 *Phys. Rev. Lett.* **108**(12) 127204
[16] Singh Y *et al.* 2012 *Phys. Rev. Lett.* **108**(12) 127203
[17] Plumb K W *et al.* 2014 *Phys. Rev. B* **90** 041112(R)
[18] Sears J A *et al.* 2015 *Phys. Rev. B* **91**(14) 144420
[19] Banerjee A *et al.* 2016 *Nature Mat.* **15** 733
[20] Sears J A, Zhao Y, Xu Z, Lynn J W and Kim Y J 2017 *Phys. Rev. B* **95**(18) 180411(R)
[21] Wolter A U B *et al.* 2017 *Phys. Rev. B* **96**(4) 041405(R)
[22] Janssen L and Vojta M 2019 *J. Phys.: Condens. Matter* **31** 423002
[23] Kasahara *et al.* 2018 *Nature* **559** 227–231
[24] Yokoi T *et al.* 2021 *Science* **373** 568 – 572
[25] Vinkler-Aviv Y and Rosch A 2018 *Phys. Rev. X* **8**(3) 031032
[26] Ye M, Halász G B, Savary L and Balents L 2018 *Phys. Rev. Lett.* **121**(14) 147201
[27] Gao Y H, Hickey C, Xiang T, Trebst S and Chen G 2019 *Phys. Rev. Research* **1**(1) 013014
[28] Balz C *et al.* 2019 *Phys. Rev. B* **100**(6) 060405(R)
[29] Gass S *et al.* 2020 *Phys. Rev. B* **101**(24) 245158
[30] Liang S, Jiang M H, Chen W, Li J X and Wang Q H 2018 *Phys. Rev. B* **98**(5) 054433
[31] Nasu J, Kato Y, Kamiya Y and Motome Y 2018 *Phys. Rev. B* **98**(6) 060416(R)
[32] Zhu Z, Kimchi I, Sheng D N and Fu L 2018 *Phys. Rev. B* **97**(24) 241110(R)
[33] Gohlke M, Moessner R and Pollmann F 2018 *Phys. Rev. B* **98**(1) 014418
[34] Hickey C and Trebst S 2019 *Nat. Commun.* **10** 530
[35] Patel N D and Trivedi N 2019 *PNAS* **116** 12199–12203 ISSN 0027-8424
[36] Hickey C, Gohlke M, Berke C and Trebst S 2021 *Phys. Rev. B* **103**(6) 064417
[37] Czajka P *et al.* 2021 *Nat. Phys.* **17** 915–919
[38] Villadiego I S 2021 Pseudoscalar U(1) spin liquids in α-$RuCl_3$ (*Preprint* 2106.05290)
[39] Jin H K, Tu H H and Zhou Y 2021 *Phys. Rev. B* **104**(2) L020409
[40] Krüger W G F and Janssen L 2021 Nesting instability of gapless U(1) spin liquids with spinon Fermi pockets in two dimensions (*Preprint* cond-mat.str-el/2107.00661)
[41] Janssen L, Andrade E C and Vojta M 2016 *Phys. Rev. Lett.* **117**(27) 277202
[42] Janssen L, Andrade E C and Vojta M 2017 *Phys. Rev. B* **96**(6) 064430
[43] Chern G W, Sizyuk Y, Price C and Perkins N B 2017 *Phys. Rev. B* **95**(14) 144427
[44] Cônsoli P M, Janssen L, Vojta M and Andrade E C 2020 *Phys. Rev. B* **102**(15) 155134
[45] Ronquillo D C, Vengal A and Trivedi N 2019 *Phys. Rev. B* **99**(14) 140413
[46] Jiang Y F, Devereaux T P and Jiang H C 2019 *Phys. Rev. B* **100**(16) 165123
[47] Jiang H C, Gu Z C, Qi X L and Trebst S 2011 *Phys. Rev. B* **83**(24) 245104

Magnetic properties of rare-earth zigzag chain systems $RAgSe_2$ (R = Ho, Er, Tm, and Yb)

S. Mizutani[1], Y. Ohmagari[1], T. Onimaru[1], Y. Shimura[1], R. Yamamoto[1], K. Umeo[2], and T. Takabatake[1]

[1]Department of Quantum Matter, Graduate School of Advanced Science and Engineering, Hiroshima University, Higashi-Hiroshima, Hiroshima 739-8530, Japan
[2]Department of Low Temperature Experiment, Integrated Experimental Support/Research Division, N-BARD, Hiroshima University, Higashi-Hiroshima, Hiroshima 739-8526, Japan

E-mail: m210279@hiroshima-u.ac.jp

Abstract. In an antiferromagnetic zigzag chain, competition between the nearest and next-nearest neighbor interactions could give rise to magnetic frustration. Magnetic semiconductors $RAgSe_2$ (R = Ho, Er, Tm, and Yb) crystallize in the $ErAgSe_2$-type orthorhombic structure, where the R ions form a zigzag chain along the orthorhombic a-axis. The magnetic susceptibility data for all the samples follow the Curie–Weiss law between 40 and 300 K. The values of the effective magnetic moment $\mu_{\rm eff}$ are close to those expected for the free R^{3+} ions. Negative values of the paramagnetic Curie temperature $\theta_{\rm p}$ indicate antiferromagnetic interactions. For R = Ho and Tm, the specific heat $C(T)$ data exhibit no anomaly down to 0.4 K, which is ascribed to the nonmagnetic singlet ground states under the crystalline electric fields. On the other hand, for R = Er, $C(T)$ shows peaks at T_1 = 1.3 K and T_2 = 0.9 K, indicating successive antiferromagnetic transitions. For R = Yb, $C(T)$ shows a lambda-type anomaly at $T_{\rm m}$ = 1.8 K. The magnetic entropy at $T_{\rm m}$ is only 30% of $R\ln 2$ expected for the ground state doublet, suggesting magnetic fluctuations above $T_{\rm m}$.

1. Introduction

An antiferromagnetic spin-1/2 zigzag chain is equivalent to a one-dimensional spin-1/2 chain with a nearest neighbor interaction J_1 and a next-nearest neighbor interaction J_2. In case J_2 strongly competes with J_1, a long-range antiferromagnetic order could be suppressed to give rise to magnetic frustration. Theoretical simulations with the antiferromagnetic spin-1/2 zigzag chain model revealed the manifestation of a variety of nontrivial quantum ground states in magnetic fields such as spin density wave, vector chirality, and 1/3 magnetization plateau [1, 2].

Recently, a magnetic semiconductor $YbCuS_2$ with an effective spin-1/2 Yb zigzag chain was found to exhibit a peculiar antiferromagnetic transition and magnetic-field induced phases [3]. $YbCuS_2$ crystallizes in an orthorhombic structure with the space group of $P2_12_12_1$, where the Yb ions located in deformed sulfur-octahedrons form a zigzag chain along the a-axis [4]. The magnetic specific heat $C_{\rm m}$ exhibits a sharp peak at $T_{\rm m}$ = 0.95 K, where the magnetic entropy is 20% of $R\ln 2$ expected for the Kramers doublet ground state [3]. $T_{\rm m}$ is robust against the magnetic fields for $B \leq 3$ T, above which the 1/3 magnetization plateau appears, being a characteristic of the up-up-down spin configuration. The B–T phase diagram suggests nontrivial phases arising from the magnetic frustration in the Yb zigzag chain.

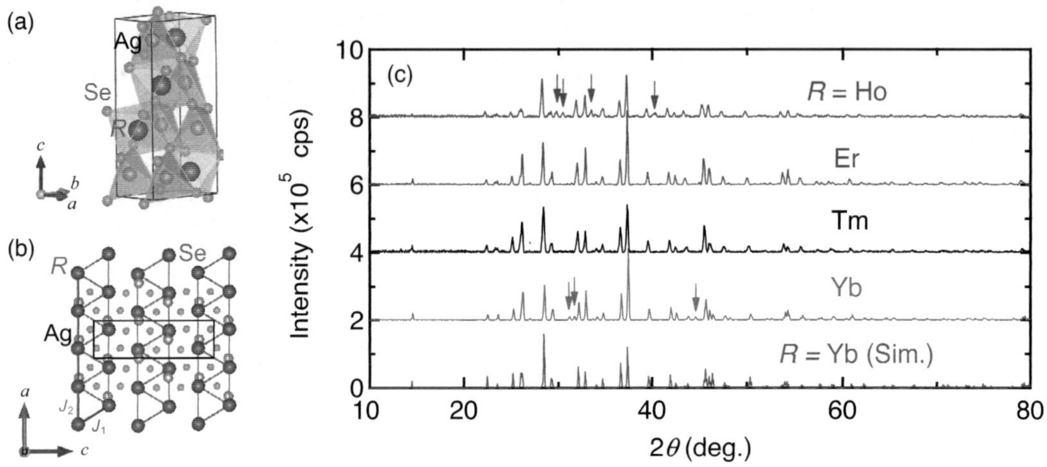

Figure 1. (a) Orthorhombic $ErAgSe_2$-type crystal structure of $RAgSe_2$ [5]. (b) R zigzag chains along the a-axis with magnetic exchange interactions J_1 and J_2. (c) Powder X-ray diffraction patterns of $RAgSe_2$ (R = Ho, Er, Tm, and Yb) and a simulated one for R = Yb with the orthorhombic structure. The arrows indicate impurity peaks.

In $RAgSe_2$ (R = Ho, Er, Tm, and Yb), the R ions are located in slightly deformed sulfur octahedrons [5] and form a zigzag chain along the a-axis as shown in Fig. 1(a) and 1(b). The magnetic susceptibility and the magnetization up to 14 T were previously reported [6]. The magnetic susceptibility data follow the Curie–Weiss law of the trivalent state of the R ions. The paramagnetic Curie temperatures are negative, indicating antiferromagnetic interactions. For $T > 4.2$ K, no magnetic transition was observed. In the present work, we synthesized polycrystalline samples of these compounds and measured the magnetization M for $T \geq 1.8$ K and the specific heat C for $T \geq 0.4$ K.

2. Experimental

Polycrystalline samples of $RAgSe_2$ (R = Ho, Er, Tm, and Yb) were synthesized by melting the constituent elements in evacuated quartz ampoules. The samples were characterized by means of the powder X-ray diffraction analysis and electron-probe microanalysis. Fig. 1(c) shows the X-ray diffraction patterns and a simulation for R = Yb with the $ErAgSe_2$-type orthorhombic structure [4]. Most peaks can be indexed by the orthorhombic structure [4], while impurity peaks were observed for R = Tm and Yb as indicated by the arrows.

The magnetization was measured between 1.8 K and 300 K in magnetic fields of $B \leq 5$ T by using a commercial SQUID magnetometer (Quantum Design, MPMS). The specific heat was measured by the thermal relaxation method using a commercial calorimeter (Quantum Design, PPMS) for $0.4 \leq T \leq 300$ K and $B \leq 14$ T.

3. Results

Figure 2(a) shows the temperature dependence of the magnetic susceptibility M/B of $RAgSe_2$ for R = Ho, Er, Tm (left-hand scale), and Yb (right-hand scale). The M/B data for R = Ho and Er at $T = 1.8$ K are much larger than those for R = Tm and Yb. As shown with the open arrow for R = Yb in the inset, a maximum appears at around 3 K. Fig. 2(b) shows the inverse magnetic susceptibility B/M. All the data follow the Curie–Weiss law for $40 \leq T \leq 300$ K. As shown in Table 1, the values of the effective magnetic moment μ_{eff} are close to those calculated

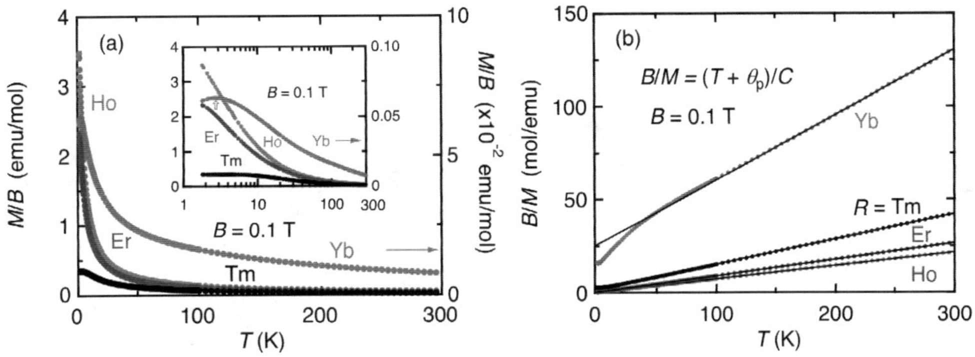

Figure 2. Temperature dependences of (a) the magnetic susceptibility $M(T)/B$ of $R\mathrm{AgSe_2}$ for R = Ho, Er, Tm (left-hand scale) and Yb (right-hand scale) in a magnetic field of B = 0.1 T. The inset shows the $M(T)/B$ data plotted against a logarithmic scale of the temperature. (b) The inverse magnetic susceptibility B/M. The solid lines are the Curie–Weiss fits.

Table 1. Effective magnetic moment $\mu_{\rm eff}$, the calculated value of $\mu_{\rm cal}$ for a free R^{3+} ion, and paramagnetic Curie temperature $\theta_{\rm p}$ of $R\mathrm{AgSe_2}$ for R = Ho, Er, Tm, and Yb.

	R = Ho	Er	Tm	Yb
$\mu_{\rm eff}$ ($\mu_{\rm B}$/f.u.)	10.59	9.54	7.53	4.77
$\mu_{\rm cal}$ ($\mu_{\rm B}/R^{3+}$)	10.58	9.59	7.55	4.54
$\theta_{\rm p}$ (K)	−3.6	−2.8	−15.0	−71.9

for the free R^{3+} ions. The values of the paramagnetic Curie temperature $\theta_{\rm p}$ are negative, which indicate antiferromagnetic interactions between the magnetic moments of the R^{3+} ions. It is noted that the absolute value of $\theta_{\rm p}$ for R = Yb is one order of magnitude larger than those for R = Ho and Er. As for the possible reason, the Kondo effect is excluded because of the semiconducting nature.

Figure 3(a) shows the specific heat of $R\mathrm{AgSe_2}$ (R = Ho, Er, Tm, and Yb). For R = Ho and Tm, the $C(T)$ data exhibit no anomaly down to 0.4 K. Figure 3(b) shows the magnetic specific heat $C_{\rm m}(T)$ which was evaluated by subtracting the $C(T)$ data for R = Lu as a phonon contribution from the total $C(T)$ data. $C_{\rm m}(T)$ for R = Ho and Tm are fitted by using a singlet-singlet two-level model. Thereby, we adopted the energy separation of 3.3 and 25 K, respectively, from the crystalline electric field (CEF) nonmagnetic singlet to the excited singlet.

On the other hand, for R = Er, two peaks appear at T_1 = 1.3 K and T_2 = 0.9 K. The inset shows the B–T phase diagram obtained from the $C(T)$ measurements. As shown in the inset of Fig. 3(a), the two peaks shift to lower temperatures with increasing B, indicating successive antiferromagnetic transitions. For R = Yb, a lambda-type peak appears at $T_{\rm m}$ = 1.8 K. The magnetic entropy at $T_{\rm m}$ is only 30% of $R\ln 2$ expected for the Kramers doublet ground states, indicating magnetic fluctuations for $T > T_{\rm m}$. Therefore, the reduction of the magnetic entropy is probably ascribed to magnetic frustration in the Yb zigzag chain as was observed in $\mathrm{YbCuS_2}$ [3].

4. Conclusion

We have prepared polycrystalline samples of magnetic semiconductors $R\mathrm{AgSe_2}$ (R = Ho, Er, Tm, and Yb) and measured the magnetic susceptibility for $T \geq 1.8$ K and the specific heat for

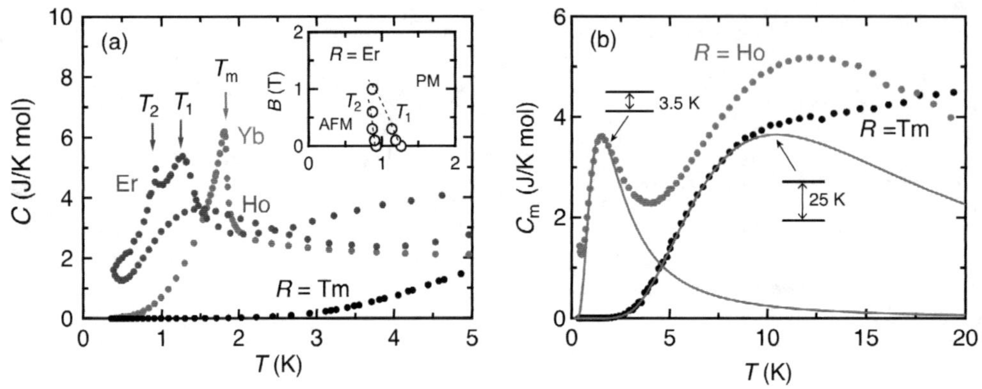

Figure 3. (a) Temperature dependence of the specific heat $C(T)$ of RAgSe$_2$ for R = Ho, Er, Tm, and Yb. The inset shows the B–T phase diagram for R = Er. (b) The magnetic specific heat $C_m(T)$ for R = Ho and Tm. The solid lines are calculations with a singlet-singlet two-level model with energy gaps of 3.5 and 25 K for R = Ho and Tm, respectively.

$T \geq 0.4$ K. The Curie–Weiss fits to the magnetic susceptibility yielded values of the effective magnetic moment $\mu_{\rm eff}$ which are close to those expected for the free R^{3+} ions. The negative values of θ_p indicate antiferromagnetic interactions between the magnetic moments of the R^{3+} ions. For R = Ho and Tm, no phase transition was observed down to 0.4 K because of the nonmagnetic crystalline electric field (CEF) singlet ground state. On the other hand, magnetic transitions were observed at $T_1 = 1.3$ K and $T_2 = 0.9$ K for R = Er and $T_m = 1.8$ K for R = Yb.

Acknowledgments

The authors thank Y. Shibata for the electron-probe microanalysis carried out at N-BARD, Hiroshima University. The measurements of magnetization and specific heat were performed at N-BARD, Hiroshima University. We acknowledge support from the Center for Emergent Condensed-Matter Physics (ECMP), Hiroshima University. This work was financially supported by Grants in Aid from MEXT/JSPS of Japan, Grants No. JP18H01182.

References

[1] Okunishi K and Tonegawa T 2003 *J. Phys. Soc. Jpn.* **72** 479
[2] Hikihara T, Momoi T, Furusaki A and Kawamura H 2010 *Phys. Rev. B* **81** 224433
[3] Ohmagari Y, Onimaru T, Yamane Y, Shimura Y, Umeo K, Takabatake T, Sato H, Kikugawa N, Terashima T, Hirose T and Uji S 2020 *J. Phys. Soc. Jpn.* **89** 093701
[4] Gulay L D, Hucha M, Olekseyuk I, Stepien-Damm J and Pietraszko A 2007 *J. Alloys Compd.* **428** 139
[5] Julien-Pouzol M and Laruelle P 1977 *Acta Cryst.* **B33** 1510
[6] Duczmal M and Pokrzywnicki S 2001 *J. Alloys Compd.* **323-324** 513

Magnetic and vibrational properties of the covalent chain antiferromagnet $RbFeS_2$

A Kiiamov[1], Z Seidov[2,3], D Croitori[4], V Tsurkan[3,4], H-A Krug von Nidda[3], A Günther[3], L Tagirov[1,5], and D Tayurskii [1]

[1]Institute of Physics, Kazan Federal University, 420008 Kazan, Russia
[2]Institute of Physics, Azerbaijan National Academy of Sciences, AZ-1143 Baku, Azerbaijan
[3]Experimental Physics V, Center for Electronic Correlations and Magnetism, Institute of Physics, University of Augsburg, D-86135 Augsburg, Germany
[4]Institute of Applied Physics, MD-20208 Chisinau, Moldova
[5]Zavoisky Physical-Technical Institute, FRC Kazan Scientific Center of RAS, 420029 Kazan, Russia

AiratPhD@gmail.com

Ternary rubidium-iron sulfide, $RbFeS_2$, belongs to a family of quasi-one-dimensional compounds with the general chemical composition $AFeCh_2$ (where A – K, Rb, Cs, Tl; Ch – S, Se). Understanding the magnetic properties of these compounds is a challenge. The controversy concerning the spin-state of the iron ion needs to be resolved to build the proper model of magnetism. Single crystals of $RbFeS_2$ were grown and characterized by powder x-ray diffraction. QD MPMS-5 SQUID magnetometry was used to measure the magnetic susceptibility, and specific heat was measured utilizing QD PPMS-9 setup. Above the transition to three-dimensional antiferromagnetic order at the Néel temperature of T_N = 188 K, the susceptibility exhibits unusual quasi-linear increase up to the highest measured temperature of 500 K. The specific heat was measured in the temperature range 1.8 – 300 K. *Ab initio* phonon dispersion and density-of-states calculations were performed by means of density functional theory (DFT), and the calculated lattice specific heat was subtracted from the measured one giving the magnetic contribution to the specific heat. Our results suggest that the features of the magnetic specific heat are general for the whole family of the covalent chain ternary iron chalcogenides of the $AFeCh_2$ structure and indicate an intermediate $S = 3/2$ spin state of the iron ion.

1. Introduction

Ternary rubidium-iron sulfide, $RbFeS_2$, is a representative of a family of quasi-one-dimensional compounds with the chemical composition $AFeCh2$ (where A – K, Rb, Cs, Tl; Ch – S, Se) [1]. Understanding the magnetic properties of these compounds meets with a controversy concerning the spin-state of the iron ion important for building the proper model of magnetism [3]. $RbFeS_2$ consists of linear chains of edge-sharing [FeS4] tetrahedra along the crystallographic c direction, separated by Rb atoms (see Fig. 1a). These chains are cross-linked by alkali atoms to form a three-dimensional structure. Concerning the question about the relation of superconductivity and magnetism in iron-based superconductors, these linear chain compounds serve as a useful model system, because the small Fe-Fe intra-chain separation (2.71 Å [2]) gives rise to strong covalence effects and is expected to promote spin reduction and charge-carrier delocalization on the verge to 1D metallic behavior [2].

In the present study, we focus on the magnetic properties of $RbFeS_2$. The aim of the present report is to obtain the temperature dependence of the specific heat of the magnetic subsystem of the compound under study. The temperature behavior of the magnetic specific heat can shed light on the magnetic intra-chain and inter-chain interaction between iron ions in $RbFeS_2$. While the magnetic interactions in $RbFeS_2$ are essential to build the correct model of the magnetic subsystem of $RbFeS_2$. We assume that the total specific heat originates from two contributions, a lattice specific heat due to acoustic and optical phonon, and a magnetic one determined by the thermal population of excited magnetic states. We estimate the magnetic specific heat as a difference between the total experimentally measured heat capacity and the calculated lattice specific heat. The lattice specific heat we use in the present study was obtained within the quasi-harmonic approach by using *ab initio* calculated phonon density of states obtained in our previous study [4].

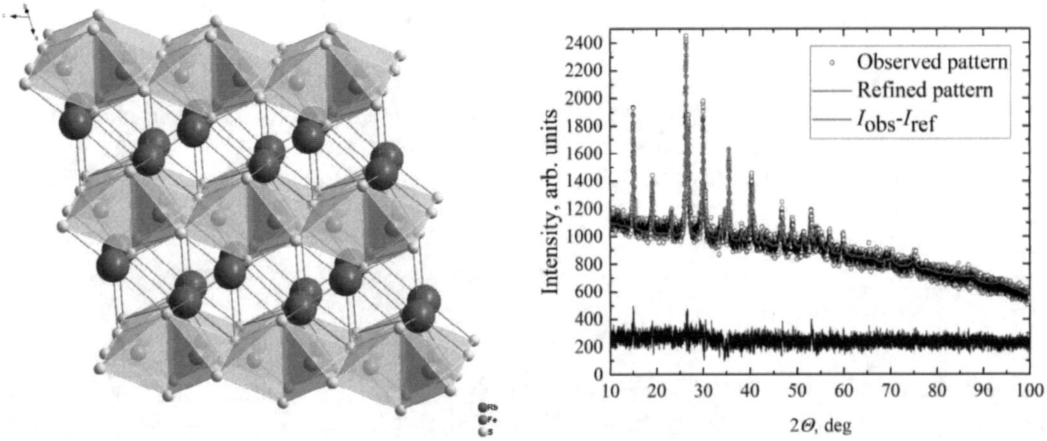

FIG. 1. *Left(a):* crystal structure of RbFeS2. The FeS4 tetrahedra, with Fe drawn as the green sphere in the center and S as the yellow sphere at the corners, are highlighted in transparent green color. Large blue spheres denote Rb ions. *Right(b):* Powder-diffraction pattern of $RbFeS_2$. The empty circles represent the measured intensities. The red solid line shows the refined pattern. The difference pattern $I_{obs.}-I_{cal}$ is indicated by the solid blue line.

2. Sample preparation and methods

Single crystals of $RbFeS_2$ were grown by the Bridgman method. The structural details of the crystals were investigated by conventional x-ray diffraction on powdered single crystals at room temperature using a STOE STADI P diffractometer with CuKα radiation. The data were analyzed by standard Rietveld refinement using the program FULLPROF [5] (Fig. 1b).

We could not detect any impurity phases above the background. The structural analysis confirmed the C2/c monoclinic structure for $RbFeS_2$ with lattice parameters $a = 7.234$ Å, $b = 11.739$ Å, $c = 5.441$ Å, and $\beta = 112.005°$[6]. The specific heat was measured by a relaxation method using a PPMS (Quantum Design) in the temperature range 1.8-300 K indicating an anomaly is at $T_N = 188$ K evidencing long-range AFM order[7].

3. Magnetic specific heat

We assume that the total specific heat consists of two contributions, a lattice specific heat due to acoustic and optical phonon, and a magnetic one determined by the thermal population of excited magnetic states. We estimate the magnetic specific heat as a difference between the total experimental heat capacity and the calculated vibrational part. The phonon density of states enables to calculate directly the lattice contribution to the specific heat by using the harmonic approximation [8]. In the harmonic approximation the lattice heat capacity can be determined as follows [9]:

$$C_V(T) = DNk_B \int \left[\frac{\hbar\omega / 2k_B T}{\sinh\left(\hbar\omega / 2k_B T\right)} \right]^2 g_T(\omega) d\omega,$$

where D is the number of degrees of freedom in the unit cell (3 in our case), and $g_T(\omega)$ is the total phonon density of states. In the present study, we use *ab initio* calculated phonon density of states [4]. We obtained the magnetic specific heat as a difference between the experimentally measured total specific heat and the calculated lattice contribution. The magnetic specific heat is presented in Fig. 2.

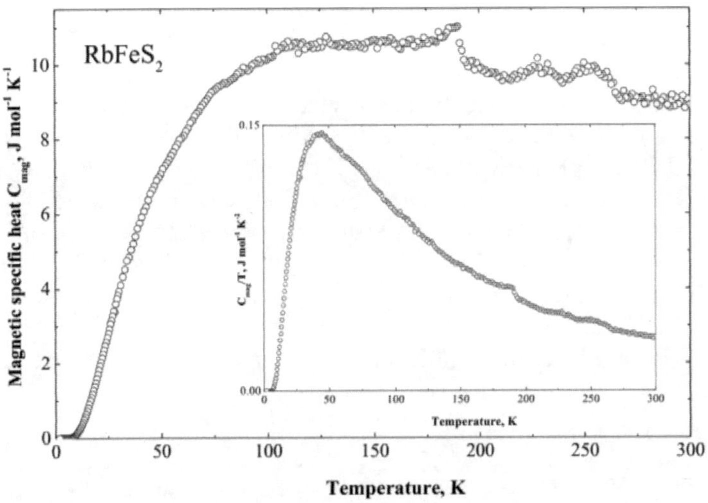

FIG. 2. The temperature dependence of the magnetic heat capacity of $RbFeS_2$ was obtained as a difference between the experimentally measured specific heat and the calculated lattice contribution. Inset represents the magnetic specific heat drawn as *C*(*T*)/*T*.

Despite the fact of the structural similarity of the $RbFeS_2$ with a related RbFeSe2 compound, the temperature dependencies of their magnetic specific heats are not similar to each other. While magnetic specific heat of $RbFeSe_2$ shows complicated temperature dependence which could not be described within existing models of spin chains, the temperature behavior of magnetic specific heat of $RbFeS_2$ corresponds to the coupled dimer model [10]. Both compounds have a similar structure, the linear dimensions are only the feature that is different [2]. In the $RbFeSe_2$, $d_{Fe\text{-}Fe}$ distance between iron ions is 2.85 Å, while in $RbFeS_2$ Å it is 2.71 [2]. It is notable that a 3-4 percent difference in the iron ions interaction range dramatically changes the magnetic properties of the compound. It should be noted, that the current results should be considered preliminary because the used *ab initio* phonon density of states was calculated within the density functional theory without taking into account DFT+*U* approach [4]. It could be necessary for the compounds with *d*-electrons and electronic delocalization.

4. Conclusion

To summarize, the crystalline $RbFeS_2$ samples were grown by the Bridgman method. The powder X-ray diffraction pattern was analyzed by the standard Rietveld refinement using the FULLPROF program [5]. Crystal structure parameters were calculated using the Rietveld refinement technique. The total specific heat was measured by a relaxation method using a PPMS (Quantum Design) in the temperature range 1.8-300 K. The lattice specific heat we use in the present study was obtained within the quasi-harmonic approach by using the *ab initio* calculated phonon density of states realized in our previous study. The magnetic specific heat was obtained as a difference between the experimentally measured total specific heat and the calculated lattice contribution. The temperature dependence of the magnetic specific heat of $RbFeS_2$ shows behavior similar to the coupled dimer spin chains.

5. Acknowledgment

This work was supported by the joint RFBR-DFG project No 19-52-12068 (RFBR) and No KR 2254/3-1 (DFG).

References

[1] M.R. Harrison and M.G. Francesconi, *Coord. Chem. Rev.* 255, 451 (2011).
[2] Z. Seidov, H.-A. Krug von Nidda, V. Tsurkan, *et al.*, *Phys. Rev. B* 94, 134414 (2016).
[3] A.G. Kiiamov, Y. V. Lysogorskiy, F.G. Vagizov, et al., *Phys. Rev. B* 98, 214411 (2018).
[4] A Kiiamov, YLysogorskiy, Z Seidov, et al, *AIP Conference Proceedings*, 2041, 1 (2018).
[5] Rodriguez-Carvajal *Physica* B 1993 **192** 55.
[6] W. Bronger, A. Kyas, and P. Muller, *Journal of Solid State Chemistry* 70, 262–270 (1987).
[7] W. Bronger, A. Kyus, and P. Müller, J. Solid State Chem. **70**, 262 (1987).
[8] C. Kittel, Introduction to Solid State Physics (Wiley, New York, 1996, 7-th Edition), Ch. 4-5.
[9] J.M. Ziman, Electrons and Phonons, the International Series of Monographs on Physics (Clarendon Press, Oxford, 1960), Ch. 1.
[10] R.M. Eremina, T.P. Gavrilova, A. Gunther, et al, *Eur. Phys. J. B* 84, 391–395 (2011).

Journal of Physics: Conference Series **2164** (2022) 012027 doi:10.1088/1742-6596/2164/1/012027

Impurity-Robust Bulk Gapless Excitation in the Yb-Based Zigzag Chain Compound $YbCuS_2$

Fumiya Hori[1], Katsuki Kinjo[1], Shunksaku Kitagawa[1], Kenji Ishida[1], Yudai Ohmagari[2], and Takahiro Onimaru[2]

[1]Depertment of Physics, Kyoto University, Kyoto 606-8502, Japan
[2]Department of Quantum Matter, Graduate School of Advanced Science and Engineering, Hiroshima University, Higashihiroshima 739-8530, Japan

E-mail: hori.fumiya.36s@st.kyoto-u.ac.jp

Abstract. We have performed ^{63}Cu-nuclear quadrupole resonance (NQR) measurements using a lump sample of the Yb zigzag-chain compound $YbCuS_2$ with a small surface area to investigate the sample dependence of low-temperature magnetic properties in $YbCuS_2$ by comparing with the previous study with different powdered sample. The line width of NQR signals in the present lump sample is larger than that in the previous powdered sample. In addition, the transition temperature $T_N \sim 0.92$ K in the present lump sample is lower than that in the previous powdered sample (~ 0.95 K). These results suggest that the quality of the present lump sample is worse than that of the previous powdered sample. However, the T-linear behavior of the nuclear spin-lattice relaxation rate $1/T_1$ was observed below 0.5 K and the value of $1/T_1T$ in both samples is almost the same even though the sample quality and sample geometry are different. This suggests that T-linear behavior in $1/T_1$ arises from the impurity-robust bulk gapless excitation inherent in $YbCuS_2$ rather than from sample issues such as the sample quality or geometry.

1. Introduction

Rare-earth-based frustrated spin systems have been attracted much attention [1–5]. This is because the strong spin-orbit coupling and crystalline electric field associated with the rare-earth ion lead to anisotropic exchange interactions, which may induce strong quantum fluctuations and host novel quantum spin states.

$YbCuS_2$ is a magnetic semiconductor with Yb zigzag chains. This compound has an orthorhombic structure with the space group $P2_12_12_1$, and Yb ions form the zigzag chains along the a-axis [6]. At zero magnetic field $H = 0$, the magnetic specific heat shows a peak at $T_N \sim 0.95$ K [7,8]. In the previous study, we performed $^{63/65}$Cu nuclear magnetic resonance and nuclear quadrupole resonance (NQR) measurements on a powdered sample of $YbCuS_2$ [9] and revealed the occurrence of the antiferromagnetic (AFM) phase transition at T_N. In addition, the nuclear spin-lattice relaxation rate $1/T_1$ was proportional to T below 0.5 K, which is generally observed in metals. The value of $1/T_1T$ below 0.5 K in $YbCuS_2$ is by more than one order of magnitude larger than that in Cu metals [10]. This indicates the presence of the novel gapless spin excitation at low temperatures in $YbCuS_2$ with the magnetic frustration based on the Yb zigzag chain.

In this study, to investigate the origin and characteristics of the peculiar T-linear behavior of $1/T_1$ in $YbCuS_2$, we performed the ^{63}Cu-NQR measurements by changing the following two conditions from the previous study [9]. First, by using different quality of a sample as the previous study, we investigated the sample quality dependence of the low-temperature magnetic excitations. Second, by using a lump sample with a small surface area, which is different sample geometry from the previous powdered sample, we checked whether the low-temperature magnetic excitation comes from the surface or the bulk. We observed the same T-linear behavior of $1/T_1$ below 0.5 K and we confirmed the value of $1/T_1T$ in both samples is almost the same even though the sample quality and geometry were different. This suggests that T-linear behavior of $1/T_1$ arises from the impurity-robust bulk gapless excitation inherent in $YbCuS_2$ rather than from extrinsic sample issues such as the sample quality or sample geometry.

2. Experiments

Polycrystalline samples of $YbCuS_2$ were synthesized by the melt-growth method [8]. ^{63}Cu-NQR measurements were performed on a cuboid lump sample of $6 \times 6 \times 5$ mm^3 size. A conventional spin-echo technique was used for the ^{63}Cu-NQR measurements (^{63}Cu nucleus have nuclear spin $I = 3/2$ with nuclear gyromagnetic ratios of $^{63}\gamma/2\pi = 11.289$ MHz/T, and there are nuclear quadrupolar interactions). A ^{3}He-^{4}He dilution refrigerator was used down to 0.075 K. The NQR spectra were obtained by the frequency-swept method without an external field. $1/T_1$ was evaluated by fitting the relaxation curve of the nuclear magnetization after the saturation to a theoretical function for the nuclear spin $I = 3/2$. While $1/T_1$ was determined by a single relaxation component down to T_N, the whole of the relaxation curve cannot be fitted to the single relaxation component below T_N. Thus, we picked up the slowest components below T_N in the this study as well as the previous study [9].

3. Results and Discussion

Figure 1(a) shows the frequency-swept ^{63}Cu-NQR spectra measured at 4.2 K (upper panel) and 0.3 K (lower panel). Similarly to the previous study [9], a single ^{63}Cu-NQR signal was observed above T_N, and the NQR peak splits into multi peaks due to the internal magnetic fields arising from the AFM ordered moments below T_N. The NQR-peak positions in the present lump sample are almost the same with those in the previous powdered samples. On the other hand, the line width of NQR signals in the present lump sample is larger than that of the previous powdered sample, suggesting that the quality of the present lump sample is worse than that of the previous powdered sample.

The difference of the sample quality in the present sample and previous one was also confirmed from the comparison of the transition temperatures determined by the decrease in the NQR peak intensity. Figure 1(b) shows the temperature dependence of the products of the NQR peak intensity and the temperature $IT(T)$ at $f = 9.3$ MHz. We define T_N as the temperature when $IT(T)$ decreases rapidly with decreasing temperature. In the present lump sample, $IT(T)$ decreases rapidly at $T_N \sim 0.92$ K, which is slightly lower than that in the previous powdered sample (~ 0.95 K).

In contrast to the NQR spectra and T_N, the behaviors of the nuclear spin-lattice relaxation rate $1/T_1$ were almost the same in the previous powdered sample and the present lump sample as shown in Fig. 2. Due to the difference of T_N, the value of $1/T_1$ in the present lump sample is slightly larger than that in the previous powdered sample for 0.5 K $\leq T \leq T_N$. On the other hand, T-linear behavior of $1/T_1$ was observed below T_N and the value of $1/T_1T$ at lowest temperatures is almost the same in both samples, suggesting that T-linear behavior of $1/T_1$ is independent of the sample quality and geometry.

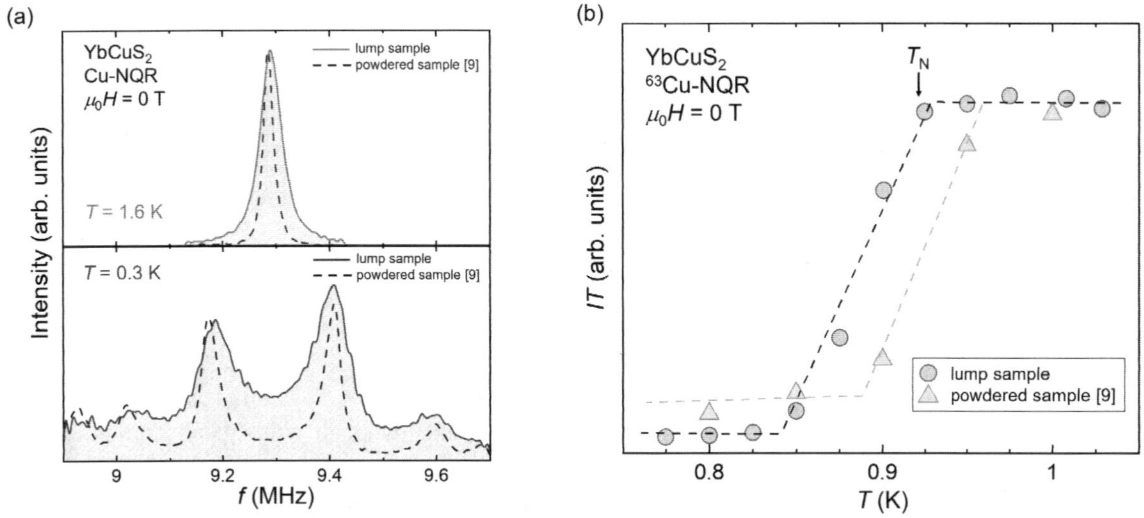

Figure 1. (Color online) (a) ^{63}Cu-NQR spectra obtained by the frequency-swept method at 4.2 K (upper panel) and 0.3 K(lower panel): the (red and blue) solid lines and (black) dashed line represents ^{63}Cu-NQR spectra in present lump sample and the previous powdered sample, respectively. (b) The temperature dependence of the products of the NQR peak intensity and the temperature $IT(T)$: the circles and triangles denote $IT(T)$ in the present lump sample and the previous powdered sample, respectively.

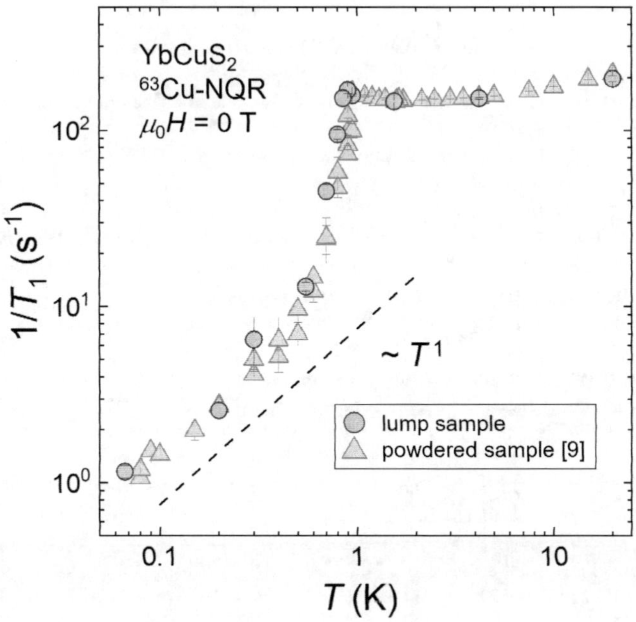

Figure 2. Temperature dependence of the ^{63}Cu-NQR nuclear spin-lattice relaxation rates $1/T_1$ in $YbCuS_2$: the circles and triangles denote $1/T_1$ in the present lump sample and the previous powdered sample, respectively.

As mentioned in the previous study [9], the similar T-linear behavior of $1/T_1$ below transition temperature was reported in several kagomé systems, and spinon excitations were proposed in these systems [11–13]. Recently, the ground states of several Yb triangular lattices such as $YbMgGaO_4$ [1–3] and $YbNaSe_2$ [4,5] were also reported to be quantum spin liquid states with a spinon Fermi surface. However, $YbMgGaO_4$ has an inhomogeneous charge environment arising from a site mixing of Mg^{2+} and Ga^{3+}, which gives rise to the orientational spin disorder and mimic a spin-liquid-like state as different interpretation of the ground state. [14,15]. In contrast, the novel gapless excitation in $YbCuS_2$ is not related to sample extrinsic effect such as the sample quality or sample geometry.

4. Conclusion

In conclusion, we have performed ^{63}Cu-NQR measurements using a lump sample of $YbCuS_2$ with a small surface area to investigate the sample dependence of low-temperature magnetic properties in $YbCuS_2$ by comparing with the previous study with different powdered sample. The line width of NQR signals in the present lump sample is larger than that of the previous powdered sample. Moreover, the transition temperature $T_N \sim 0.92$ K in the present lump sample is slightly smaller than that in previous one (~ 0.95 K). These results suggest that the quality of the present lump sample is worse than that of the previous powdered sample. In contrast, the novel T-linear behavior of $1/T_1$ was observed below 0.5 K and the value of $1/T_1T$ in both samples is almost the same even though the sample quality and sample geometry are different. This suggests that the behavior of $1/T_1$ at low temperature arises from the impurity-robust bulk gapless excitation inherent in $YbCuS_2$ and is not related to sample issues such as the sample quality or sample geometry. The properties of this novel gapless magnetic excitation deserve further investigation.

References

[1] Y. Li, H. Liao, Z. Zhang, S. Li, F. Jin, L. Ling, L. Zhang, Y. Zou, L. Pi, and Z. Yang, Sci. Rep. **5**, 16419 (2015).
[2] Y. Li, G. Chen, W. Tong, L. Pi, J. Liu, Z. Yang, X. Wang, and Q. Zhang, Phys. Rev. Lett. **115**, 167203 (2015).
[3] Y. Li, D. Adroja, P. K. Biswas, P. J. Baker, Q. Zhang, J. Liu, A. A. Tsirlin, P. Gegenwart, and Q. Zhang, Phys. Rev. Lett. **117**, 097201 (2016).
[4] K. M. Ranjith, S. Luther, T. Reimann, B. Schmidt, Ph. Schlender, J. Sichelschmidt, H. Yasuoka, A. M. Strydom, Y. Skourski, J. Wosnitza, H. Kühne, T. Doert, and M. Baenitz, Phys. Rev. B **100**, 224417 (2019).
[5] P.-L. Dai, G. Zhang, Y. Xie, C. Duan, Y. Gao, Z. Zhu, E. Feng, Z. Tao, C.-L. Huang, H. Cao, A. Podlesnyak, G. E. Granroth, M. S. Everett, J. C. Neuefeind, D. Voneshen, S. Wang, G. Tan, E. Morosan, X. Wang, H.-Q. Lin, L. Shu, G. Chen, Y. F. Guo, X. Lu, and P. Dai, Phys. Rev. X **11**, 021044 (2021).
[6] L. D. Gulay and I. D. Olekseyuk, J. Alloys Compd. **402**, 89 (2005).
[7] Y. Ohmagari, Y. Yamane, T. Onimaru, K. Umeo, Y. Shimura, and T. Takabatake, JPS Conf. Proc. **30**, 011167 (2020).
[8] Y. Ohmagari, T. Onimaru, Y. Yamane, Y. Shimura, K. Umeo, T. Takabatake, H. Sato, N. Kikugawa, T. Terashima, H. T. Hirose, and S. Uji, J. Phys. Soc. Jpn. **89**, 093701 (2020).
[9] F. Hori, K. Kinjo, S. Kitagawa, K. Ishida, Y. Ohmagari, and T. Onimaru, submitted.
[10] G. C. Carter, L. H. Bennett, and D. J. Kahan, *Metallic Shifts in NMR*, Part I (Pergamon, London, 1977).
[11] M. Yoshida, M. Takigawa, H. Yoshida, Y. Okamoto, and Z. Hiroi, Phys. Rev. Lett. **103**, 077207 (2009).
[12] F. Bert, D. Bono, P. Mendels, F. Ladieu, F. Duc, J.-C. Trombe, and P. Millet, Phys. Rev. Lett. **95**, 087203 (2005).
[13] M. Gomilšek, M. Klanjšek, R. Žitko, M. Pregelj, F. Bert,P. Mendels, Y. Li, Q. M. Zhang, and A. Zorko, Phys. Rev. Lett. **119**, 137205 (2017).
[14] Z. Zhu, P. A. Maksimov, S. R. White, and A. L. Chernyshev, Phys. Rev. Lett. **119**, 157201 (2017).
[15] I. Kimchi, A. Nahum, and T. Senthil, Phys. Rev. X **8**, 031028 (2018).

Majorana excitations in the anisotropic Kitaev model with an ordered-flux structure

Akihiro Hashimoto, Yuta Murakami, and Akihisa Koga
Department of Physics, Tokyo Institute of Technology, Meguro, Tokyo 152-8551, Japan
E-mail: akihiro.hashimoto@stat.phys.titech.ac.jp

Abstract. We investigate the anisotropic $S = 1/2$ Kitaev model on the honeycomb lattice with the ordered-flux structure. By diagonalizing the Majorana Hamiltonian for the flux configuration, we find two distinct gapped quantum spin liquids. One of them is the gapped state realized in the large anisotropic case, where low energy properties are described by the toric code. On the other hand, when the system has small anisotropy, the other gapped quantum spin liquid is stabilized by the ordered-flux configuration. Since these two gapped quantum spin liquids are separated by the gapless region, these are not adiabatically connected to each other.

1. Introduction

Quantum spin liquids in the Kitaev model [1, 2] have been studied intensively in recent years because of their potential applications in topological quantum computation and spintronics. It is known that in the Kitaev model, spin degrees of freedom are split into itinerant Majorana fermions and localized fluxes due to the spin fractionalization. The ground state belongs to the flux-free state and low energy properties are described by itinerant Majorana fermions. In fact, the Majorana edge current has been observed in the half-quantized plateau in the thermal quantum Hall effect [3, 4]. Furthermore, the Majorana-mediated spin transport has been theoretically suggested [5], which should stimulate further investigation to realize quantum devices using Majorana fermions.

In our previous study [6], we have found that the gapped quantum spin liquid is realized in a certain flux configuration. This indicates that the flux configuration is a key role in controlling Majorana excitation. On the other hand, it is known that the large anisotropy in the exchange couplings stabilizes the gapped quantum spin liquid, which should be described by the toric code [7]. Then a question arises. Are the above gapful states with distinct origins are adiabatically connected to each other? To answer this question, we focus on a flux configuration to investigate how the anisotropy in the exchange couplings affects the Majorana excitations.

2. Model and Hamiltonian

We consider the Kitaev model on the honeycomb lattice, which is given by

$$H = -J_x \sum_{\langle i,j\rangle_x} S_i^x S_j^x - J_y \sum_{\langle i,j\rangle_y} S_i^y S_j^y - J_z \sum_{\langle i,j\rangle_z} S_i^z S_j^z, \tag{1}$$

where $\langle i,j\rangle_\alpha$ shows the nearest-neighborhood pair on the $\alpha(= x, y, z)$-bonds. $S_i^\alpha (= \frac{1}{2}\sigma_i^\alpha)$ is the α-component of the $S = 1/2$ spin at the ith site and σ^α is the α-component of the Pauli matrix.

$J_\alpha (> 0)$ is the exchange coupling on the α-bonds. The model Hamiltonian is schematically shown in Fig. 1(a). There is a local conserved quantity on each hexagonal plaquette in the

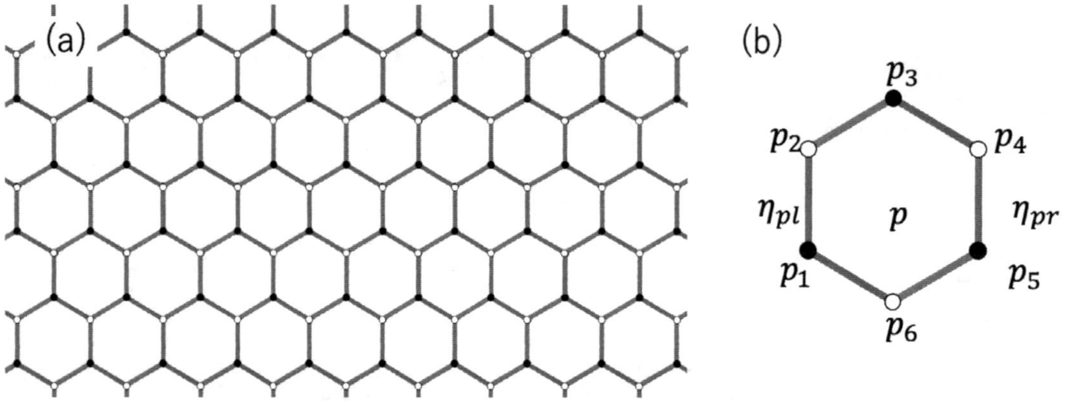

Figure 1. (a) $S = 1/2$ Kitaev model on the honeycomb lattice. Red, blue and green lines represent x-, y- and z-bonds. (b) Hexagonal plaquette with sites $p_1, p_2, \cdots, p_6$ shown for the local operator W_p. η_{pl} and η_{pr} denote the left and right z bonds on the plaquette.

honeycomb lattice. The operator W_p is defined on a plaquette p as

$$W_p = \sigma_{p1}^x \sigma_{p2}^y \sigma_{p3}^z \sigma_{p4}^x \sigma_{p5}^y \sigma_{p6}^z, \tag{2}$$

where $p_i (i = 1, 2, \cdots, 6)$ is the site on the plaquette p [see Fig. 1(b)]. This satisfies $\left[W_p, W_{p'}\right] = 0, [W_p, H] = 0$, and $W_p^2 = 1$. The corresponding eigenvalue w_p takes ± 1. The eigenstate of the Hamiltonian can be specified by the set of the eigenvalues w_p. It is known that the ground state is realized with $w_p = +1$ for all plaquettes. Therefore, a plaquette with $w_p = -1$ can be regarded as an excited flux.

In the study, we discuss Majorana excitations in the Kitaev system with certain flux-configurations. To this end, we first use the Jordan-Wigner transformation, $S_i^+ = \prod_{i'}^{i-1}(1 - 2n_{i'})c_i^\dagger$, $S_i^- = \prod_{i'}^{i-1}(1 - 2n_{i'})c_i$, $S_i^z = n_i - \frac{1}{2}$, where $c_i^\dagger$ and c_i are the creation and annihilation operators of the fermion at the ith site. The Hamiltonian is rewritten into

$$\begin{aligned} H = &-\frac{J_x}{4} \sum_{\langle rb, r'w \rangle_x} \left(c_{rb} + c_{rb}^\dagger\right)\left(c_{r'w} - c_{r'w}^\dagger\right) - \frac{J_y}{4} \sum_{\langle rb, r'w \rangle_y} \left(c_{rb} + c_{rb}^\dagger\right)\left(c_{r'w} - c_{r'w}^\dagger\right) \\ &- \frac{J_z}{4} \sum_r (2n_{rb} - 1)(2n_{rw} - 1), \end{aligned} \tag{3}$$

where $c_{rb} (c_{rw})$ is an annihilation operator of the fermion at the black (white) site on the rth z-bond. $\langle rb, r'w \rangle_\alpha$ indicates the nearest-neighbor pair connected by the α-bond. Here, we define Majorana fermion operators $\gamma, \bar{\gamma}$ [8, 9, 10] as

$$\left\{ \begin{array}{ll} i\gamma_{rw} & = c_{rw} - c_{rw}^\dagger \\ \bar{\gamma}_{rw} & = c_{rw} + c_{rw}^\dagger \end{array} \right., \quad \left\{ \begin{array}{ll} \gamma_{rb} & = c_{rb} + c_{rb}^\dagger \\ i\bar{\gamma}_{rb} & = c_{rb} + c_{rb}^\dagger \end{array} \right., \tag{4}$$

where Majorana operators satisfy $\gamma_i^\dagger = \gamma_i, \bar{\gamma}_i^\dagger = \bar{\gamma}_i, \{\gamma_i, \gamma_j\} = \{\bar{\gamma}_i, \bar{\gamma}_j\} = 2\delta_{ij}$. The Hamiltonian is given as

$$H = -\frac{iJ_x}{4} \sum_{\langle rb, r'w \rangle_x} \gamma_{rb}\gamma_{r'w} - \frac{iJ_y}{4} \sum_{\langle rb, r'w \rangle_y} \gamma_{rb}\gamma_{r'w} - \frac{iJ_z}{4} \sum_r \eta_r \gamma_{rb}\gamma_{rw}, \tag{5}$$

where $\eta_r = i\bar{\gamma}_{rb}\bar{\gamma}_{rw}$. Since η_r satisfies $[H, \eta_r] = 0$, $[\eta_r, \eta_{r'}] = 0$, and $\eta_r^2 = 1$, η_r is a local conserved quantity and the corresponding eigenvalue takes ± 1. The local operator W_p is represented as $W_p = \eta_{pl}\eta_{pr}$, where η_{pl} and η_{pr} are defined on the left and right z-bonds on the plaquette p [see Fig. 1(b)].

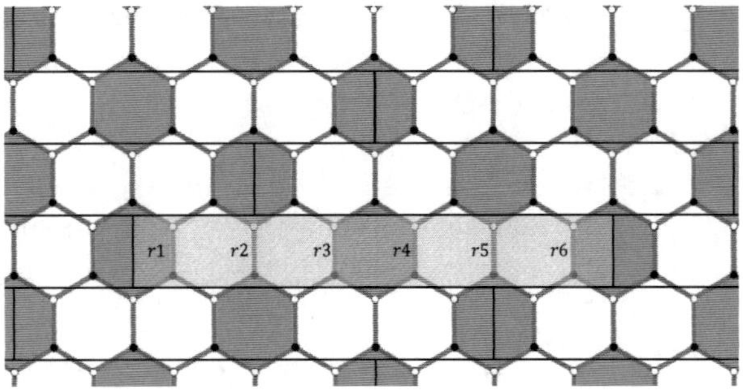

Figure 2. Flux configuration in the Kitaev model studied here. Each plaquette with orange shade shows $w_p = -1$ and that without shade shows $w_p = +1$. Each rectangle including 6 z-bonds represents the unit cell, which is marked with thin yellow shade.

In the present study, we focus on an ordered-flux structure shown in Fig. 2 as a simple example. This flux-configuration is represented by the set $\{\eta_{r_1}, \eta_{r_2}, \cdots, \eta_{r_6}\} = \{1, 1, 1, -1, -1, -1\}$ and its periodic arrangements. From the previous study [6], it is known that in the isotropic case with $J_x = J_y = J_z$, the excitation gap appears in the Majorana excitations. Here, we consider the Majorana excitations in the anisotropic Kitaev model under the conditions $J_x + J_y + J_z = 3$, and $J_x = J_y$.

3. Result

By diagonalizing the Hamiltonian of the Kitaev model with the flux configuration, we obtain the Majorana excitations. The results for the Majorana gap are shown in Fig. 3. When $J_z = 0$,

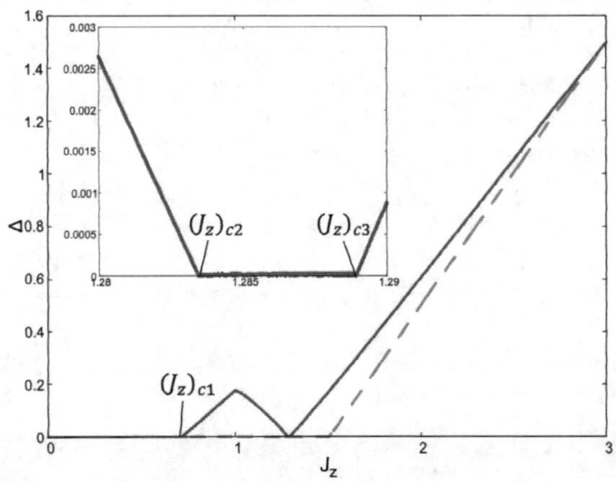

Figure 3. Gap magnitude Δ as a function of the coupling J_z. The purple solid (green dashed) line represents the ordered-flux configuration (the ground state). Inset shows magnified image around $J_z = 1.285$. The isotropic case corresponds to $J_z = 1$.

the system is reduced to the quantum spin chain composed of x- and y-bonds. In this limit, the flux configuration never affects the Majorana excitation, and $\Delta = 0$. This gapless state is stable against the introduction of the exchange J_z, as shown in Fig. 3. Beyond $(J_z)_{c1} \simeq 0.7129$, the Majorana gap appears, takes a maximum at $J_z = 1$, and finally closes at $(J_z)_{c2} \simeq 1.2834$. Thus, we can conclude that the gapped quantum spin liquid is stabilized by the flux-configuration since the Kitaev system in the flux-free state is gapless in this parameter regime, as shown in Fig. 3. When $(J_z)_{c2} < J_z < (J_z)_{c3} \simeq 1.2889$, the gapless quantum spin liquid state is realized, although the region is very narrow, as shown in the inset of Fig. 3. For $J_z > (J_z)_{c3}$, another gapped quantum spin liquid state emerges with $\Delta \simeq J_z$. In this case, the flux configuration plays no essential role and the ground state should be described by the toric code. Therefore, our results suggest that origins for these two gapful quantum spin liquids are distinct from each other.

4. Summary

We have studied the anisotropic $S = 1/2$ Kitaev model on the honeycomb lattice with the ordered-flux structure. In this flux configuration, two distinct gapped quantum spin liquids appear when J_z varies. With large anisotropy, the system is well described by the toric code. On the other hand, when the system has small anisotropy, another gapped quantum spin liquid is realized. These two gapped quantum spin liquids are separated by the gapless region. Therefore, these are not adiabatically connected to each other and thus we conclude that they have different origins. In the sense, one can potentially control the motion of the Majorna excitations using both the flux configuration and anisotropy in the exchanges [11]. The systematic study for other ordered-flux structures are left for future work.

References

[1] Kitaev A 2006 *Ann. Phys.* **321** 2–111
[2] Motome Y and Nasu J 2020 *J. Phys. Soc. Jpn.* **89** 012002
[3] Kasahara Y, Ohnishi T, Mizukami Y, Tanaka O, Ma S, Sugii K, Kurita N, Tanaka H, Nasu J, Motome Y, Shibauchi T and Matsuda Y 2018 *Nature* **559** 227–231
[4] Yokoi T, Ma S, Kasahara Y, Kasahara S, Shibauchi T, Kurita N, Tanaka H, Nasu J, Motome Y, Hickey C, Trebst S and Matsuda Y 2021 *Science* **373** 568572
[5] Minakawa T, Murakami Y, Koga A and Nasu J 2020 *Phys. Rev. Lett.* **125**(4) 047204
[6] Koga A, Murakami Y and Nasu J 2021 *Phys. Rev. B* **103** 214421
[7] Kitaev A Y 2003 *Ann. Phys.* **303** 2–30
[8] Chen H D and Hu J 2007 *Physical Review B* **76** 193101
[9] Feng X Y, Zhang G M and Xiang T 2007 *Phys. Rev. Lett.* **98** 087204
[10] Chen H D and Nussinov Z 2008 *Journal of Physics A: Mathematical and Theoretical* **41** 075001
[11] Taguchi H, Murakami Y, Koga A and Nasu J 2021 *Phys. Rev. B* **104** 125139

Discord effects of inter-cluster interactions on a cluster-based Haldane state in a triangular spin tube

Takanori Sugimoto

E-mail: sugimoto.takanori@rs.tus.ac.jp
Department of Applied Physics, Tokyo University of Science, Katsushika, Tokyo 125-8585, Japan
Advanced Science Research Center, Japan Atomic Energy Agency, Tokai, Ibaraki 319-1195, Japan

Takami Tohyama

Department of Applied Physics, Tokyo University of Science, Katsushika, Tokyo 125-8585, Japan

Abstract. We have recently proposed a new concept of cluster-based Haldane states supported by an inherent chirality in a triangular spin tube with equivalent inter-cluster interactions. In this state, there appear spin-1/2 degrees of freedom consisting of a real spin and a scalar chirality as edge states. With applying a magnetic field, we can observe approximately a half of magnetization of $S = 1/2$ spin, i.e., a quarter spin magnetization, due to symmetrization of the real spin and the chirality. Here, we consider effects of discord between two types of inter-cluster interactions, inducing anisotropic biquadratic interactions of effective $S = 1$ spins. The discord brings the edge magnetization nearer to an exact value of quarter spin magnetization, whereas localization of the edge states becomes worse. These effects are also confirmed in a corresponding spin-1 model.

1. Introduction

Fractionalization of quantum degrees of freedom has opened the ways to new phenomena and effects in modern physics, e.g., quark model in particle physics and spin-charge separation in condensed-matter physics. Particularly, in condensed-matter physics, concepts of fractionalization have been applied to understand complicated many-body states, such as the fractional quantum Hall effects [1] and Majorana fermions in topological superconductors [2]. The Affleck–Kennedy–Lieb–Tasaki (AKLT) model [3] is an important example using fractionalization of quantum spin to give an intuitive picture of the Haldane state [4], which is the ground state of an $S = 1$ spin chain. Moreover, the Haldane state has recently attracted broader attention because of a possibility of holographic quantum computing [5].

To flexibly design the Haldane state with $S = 1/2$ spins, we have considered spin cluster chains (SCCs) where clusters of strongly interacting $S = 1/2$ spins line up and weakly interact with neighboring clusters [6, 7]. In addition, we have clarified general conditions for the Haldane state in SCCs, which we call cluster-based Haldane states (CBHS): (I) the ground state of a cluster is a triplet, (II) the inter-cluster interaction is weak enough as compared with any intra-cluster interactions, and (III) the effective inter-cluster interaction between clusters' ground states is mainly antiferromagnetic and XXZ-like (XY, Heisenberg, and weakly Ising types). The Kramers theorem allows the triplet ground state of a cluster (Condition I), composed by an even number of spins only with Hamiltonian terms keeping the time-reversal symmetry (TRS). However, the condition can be extended to SCCs whose clusters contain an odd number of spins, if there is a term breaking the TRS, e.g., magnetic fields and a scalar spin chirality. Both cases of the magnetic field and scalar chirality have been demonstrated with explicit spin Hamiltonians in our preceding works [7, 8]. Particularly, in the latter case, we have also shown an exotic edge state consisting of a real spin and a scalar chirality, which exhibits a quarter spin magnetization as a response to a magnetic field.

2. Model and Method

In this paper, we consider a CBHS supported by a chirality appearing in triangular spin tube (TST) defined by

$$\mathcal{H}_{\rm TST} = \sum_{j=1}^{L}\left(J\sum_{i<i'}\boldsymbol{s}_{i,j}\cdot\boldsymbol{s}_{i',j} - K\boldsymbol{s}_{{\rm tot},j}\cdot\boldsymbol{\chi}_j\right) + \sum_{j=1}^{L-1}\left(J'_1\sum_{i}\boldsymbol{s}_{i,j}\cdot\boldsymbol{s}_{i,j+1} + J'_2\sum_{i\neq i'}\boldsymbol{s}_{i,j}\cdot\boldsymbol{s}_{i',j+1}\right), \tag{1}$$

where L is the number of clusters. The first (second) term represents intra-cluster (inter-cluster) interactions. The $S = 1/2$ local spin (the total spin) operator in a cluster is denoted by $\boldsymbol{s}_{i,j}$ ($\boldsymbol{s}_{{\rm tot},j} = \sum_i \boldsymbol{s}_{i,j}$), where $i = 1, 2, 3$ ($j = 1, 2, \cdots, L$) denotes the site (cluster) index. The $S = 1/2$ pseudo spin operator $\boldsymbol{\chi}_j$ is constructed by the scalar chirality: $\chi_j^z = (2/\sqrt{3})\boldsymbol{s}_{1,j}\cdot\boldsymbol{s}_{2,j}\times\boldsymbol{s}_{3,j}, \chi_j^x = \left(\chi_j^+ + \chi_j^-\right)/2, \chi_j^y = \left(\chi_j^+ - \chi_j^-\right)/(2\imath)$, where $\chi_j^\pm$ represents the chirality ladder operators defined by $\chi_j^\pm = \sum_{\sigma=\pm}|\sigma,\pm\rangle_j\langle\sigma,\mp|_j$ with the local eigenstates of the total spin and scalar chirality of a cluster, $|s_{{\rm tot},j}^z, \chi_j^z\rangle_j$.

In this model, we assume the SCC condition (II), i.e., $J, K \gg J'_i$ (i=1,2), resulting in a triplet ground state obtained by symmetrization of the total spin and scalar chirality in a cluster. The low-temperature behavior is dominated by the triplet, corresponding to a pseudo spin-1 operator $\boldsymbol{T}_j$. Therefore, we can obtain an effective Hamiltonian for the $S = 1$ pseudo spins as follows,

$$\mathcal{H}_{\rm eff} = J'_1\sum_{j=1}^{L-1}\left\{{\rm Hs}_j\left(\frac{5}{3}-\frac{2}{3}\alpha_{J'}\right) + \frac{8}{3}(1-\alpha_{J'})\,{\rm Hs}_j^2\left(\frac{1}{2}\right) - \frac{8}{3}(1-\alpha_{J'})\,{\rm Iz}_j\right\} \tag{2}$$

with the anisotropic Heisenberg (XXZ) interaction and an additional z component interaction,

$${\rm Hs}_j(\Delta) = T_j^x T_{j+1}^x + T_j^y T_{j+1}^y + \Delta T_j^z T_{j+1}^z,\ {\rm Iz}_j = \left[1-(T_j^z)^2/2\right]\left[1-(T_{j+1}^z)^2/2\right], \tag{3}$$

where we use the ratio of the inter-cluster interactions $\alpha_{J'} = J'_2/J'_1$. Since an equivalent case $\alpha_{J'} = 1$ has been presented in the preceding paper [8], we concentrate on a discord case $\alpha_{J'} \neq 1$.

3. Discord effects of inter-cluster interactions

In this section, we show effects of the discord of two inter-cluster interactions $J'_1 \neq J'_2$ ($\alpha_{J'} \neq 1$). As explained above, the equivalence of the inter-cluster interactions $\alpha_{J'} = 1$ effectively gives the Heisenberg spin chain of the $S = 1$ pseudo spins at low temperatures. On the other hand, if there is the discord $\alpha_{J'} \neq 1$, additional interactions appear in the effective Hamiltonian (2) such as an anisotropy and biquadratic interactions. To clarify effects of the additional terms as compared with a spin-1 chain, we consider an extended bilinear-biquadratic (eBLBQ) model of

$S = 1$ operator $\boldsymbol{S}_j$ defined by,

$$\mathcal{H}_{\text{eBLBQ}} = J \sum_{j=1}^{L-1} \left[\text{Hs}_j(\Delta) + Q\,\text{Hs}_j^2(\Delta') - A\,\text{Iz}_j \right]_{\boldsymbol{T}_j \to \boldsymbol{S}_j} . \quad (4)$$

It is well known that the Haldane state is stabilized by introducing an isotropic biquadratic interaction $Q > 0$ with $\Delta = \Delta' = 1$ in the spin-1 eBLBQ model ($A = 0$). AKLT shows that at $Q = 1/3$, the Haldane gap is maximized and the edge states are most localized [3]. Therefore, in our model, the stabilization of the CBHS can be enhanced by the discord of the inter-cluster interactions $\alpha_{J'} \neq 1$. To confirm the speculation, we have performed the variational matrix-product state (VMPS) calculation with $L = 60$, $K = J/2$, and $\sqrt{(J_1')^2 + (J_2')^2} = J/10$ in (1), and obtained the Haldane gap Δ_2 as a function of the J' ratio defined by $\theta_{J'} = \tan^{-1}\alpha_{J'}$.

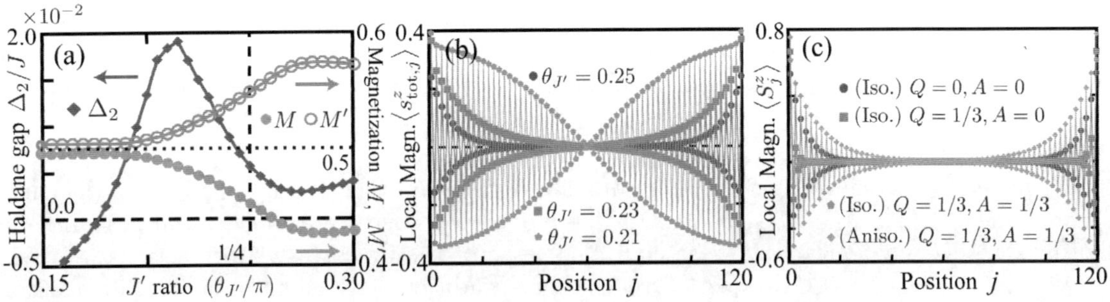

Figure 1. (a) The Haldane gap Δ_2 corresponding to the energy difference between the first ($\widetilde{M} = 1$) and second ($\widetilde{M} = 2$) excited states, and real (pseudo) magnetization M (M') of the $\widetilde{M} = 1$ state, as a function of the J' ratio, $\theta_{J'} = \tan^{-1}\alpha_{J'}$. The vertical dashed lines denote the equivalent condition $\alpha_{J'} = 1$. The horizontal dashed line represents the zero-energy level, corresponding to the stable-unstable boundary of the Haldane state. The horizontal dotted line denotes the symmetric case $M = M'$. (b) Local magnetization $\left\langle s^z_{\text{tot},j} \right\rangle$ in the $\widetilde{M} = 1$ state of the TST model. (c) Local magnetization $\left\langle S^z_j \right\rangle$ in the $M = 1$ state of the spin-1 chain model. In (b), the J' ratio dependency is shown for $\theta_{J'} = \tan^{-1}\alpha_{J'} = 0.25, 0.23$, and 0.21 for a TST (1). In (c), effects of the Q and A terms including an anisotropy are presented for a spin-1 chain (4), where "Iso." and "Aniso." represent the isotropic case ($\Delta = \Delta' = 1$) and an anisotropic case ($\Delta = 13/12, \Delta' = 1/2$) for the bilinear-biquadratic term, respectively. The parameters of the orange diamonds in (c) correspond to the $\alpha_{J'} = 7/8$ ($\theta_{J'} \cong 0.229\pi$) in the effective model (4).

Figure 1(a) shows the Haldane gap defined by the energy difference between the $\widetilde{M} = 1$ and $\widetilde{M} = 2$ ground states, where the total pseudo magnetization is defined by $\widetilde{M} = M + M'$ with the real and pseudo magnetizations, $M = \sum_j \left\langle s^z_{\text{tot},j} \right\rangle$ and $M' = \sum_j \left\langle \chi^z_j \right\rangle$. As we can see, the Haldane gap has a peak at $\theta_{J'} \cong 0.215\pi$, whereas the $\theta_{J'} = \pi/4$ represents the equivalent point of $J_1' = J_2'$ denoted by the vertical dashed line. In the effective Hamiltonian (2), we cannot find the AKLT condition [$Q = 1/3$, $A = 0$, $\Delta = \Delta' = 1$ in (4)] due to a necessarily finite anisotropy and another term. If we ignore effects of the anisotropy and the additional term, an AKLT-like condition $\alpha_{J'} = 7/8$ corresponds to $\theta_{J'} = \tan^{-1}\alpha_{J'} \cong 0.229\pi$, which is not so far from the point $\theta_{J'} \cong 0.215\pi$ showing the peak of the Haldane gap in Fig. 1(a).

Next, we have also investigated the difference between the real and pseudo magnetizations, M and M', at the $\widetilde{M} = 1$ ground state as a function of the J' ratio [see Fig. 1(a)].

Interestingly, symmetrization of the real and pseudo magnetizations recovers with decreasing $\theta_{J'}$, i.e., approaching the AKLT-like point. Since the real magnetization corresponds to the sum of magnetization of the two edge states, the quasi-fractionalization of $S = 1/2$ spin magnetization shows nearly an exact half value, that is, an $S = 1/4$ spin magnetization for $\theta_{J'} \lesssim 0.2\pi$, while the CBHS becomes unstable for $\theta_{J'} \lesssim 0.175\pi$ due to a negative Haldane gap.

Furthermore, to check the localization of the edge state, we have calculated the local magnetization, $\left\langle s^z_{\mathrm{tot},j} \right\rangle$ in the $\widetilde{M} = 1$ state of the TST model with various J' ratios [see Fig. 1(b)]. With approaching the most stable point $\theta_{J'} \cong 0.215\pi$, the edge state surprisingly extends to the bulk, so that the localization becomes worse. To clarify which term dominates the extension in the effective Hamiltonian (2), we have also obtained the local magnetization in the spin-1 eBLBQ model (4). Figure 1(c) shows effects of the additional terms and anisotropy. We can confirm that the isotropic additional terms prevent the edge state from extending, while the anisotropy automatically appearing in the effective Hamiltonian (4) makes the localization worse. Therefore, we conclude that in the TST model, the discord of the inter-cluster interactions stabilizes the CBHS and brings the edge magnetization nearer to the exact value of the quarter spin magnetization, whereas the localization of the edge states simultaneously becomes worse.

4. Summary

In this paper, we have investigated the discord effects of the inter-cluster interactions on the edge state of the CBHS in the TST model. In this model, the edge state shows an approximately half of spin-1/2 magnetization in the equivalent case of the inter-cluster interactions [8]. On the other hand, if the equivalence is broken, some additional terms such as the biquadratic interactions appear in the effective Hamiltonian. To confirm the effects, we have performed the VMPS calculations for the TST model, together with the spin-1 eBLBQ model for comparison. In conclusion, we have clarified that the biquadratic interactions certainly stabilize the CBHS and brings the edge magnetization nearer to the exact value of the quarter spin magnetization, while the localization of the edge states simultaneously becomes worse due to the anisotropy automatically appearing with the biquadratic terms.

Acknowledgments

This work was partly supported by a Grant-in-Aid for Scientific Research (C) (Grant No. 20K03840). Numerical computation was carried out on the supercomputers at JAEA and the Supercomputer Center at the Institute for Solid State Physics, University of Tokyo.

References

[1] Stormer H L 1999 Fractional quantum hall effect today *Rev. Mod. Phys.* **71** S298.
[2] Qi X L and Zhang S C 2011 Topological insulators and superconductors *Rev. Mod. Phys.* **83** 1057.
[3] Affleck I, Kennedy T, Lieb E H, and Tasaki H 1988 Valence bond ground states in isotropic quantum antiferromagnets *Commun. Math. Phys.* **115** 477
[4] Haldane F D M 1983 Nonlinear field theory of large-spin Heisenberg antiferromagnets: Semiclassically quantized solitons of the one-dimensional easy-axis Néel state *Phys. Rev. Lett.* **50** 1153
[5] Miyake A 2010 Quantum computation on the edge of a symmetry-protected topological order *Phys. Rev. Lett.* **105** 040501.
[6] Fujihala M, Sugimoto T, Tohyama T, Mitsuda S, Mole R A, Yu D H, Yano S, Inagaki Y, Morodomi H, Kawae T, Sagayama H, Kumai R, Murakami Y, Tomiyasu K, Matsuo A, and Kindo K 2018 Cluster-Based Haldane State in an Edge-Shared Tetrahedral Spin-Cluster Chain: Fedotovite $K_2Cu_3O(SO_4)_3$ *Phys. Rev. Lett.* **120** 077201
[7] Sugimoto T, Morita K, and Tohyama T 2020 Cluster-based Haldane states in spin-1/2 cluster chains *Phys. Rev. Res.* **2** 023420
[8] Sugimoto T and Tohyama T 2021 Quasi-fractionalization of spin in a cluster-based Haldane state supported by chirality of a triangular spin tube *Preprint* cond-mat/2108.00681

Field-Induced Quantum Spin Nematic Liquid Phase in the S=1 Antiferromagnetic Heisenberg Chain with Additional Interactions

Tôru Sakai[1,2], Hiroki Nakano[1], Rito Furuchi[1] and Kiyomi Okamoto[1]

[1]Graduate School of Science, University of Hyogo, Hyogo 678-1297, Japan
[2]National Institutes for Quantum Science and Technology, SPring-8, Hyogo 679-5148, Japan

E-mail: sakai@spring8.or.jp

Abstract.
The magnetization process of the $S = 1$ antiferromagnetic chain with the single-ion anisotropy D and the biquadratic interaction is investigated using the numerical diagonalization. Both interactions stabilize the 2-magnon Tomonaga-Luttinger liquid (TLL) phase in the magnetization process. Based on several excitation gaps calculated by the numerical diagonalization, some phase diagrams of the magnetization process are presented. These phase diagrams reveal that the spin nematic dominant TLL phase appears at higher magnetizations for sufficiently large negative D.

1. Introduction

The quantum spin nematic state is one of interesting topics in the field of the strongly correlated electron systems. It was theoretically predicted to be realized in several frustrated systems; the square-lattice model with the ferromagnetic nearest-neighbor and antiferromagnetic next-nearest-neighbor interactions[1], the triangular-lattice antiferromagnetic with multi spin exchange interactions[2], and the ferromagnetic and antiferromagnetic zigzag chain[3, 4]. The high-magnetic field measurement[5] on the zigzag chain compound $LiCuVO_4$ detected an anomalous behavior of the magnetization curve just below the saturation magnetization, which was supposed to be the spin nematic phase. The quantum spin liquid like behavior of the $S = 1$ triangular-lattice compound $NiGa_2S_4$ was theoretically explained as the spin nematic state[6]. A density matrix renormalization group (DMRG) analysis[7] indicated that the field-induced spin nematic liquid phase appears in the $S = 1$ antiferromagnetic chain with the biquadratic interaction. It was supported by the numerical diagonalization study[8].

On the other hand, the 2-magnon bound state similar to the spin nematic state was predicted to appear in the magnetization process of the $S = 1$ antiferromagnetic chain with the easy axis anisotropy based on the numerical diagonalization analysis[9]. It would possibly correspond to the spin nematic state. In the present paper, we investigate the $S = 1$ antiferromagnetic chain with the easy axis anisotropy and the biquadratic interaction using the numerical diagonalization, in order to consider the relation between this 2-magnon bound state and the spin nematic liquid phase.

2. Model

We consider the magnetization process of the $S = 1$ antiferromagnetic chain with the easy axis single ion anisotropy D and the biquadratic interaction $J_{\rm BQ}$. The system under the magnetic field is described by the Hamiltonian

$$
\begin{aligned}
\mathcal{H} &= \mathcal{H}_0 + \mathcal{H}_Z \\
\mathcal{H}_0 &= J\sum_{j=1}^{L} \vec{S}_j \cdot \vec{S}_{j+1} + D\sum_{j=1}^{L}(S_j^z)^2 + J_{\rm BQ}\sum_{j=1}^{L}(\vec{S}_j \cdot \vec{S}_{j+1})^2 \\
\mathcal{H}_Z &= -H\sum_{j=1}^{L} S_j^z. \qquad (1)
\end{aligned}
$$

In this paper we consider the case of $J_{\rm BQ} < 0$, $D < 0$ and fix $J = 1$. The first term of $\mathcal{H}_0$ is the bilinear exchange interaction which stabilizes the antiparallel configuration of the nearest neighbor spin pair. On the other hand, the third term, the biquadratic interaction does not only stabilize the antiparallel configuration, but also the parallel one. Thus the spin nematic liquid phase would possibly occur for sufficiently large $|J_{\rm BQ}|$. A numerical diagonalization and DMRG study indicated that the biquadratic interaction stabilizes the spin nematic correlation[10]. If the spin nematic liquid phase is realized, the four spin correlation function would exhibit the power-law decay, while the two spin one decays exponentially. Then in this phase the single magnon excitation should be gapped, but the 2-magnon excitation should be gapless. Therefore the 2-magnon bound state would be realized there. If the spin nematic liquid phase appears in the magnetization process described by the Hamiltonian (1), each step of the magnetization curve for the finite-size systems should be $\delta S^z_{\rm total} = 2$, while $\delta S^z_{\rm total} = 1$ in the conventional Tomonaga-Luttinger liquid (TLL) phase. The negative D also yields a similar 2-magnon bound state, because the $S^z_j = \pm 1$ states are stabilized and the $S^z_j = 0$ state is skipped[9]. In order to clarify the feature of these 2-magnon bound states, we investigate the model (1) using the numerical diagonalization of finite clusters under the periodic boundary condition. The calculated lowest energy for each magnetization $M = (\sum_{j=1}^{L} S_j^z)$ and each momentum k is denoted as $E_k(M)$ in the following sections.

3. Spin Nematic Liquid

The negative biquadratic interaction stabilizes the TLL phase where the 2-magnon bound state is realized, as well as the negative D. In this 2-magnon TLL phase the spin correlation functions are expected to have the following asymptotic forms:

$$\langle (S_0^+)^2 (S_r^-)^2 \rangle \sim r^{-\eta_2}, \quad \langle S_0^z S_r^z \rangle \sim \cos(2k_F r) r^{-\eta_z} \quad (r \to \infty), \qquad (2)$$

where $2k_F$ is given by $2k_F = (1 - M/M_s)\pi$ and M_s is the saturation magnetization ($M_s = L$). $\langle (S_0^+)^2 (S_r^-)^2 \rangle$ is the nematic (quadrapole) correlation perpendicular to the magnetic field and $\langle S_0^z S_r^z \rangle$ is the SDW correlation parallel to it. In a DMRG work[7] the region with $\eta_2 < \eta_z$ is called the nematic phase and the region with $\eta_2 > \eta_z$ is called the SDW phase, although the boundary between them is just a crossover line. Thus in order to find the spin nematic liquid phase, we should look for the parameter region where the 2-magnon TLL is realized and $\eta_2 < \eta_z$ is satisfied.

4. Excitation Gaps

In order to determine the boundary between the 1-magnon (conventional) and 2-magnon TLL phases, we consider several excitation gaps. The 2-magnon excitation gap Δ_2 is gapless in both phases. The 1-magnon excitation gap Δ_1 (the $2k_F$ excitation gap of the 2-magnons Δ_{2k_F}) is

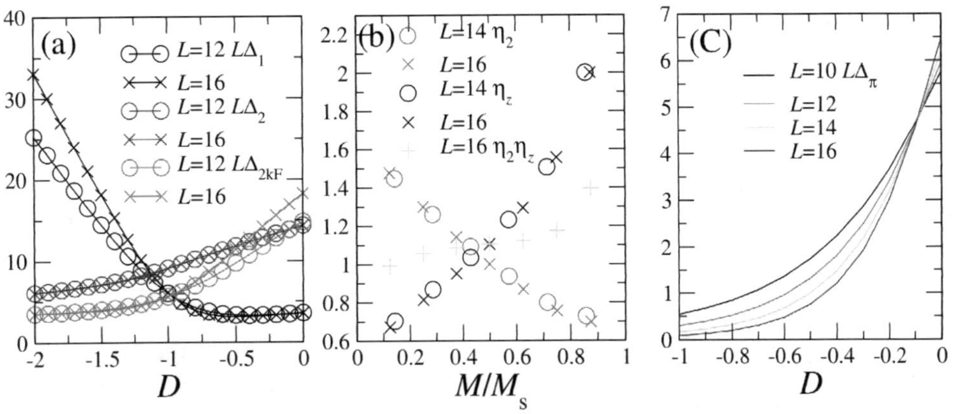

Figure 1. (a)Scaled gaps $L\Delta_1$, $L\Delta_2$ and $L\Delta_{2k_F}$ at $M = M_s/2$ for $J_{\rm BQ}$=-0.2 are plotted versus D for $L = 12$ and 16, (b)Exponents η_2 and η_z estimated by (3) plotted versus M/M_s for L =14 and 16 ($J_{\rm BQ}$=-0.2, D=-1.5). $\eta_2\eta_z = 1$ is well satisfied around the cross of η_2 and η_z. (c)Scaled gaps $L\Delta_\pi$ for $J_{\rm BQ} = -0.2$. It indicates the phase boundary about $D \sim -0.1$.

gapless (gapped) in the 1-magnon TLL phase, while gapped (gapless) in the 2-magnon one. D dependence of the scaled excitation gaps $L\Delta_1$, $L\Delta_2$ and $L\Delta_{2k_F}$ at $M = M_s/2$ for $J_{\rm BQ} = -0.2$ are shown in Fig. 1(a). Fig. 1(a) indicates the above behaviors of these gaps. The cross point of Δ_1 and Δ_{2k_F} is one of good estimations of the phase boundary between the 1-magon and 2-magon TLL phases. Our analysis of the L dependence indicates that this phase boundary converges faster than $1/L^2$ with respect to L. Thus we adopt this method to determine the boundary between the 1-magnon and 2-magnon TLL phases.

5. Spin Correlation Exponents

According to the conformal field theory, the critical exponents η_2 and η_z can be estimated the forms

$$\eta_2 = \frac{E_0(M+2) + E_0(M-2) - 2E_0(M)}{E_{k_1}(M) - E_0(M)}, \quad \eta_z = 2\frac{E_{2k_F}(M) - E_0(M)}{E_{k_1}(M) - E_0(M)}, \tag{3}$$

for each magnetization M, where k_1 is defined as $k_1 = L/2\pi$. η_2 and η_z estimated for L=14 and 16 are plotted versus M/M_s in Fig. 1 (b). It suggests that the SDW correlation is dominant for small M, while the nematic one is for large M. Since the cross point of η_2 and η_z is not so strongly dependent of L, the cross point of $L = 16$ is used as the crossover point between the SDW and nematic TLL phases in Fig. 2(a), (b) and (c), for $J_{\rm BQ}$=0, -0.2 and -0.5, respectively. At least around the cross point the system well holds the relation $\eta_2\eta_z = 1$, which should be satisfied for TLL, as shown in Fig. 1(b).

6. Phase diagrams

The phase diagrams on the D-M/M_s plane are presented for $J_{\rm BQ}$=0, -0.2, and -0.5 in Figs. 2 (a), (b), and (c), respectively. The phase boundary at $M = 0$ between the Haldane and Néel ordered phases, which is denoted as the green triangle in Fig. 2, is determined by the phenomenological renormalization using the excitation gap Δ_π with $k = \pi$ for L=10, 12, 14, and 16. The scaled gap $L\Delta_\pi$ is plotted versus D for $J_{\rm BQ} = -0.2$ in Fig. 1(c). The L dependent critical point

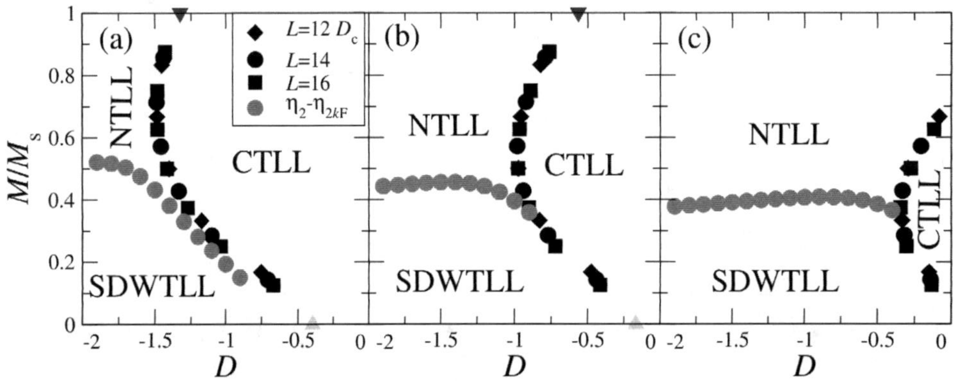

Figure 2. Phase diagrams with respect to D and M/M_s for J_{BQ}=0(a), -0.2(b) and -0.5(c).

obtained by $L\Delta_\pi(L, D_c) = (L+2)\Delta_\pi(L+2, D_c)$ can be easily extrapolated to the infinite L limit. The critical point at $M = M_s$ is determined as the point where the 1-magnon and 2-magnon excitation gaps are equal, which is the blue triangle in Fig. 2. Here, NTLL, SDWTLL and CTLL denote the nematic TLL, the SDW TLL, and the conventional TLL, respectively. Fig.2(a) indicates that the NTLL can appear even for $J_{\mathrm{BQ}} = 0$. Figs. 2(b) and (c) suggest that the biquadratic interaction enhances both of NTLL and SDWTLL phases. The crossover boundary between NTLL and SDWTLL is always about the half the saturation magnetization. It is not so strongly dependent on J_{BQ}.

7. Summary

The magnetization process of the $S = 1$ antiferromagnetic chain with the easy-axis single-ion anisotropy and the biquadratic interaction is investigated by the numerical diagonalization for finite-size systems. It is found that the nematic TLL phase appears in higher magnetization region for sufficiently large negative D, even for $J_{\mathrm{BQ}} = 0$.

Acknowledgments

This work was partly supported by JSPS KAKENHI, Grant Numbers JP20K03866, JP18H04330 (J-Physics) and JP20H05274. A part of the computations was performed using facilities of the Supercomputer Center, Institute for Solid State Physics, University of Tokyo, and the Computer Room, Yukawa Institute for Theoretical Physics, Kyoto University.

References

[1] N. Shannon, T. Momoi and P. Sindzingre, Phys. Rev. Lett. **96**, 027213 (2006).
[2] T. Momoi, P. Sindzingre and N. Shannon, Phys. Rev. Lett. **97**, 257204 (2006).
[3] A. V. Chubukov, Phys. Rev. B **44**, R4693 (1991).
[4] T. Hikihara, L. Kecke, T. Momoi and A. Furusaki, Phys. Rev. B **78**, 144404 (2008).
[5] N. Büttgen et al., Phys. Rev. B **90**, 134401 (2014).
[6] S. Nakatsuji et al., Science **309**, 1697 (2005).
[7] S. R. Manmana, A. M, Lauchli, F. H. L. Essler and F. Mila, Phys. Rev. B **83**, 184433 (2011).
[8] T. Sakai, AIP Advances **11**, 015306 (2021).
[9] T. Sakai, Phys. Rev. B **58**, 6268 (1998).
[10] A. Lauchli, G. Schmid and S. Trebst, Phys. Rev. B **74**, 144426 (2006).

Correlations among T_C, A, and $\rho_\circ$ within FL region of T-p phase diagram of heavy-Fermion superconductors

M. ElMassalami[1], **P. B. Castro**[1], **M. B. Silva Neto**[1]
[1] Instituto de Física, Universidade Federal do Rio de Janeiro, 21941-972 Rio de Janeiro RJ, Brazil
E-mail: massalam@if.ufrj.br, CASTRO.Pedro@nims.go.jp, mbsn@if.ufrj.br

Abstract. Extensively reported experimental observations indicate that on varying a control parameter, such as pressure p, within the phase diagram of most quantum critical heavy fermion HF superconductors, one identifies a cascade of distinct electronic states which may be magnetic, of Kondo-type, non-conventional superconducting, Fermi Liquid, FL, or non-FL character. Of particular interest is the part of the phase diagram wherein superconductivity emerges from a strongly renormalized FL state. This region resembles the overdoped region of the *T-doping* phase diagram of cuprate superconductors. Remarkably, within this highly nontrivial region, one identifies a universal correlation among T_C and A: $\ln\frac{T_C}{\theta} \propto A^{-\frac{1}{2}}$ (θ is a characteristic energy scale and A is the coefficient of T^2 resistivity term). Commonly, these features are considered to be driven by a Spin-Fluctuation-mediated electron-electron scattering channel. On adopting such a channel and applying standard theories of Migdal-Eliashberg (superconductivity) and Boltzmann (transport), we derive analytic expressions that satisfactorily reproduce the aforementioned empirical correlations.

1 Introduction

Recently, much attention has been focused on the overdoped region of high-Tc cuprate superconductors wherein only superconductivity and Fermi liquid, **FL**, state are involved [1]. This region offers a welcome simplification since one avoids complications related to competing states such as antiferromagnetism, spin glass, stripes, or pseudogap. More importantly, the continuous and monotonic evolution of T_C suggests that the same pairing channel is involved: Then, as that the trend of $T_C \to 0$ within this region is considered to be related to a decrease in the strength of the spin-fluctuation, **SF**, mediated pairing channel [1] (a)], it is tempting to correlate the shape of T_C dome to the rise and fall of the strength of this channel.

In spite of this gain in simplification, analysis of this overdoped region is plagued with doping-induced disorder [1]. But, there is a similar region within the T-p phase diagrams of the closely related heavy fermion, **HF**, superconductors which does not manifest disorder effects [2]. Indeed, the similarity is evidenced as: (i) Within this HF region, the pairing is considered to be mediated by SF modes [3]; (ii) the superconductivity emerges from a strongly renormalized FL state; and (iii) $T_C \to 0$. Interestingly, in addition to avoiding the influence of competing states (antiferromagnetism, spin glass phase, non-FL state, etc.), here one also avoids disorder effects: a much better simplification, electing HFs as suitable candidates for such an investigation.

It is remarkable that, in spite of an absence of a deliberately introduced disorder, the residual resistivity, $\rho_\circ$, of HF superconductors is considerably high and manifest nonmonotonic baric evolution [see Fig. 1(d)]. We consider this resistivity excess to be due to a scattering from SF-related non-homogeneous spacial distribution within a HF superconducting sample: at each instant, only some randomly-distributed micro-sized domains of a sample are assumed to sustain SF modes and that their average dimension is longer than the mean free path and coherent length. The concentration of these domains (a measure of the strength of the SF-mediated channel) is reflected in the excess of $\rho_\circ$. Moreover, we consider that the very same SF-mediated e-e coupling is responsible for the emergence of both superconductivity and FL character: This is reminiscent to that the electron-phonon interaction is responsible for both the BCS superconductivity and Bloch-Gruneisen βT^5 term. Accordingly, within the FL region, an increase in a control parameter (e.g. pressure) leads to a decrease in the size and concentration of the SF-bearing domains, till eventually one reaches, at very high pressure, the SF-free state wherein no modes are available for mediation and thus to a removal of SF-related features: quench of superconductivity ($T_c \rightarrow 0$), reduction of FL behavior ($A_{tot} \rightarrow A_\circ$), and reduction of residual resistivity ($\rho_\circ^{tot} \rightarrow \rho_\circ^\circ$). Below, two FL contributions are empirically identified: (i) a conventional FL contribution manifested as $A_\circ$; (ii) another, driven by SF-electron scattering channel, is manifested as $A_{tot} - A_\circ$.

In this work, we are interested in analyzing the involved SF-mediated interaction channel and how this channel influences the evolution of A, and T_c within the FL region of a typical archetype HF superconductor (analysis of further HF examples are given in Ref. [4]). Our analysis identified universal correlations among $\rho_\circ$, A, and T_c. As mentioned above, we argue that the SF-electron scattering channel is responsible for the emergence of the superconductivity, the FL character, and the correlations among them. On adopting such a mechanism and applying standard Migdal-Eliashberg theory for description of superconductivity and Boltzmann theory for description of transport properties, we derive analytic expressions that compare favorably with the empirically-obtained correlations.

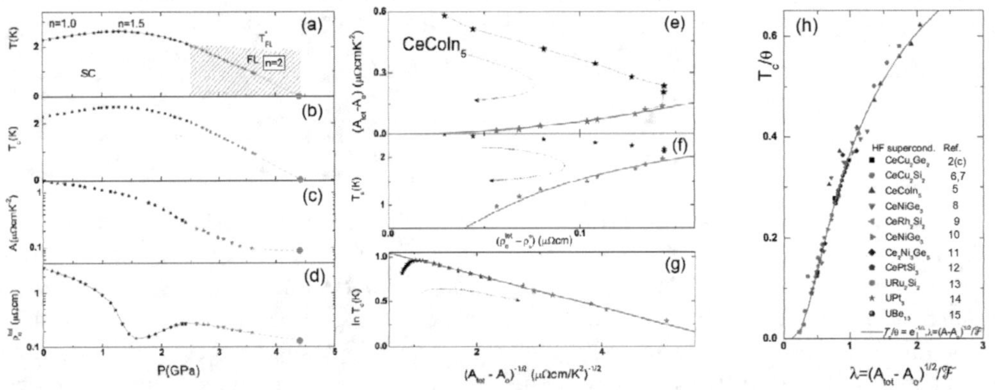

Figure 1. (a) *T-p* phase diagram of $CeCoIn_5$. Also shown are the baric evolution of $T_c(p)$ in (b), $A_{tot}(p)$ in (c), and $\rho_\circ^{tot}(p)$ in (d). All curves were taken from the continuous line (the visual guide) of Ref. [5]. The red symbols and hatched area identify the FL region while the dashed lines are visual guides. For p $\rightarrow$4.4 GPa, $T_c \rightarrow 0$, $A \rightarrow A_\circ = 0.082(4)\mu\Omega$cm and $\rho_\circ \rightarrow \rho_\circ^\circ = 0.130(5)\mu\Omega$cm (all shown as green large filled circle). (e) A plot of $A_{tot} - A_\circ$ *versus* $\rho_\circ^{tot} - \rho_\circ^\circ$. The solid curve is the best fit to Eq. 3 (A_1=0, $A_2 = 6.0(2) \times 10^6\Omega^{-1}cm^{-1}/K^2$). (f) T_c *versus* $(\rho_\circ^{tot} - \rho_\circ^\circ)$: The solid calculated line is based on Eq. (5). (g) $\ln(T_c)$ *versus* $(A - A_\circ)^{-\frac{1}{2}}$. The best fit to the linearization of Eq. (4) (Fit parameters $\theta = 3.2(1)$K, $\mathcal{F} = 1.8(1) \times 10^{-4}(\Omega cm/K^2)^{1/2}$). (h) The generalized plot of $\frac{T_c(p)}{\theta}$ *versus* $\lambda = \frac{(A - A_\circ)^{\frac{1}{2}}}{\mathcal{F}}$ of the tabulated HF superconductors [2, (c)], [6, 7, 5, 8, 9, 10, 11, 12, 13, 14, 15]. Each pair of sample-dependent θ and $\mathcal{F}$ was evaluated within the pressure range wherein Eq. (4) is valid [see Figs. 1(g)]. As evident, the distribution of (T_c/θ, λ) extends up to $\approx (0.6, 2.1)$. This is similar to the limit reported for strong superconductors, see Ref. [16] and Fig. 4 of Ref. [17].

2 Results and Analysis

2.1 Empirical analysis: Identification of correlations among $\rho_\circ$, A, *and* T_c

The *T-p* phase diagram of $CeCoIn_5$ [5, 18] is shown in Fig. 1(a) while the baric evolution of $T_c(p)$, $A_{tot}(p)$ and $\rho_\circ^{tot}(p)$ are shown in Figs. 1(b), 1(c) and 1(d), respectively All data of Figs. 1 were taken from Ref. [5]. As shown in the figures, the FL character is manifested at sufficiently higher pressure and is evidenced as a *quadratic-in-T* resistivity contribution, marked by the hatched area and the *n=2* notation. We are interested in the FL region which starts at the event of switching from NFL into FL character and ends when $T_c \to 0$. Specifically, the $T_c \to 0$ event at 4.4 GPa is taken as signaling the pressure-induced quench of the SF-related FL contributions and that all other parameters extrapolate to non-SF-bearing limits: namely, $A_{tot} \to A_\circ$ and $\rho_\circ^{tot} \to \rho_\circ^\circ$. Then, based on Matthiessen's rule, we identify two residual resistivities: $\rho_\circ^\circ$ and $(\rho_\circ^{tot} - \rho_\circ^\circ)$. In fact, as mentioned above, we identify two contributions: (i) The non-SF-bearing contribution leading to $\rho_\circ^\circ$, $A_\circ$, a weak coupling constant $\lambda_\circ$, and a mean free path $\ell_\circ$ ($\ell_\circ \propto \frac{1}{\rho_\circ^\circ}$). (ii) The SF-bearing contribution leading to $\rho_\circ = \rho_\circ^{tot} - \rho_\circ^\circ$, $A = A_{tot} - A_\circ$, $\lambda = \lambda_{tot} - \lambda_\circ$, and $\ell = 1/(\frac{1}{\ell_{tot}} - \frac{1}{\ell_\circ})$. Based on such a clear distinction, we were able to identify three correlations among T_c, A, and $\rho_\circ$. As shown in Fig. 1(e,f,g), these are, respectively, $[A_{tot} - A_\circ] \propto [\rho_\circ^{tot} - \rho_\circ^\circ]^2$, $\ln\frac{T_c}{\theta} \propto [\rho_\circ^{tot} - \rho_\circ^\circ]^{-1}$, and $\ln\frac{T_c}{\theta} \propto (A_{tot} - A_\circ)^{-\frac{1}{2}}$.

2.2 Theoretical analysis: interpretation of the correlations among $\rho_\circ$, A, *and* T_c

It is taken that within a HF superconducting sample, the SF-related quasiparticles can be created or annihilated and that their mediation of the e-e interaction, V_{ee}^f [3], is the driving mechanism behind the surge of the features understudy. As usual, the $I^2\chi(\omega)$ [I is the exchange coupling while $\chi(\omega)$ is the dynamic susceptibility] plays the same role as that of Eliashberg's electron-phonon spectral function $\alpha^2\mathcal{F}(\omega)$. V_{ee}^{sf} can then be obtained from the integral over frequencies of such an Eliashberg's type spectral function. Therefore, any modification in the interaction strength [1, (a)] or redistribution of the spectral weight [1, (c)] would modify both T_c and A as well as their correlation.

On adopting the traditional strategy for calculating $T_c(\ell)$, we solved the set of coupled linearized Eliashberg's equations within the BCS limit. In this limit, $T_c(\ell)$ is determined by the superconducting coupling constant $\lambda(\ell) = N(\epsilon_F)V_{ee}^{sf}(\ell) = 2\int d\omega \frac{I^2\chi(\omega)}{\omega_{opt}^2 + \omega^2}$, where ω_{opt} is an optimal frequency at which λ is maximal, and can thus be written as [19]:

$$T_c(\ell) = \theta \exp\left(\frac{-1 - \lambda(\ell)}{\lambda(\ell) - \mu^*}\right), \tag{1}$$

where μ^* is the Coulomb pseudopotential and θ is an energy scale that depends on the cut-off bandwidth ω_c^{sf}. On applying a variational approach to the linearized version of Boltzmann's transport equations within the relaxation time approximation (wherein the inverse scattering time is calculated by the use of Fermi's golden rule) we obtain [19] $\Delta\rho = A(\ell)T^2$ and

$$A(\ell) = F_\ell^2 \left|\frac{\lambda(\ell) - \mu^*}{1 + \lambda(\ell)}\right|^2. \tag{2}$$

Here, F_ℓ represents the efficiency of momentum relaxation and the availability of phase space for scattering [19]. Within the FL phase (where ℓ is long, $1 \ll k_F\ell_p < \infty$, $\rho_\circ$ is small, and k_F is the Fermi wave number), Eq. (2) can be expanded to give the following relation [19]:

$$A(\ell) \simeq A_\circ + A_1(\rho_\circ) + A_2(\rho_\circ)^2, \tag{3}$$

where $A_i(i = 0, 1, 2)$ are functions of F_ℓ, μ^*, and $\lambda_\circ = \lambda(\ell \to \infty)$. The finding of Fig. 1(e) indicates that Eq. 3 is reduced to $(A - A_\circ) \propto (\rho_\circ)^2$: a contribution with a dominant A_2 but a negligible A_1 (no Koshino-Taylor contribution).

Finally, on combining the expressions for $T_c(\ell)$ in Eq. 1 and $A(\ell)$ in Eq. 2, we arrive at

$$T_c(\ell) = \theta\, e^{-\mathcal{F}/\sqrt{A(\ell)}}. \tag{4}$$

This universal and exact kinematic scaling relation (valid for long ℓ, low $\rho_\circ$, within the Fermi-liquid region) is the essence of the the fits in Figs. 1(g,h).

Finally, a simplified but approximate relation between T_c and $(\rho_\circ^{tot} - \rho_\circ^\circ)$ can be obtained in the specific case wherein the *quadratic-in-*$\rho_\circ$ term in Eq. 3 is dominant leading to

$$T_c(\ell) \approx \theta \exp\Big(\frac{-\mathcal{F}}{\sqrt{A_2}\left[\rho_\circ^{tot}(\ell) - \rho_\circ^\circ\right]}\Big) \tag{5}$$

On substituting θ, $\mathcal{F}$, A_2, and $\rho_\circ^\circ$, the solid curve of Fig. 1(f) is obtained with no adjustment.

3 Discussion and Conclusion

Analysis of Fig. 1 indicates that on moving out of the FL region towards the quantum critical point, one observes a monotonic and continuous enhancement in $\rho_\circ$, A, and T_c: an indication that the same interaction channel is in operation and that its strength is enhanced on reducing the pressure. However, Fig. 1 also indicates that below the FL region our analytical expressions do not reproduce the evolution of A(p) or T_c(p). This shortcoming is related to the breakdown of the long-ℓ condition and as such does not invalidate our analysis within the FL region.

In summary, we identified three empirical correlations among T_c, A, and ρ_o within the FL region of the investigated *T-p* phase diagrams. These correlations are attributed to a SF-mediated electron-electron interaction. Theoretically, on adopting many-body techniques, we derive analytic expressions for T_c and A and their correlations that reproduce satisfactorily the aforementioned empirical correlations. It is worth adding that our internally-consistent empirical and theoretical analyses are much helped by the clarity in identifying the difference between the normal (non-SF-related) and the SF-related contributions.

References

[1] (a) Maier T A, *et al.* 2020 *Phys Rev Res* **2** 033132; (b) Li Z, *et al.* 2021 *npj Quantum Mater* **6** 36; (c) Dean M, *et al.* 2013 *Nat Mater* **12** 1019; (d) Lee-Hone N, *et al.* 2020 *Phys Rev Res* **2** 013228
[2] (a) Gegenwart P, *et al.* 2008 *Nat. Phys.* **4** 186; (b) Coleman P 2007 *Handbook of Magnetism and Advanced Magnetic Materials*; (c) Jaccard D, *et al.* 1999 *Physica B* **259-261** 1; (d) Yuan H Q, *et al.* 2003 *Science* **302** 2104; (e) Löhneysen H v, *et al.* 2007 *Rev. Mod. Phys.* **79** 1015
[3] Mathur N D, *et al.* 1998 *Nature* **394** 39
[4] ElMassalami M, Castro P B and Neto M B S 2021 arXiv:2109.14032
[5] Sidorov V, *et al.* 2002 *Phys. Rev. Lett.* **89** 157004
[6] Holmes A T, Jaccard D and Miyake K 2004 *Phys. Rev. B* **69** 024508
[7] Sheikin I, Braithwaite D, Brison J P, Assmus W and Flouquet J 2000 *J. Low Temp. Phys.* **118** 113
[8] Kotegawa H,*et al.* 2006 *J. Phys. Soc. Jpn.* **75** 044713
[9] Araki S, Nakashima M, Settai R, Kobayashi T C and Onuki Y 2002 *J. Phys.: Condens. Matter* **14** L377
[10] Nakashima M, *et al.* 2004 *J. Phys.: Condens. Matter* **16** L255
[11] Nakashima M, *et al.* 2006 *Physica B: Cond Matter* **378** 402–403
[12] Onuki Y*et al.* 2008 *J. Phys. Soc. Jpn.* **77** 37
[13] Hassinger E, Knebel G, Izawa K, Lejay P, Salce B and Flouquet J 2008 *Phys. Rev. B* **77** 115117
[14] de Visser A, Franse J and Menovsky A 1984 *J. Magn. Magn. Mater.* **43** 43
[15] Ott H R, Rudigier H, Fisk Z and Smith J L 1983 *Phys. Rev. Lett.* **50** 1595
[16] Gurvitch M 1986 *Phys. Rev. Lett.* **56** 647
[17] Nunez-Regueiro M, Garbarino G and Nunez-Regueiro M D 2012 *J Phys: Conf. Series* **400** 022085
[18] (a) Ronning F, *et al.* 2006 *Phys. Rev. B* **73** 064519; (b) Paglione J, *et al.* 2003 *Phys. Rev. Lett.* **91** 246405; (c) Bianchi A, *et al.* 2003 *Phys. Rev. Lett.* **91** 257001
[19] ElMassalami M and Neto M B S 2021 *Phys. Rev. B* **104** 014520

Determination of the magnetic structures in the heavy fermion superconductor Ce_3PtIn_{11}

Naoki Shioda[1], Kazuki, Kumeda[1], Hideto Fukazawa[1], Tetsuo Ohama[1], Yoh Kohori[1], Debarchan Das[2,3], Joanna Bławat[2], Dariusz Kaczorowski[2], Koudai Sugimoto[4]

[1]Department of Physics, Chiba University, Chiba 263-8522, Japan
[2]Institute of Low Temperature and Structure Research, Polish Academy of Science, P.O. Box 1410, PL-50-950 Wrocław, Poland
[3]Laboratory of Muon Spin Spectroscopy, Paul Scherrer Institute, CH-5232 Villigen PSI, Switzerland
[4]Department of Physics, Keio University, Yokohama 223-8522, Japan

E-mail: `afga6716@chiba-u.jp`

Abstract. Comprehensive analyses of the ^{115}In-NQR spectra employing a magnetic dipolar model and a transferred hyperfine model are performed to determine the magnetic structures of Ce_3PtIn_{11}. Based on this analysis, we confirmed that the magnetic moment at the Ce(1) site is very small ($\sim$ 4% of that at the Ce(2) site) but finite. Further, for $T_{N2} < T < T_{N1}$ and $T < T_{N2}$, the magnetic moments at the Ce(1) and Ce(2) sites are parallel to the c-axis, and the in-plane propagation vectors in the two Ce atoms sublattices are $(q_a,\ q_b) = (1/2,\ 1/2)$. In the temperature interval $T_{N2} < T < T_{N1}$, for example, both of them have a form of $\boldsymbol{q} = (1/2, 1/2, 0)$. For $T < T_{N2}$, the magnetic structure in the Ce(2) sublattice is defined by the propagation vector $\boldsymbol{q} = (1/2, 1/2, 1/2 \text{ or } 0)$, whereas the Ce(1) sublattice forms a six-fold period structure given by $\boldsymbol{q} = (1/2, 1/2, 1/6)$.

1. Introduction

In regards of quantum criticality and quantum phase transition in rare earth based strongly correlated systems, $Ce_nM_mIn_{3n+2m}$(M = Co, Rh, Pd, Ir, Pt) systems are quite promising because of the interplay between heavy-fermion superconductivity and magnetism in close proximity with quantum critical point [1, 2, 3, 4, 5, 6, 7, 8]. Ce_3PtIn_{11} (n = 3, m = 1, M = Pt) is a relatively new member of the $Ce_nM_mIn_{3n+2m}$ family [9]. Figure 1 shows the tetragonal crystal structure of Ce_3PtIn_{11}, which consists of periodically stacked two layers of $CeIn_3$ and one layer of $CePtIn_5$. In contrast with $CeMIn_5$, Ce_2MIn_8, and $CeIn_3$, this material has two inequivalent Ce sites, Ce(1) and Ce(2), in the unit cell. Ce_3PtIn_{11} undergoes two successive magnetic transitions at T_{N1} = 2.2 K and T_{N2} = 2.0 K followed by superconducting transition at a low temperature (T_c = 0.32 K), clearly observed in thermodynamic quantities such as specific heat and magnetic susceptibility [10, 11, 12]. Temperature dependence of the magnetic entropy shows that one Ce site is magnetic and the other is nearly non-magnetic. It suggests that the magnetic site is the Ce(2), whereas the non-magnetic one is the Ce(1) when considering the crystal structure [11]. Hence, the duality of the Ce sites may play an essential role in governing such a complex ground state exhibiting co-occurrence of antiferromagnetic (AFM) ordering and

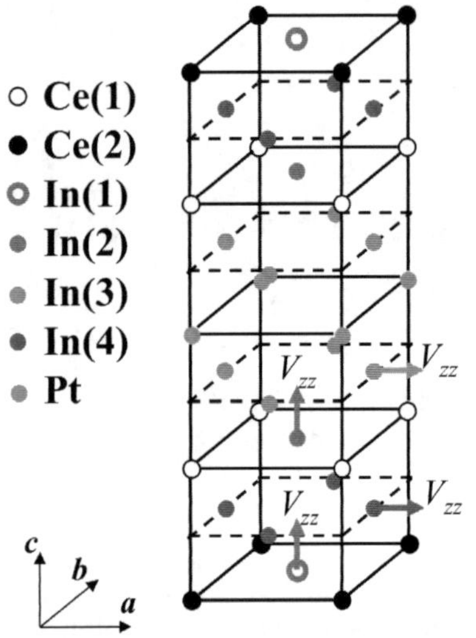

Figure 1. Crystal structure of Ce_3PtIn_{11}. Arrows represent maximum principal axes of the electric field gradient at respective In sites.

superconductivity. However, from the specific heat alone, it is not clear which Ce site is magnetic. Furthermore, direct determination of the magnetic structure by neutron scattering is difficult because of the very high neutron absorption cross-section of the ^{115}In nucleus. Therefore, indirect determination or deduction of the magnetic structures using nuclear quadrupole resonance (NQR) is required as for other members of $Ce_nM_mIn_{3n+2m}$ family [13, 14, 15].

2. Analysis with Dipolar Model

We previously performed ^{115}In-NQR of Ce_3PtIn_{11} [16]. The temperature dependencies of the representative frequencies in the NQR spectra around T_{N1} or T_{N2} were classified into three types: no change, shift, and split, as summarized in Table 1 [16, 17]. To determine the magnetic structures of Ce_3PtIn_{11}, we performed a spectral analysis with a magnetic dipolar model. In this analysis, we confirmed that the magnetic moment at the Ce(1) site is essential to explain the splittings of the In(3)-ν_1, -ν_2 lines. However, the magnitude of the magnetic moment at the Ce(1) site is nearly 4% of that at the Ce(2) site. This is consistent with the specific heat results [11]. For $T_{N2} < T < T_{N1}$ and $T < T_{N2}$, the magnetic moments at the Ce(1) and

Table 1. T-dependence of NQR lines. Spectra in T regions, $T_{N2} < T < T_{N1}$ and $0 < T < T_{N2}$ [16, 17].

site	line	T_{N1}	T_{N2}
In(1)	ν_4	No change	Not observed
In(2)	ν_2	No change	No change
	ν_3	No change	No change
	ν_4	No change	No change
In(3)	ν_1	Split into 2 lines	Split into 4 lines
	ν_2	Split into 2 lines	Split into 6 lines
	ν_3	No change	No change
	ν_4	No change	No change
In(4)	ν_4	Shift	Shift

Ce(2) sites are parallel to the c-axis, and the in-plane propagation vectors in the two Ce atoms sublattices are $(q_a,\ q_b) = (1/2,\ 1/2)$. Magnetic structures such as the propagation vector are analogous to those of other Ce-based materials [14, 18, 19, 20]. The results showed that it has a commensurate magnetic structure when $T_{\mathrm{N1}} < T < T_{\mathrm{N2}}$. In addition, a qualitative magnetic structure was proposed for $T < T_{\mathrm{N2}}$. The structure is commensurate at the Ce(2) sublattice and incommensurate on Ce(1). However, we could not quantitatively reproduce the representative frequency evolution and the intensity ratio of the peaks using the magnetic dipolar model.

3. Analysis with Hyperfine Transferred Model

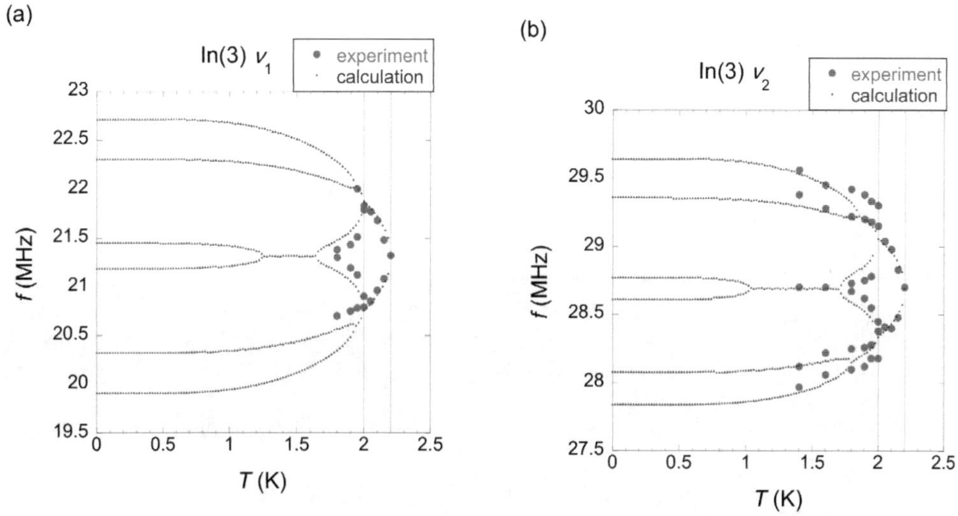

Figure 2. Calculated results for In(3) -ν_1 and -ν_2 lines are plotted in figures. Magnetic structures are $\boldsymbol{q}_{\mathrm{Ce(2)}} = \left(\frac{1}{2},\ \frac{1}{2},\ 0\right)$ for $T_{\mathrm{N2}} < T < T_{\mathrm{N1}}$ and $\boldsymbol{q}_{\mathrm{Ce(2)}} = \left(\frac{1}{2},\ \frac{1}{2},\ \frac{1}{2}\right)$ and $\boldsymbol{q}_{\mathrm{Ce(1)}} = \left(\frac{1}{2},\ \frac{1}{2},\ \frac{1}{6}\right)$ for $T < T_{\mathrm{N2}}$. (color online)

To address these issues and to determine the magnetic structure of this compound with better resolution, we performed spectral analysis beyond the dipolar model to reproduced the ^{115}In-NQR spectra of $\mathrm{Ce_3PtIn_{11}}$ with improved precision. We have obtained the magnetic structure of $\mathrm{Ce_3PtIn_{11}}$ by conducting analysis considering the transferred hyperfine field. The calculations of the hyperfine fields are based on those performed by Curro for the CeMIn$_5$ [15]. We described the details of the calculation method in ref. [21].

Fig. 3 shows the fitting results of ν_1 and ν_2 lines of the In(3) site. As an example of a magnetic structure which reproduces the experimental results, we propose the following structure: $\boldsymbol{q}_{\mathrm{Ce(1)}} = \left(\frac{1}{2},\ \frac{1}{2},\ \frac{1}{6}\right)$, $\boldsymbol{q}_{\mathrm{Ce(2)}} = \left(\frac{1}{2},\ \frac{1}{2},\ \frac{1}{2}\right)$. Specific heat measurement suggests that the magnetic structure of the Ce(2) site is responsible for the magnetic transition at T_{N2} [11]; hence, $\boldsymbol{q}_{\mathrm{Ce(2)}} = \left(\frac{1}{2},\ \frac{1}{2},\ 0\right) (= \boldsymbol{q}_1)$ is required in the intermediate-temperature region. The fittings reproduce the splittings and merges of the lines well, particularly in the low-temperature region. The ratio of the magnetic moments at the Ce(1) and Ce(2) sites at $T = 0$ is $\frac{m_1}{m_2} \simeq \frac{1}{40}$ in both models, which indicates that the magnetic moment at the Ce(1) site is very small even in the present model, similar to the magnetic dipolar model [16].

4. Summary

We analyzed the temperature dependence of the NQR spectra of Ce_3PtIn_{11} by two methods, magnetic dipolar model and transferred hyperfine model. With the dipolar model analysis, we revealed that both the Ce(1) and Ce(2) sites must have $\boldsymbol{m}_{\mathrm{Ce}(\kappa)} \parallel \boldsymbol{c}$, $\boldsymbol{q}_{\mathrm{Ce}(\kappa)} = \left(\frac{1}{2}, \frac{1}{2}, 0 \text{ or } \frac{1}{2}\right)$, and ($\kappa$ = 1, 2) below T_{N1}. Furthermore, with the transferred hyperfine model analysis, we fitted for the ν_1 and ν_2 lines of the In(3) site by a magnetic structure with propagation vectors $\boldsymbol{q}_{\mathrm{Ce}(1)} = \left(\frac{1}{2}, \frac{1}{2}, \frac{1}{6}\right)$. The ratio of the magnitudes of the magnetic moments at the Ce(1) and Ce(2) sites was evaluated as $\frac{m_1}{m_2} \sim \frac{1}{40}$. This is consistent with the specific heat results. We conclude that the Ce(2) site plays a dominant role in the magnetism of Ce_3PtIn_{11}. Concurrently, the magnetic moment at the Ce(1) site, which is very small but finite, may play an essential role in the successive magnetic transitions. Therefore, the transitions of Ce_3PtIn_{11} are considered to be a unique property among the $Ce_nM_mIn_{3n+2m}$ family.

Acknowledgments

This work was supported by JSPS KAKENHI Grant (No. 18K03505, No. 19K14644 and No. 21K03439). The work in Poland was supported by the National Science Centre (Poland) under Research Grant No. 2015/19/B/ST3/03158.

References

[1] Hegger H, Petrovic C, Moshopoulou E G, Hundley M F, Sarrao J L, Fisk Z and Thompson J D 2000 *Phys. Rev. Lett.* **84** 4986.
[2] Petrovic C, Pagliuso P G, Hundley M F, Movshovich R, Sarrato J L, Thompson J D, Fisk Z and Monthoux P 2001 *J. Phys.: Condens. Matter* **13** L377.
[3] Kohori Y, Yamato Y, Iwamoto Y, Kohara T, Bauer E D, Maple M B and Sarrao J L 2001 *Phys. Rev. B* **64** 134526.
[4] Kawasaki S, Mito T, Zheng G-q, Thessieu C, Kawasaki Y, Ishida K, Kitaoka Y, Muramatsu T, Kobayashi T C, Aoki D, Araki S, Haga Y, Settai R and Onuki Y 2001 *Phys. Rev. B* **65** 020504(R).
[5] Fukazawa H, Okazaki T, Hirayama K, Kohori Y, Chen G, Ohara S, Sakamoto I and Matusmoto T 2007 *J. Phys. Soc. Jpn.* **76** 124703.
[6] Kaczorowski D, Gnida D, Pikul A P and Tran V H 2010 *Solid State Commun.* **150** 411.
[7] Yashima M, Taniguchi S, Miyazaki H, Mukuda H, Kitaoka Y, Shishido H, Settai R and Onuki Y 2010 *J. Phys.: Conf. Series* **200** 012238.
[8] Fukazawa H, Nagashima R, Shimatani S, Kohori Y and Kaczorowski D 2012 *Phys. Rev. B* **86** 094508.
[9] Das D, Bławat J, Gnida D and Kaczorowski D 2021 *Physica B: Condensed Matter* **603** 412724.
[10] KratochvílováM, Dušek M, Uhlířová K, Rudajevová A, Prokleška J, Vondráčková V, Custers J and Sechovský V 2014 *J. Cryst. Growth* **397** 47.
[11] Prokleška V, Kratochvílová M, Uhlířová K, Sechovský V and Custers J 2015 *Phys. Rev. B* **92** 161114(R).
[12] Das D, Gnida D, Bochenek Ł, Rudenko A, Daszkiewicz A and Kaczorowski D 2018 *Sci. Rep.* **8** 16703.
[13] Sakai H, Tokunaga Y, Kambe S, Lee H -O, Sidorov V A, Tobash P H, Ronning F, Bauer E D and Thompson J D 2011 *Phys. Rev. B* **83** 140408(R).
[14] Curro N J, Hammel P C, Pagliuso P G, Sarrao J L, Thompson J D and Fisk Z 2000 *Phys. Rev. B* **62** R6100.
[15] Curro N J 2006 *New J. Phys.* **8** 173.
[16] Fukazawa H, Kumeda K, Shioda N, Lee Y S, Kohori Y, Sugimoto K, Das D, Bławat J and Kaczorowski D 2020 *Phys. Rev. B* **102** 165124.
[17] Kambe S, Sakai H, Tokunaga Y, Walstedt R E, Kratochvílová M, Uhlířová K and Custers J 2020 *Phys. Rev. B* **101** 081103(R).
[18] Benoit A, Boucherle J X, Convert P, Flouquet J, Palleau J and Schweizer J 1980 *Solid State Commun.* **34** 293.
[19] Bao Wei, Pagliuso P G, Sarrao J L, Thompson J D, Fisk Z, Lynn J W and Erwin R W 2000 *Phys. Rev. B* **62** R14621.
[20] Bao W, Pagliuso P G, Sarrao J L, Thompson J D, Fisk Z and Lynn j W 2001 *Phys. Rev. B* **64** 020401(R).
[21] Shioda N, Kumeda K, Fukazawa H, Ohama T, Kohori Y, Das D, Bławat J, Kaczorowski D and Sugimoto K 2021 *arXiv*: 2109.06454v1.

Journal of Physics: Conference Series **2164** (2022) 012033 doi:10.1088/1742-6596/2164/1/012033

High-field magnetization and magnetoresistance of the honeycomb Kondo lattice alloys $Ce(Pt_{1-x}Pd_x)_6Al_3$

R. Oishi[1], A. Miyake[2], M. Tokunaga[2], Y. Shimura[1], K. Umeo[3], T. Onimaru[1], and T. Takabatake[1]

[1] *Department of Quantum Matter, Graduate School of Advanced Science and Engineering, Hiroshima University, Higashi-Hiroshima 739-8530, Japan*

[2] *Institute for Solid State Physics, University of Tokyo, Kashiwa, Chiba 277-8581, Japan*

[3] *Department of Low Temperature Experiment, Integrated Experimental Support/Research Division, N-BARD, Hiroshima University, Higashi-Hiroshima 739-8526, Japan*

E-mail: m205560@hiroshima-u.ac.jp

Abstract. $CePt_6Al_3$ is a rare example of the Ce-based honeycomb lattice compound. Substitution of the isovalent Pt for Pd transforms the ground state from a paramagnetic heavy-fermion state to an antiferromagnetic (AFM) ordered state. To gain insight into the role of magnetic frustration in the honeycomb Kondo lattice, we have measured the magnetization $M(B)$ and electrical resistivity $\rho(B)$ of $Ce(Pt_{1-x}Pd_x)_6Al_3$ ($x \leq 0.3$) in pulsed magnetic fields up to 60 T. For the single crystal with $x = 0$, $M(B)$ for B along the easy c axis exhibits a metamagnetic increase and $\rho(B \parallel c)$ bends at $B_m = 28$ T. Taking the temperature at a shoulder in the magnetic susceptibility $\chi(T)$ for $B \parallel c$ as $T_{\chi m} = 17$ K, the ratio $\mu_B B_m / k_B T_{\chi m}$ is estimated to be 1.1. This value agrees with those reported for Ce-based paramagnetic heavy-fermion compounds with similar magnitudes of B_m. With increasing x from 0 to 0.3 in polycrystalline samples, the magnitude of $M(B = 60\text{ T})$ at 1.4 K increases from 0.75 to 2.0 μ_B/Ce, indicating the recovery of localized 4f moments by the Pd substitution for Pt. Using the data of $M(B)$ and $\rho(B)$ for x = 0.1, 0.2, 0.3, we constructed B-T phase diagrams, which show a trend from a small moment AFM ordered state for $x = 0.1$ to a large moment AFM ordered state for $x = 0.3$.

1. Introduction

The cerium atoms in $CePt_6Al_3$ are arranged on a honeycomb lattice [1] which is expected to host magnetic frustration [2]. At low temperatures, a paramagnetic heavy-fermion state is realized, as manifested by the specific heat divided by temperature C/T of 0.5 J/K^2 mol at 0.5 K [3]. Thereby, the Kondo effect and the magnetic frustration likely work cooperatively. In fact, the magnetization of 0.13 μ_B/Ce for $M(B \parallel c = 5\text{ T})$ at 1.8 K is one order of magnitude smaller than 2.14 μ_B/Ce expected for the full moment of a free Ce^{3+} ion. Application of high magnetic fields has significant effects on frustrated Kondo lattice systems by relieving both the Kondo effect and magnetic frustration. For example, in the kagome antiferromagnet CePdAl with T_N = 2.7 K, both $M(B)$ and electrical resistivity $\rho(B)$ show multi-step transitions due to the successive transformations of the ground state [4–6].

We recently found the development of an antiferromagnetic (AFM) order in partially Pd-substituted $Ce(Pt_{1-x}Pd_x)_6Al_3$ ($x \leq 0.3$) without carrier doping and chemical pressure [7]. For $x = 0.05$, C/T and $\rho(T)$ jump at T_N = 1.8 K. With increasing x to 0.2, the λ-type anomaly in C/T becomes pronounced and T_N increases to 3.8 K. The value of magnetic entropy at T_N increases by three times with increasing x and reaches 3.6 J/K mol for $x = 0.3$. It was suggested that the ground state transforms from a spin density wave state for $x \leq 0.15$ to an AFM ordered state of more localized moments for $x \geq 0.2$. In this paper, we intend to reveal the characteristic behaviour of the honeycomb Kondo lattice compounds $Ce(Pt_{1-x}Pd_x)_6Al_3$ by measuring $M(B)$ and $\rho(B)$ in pulsed magnetic fields up to 60 T.

2. Experiments

For the measurements of $M(B)$ and $\rho(B)$ of $Ce(Pt_{1-x}Pd_x)_6Al_3$, we used single crystalline samples for x = 0 and polycrystalline ones for $x \neq 0$. The single crystal for $x = 0$ was grown by the Czochralski method with a radio-frequency heating furnace [8], whereas polycrystalline samples with x = 0.1, 0.2, and 0.3 were prepared by the arc-melting and annealing [7]. Pulsed magnetic fields up to 60 T were generated using nondestructive magnets with durations of ~ 36 ms at the International MegaGauss Science Laboratory of the Institute for Solid State Physics at the University of Tokyo [9]. $M(B)$ was measured by the standard induction method using coaxial pick-up coils. For the measurement of $\rho(B)$, the four-probe AC method was employed in the longitudinal configuration $B \parallel I$ for the direction of electric current I to avoid the effect of cyclotron motion. We used samples with thickness less than 0.3 mm to reduce self-heating due to eddy currents induced by pulsed magnetic fields.

3. Results and discussion

Figures 1 (a) and 1(b) show, respectively, the field dependences of $M(B)$ and normalized magnetoresistance, $MR(B) = \{\rho(B) - \rho(0)\}/\rho(0)$, for single-crystalline $CePt_6Al_3$. A metamagnetic behaviour manifests in $M(B \parallel c, T = 1.4\text{ K})$ at around B_m = 28 T, above which $M(B \parallel c)$ continues to increase to 1.4 μ_B/Ce with increasing B to 60 T. At 4.2 K, the metamagnetic behaviour is blurred. On the contrary, $M(B \perp c)$ increases linearly with B in the whole range. As shown in Fig. 1(b), $MR(B \parallel c, T = 1.4\text{ K})$ displays a bend at around B_m = 28 T, decreases with further increasing B, and reaches −36% at 55 T. For $B \perp c$, however, MR(B) increases by 5 % up to 15 T and becomes flat.

We recall that most of paramagnetic heavy fermion compounds exhibit metamagnetic behaviours in $M(B)$. It is empirically known that the metamagnetic field B_m is scaled by the temperature at the maximum and/or shoulder in the magnetic susceptibility $\chi(T)$, $T_{\chi m}$, by the relation $\mu_B B_m / k_B T_{\chi m}$ = 0.54 ~ 1.1 [10]. For example, Ce-based tetragonal compounds $CeFe_2Ge_2$ [11] and $CeNi_2Ge_2$ [12] exhibit metamagnetic increases in $M(B \parallel c)$ at B_m =

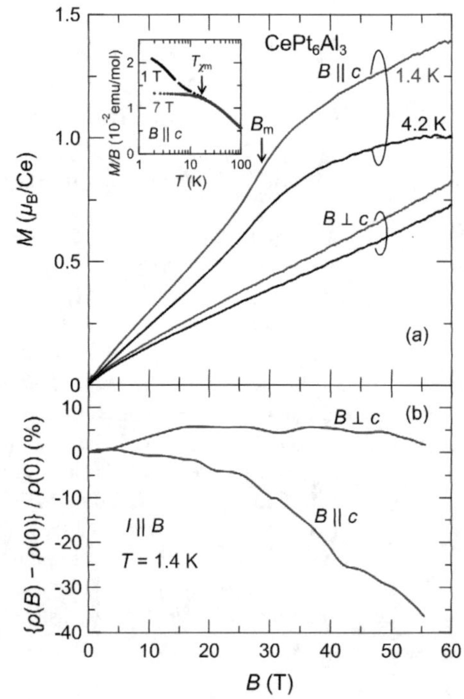

Figure 1. Isothermal (a) magnetization and (b) longitudinal magnetoresistance of single-crystalline $CePt_6Al_3$ samples for $B \parallel c$ and $B \perp c$ in pulsed magnetic fields. The arrow indicates the magnetic field B_m where $M(B \parallel c)$ exhibits a metamagnetic increase. The inset displays the temperature dependence of M/B measured in $B \parallel c$ = 1 and 7 T. The arrow indicates the shoulder.

30 and 45 T, respectively, yielding the ratio $\mu_B B_m / k_B T_{\chi m}$ = 1.1. For $CePt_6Al_3$, as shown in the inset of Fig. 1(a), $\chi(T) = M(T)/B$ measured in $B \parallel c$ = 7 T exhibits a shoulder at 17 K. Using this temperature as $T_{\chi m}$, we obtain the ratio $\mu_B B_m / k_B T_{\chi m}$ as 1.1, which agrees with the above value. However, it is not clear yet how the metamagnetic increase in $M(B)$ is associated with the relief of magnetic frustration in $CePt_6Al_3$.

Let us discuss the result of MR($B \parallel c$) showing the bend at B_m = 28 T. The large and negative MR($B \parallel c$) up to −36% may have two origins. One is the decrease in Kondo scattering due to the development of localized 4*f* moments. The other is the decrease in magnetic scattering from the frustrated state due to the relief of magnetic frustration. In any case, the large and negative value of MR indicates that most of the residual resistivity in zero field arises from scattering of conduction electrons from 4*f* electrons.

We turn our attention to the data of $M(B)$ for polycrystalline samples $Ce(Pt_{1-x}Pd_x)_6Al_3$ with x = 0.1, 0.2, and 0.3, in which AFM transitions occur at 3.2, 3.75, and 3.3 K, respectively [7]. Figure 2 displays the isothermal magnetization curves $M(B)$ up to 60 T at 1.4 K for x = 0.1, 0.2, and 0.3. For comparison, we added the data for x = 0 which are calculated as $\{M(B \parallel c) + 2M(B \perp c)\}/3$ from the data in Fig. 1(a). Upon increasing B up to 60 T, the $M(B)$ curves for x = 0.1, 0.2, and 0.3 gradually increase without showing any clear anomaly. The magnitude of M at 60 T increases with x by three times from 0.76 μ_B/Ce for x = 0 to 2.0 μ_B/Ce for x = 0.3. The latter value is close to the saturation value expected for a free Ce^{3+} ion, suggesting the recovery of a localized magnetic moment. However, the continuous increase of $M(B)$ with B even at 60 T is unlike the saturation behavior expected for fully localized magnetic moments.

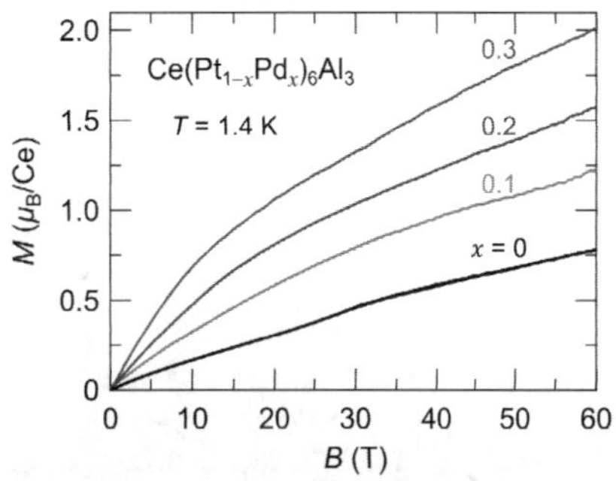

Figure 2. Isothermal magnetization $M(B)$ curves at 1.4 K for polycrystalline samples of $Ce(Pt_{1-x}Pd_x)_6Al_3$ ($x \leq 0.3$).

Note that the temperature 1.4 K for the measurements is lower than T_N for x = 0.1, 0.2, and 0.3. Thereby, we expect some anomaly to appear in $M(B)$ and $\rho(B)$ at the field where the AFM ordered structure is changed and/or destroyed by external fields. To detect such anomalies, we take the differential with respect to the data at 4.2 K where the systems are in paramagnetic states. As an example, $M(B, T)$ and $\Delta M(B, T) = M(B, T) - M(B, 4.2\ K)$ for x = 0.2 are represented in Figs. 3(a) and 3(b), respectively. In the $\Delta M(B, T)$ curves at T = 1.4 K and 2.3 K, there are bends at around B_M = 15 T and 12 T, respectively. We treated the $\rho(B)$ data similarly as shown in Fig. 4 for x = 0.2. The results of $\Delta\rho(B, T) = \rho(B, T) - \rho(B, 4.2\ K)$ are displayed in Fig. 4(b). In the $\Delta\rho(B, T)$ curves at 1.4 K and 2.3 K, we notice bends at B_ρ = 14 T and 12 T, respectively, which are close to the values of B_M. For x = 0.1 and 0.3, we also obtained the values of B_M and B_ρ, respectively, from the data of $\Delta M(B, T)$ and $\Delta\rho(B, T)$.

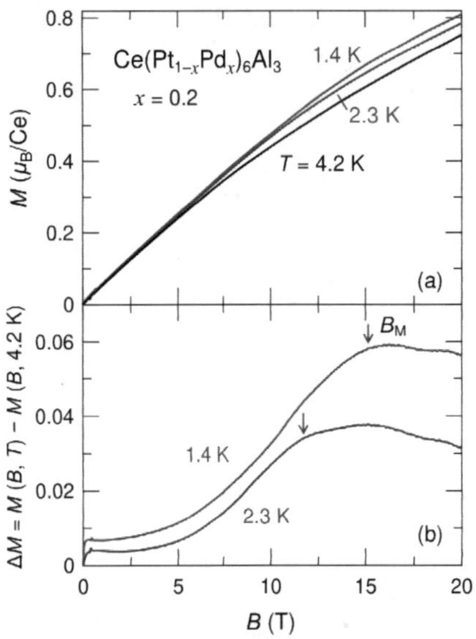

Figure 3. (a) Isothermal magnetization curves $M(B)$ at 1.4, 2.3, and 4.2 K for a polycrystalline $Ce(Pt_{0.8}Pd_{0.2})_6Al_3$ sample. (b) The differential $\Delta M(B) = M(B, T) - M(B, 4.2\ K)$ at T = 1.4 and 2.3 K. The arrows mark the field B_M where the differential $\Delta M(B)$ bends.

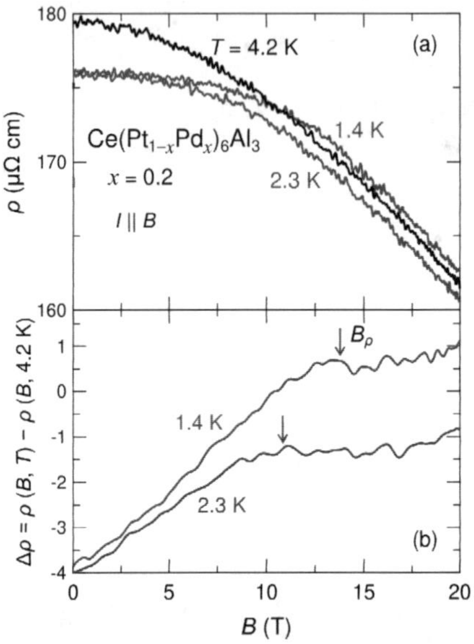

Figure 4. (a) Field dependence of the electrical resistivity $\rho(B)$ in a longitudinal configuration $I \parallel B$ at 1.4, 2.3, and 4.2 K for a polycrystalline $Ce(Pt_{0.8}Pd_{0.2})_6Al_3$ sample. (b) The differential $\Delta\rho(B) = \rho(B, T) - \rho(B, 4.2\ K)$ at T = 1.4 and 2.3 K. The arrows mark the field B_ρ where the differential $\Delta\rho(B)$ bends.

In our previous paper [7], we presented a B-T phase diagram of $Ce(Pt_{1-x}Pd_x)_6Al_3$ in a range $B < 13$ T, as shown in Fig. 5. The open squares, triangles, and circles denote the anomalies observed in the measurements of $\chi(T)$, $\rho(B)$, and specific heat $C(T)$ in steady magnetic fields. The data obtained from the present experiments in pulsed fields are plotted with closed marks in the B-T phase diagram. The closed marks are on the lines extrapolated from phase boundaries between the paramagnetic state and AFM ordered state. It is noteworthy that the boundary line for $x = 0.3$ bends more markedly than that for $x = 0.1$ despite having almost equal value of T_N at $B = 0$. Looking at Fig. 2, we recall that the value of $M(B = 60\ T)$ at 1.4 K is increased from 1.25 to 2.0 μ_B/Ce on going from $x = 0.1$ to 0.3. The elongation of magnetic moment per Ce and strong bending of the boundary line for $x = 0.3$ are consistent with the idea that the Kondo effect and magnetic frustration coexisting for $x = 0.1$ are relieved for $x = 0.3$. However, we did not observe any indication of multiple metamagnetic transitions as a result of partial release of frustration, as was reported in the quasikagome Kondo lattice CePdAl [4–6].

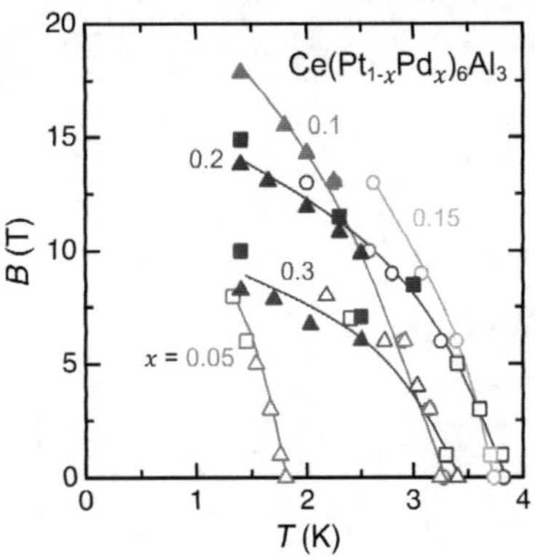

Figure 5. Magnetic field vs temperature phase diagrams for $Ce(Pt_{1-x}Pd_x)_6Al_3$ ($x \leq 0.3$) constructed from measurements of magnetization (closed squares) and magnetoresistance (closed triangles) in pulsed fields together with the data obtained by magnetic susceptibility, magnetoresistance, and specific heat measurements in steady fields as marked by open squares, triangles, and circles, respectively [7] .

4. Summary

To gain insight into the role of magnetic frustration in the Ce-based honeycomb compounds $Ce(Pt_{1-x}Pd_x)_6Al_3$, we have measured field dependences of $M(B)$ and $\rho(B)$ up to 60 T for the single-crystalline sample for $x = 0$ with the paramagnetic heavy-fermion state and polycrystalline samples for $x = 0.1$, 0.2, and 0.3 with AFM ordered states. For the single crystal with $x = 0$, $M(B)$ in B along the easy c axis exhibits the metamagnetic increase and $\rho(B \parallel I)$ bends at around $B_m = 28$ T. Taking the temperature of 17 K at the shoulder in $\chi(T)$ for $B \parallel c$ as $T_{\chi m}$, we obtain the ratio $\mu_B B_m / k_B T_{\chi m} = 1.1$, in agreement with the values for Ce-based paramagnetic heavy-fermion compounds $CeNi_2Ge_2$ and $CeFe_2Ge_2$. Upon going from $x = 0$ to 0.3, the magnitude of $M(B = 60$ T) at 1.4 K increases from 0.75 to 2.0 μ_B/Ce, indicating the recovery of localized 4f moment by Pd substitution for Pt. Using the data of $M(B)$ and $\rho(B)$ for $x =$ 0.1, 0.2, and 0.3, we constructed B-T phase diagrams, which show a trend from a small moment AFM state for $x = 0.1$ to a large moment AFM state for $x = 0.3$. However, we did not observe multiple metamagnetic transitions characterizing frustrated magnetic systems. To find such transitions, further experiments using the Pd-substituted single-crystalline samples are required.

5. Acknowledgments

This work was supported by JSPS KAKENHI Grant Number No. JP21K03473. Magnetization and magnetoresistance measurements were carried out at the International MegaGauss Science Laboratory under the joint-research in the Institute for Solid State Physics, the University of Tokyo (Proposal No. 202012-HMBXX-0003).

References

[1] Eusterman F, Stegemann F, Renner K and Janka O 2017 *Z. Anorg. Allg. Chem.* **643** 1836
[2] Ferrari F and Becca F 2020 *J. Phys.: Condens. Mater* **32** 274003
[3] Oishi R, Ohmagari Y, Kusanose Y, Yamane Y, Umeo K, Shimura Y, Onimaru T and Takabatake T 2020 *J. Phys. Soc. Jpn.* **89**, 104705
[4] Zhao H, Zhang J, Hu S, Isikawa Y, Luo J, Steglich F and Sun P 2016 *Phys. Rev B* **94** 235131
[5] Lucas S, Grube K, Hung L. C, Sakai A, Wunderlich S, Green L E, Wosnitza J, Fritsch V, Gegenwart P, Stockert O and Löhneysen v H 2017 *Phys. Rev. Lett.* **118** 107204
[6] Mochizuki K, Shimizu Y, Kondo A, Nakamura S, Kittaka S, Kono Y, Sakakibara T, Ikeda Y, Isikawa Y and Kindo K 2017 *J. Phys. Soc. Jpn.* **86** 034709
[7] Oishi R, Umeo K, Shimura Y, Onimaru T, Strydom M A and Takabatake T 2021 *Phys. Rev B* **104** 104411
[8] Kim S M, Echizen Y, Umeo K, Kobayashi S, Sera M, Salamakha S P, Sologub L O, Takabatake T, Chen X, Tayama T, Sakakibara T and Maple B M 2003 *Phys. Rev. B* **68** 054416
[9] Kindo K 2001 *Physica B*, **294** 585
[10] Aoki D, Knafo W and Sheikin I 2013 *C. R. Phys.* **14** 53
[11] Sugawara H, Aoki Y, Sato H, Mushnikov N and Goto T 1999 *J. Phys. Soc. Jpn.* **68** 1094
[12] Miyake A, Sato Y, Tokunaga M, Jatmika J and Ebihara T 2017 *Phys. Rev. B* **96** 085127

Thermal Expansion in a diluted Ce system $La_{1-x}Ce_xCu_6$

Y. Yamane[1], **T. Maeda**[2], **A. Yamaguchi**[1], **A. Sumiyama**[1], **and Y. Ōnuki**[3]
[1]Graduate School of Science, University of Hyogo, Kamigori, Hyogo 678-1297, Japan
[2]School of Science, University of Hyogo, Kamigori, Hyogo 678-1297, Japan
[3]Faculty of Science, University of the Ryukyus, Nishihara, Okinawa 903-0213, Japan

E-mail: y-yamane@sci.u-hyogo.ac.jp

Abstract. The thermal expansion of a diluted Ce system $La_{1-x}Ce_xCu_6$ for $(0.6 \leq x \leq 1)$ has been measured between 10 and 150 K to reveal the change from the coherent heavy Fermion state $(0.9 \leq x \leq 1)$ to the incoherent Kondo state $(0 < x \leq 0.73)$. The large Ce concentration x dependence of the linear thermal expansion coefficient along b-axis $\alpha_b(T)$ suggests that the coupling between the $4f^1$ electron and the lattice strain is the largest along the b-axis in the three crystallographic axes. The maximum of the magnetic contribution to the volume thermal expansion coefficient $\beta_m(T)$ at $T = 50$ K is retained in the x range of $0.6 \leq x \leq 1$, suggesting the crystalline electric field (CEF) level for $x = 1$ doesn't change by the substitution. Furthermore, the upturn in $\beta_m(T)$ below 25 K, which should be a precursor of the maximum at $T = 2.5$ K reported for $x = 1$, is retained when we decrease x from 1 to 0.6. Because the ground state for $x = 0.6$ is the incoherent Kondo state, the robustness of the maximum at $T = 50$ K and upturn in the current x value implies that $\beta_m(T)$ in $10 \leq T \leq 150$ K is attributed to the CEF and Kondo effects rather than the formation of the heavy Fermion state.

1. Introduction

Cerium-based intermetallic compounds show a variety of phenomena arising from the strong hybridization between the $4f^1$ electron and the conduction electrons ($c - f$ hybridization), i.e., heavy Fermion states, non-Fermi-liquid behavior, and unconventional superconductivity [1, 2, 3]. Among them, an orthorhombic $CeCu_6$ is a well-known heavy Fermion compound exhibiting the large Sommerfeld coefficient $\gamma \sim 1.5$ J/K^2-mol, the T^2 dependence in the electrical resistivity, and a maximum in the magnetic susceptibility, which is accompanied by the Kondo temperature $T_K \sim 3$ K [4, 5, 6, 7]. While $CeCu_6$ crystallizes in the orthorhombic structure with the space group *Pnma* (#62, D_{2h}^{16}) at room temperature, it shows a structural transition at $T_s \sim 220$ K, below which the space group changes to the monoclinic $P2_1/c$ (#14, C_{2h}^5) with a small angle modification $\delta\beta = 1.36°$. The Au-substituted system $CeCu_{6-x}Au_x$ shows the non-Fermi-liquid behaviors, which have revealed the quantum criticality originating from the competition between the onsite Kondo effect and the intersite Ruderman-Kittel-Kasuya-Yosida (RKKY) interaction [8]. The observation of the quantum criticality in $CeCu_{6-x}Au_x$ indicates that the non-substituted $CeCu_6$ is located near the quantum critical point. On the other hand, the electrical resistivity and magnetic susceptibility measurements on the diluted Ce system $La_{1-x}Ce_xCu_6$ have revealed the consecutive change from the coherent heavy Fermion state to the incoherent Kondo state with a crossover concentration $x \sim 0.7$ [9].

The thermal expansion of $CeCu_6$ have been reported in Refs. [10, 11]. The volume thermal expansion coefficient $\beta(T)$ shows a maximum at $T = 2.5$ K, which temperature is close to $T_K \sim 3$ K. Because the maximum is observed in the other heavy Fermion compounds such as $CeRu_2Si_2$ [12] and $EuNi_2P_2$ [13], it would be attributed to formation of the heavy Fermion state. On the other hand, in our knowledge, the thermal expansion arising from the Kondo effect has not been reported so far. Bearing this in mind, we measured the thermal expansion of the diluted Ce systems $La_{1-x}Ce_xCu_6$ ($0.6 \le x \le 1$) down to 10 K in order to reveal the change of $\beta(T)$ from the heavy Fermion state ($0.9 \le x \le 1$) to the incoherent Kondo state ($0 < x \le 0.73$).

2. Experiments

Single-crystalline samples of $La_{1-x}Ce_xCu_6$ were prepared by the Czochralski method with boron nitride or tungsten crucibles as described in the previous papers [9, 14]. The thermal expansion was measured at 10–150 K by the active-dummy method with two strain gauges (Kyowa Dengyo, KFLB-12-120-C1-11). For the measurements, the laboratory-made dilatometer was installed on a commercial SQUID magnetometer (Quantum Design, MPMS) and operated with an external device control option.

3. Results and Discussion

Figure 1 shows the temperature dependences of the linear thermal expansion coefficients in $La_{1-x}Ce_xCu_6$ for $0.6 \le x \le 1$ along three crystallographic axes, $\alpha_i(T)$ $(i = a, b, c)$. The $\alpha_i(T)$ data for $x = 0$ ($LaCu_6$) [11] are plotted with the black solid curve for comparison. All of the $\alpha_i(T)$ data are derived by differentiating the relative length changes $dL_i(T)/L_i$ with respect to temperature. At first, we discuss the behavior of $\alpha_i(T)$ for the non-substituted system $x = 1$. The $\alpha_a(T)$ data for $x = 1$ monotonically decrease on cooling from 150 K to 12 K and the sign changes to negative below 12 K. The $\alpha_b(T)$ data, in contrast, remain constant in the temperature range of $40 \le T \le 150$ K and decrease on further cooling below 40 K. The $\alpha_c(T)$ data show a minimum at $T = 25$ K, and then an upturn in $T < 25$ K, . The magnitude and the temperature dependences of $\alpha_i(T)$ are almost in good agreement with those of the previous reports [10, 11].

Next, we turn to the data of the diluted Ce system $La_{1-x}Ce_xCu_6$ for $0.6 \le x < 1$. The $\alpha_b(T)$ curves mostly decrease with decreasing the Ce concentration x from 1 to 0.6. The $\alpha_a(T)$ and $\alpha_c(T)$ data, in contrast, hardly change in the current x region. Rather, the $\alpha_a(T)$ data for $x = 1$ are almost identical to those for the nonmagnetic $x = 0$. The anisotropic change against x would be attributed to the anisotropic coupling between the $4f^1$ electron and lattice strain. In particular, the large change in $\alpha_b(T)$ suggests that the coupling along the b-axis is the largest in the three crystallographic axes.

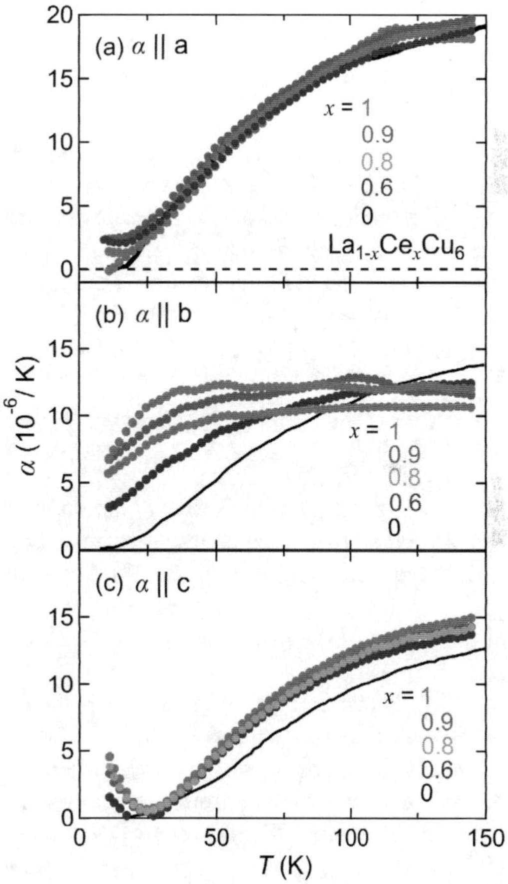

Figure 1. Linear thermal expansion coefficients $\alpha(T)$ of $La_{1-x}Ce_xCu_6$ ($0.6 \le x \le 1$) along (a) a-axis, (b) b-axis, and (c) c-axis. Data for $x = 0$ ($LaCu_6$) [11] are plotted with a black solid curve for comparison.

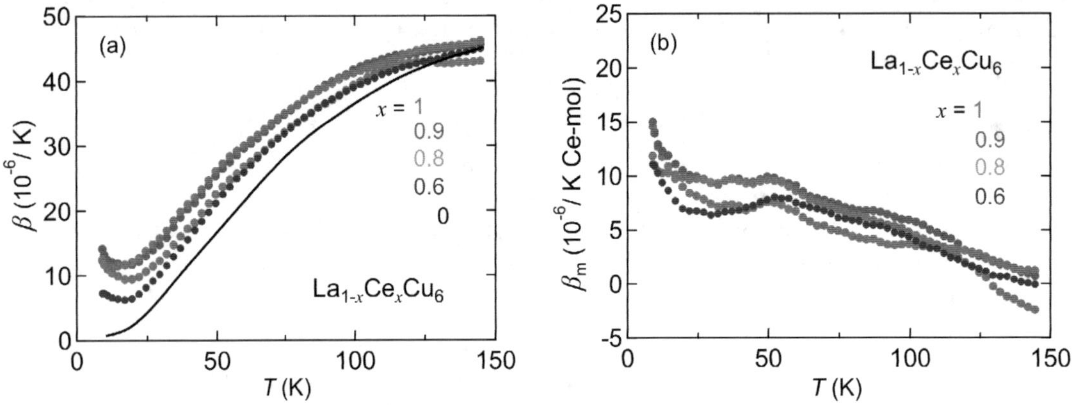

Figure 2. (a) Volume thermal expansion coefficients $\beta(T)$ of $La_{1-x}Ce_xCu_6$ ($0 \leq x \leq 1$). (b) Magnetic contribution to $\beta(T)$, $\beta_m(T)$, per Ce-mol.

Figure 2(a) shows the temperature dependences of the volume thermal expansion coefficient $\beta(T)$ in $La_{1-x}Ce_xCu_6$ for $0 \leq x \leq 1$, where $\beta(T)$ is derived by $\beta(T) = \alpha_a(T) + \alpha_b(T) + \alpha_c(T)$. Here, since the angle modification at T_s is small as described in the introduction, we approximate that the system remains in the orthorhombic structure. All the $\beta(T)$ data monotonically decrease on cooling from 150 K. For $x = 1$, the $\beta(T)$ data exhibit a minimum at $T = 12$ K and an upturn on further cooling below 12 K. With decreasing x from 1 to 0, the $\beta(T)$ curves gradually decrease, indicating the reduced magnetic contribution. At $x = 0$ [11], the $\beta(T)$ data monotonically decrease down to the lowest temperature of 10 K with no minimum.

In order to discuss the thermal expansion arising from the $4f^1$ electron, we estimated the magnetic contribution to $\beta(T)$, $\beta_m(T)$, by subtracting the data for $x = 0$ from the $\beta(T)$ data for $0.6 \leq x \leq 1$ as a phonon contribution, i.e., $\beta_m(T) = \beta_{x\neq0}(T) - \beta_{x=0}(T)$. Figure 2(b) shows $\beta_m(T)$ per molar cerium. For $x = 1$, the $\beta_m(T)$ data show a maximum at $T = 50$ K. The maximum was observed in the previous study as well [10], in which it was proposed that it may be due to crystal electric field (CEF) excitations with the level scheme 0–64–128K. In fact, similar maximum has been observed in the magnetic specific heat, which could be reproduced by the Schottky anomaly with the above CEF scheme. In the x range of $0.6 \leq x \leq 1$, the maximum retains at the same temperature and magnitude, suggesting that the CEF scheme doesn't change by the La-substitution.

Another feature in $\beta_m(T)$ for $x = 1$ is the upturn at $T < 25$ K. Since the upturn should be a precursor of the maximum at $T = 2.5$ K, we had initially expected that it might be attributed to the formation of the heavy Fermionic state. However, when we decrease the Ce concentration x from 1 to 0.6 in which the ground state changes from the heavy Fermion state to the incoherent Kondo state, the upturn are retained. Therefore, the retaining of the upturn for $x = 0.6$ implies that it arises from the Kondo effect rather than the formation of the heavy Fermion state. To clarify the origin of the upturn accompanied by the maximum at $T = 2.5$ K, further low temperature measurements down to 1.8 K are desired. In particular, it is needed to confirm whether the maximum at $T = 2.5$ K exists or not at $x = 0.6$.

4. Summary

To reveal the change of the volume thermal expansion coefficients $\beta(T)$ from the coherent heavy Fermion state to the incoherent Kondo state, we performed the thermal expansion measurements on the diluted Ce system $La_{1-x}Ce_xCu_6$ for $0.6 \leq x \leq 1$ in the temperature range of $10 \leq T \leq 150$ K. The anisotropic Ce concentration x dependence of the linear thermal expansion coefficients $\alpha_i(T)$ ($i = a, b, c$) suggests that the coupling between the $4f^1$ electron and the lattice strain along the b-axis is the largest in the three crystallographic axes. The maximum in the magnetic contribution to the volume thermal expansion, $\beta_m(T)$, at $T = 50$ K is retained in the x range of $0.6 \le x \le 1$, indicating that the crystalline electric field (CEF) levels for $x = 1$ doesn't change with decreasing x. Moreover, the upturn below 25 K in $\beta_m(T)$, which should be accompanied by the maximum at $T = 2.5$ K reported for $x = 1$, is retained when x is reduced from 1 to 0.6, which implies that $\beta_m(T)$ in $10 \le T \le 150$ K is attributed to the CEF and incoherent Kondo effects rather than the formation of the heavy Fermion state.

5. Acknowledgments

The authors would like to thank K. Wakiya, T. Yanagisawa, E. Hayasaka, and H. Hidaka for helpful discussion. This work was financially supported by grants in aid from JSPS KAKENHI Grant No. 20K22332.

References

[1] Stewart G R, 1984 Rev. Mod. Phys. **56** 755
[2] Stewart G R, 2001 Rev. Mod. Phys. **73** 797
[3] Pfleiderer C, Rev. Mod. Phys. **81** 1551
[4] Ōnuki Y, Shimizu Y, and Komatsubara T, 1984 J. Phys. Soc. Jpn. **53** 1210
[5] Stewart G R, Fisk Z, and Wire M S, 1984 Phys. Rev. B **30**, 482(R)
[6] Ōnuki Y, Shimizu Y, and Komatsubara T, 1985 J. Phys. Soc. Jpn. **54** 304
[7] Ōnuki Y, Settai R, Sugiyama K, Takeuchi T, Kobayashi T C, Haga Y, and Yamamoto E, 2004 J. Phys. Soc. Jpn. **73** 769
[8] Löhneysen H v, Pietrus T, Portisch G, Schlager H G, Schröder A, Sieck M, and Trappmann T, 1994 Phys. Rev. Lett. **72**, 3262
[9] Sumiyama A, Oda Y, Nagano G, Ōnuki Y, Shibutani K, and Komatsubara T, 1986 J. Phys. Soc. Jpn. **55** 1294
[10] De Visser A, Lacerda A, Haen P, Flouquet J, Kayzel F E, and Franse J J M, 1989 Phys. Rev. B **39** 11301
[11] Oomi G, Ōnuki Y, and Komatsubara T, 1990 J. Phys. Soc. Jpn. **59** 803
[12] Lacerda A, De Visser A, Haen P, Lejay P, and Flouquet J, 1989 Phys. Rev. B **40** 8759
[13] Hiranaka Y, Nakamura A, Hedo M, Takeuchi T, Mori A, Hirose Y, Mitamura K, Sugiyama K, Hagiwara M, Nakama T, and Ōnuki Y, 2013 J. Phys. Soc. Jpn. **82**, 083708.
[14] Ōnuki Y, Shibutani K, Hirai T, Komatsubara T, Sumiyama A, Oda Y, Nagano H, Sato H, and Yonemitsu K, 1985 J. Phys. Soc. Jpn. **54**, 2804

Low-temperature physical properties of a new cubic compound $CeMgZn_2$

E Matsuoka[1], T Yoshimoto[2], H Sugawara[1], T Sakurai[3] and H Ohta[4]

[1]Department of Physics, Graduate School of Science, Kobe University, Kobe, Japan
[2]Department of Physics, Faculty of Science, Kobe University, Kobe, Japan
[3]Research Facility Center for Science and Technology, Kobe University, Kobe, Japan
[4]Molecular Photoscience Research Center, Kobe University, Kobe, Japan

E-mail: matsuoka@crystal.kobe-u.ac.jp

Abstract. Magnetic and transport properties of a new cubic compound $CeMgZn_2$ have been examined by measuring the magnetic susceptibility, the magnetization, the electrical resistivity, and the specific heat. $CeMgZn_2$ is a Kondo-lattice compound with trivalent Ce ions. The magnetic susceptibility measured at 0.1 T exhibits a shoulder and a cusp at T_{N1} = 5.4 K and T_{N2} = 3.1 K, respectively. T_{N1} and T_{N2} correspond to the antiferromagnetic-transition temperatures since these temperatures decrease with increasing magnetic field. The large value of the paramagnetic Curie temperature divided by T_{N1} (13.5) implies that T_{N1} is suppressed by geometrical frustration on a face-centered cubic Ce sublattice. The geometrical frustration may also be responsible for the appearance of many magnetic phases in magnetic fields.

1. Introduction

Magnetic ground states of Ce- or Yb-based compounds have been understood by a concept proposed by Doniach [1], i.e., if the Ruderman-Kittel-Kasuya-Yosida (RKKY) interaction is dominant, a magnetically ordered state is realized, whereas the Kondo effect is dominant, a paramagnetic heavy fermion state is realized. In particular, a quantum critical point (QCP) where the ground state changes from a magnetically ordered state to a heavy fermion state has been a subject of intensive studies. This is because quantum fluctuation in the vicinity of a QCP leads some anomalous phenomena such as unconventional superconductivity. Recently, the other view point of quantum criticality has been discussed theoretically, i.e., quantum zero-point motion of the spins induced by magnetic frustration [2]. The magnetic ground states are described by two-axis diagram with the Kondo screening axis (K) and quantum-zero point motion axis (Q) [2]. According to the QK diagram, Ce and Yb compounds possessing magnetic frustration can exhibit a variety of magnetic ground states as a result of competition between the RKKY interaction, the Kondo effect, and the quantum-zero point motion. Although the simplest way to introduce magnetic frustration is to locate Ce or Yb ions on geometrically frustrated lattices, such lattices are rather rare compared with those for transition-metal ions. Therefore, searching for new Ce or Yb compounds with geometrical frustration is important to explore a variety of magnetic ground states and concomitant anomalous phenomena.

We have focused on a new Heusler-type cubic compound $CeMgZn_2$ (space group $Fm\bar{3}m$, O_h^5, No. 225) [3]. The Ce ions are located on the geometrically frustrated lattice since they are on a face-centered cubic lattice and form a network of edge-sharing tetrahedra. No physical properties of this compound have been reported thus far. In this study, polycrystalline samples of $CeMgZn_2$ have been grown and their magnetic and transport properties have been examined by measuring the magnetic susceptibility, the magnetization, the electrical resistivity, and the specific heat.

2. Experimental methods

Polycrystalline samples of $CeMgZn_2$ and a nonmagnetic reference compound $LaMgZn_2$ were grown by melting the stoichiometric constituent elements in Mo crucibles at 1320 K for 12 hours and subsequent annealing at 1220 K for 50 hours. The X-ray diffraction experiments revealed that all of the Bragg peaks in the diffraction pattern of $CeMgZn_2$ can be indexed on the basis of the reported structure, although small peaks of unidentified impurity phases are discernible for $LaMgZn_2$. The lattice constants determined by X-ray diffraction experiments at room temperature are a = 7.023(1) Å for $CeMgZn_2$ and a = 7.101(1) Å for $LaMgZn_2$. The magnetization $M(T, H)$ was measured using a superconducting quantum interference device magnetometer (Quantum Design, MPMS) as functions of the temperature T and magnetic field H between 1.8 and 300 K up to 5 T. The electrical resistivity $\rho(T)$ was measured with a dc four-probe method between 0.4 and 300 K in a laboratory-built ^{3}He cryostat. The specific heat $C(T)$ was measured with a thermal relaxation method between 0.6 and 9 K in a ^{3}He cryostat.

3. Results and Discussion

The inverse magnetic susceptibility H/M of $CeMgZn_2$ (not shown) above 100 K can be fitted by the Curie-Weiss law. The effective magnetic moment μ_{eff} and the paramagnetic Curie temperature θ_P derived by Curie-Weiss fitting were 2.75 μ_B/Ce (μ_B: Bohr magneton) and −73 K, respectively. The Ce ions in this compound are trivalent because the effective magnetic moment is close to that of the free Ce^{3+} ion (2.54 μ_B/Ce). The negative θ_P indicates antiferromagnetic correlation between Ce moments.

Figure 1 shows the temperature dependences of the magnetic susceptibilities M/H of $CeMgZn_2$ measured in various magnetic fields. The M/H measured at 0.1 T shows a shoulder and a cusp at T_{N1} = 5.4 K and T_{N2} = 3.1 K, respectively. T_{N1} and T_{N2} decrease with increasing magnetic fields as indicated by solid and dashed arrows, respectively. The decrease in T_{N1} and T_{N2} by magnetic fields and the negative θ_P indicate that these temperatures are the antiferromagnetic-transition temperatures. Above 2.5 T, only T_{N1} was observed. No anomalies were observed above 3 T.

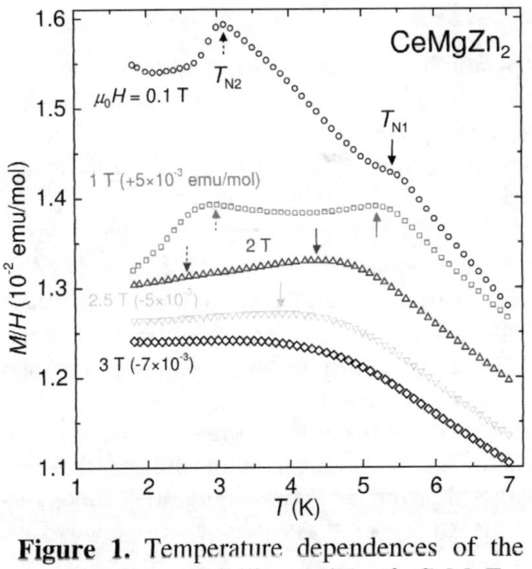

Figure 1. Temperature dependences of the magnetic susceptibility M/H of $CeMgZn_2$ measured in various magnetic fields. The M/H values measured at 1, 2.5 and 3 T are shifted vertically by the values in parentheses.

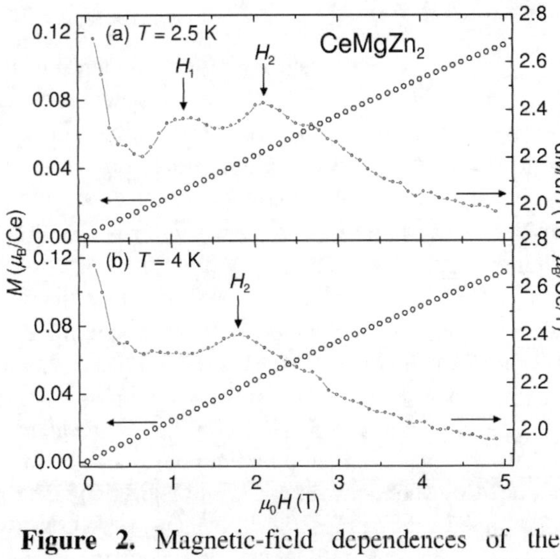

Figure 2. Magnetic-field dependences of the magnetization M and their field derivatives dM/dH of $CeMgZn_2$ measured at (a) 2.5 K and (b) 4 K.

The magnetic-field dependence of the magnetization M at 2.5 K ($T < T_{N2}$) has two anomalies at H_1 and H_2 as shown in the figure 2(a). These anomalies are clearly observed as the peaks of the field derivative dM/dH. On the other hand, the M at 4 K ($T_{N2} < T < T_{N1}$) has one anomaly at H_2 as shown in the figure 2(b).

The temperature dependence of the electrical resistivity ρ is shown in the figure 3. The $\rho(T)$ decreases with decreasing temperature, followed by a minimum at 20 K and shows the $-\ln T$ dependence between 8 and 15 K. This means that $CeMgZn_2$ is a Kondo-lattice compound. A steep decrease is observed below T_{N1}, while no anomalies are observed below T_{N2}.

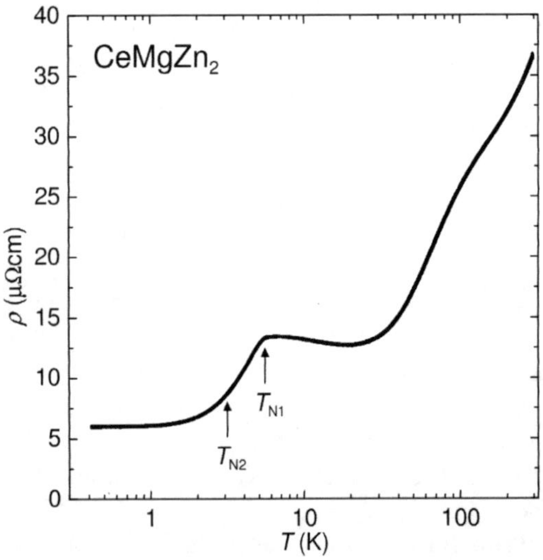

Figure 3. Temperature dependence of the electrical resistivity ρ of $CeMgZn_2$.

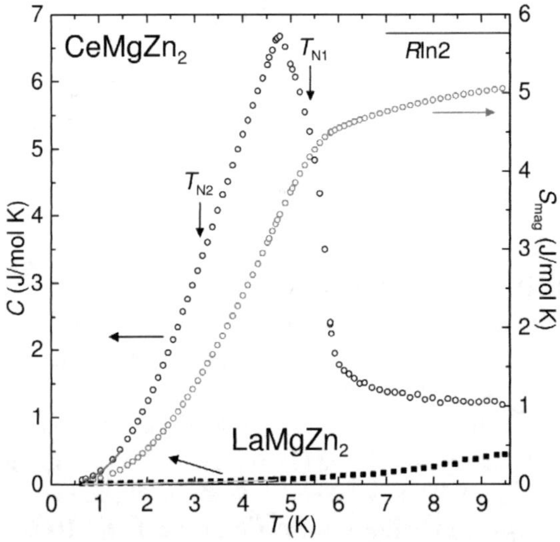

Figure 4. Temperature dependences of the specific heat C and the magnetic entropy S_{mag} of $CeMgZn_2$ (open symbols). The C of $LaMgZn_2$ is shown as the solid squares.

Figure 4 shows the temperature dependence of the specific heat C of $LaMgZn_2$ and $CeMgZn_2$. The $C(T)$ of $LaMgZn_2$ shows typical nonmagnetic behavior. On the other hand, the $C(T)$ of $CeMgZn_2$ has a λ-type anomaly at T_{N1}. This means that the second-order antiferromagnetic transition occurs at T_{N1}. In contrast, the $C(T)$ of $CeMgZn_2$ has no anomaly at T_{N2} as for the case of the resistivity. To examine the mass enhancement of the conduction electrons of $CeMgZn_2$, the $C(T)$ of these compounds were fitted by using a formula $C(T) = \gamma T + \beta T^3$. γ is the electronic specific-heat coefficient. Although only the phonon contributes to β for $LaMgZn_2$, the antiferromagnetic magnon also contributes to β for $CeMgZn_2$. The red and light-blue lines in the figure 4 are the fitted results using this formula. The γ and the β values were estimated to be $\gamma = 0.0054$ J/mol K^2 and $\beta = 3.7 \times 10^{-4}$ J/mol K^4 for $LaMgZn_2$, and $\gamma = 0.018$ J/mol K^2 and $\beta = 0.16$ J/mol K^4 for $CeMgZn_2$. The rather small γ value of $CeMgZn_2$ (only 3 times larger than that of $LaMgZn_2$) reveals that the 4f electrons have localized nature even at low temperatures.

The temperature dependence of the magnetic entropy S_{mag} of $CeMgZn_2$ is also depicted in the figure 4. The S_{mag} was calculated by integrating C_{mag}/T in T, where C_{mag} is the magnetic specific heat that was deduced by subtracting the $C(T)$ of $LaMgZn_2$. The magnetic entropy at T_{N1} is 4.2 J/mol K. Since this value is 73% of $R\ln 2$ (R: gas constant), it is considered that the ground state of 4f energy levels split under the cubic crystalline electric field is the Γ_7 doublet.

The magnetic field versus temperature phase diagram was drawn by plotting the susceptibility anomalies and the magnetization anomalies as shown in the figure 5. Below 3 T, four magnetic phases named I, II, III, and IV appear. We now consider that competing magnetic interaction due to geometrical frustration is responsible for the appearance of many magnetic phases. The frustration effect is evidenced by the reduction of T_{N1}. The value of $|\theta_P|$ divided by transition temperature is often used for evaluating the strength of suppression of transition temperature due to frustration and the value larger than 5 indicates a strong suppression [4]. The $|\theta_P|/T_{N1}$ value of $CeMgZn_2$ is 13.5. This value is comparable to $|\theta_P|/T_N = 12.6$ for CePdAl whose antiferromagnetially ordered state below T_N is strongly affected by frustration [5].

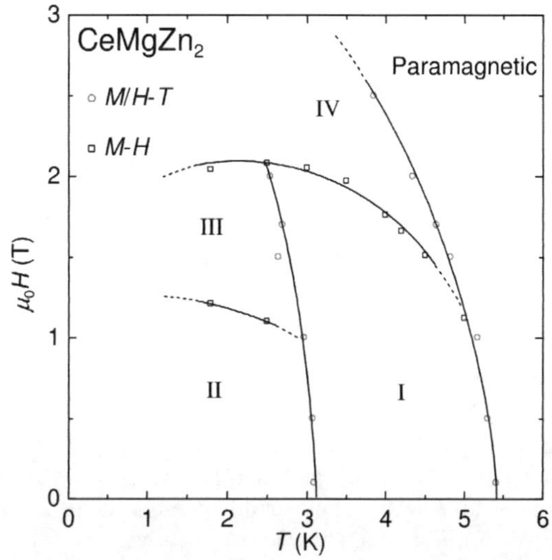

Figure 5. Magnetic field vs. temperature phase diagram of $CeMgZn_2$

4. Conclusions

In this study, polycrystalline samples of $CeMgZn_2$ and the reference compound $LaMgZn_2$ have been grown and their magnetic and transport properties have been examined. We found that $CeMgZn_2$ is a Kondo-lattice compound with the trivalent Ce ions. $CeMgZn_2$ shows antiferromagnetic transition at $T_{N1} = 5.4$ K and $T_{N2} = 3.1$ K. By constructing magnetic field versus temperature phase diagram, we found four magnetic phases below 3 T. Because the $|\theta_P|/T_{N1}$ value reaches 13.5, we consider that geometrical frustration on a face-centered cubic Ce sublattice reduces T_{N1}. The geometrical frustration is also responsible for the appearance of many magnetic phases.

Acknowledgments

This work was partly supported by JSPS KAKENHI Grant Number JP15H05882 (J-Physics) and 20K03861, and the Nippon Sheet Glass Foundation for Materials Science and Engineering.

References

[1] Doniach S 1977 *Physica B+C* **91** 231
[2] Coleman P, Nevidomskyy A H 2010 *J. Low. Temp. Phys.* **161** 182
[3] Pavlyuk V, Solokha P, Dmytriv G, Marciniak B and Paul-Boncour V 2007 *Acta. Cryst.* **E63** i161
[4] Balents L 2010 *Nature* **464** 199
[5] Dönni A, Ehlers G, Maletta H, Fischer P, Kitazawa H and Zolliker M 1996 *J. Phys.: Condens. Matter* **8** 11213

Magnetic and transport properties of a new ferromagnetic orthorhombic compound $CePtAl_2$

H HAYASHI[1], E MATSUOKA[1], H SUGAWARA[1], T SAKURAI[2] and H OHTA[3]

[1]Department of Physics, Graduate School of Science, Kobe University, Kobe, Japan
[2]Research Facility Center for Science and Technology, Kobe University, Kobe, Japan
[3]Molecular Photoscience Research Center, Kobe University, Kobe, Japan

E-mail: matsuoka@crystal.kobe-u.ac.jp

Abstract. We have synthesized polycrystalline samples of $CePtAl_2$ by arc melting method and examined their magnetic, transport and thermal properties by measuring the magnetization, the electrical resistivity, and the specific heat down to 0.4 K. As a result of these measurements, we found that $CePtAl_2$ is a ferromagnetic Ce-based compound with the Curie-temperature T_C = 2.7 K.

1. Introduction

It has been reported that ferromagnetism and superconductivity, which are generally contradictory properties, coexist in some U-based compounds [1]. In order to elucidate the coexistence mechanism, it is necessary to investigate similar phenomena in a larger number of compounds. However, no coexistence phenomena have been found except in a few U-based compounds thus far. Ce-based compounds are the promising candidate showing coexistence since they have the same strong correlation between *f*-electrons and conduction electrons as U-based compounds. We have focused on new compounds $RPtAl_2$ (R = Sc, Y, La-Nd, Sm, Gd-Tm, Lu) [2]. As shown in the figure 1, $RPtAl_2$ crystallizes in the orthorhombic $MgCuAl_2$-type structure (space group *Cmcm*, D_{2h}^{17}, No. 63) [2]. The magnetization measurements of $CePtAl_2$ down to 2.5 K have revealed that $CePtAl_2$ did not show any phase transition and the effective magnetic moment μ_{eff} and the paramagnetic Curie temperature θ_P were estimated to be 2.52 μ_B/Ce and -20.6 K, respectively. In this study, we have grown polycrystalline samples of $CePtAl_2$ and examined their magnetic and transport properties by measuring the magnetization, the electrical resistivity, and the specific heat down to 0.4 K.

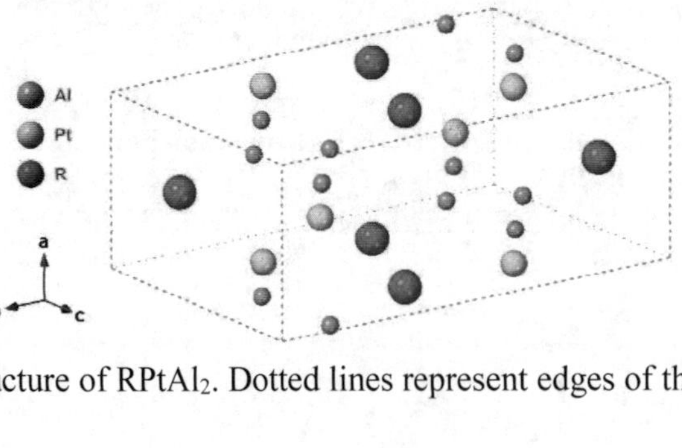

Figure 1. Crystal structure of $RPtAl_2$. Dotted lines represent edges of the chemical unit cell.

2. Experimental Methods

Polycrystalline samples of $CePtAl_2$ are synthesized as the following procedure. First, Ce (99.9%), Pt (99.99%), and Al (99.999%) were weighed in the ideal 1:1:2 atomic ratio and melted using an arc furnace under an Ar gas atmosphere. Next, the samples wrapped in Ta foils were sealed in an evacuated quartz tube to prevent oxidation and annealed in a muffle furnace at 1320 K for one week. The annealed samples were characterized by a powder X-ray diffraction experiment. Almost all of the Bragg peaks in the diffraction pattern can be indexed on the basis of the $MgCuAl_2$-type structure, although small unidentified impurity peaks are discernible. The lattice parameters were estimated to be a = 4.218 Å, b = 11.062 Å, and c = 7.032 Å, which agree with the reported values within the experimental precision.

The magnetization was measured in the temperature and magnetic-field ranges of $1.8 \leq T \leq 300$ K and $0 \leq \mu_0 H \leq 5$ T, respectively, using a superconducting quantum interference device magnetometer (Quantum Design, MPMS). The electrical resistivity was measured with a dc four-probe method in a temperature range of $0.4 \leq T \leq 300$ K using a ^{3}He cryostat. The specific heat was measured with a thermal relaxation method in the temperature range of $0.4 \leq T \leq 10$ K using a ^{3}He cryostat.

3. Results and Discussion

Figure 2 shows the temperature dependence of the inverse magnetic susceptibility H/M of $CePtAl_2$. The H/M follows the Curie-Weiss law above 100 K (see the solid line in the figure 2). From the Curie-Weiss fitting of the H/M, we obtained the paramagnetic Curie temperature θ_P = -5.23 K and the effective magnetic moment μ_{eff} = 2.42 μ_B/Ce. The smaller $|\theta_P|$ compared with that reported previously may be ascribable to the magnetic anisotropy: if the polycrystalline samples of the present and the previous studies have a crystallographic preferred orientation, θ_P value depends on the angle between the orientation direction and the applied field direction. The μ_{eff} value is comparable to that reported previously and we can suggest that the Ce ions in $CePtAl_2$ are in the trivalent state because this value is in excellent agreement with the theorical value μ_{calc} = 2.54 μ_B/Ce for a free Ce^{3+} ion. The inset of the figure 2 shows the temperature dependence of the magnetic susceptibility M/H at low temperatures. M/H increases steeply below 3 K. This behavior is reminiscent of the ferromagnetic transition.

Figure 3 shows the magnetic-field dependences of the magnetization M of $CePtAl_2$ measured at 1.8 and 5 K. In contrast to the paramagnetic behavior of M measured at 5 K, M measured at 1.8 K shows ferromagnetic behavior, i.e., M increases steeply by applying magnetic field and saturates above 2 T.

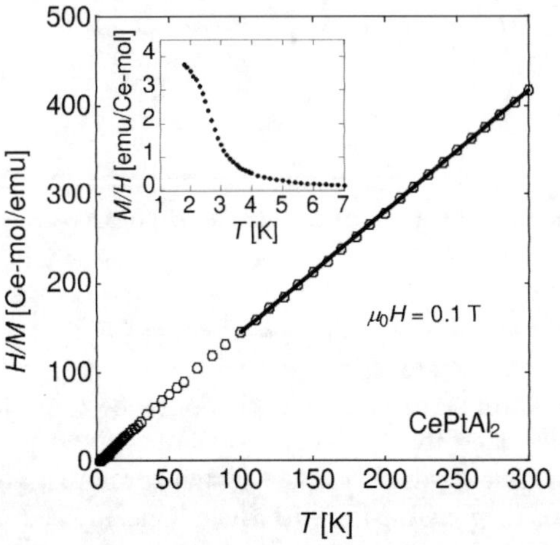

Figure 2. Temperature dependence of the inverse magnetic susceptibility H/M of $CePtAl_2$ measured at 0.1 T. The inset shows the low-temperature part of the magnetic susceptibility M/H.

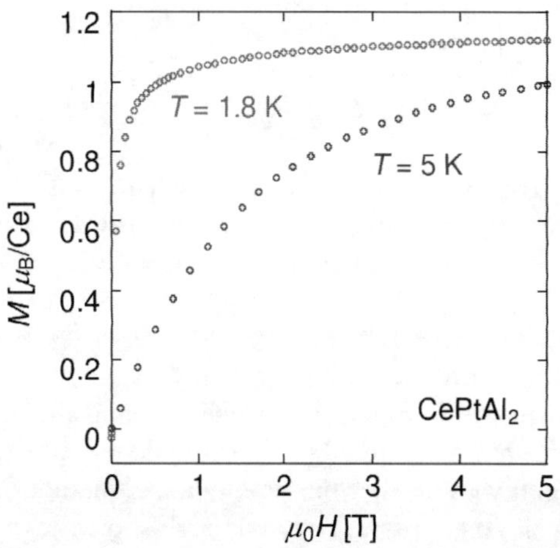

Figure 3. Magnetic-field dependences of the magnetization M of $CePtAl_2$ measured at 1.8 K and 5 K.

Figure 4 shows the temperature dependence of the electrical resistivity ρ of $CePtAl_2$. The low-temperature part of ρ is shown in the inset. The ρ decreases almost linearly with decreasing temperature in the high temperature region. However, ρ decreases rapidly below 2.7 K as indicated by an arrow in the inset. The decrease temperature corresponds to the temperature at which the magnetic susceptibility *M*/*H* increases steeply.

Figure 5 shows the temperature dependence of the specific heat *C*. The *C* has a clear λ-type jump at T_C = 2.7 K. This means that the increase in *M*/*H* and the decrease in ρ are due to the second-order ferromagnetic transition at T_C. The negative θ_P (antiferromagnetic) obtained in this study may be explained by the magnetic anisotropy described above: if we measure the *M*/*H* with the different sample angle to the magnetic field, positive θ_P (ferromagnetic) may be obtained. The temperature dependence of the total entropy *S* deduced by integrating *C*/*T* in temperature is also shown in the figure 5. Since the *S* value at T_C is 3.9 J/mol K, which corresponds to 68% of *R*ln2 (*R* : gas constant), we consider that the ground doublet of Ce^{3+} is responsible for the ferromagnetic transition. Here, the 32% reduction of *S* is not explained by the Kondo effect since the ρ does not show any signature of the Kondo effect and the localized nature of 4*f* electrons are kept down to low temperatures. By considering the fact that the *C* increases with decreasing temperature below 9 K, the reduced *S* value at T_C can be ascribable to the development of short-range ferromagnetic correlation above T_C. In fact, the *S* value reaches 98% at 9 K.

Note that ferromagnetic Ce compounds are rather rare compared with antiferromagnetic ones. Therefore, further studies of $CePtAl_2$ such as pressure effect measurements are worth performing to search for the coexistence of ferromagnetism and superconductivity.

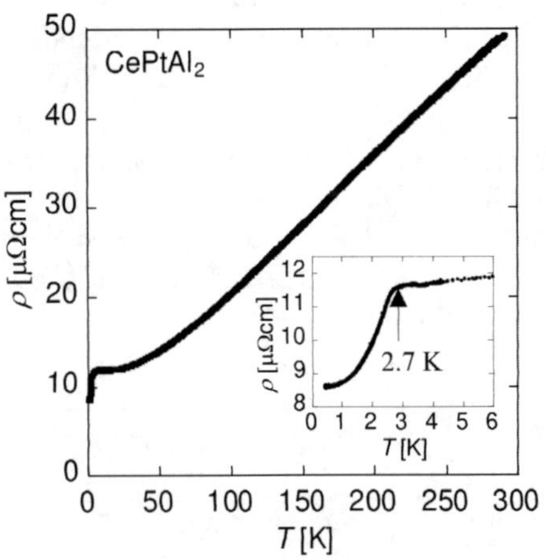

Figure 4. Temperature dependence of the electrical resistivity ρ of $CePtAl_2$. The inset shows the low-temperature part of ρ.

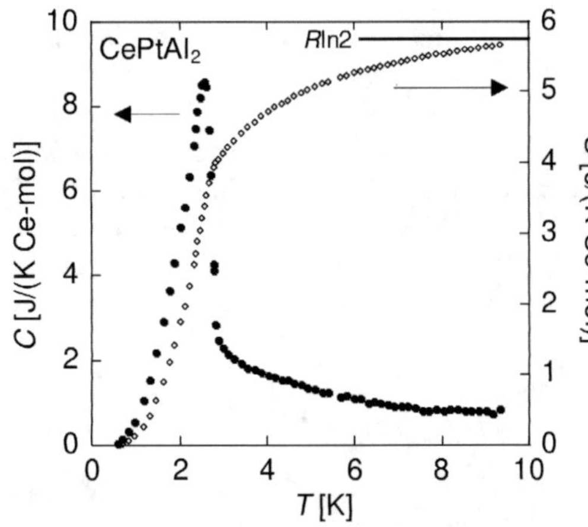

Figure 5. Temperature dependences of the specific heat *C* and the total entropy *S* of $CePtAl_2$.

4. Conclusion

In this study, we have grown polycrystalline samples of $CePtAl_2$ and measured their magnetic, transport, and thermal properties. We have found that $CePtAl_2$ is a ferromagnetic Ce-based compound with the Curie temperature T_C = 2.7 K. Since the electrical resistivity does not show any signature of the Kondo effect, we consider that the localized nature of 4*f* electrons is kept down to the low temperatures. The development of the ferromagnetic short-range correlation is responsible for the reduced entropy at T_C.

Acknowledgment
This work was supported partly by JPJS KAKENHI Grant Number JP15H05882 (J-Physics).

References

[1] Aoki D, Ishida K and Flouquet J 2019 *J. Phys. Soc. Jpn.* **88** 022001
[2] Radzieowski M, Stegemann F, Doerenkamp C, Matar F Samir, Eckert H, Dosche C, Wittstock G and Janka O 2019 *Inorg.Chem.* **58** 7010

Low-temperature physical properties of new orthorhombic compounds $RE_2Au_3Sn_6$ (RE = Ce, La)

K ONISHI[1], E MATSUOKA[1], H SUGAWARA[1], T SAKURAI[2] and H OHTA[3]

[1]Department of Physics, Graduate School of Science, Kobe University, Kobe, Japan
[2]Research Facility Center for Science and Technology, Kobe University, Kobe, Japan
[3]Molecular Photoscience Research Center, Kobe University, Kobe, Japan

E-mail: matsuoka@crystal.kobe-u.ac.jp

Abstract. We have prepared polycrystalline samples of $RE_2Au_3Sn_6$ (RE = Ce, La) and investigated their magnetic, transport and thermal properties. We found that $Ce_2Au_3Sn_6$ shows antiferromagnetic transition at T_N = 2.54 K, while $La_2Au_3Sn_6$ shows no phase transition above 0.4 K. The electronic specific heat coefficient of $Ce_2Au_3Sn_6$ is γ = 350 mJ Ce-mol K^2, which indicates that $Ce_2Au_3Sn_6$ is a heavy fermion compound. The magnetic entropy of $Ce_2Au_3Sn_6$ at T_N is estimated to be 70% of Rln2, which can be attributed to the shielding of the magnetic moment by the Kondo effect.

1. Introduction

Compounds containing Ce show a variety of interesting phenomena, such as anomalous magnetic ordering and heavy fermion state due to the interaction between 4*f* and conduction electrons. These phenomena are due to the interplay between the Ruderman-Kittel-Kasuya-Yosida interaction and the Kondo effect [1]. It is well known that Ce compounds with their magnetic states being at the vicinity of the quantum critical point exhibit some anomalous phenomena such as non-Fermi liquid behavior and unconventional superconductivity. One possible way to search for new compounds with quantum criticality is to search for compounds without magnetic order down to cryogenic temperatures. From this view point, we have focused on newly synthesized compounds $RE_2Au_3Sn_6$ (RE = La, Ce, Pr, Nd, Sm) [2]. $RE_2Au_3Sn_6$ crystallize with the orthorhombic $La_2Zn_3Ge_6$ type structure (space group *Cmcm*, D_{2h}^{17}, No. 63) as shown in the figure 1 [2]. This structure has two independent RE sites. The magnetic susceptibility measurement down to 2.5 K has revealed that $Ce_2Au_3Sn_6$ has trivalent Ce ions and no phase transition above 2.5 K [2]. In this study, we have prepared polycrystalline samples of $RE_2Au_3Sn_6$ (RE = Ce, La) and investigated their magnetic, transport and thermal properties down to 0.4 K.

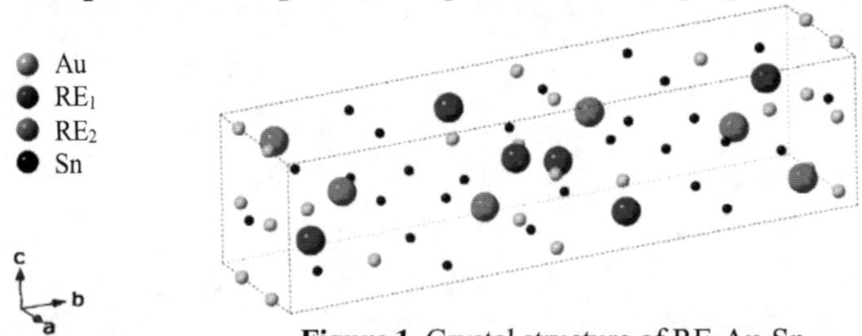

Figure 1. Crystal structure of $RE_2Au_3Sn_6$.

2. Experimental Methods

The polycrystalline samples of $RE_2Au_3Sn_6$ (RE = Ce, La) have been prepared as the following procedure. First, stoichiometric amount of RE (99.9%), Au (99.99%) and Sn (99.999%) were melted by using arc furnace in an argon atmosphere. After that, the samples wrapped in Ta foils were sealed in the evacuated quartz tubes and annealed at 600°C for one week. The resulting polycrystalline samples were analyzed by powder X-ray diffraction. For both $Ce_2Au_3Sn_6$ and $La_2Au_3Sn_6$, all Bragg peaks in the diffraction patterns are indexed on the basis of the reported structure. The lattice parameters determined from X-ray diffraction agree with those reported in the Ref. [2]. We then measured the following three properties. The magnetization M was measured from 1.8 K to 300 K and from 0 T to 5 T using a superconducting quantum interference device magnetometer (Quantum Design, MPMS). The electrical resistivity ρ was measured from 0.4 K to 300 K using a ^{3}He cryostat with the DC four-terminal method. The specific heat C was measured from 0.6 K to 10 K by the thermal relaxation method using a ^{3}He cryostat.

3. Results and Discussion

Figure 2 shows the temperature dependence of the magnetic susceptibility M/H of $Ce_2Au_3Sn_6$. The M/H above 100 K obeys the modified Curie Weiss law $M/H = C_{Curie}/(T - \theta_P) + \chi_0$ as shown by the red line in the figure 2. Here, C_{Curie}, θ_P, and χ_0 are the Curie constant, the paramagnetic Curie temperature, and the sum of the Pauli paramagnetic contribution and the core diamagnetic one, respectively. The obtained parameters are C_{Curie} = 0.769 emu K/Ce-mol, θ_P = -8.02 K, and χ_0 = -0.000437 emu/Ce-mol. The effective magnetic moment of Ce ion is calculated to be μ_{eff} = 2.48 μ_B/Ce by using C_{Curie}. This μ_{eff} value indicates that the Ce ions in $Ce_2Au_3Sn_6$ are trivalent since this value is close to that of the free Ce^{3+} ion (2.54 μ_B/Ce). The negative value of θ_P suggests that the magnetic interaction between the Ce ions is antiferromagnetic. These results are consistent with the previous report.

The low-temperature part of the M/H is shown in the inset of the figure 2. The M/H shows a cusp at T_N = 2.54 K, implying that the antiferromagnetic transition occurs at T_N.

The temperature dependence of the electrical resistivity ρ is also shown in the figure 2 and the inset. The ρ decreases with decreasing temperature and shows a broad shoulder around 70 K. The steep decrease at T_N is ascribable to the decrease in scattering of conduction electrons by Ce magnetic moments.

The magnetic-field dependence of the magnetization M of $Ce_2Au_3Sn_6$ is shown in the figure 3. At 5 K (> T_N), the M increases with increasing magnetic field and shows no anomalies. On the other hand, the M at 1.8 K (< T_N) shows the metamagnetic transition due to a spin flop at 2 T. The transition field

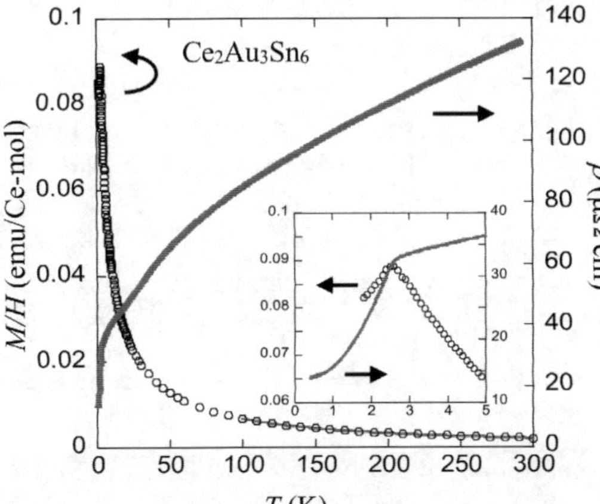

Figure 2. Temperature dependence of the magnetic susceptibility M/H measured at 0.1 T and the electrical resistivity ρ of $Ce_2Au_3Sn_6$.

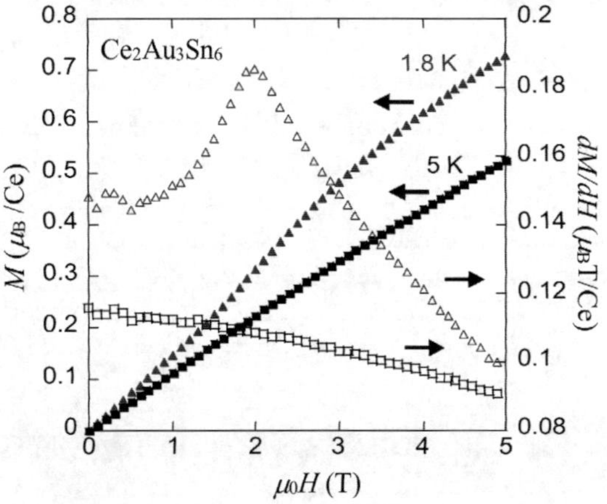

Figure 3. Magnetic-field dependence of the magnetization M and the field derivative dM/dH of $Ce_2Au_3Sn_6$.

was determined as the peak field of the field derivative dM/dH as shown in the figure 3.

Figure 4 shows the temperature dependence of the specific heat C of $Ce_2Au_3Sn_6$ and $La_2Au_3Sn_6$. The C of $Ce_2Au_3Sn_6$ shows a λ-type anomaly at T_N, which indicates that the cusp of M/H and the decrease in ρ are due to the intrinsic second-order antiferromagnetic transition. On the other hand, $La_2Au_3Sn_6$ does not show any anomaly down to 0.4 K. The electronic specific heat coefficient of $Ce_2Au_3Sn_6$ is estimated to be $\gamma = 350$ mJ/Ce-mol K^2, which is about 600 times larger than that of $La_2Au_3Sn_6$, $\gamma = 0.59$ mJ/La-mol K^2. This indicates that $Ce_2Au_3Sn_6$ is a heavy fermion compound. Here, γ and the phonon specific heat coefficient β were determined by fitting the formula $C/T = \gamma + \beta T^2$ on the experimental C/T from 6 K to 10 K for $Ce_2Au_3Sn_6$ and from 0.4 K to 4 K for $La_2Au_3Sn_6$.

Figure 5 shows the temperature dependence of the magnetic entropy S_{mag} calculated by integrating C_{mag}/T in temperature. C_{mag} is the magnetic specific heat obtained by subtracting the C of $La_2Au_3Sn_6$ from that of $Ce_2Au_3Sn_6$ under the assumption that the nonmagnetic contribution to the C of these compounds is identical. The S_{mag} at T_N is estimated to be 4.0 J/Ce-mol K which is 70% of Rln2 (R: gas constant). The reduction of S_{mag} value from Rln2 can be attributed to the shielding of the magnetic moment by the Kondo effect.

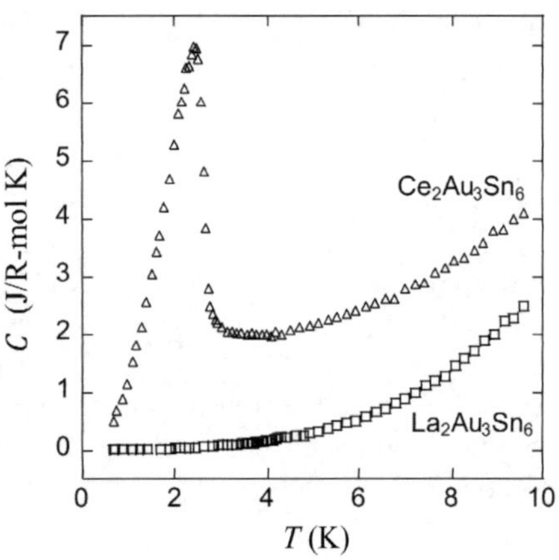

Figure 4. Temperature dependence of the specific heat C of $Ce_2Au_3Sn_6$ and $La_2Au_3Sn_6$.

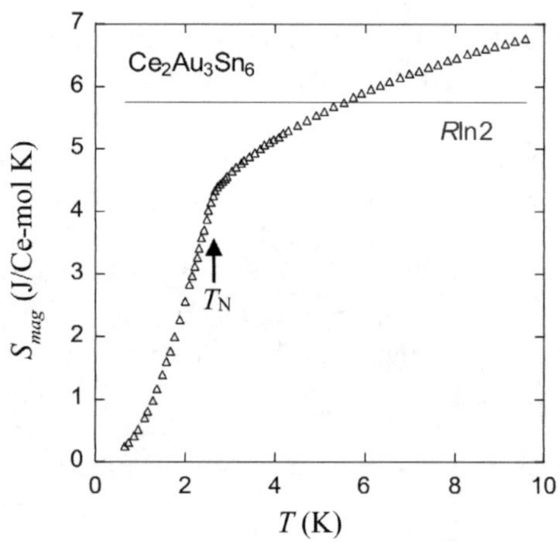

Figure 5. Temperature dependence of magnetic entropy S_{mag} of $Ce_2Au_3Sn_6$.

4. Conclusion

We have prepared polycrystalline samples of $RE_2Au_3Sn_6$ (RE = Ce, La) and investigated their magnetic, transport, and thermal properties by measuring the magnetization, the electrical resistivity, and the specific heat. $Ce_2Au_3Sn_6$ is an antiferromagnet with the transition temperature at $T_N = 2.54$ K. The large electronic specific heat coefficient $\gamma = 350$ mJ/Ce-mol K^2 indicates that $Ce_2Au_3Sn_6$ is a heavy fermion compound. The reduced magnetic entropy of $Ce_2Au_3Sn_6$ (70% of Rln2) at T_N can be attributed to the Kondo effect. The nonmagnetic reference compound $La_2Au_3Sn_6$ shows a nonmagnetic metallic behavior and does not show any phase transition down to 0.4 K.

Acknowledgment

This work was partly supported by JSPS KAKENHI Grant Number 20K03861.

References

[1] Doniach S 1977 *Physica B+C* **91** 231
[2] Engelbert S, Janka O, Klenner S and Pöttgen R 2020 *Z. Anorg. Allg. Chem.* **646** 1508

Low-temperature physical properties of new orthorhombic compounds $R_4Pt_9Al_{13}$ (R = Ce, Pr)

Y NAKAMURA[1], E MATSUOKA[1], H SUGAWARA[1], T SAKURAI[2] and H OHTA[3]

[1]Department of Physics, Graduate School of Science, Kobe University, Kobe, Japan
[2]Research Facility Center for Science and Technology, Kobe University, Kobe, Japan
[3]Molecular Photoscience Research Center, Kobe University, Kobe, Japan

E-mail: matsuoka@crystal.kobe-u.ac.jp

Abstract. Magnetic, transport, and thermal properties of new orthorhombic compounds $Ce_4Pt_9Al_{13}$ and $Pr_4Pt_9Al_{13}$ have been investigated by the magnetization, the electrical resistivity, and the specific-heat measurements. $Ce_4Pt_9Al_{13}$ is a Kondo-lattice compound and shows a ferromagnetic or ferrimagnetic transition at T_C = 0.88 K. $Pr_4Pt_9Al_{13}$ is an antiferromagnetic compound with the transition temperature at T_N = 2.6 K.

1. Introduction

Recently, coexistence of ferromagnetism and superconductivity in some U-based compounds have attracted much attention [1]. In contrast to the discovery of the U-based ferromagnetic superconductors, no other systems such as Ce-based or Yb-based ones have been discovered thus far. In order to examine the mechanism of the coexistence more precisely, it is important to examine physical properties of the other systems showing the coexistence. From this view point, we have focused on a new orthorhombic compound $Ce_4Pt_9Al_{13}$ (space group: *Pmmn*, D_{2h}^{13}, No. 59) [2]. There are three independent Ce sites in this structure as shown in the figure 1. The physical property measurements down to 1.72 K have revealed that this compound does not show any magnetic transition above 1.72 K, although the positive paramagnetic Curie temperature (41.6 K) implies ferromagnetic interaction between Ce moments [2]. In this study, magnetic and transport properties of $Ce_4Pt_9Al_{13}$ have been examined down to 0.4 K. Additionally, we have tried to prepare samples of $Pr_4Pt_9Al_{13}$ and examined their physical properties since

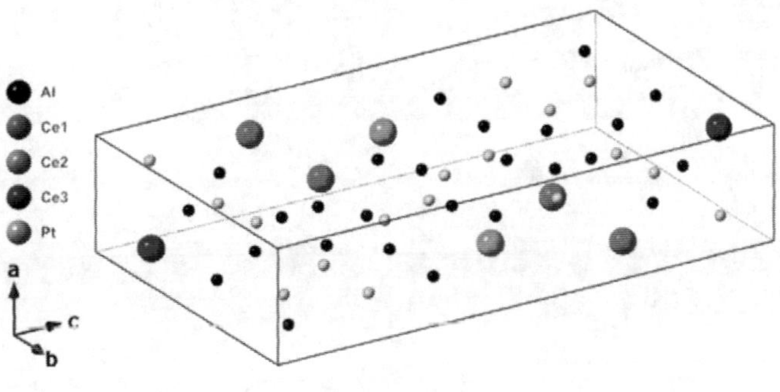

Figure 1. Crystal structure of $Ce_4Pt_9Al_{13}$. Solid lines show the edges of the chemical unit cell.

$Ce_4Pt_9Al_{13}$ is a new compound and no isomorphous compounds with the other rare earths have been synthesized.

2. Experimental methods

Polycrystalline samples of $Ce_4Pt_9Al_{13}$ and $Pr_4Pt_9Al_{13}$ were synthesized by arc melting the stoichiometric constituent elements in an Ar atmosphere. The X-ray powder diffraction experiments for these compounds revealed that all of the Bragg peaks in the diffraction patterns can be indexed on the basis of the reported structure. No impurity peaks were detected in the diffraction patterns. The lattice constants were determined to be a = 4.1739, b = 11.4373, and c = 19.8125 Å for $Ce_4Pt_9Al_{13}$, and a = 4.1615, b = 11.4162, and c = 19.7620 Å for $Pr_4Pt_9Al_{13}$. The lattice constants of $Ce_4Pt_9Al_{13}$ agree with the previously reported ones [2] within the experimental precision. The magnetization M was measured using a superconducting quantum interference device magnetometer (Quantum Design, MPMS) as functions of the temperature T and magnetic field H between 1.8 and 300 K up to 5 T. The electrical resistivity ρ was measured with a dc four-probe method between 0.4 and 300 K in a laboratory-built ^{3}He cryostat. The specific heat C was measured with a thermal relaxation method between 0.5 and 9 K in a ^{3}He cryostat.

3. Results and Discussion

3.1 $Ce_4Pt_9Al_{13}$

The inverse magnetic susceptibility H/M of $Ce_4Pt_9Al_{13}$ (not shown) above 100 K obeys the Curie-Weiss law $H/M = (T - \theta_P)/C_{Curie}$, where C_{Curie} = 0.894 emu K/Ce-mol and θ_P = -189.8 K show the Curie constant and the paramagnetic Curie temperature, respectively. The effective magnetic moment μ_{eff} calculated from C_{Curie} is 2.67 μ_B/Ce (μ_B: Bohr magneton). Since the calculated μ_{eff} is close to the theoretical one for the free Ce^{3+} ion (2.54 μ_B/Ce), we can consider that the Ce ions in this compound are trivalent. The negative θ_P indicates that the antiferromagnetic interaction acts between the Ce magnetic moments, although the positive θ_P (ferromagnetic interaction) was reported in the previous report. No anomaly of H/M has been detected above 1.8 K as already reported previously.

Figure 2 shows the temperature dependence of the electrical resistivity ρ of $Ce_4Pt_9Al_{13}$. ρ decreases

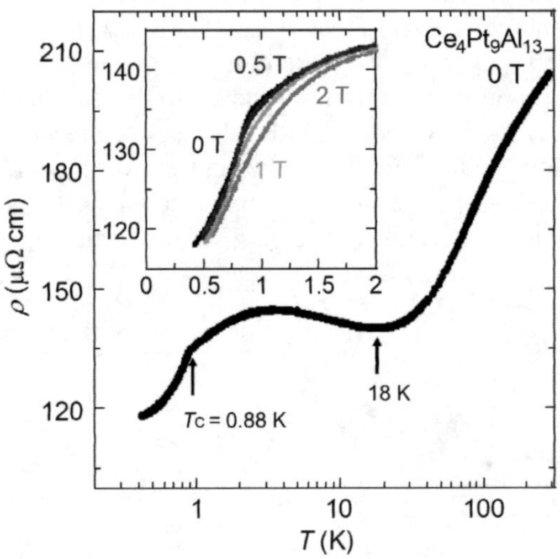

Figure 2. Temperature dependence of the electrical resistivity ρ of $Ce_4Pt_9Al_{13}$. The inset represents the low-temperature part of ρ measured in the various magnetic fields.

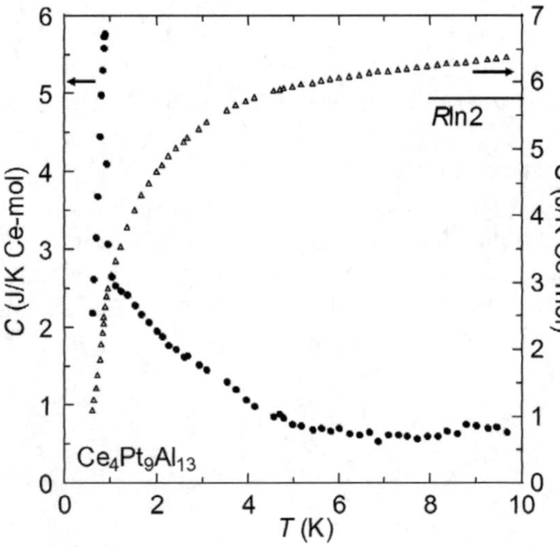

Figure 3. Temperature dependence of the specific heat C and the total entropy S of $Ce_4Pt_9Al_{13}$.

with decreasing temperature, followed by a minimum at 18 K, and shows -ln$T$ dependence between 18 K and 7 K. These $\rho(T)$ behavior indicates that $Ce_4Pt_9Al_{13}$ is a Kondo lattice compound. In addition, ρ shows a steep decrease at $T_C = 0.88$ K, indicating that some phase transition occurs at this temperature. The low-temperature part of ρ measured in various magnetic fields are shown in the inset of the figure 2. T_C shifts to higher temperatures with increasing magnetic field, indicating that the decrease in the ρ at 0.88 K corresponds to the ferromagnetic transition temperature. No superconductivity has been detected above 0.4 K.

Figure 3 shows the temperature dependence of the specific heat C of $Ce_4Pt_9Al_{13}$. The λ-type anomaly at T_C indicates that the second-order ferromagnetic transition occurs at T_C. Figure 3 also shows the temperature dependence of the total entropy S derived by integrating C/T in T. Since the symmetry of the crystalline electric field is monoclinic for Ce1 site, and orthorhombic for Ce2 and Ce3 sites, the six-fold ground multiplet of Ce^{3+} splits into three doublets for these Ce sites. In this case, it is expected that the entropy released at T_C reaches Rln2 (R: gas constant). However, the S value at T_C is only 45% of Rln2 (2.6 J/Ce-mol K). Such a large reduction of S from Rln2 cannot be explained only by the Kondo effect. The reduced S is mainly ascribable to the short-range magnetic correlation below 4 K. In fact, the C increases gradually below 4 K and the S value reaches 99% of Rln2 at 4 K.

There are three independent Ce sites in the crystal structure of $Ce_4Pt_9Al_{13}$ as described above. In such structure, the magnitude of Ce magnetic moments and/or the magnetic interaction are not necessarily identical among these Ce sites. We therefore cannot decide from the present study if the magnetic structure is simple ferromagnetic or ferrimagnetic. As the next step, we have to determine magnetic structure by neutron diffraction experiment.

3.2 $Pr_4Pt_9Al_{13}$

Figure 4 shows the temperature dependence of the inverse magnetic susceptibility H/M of $Pr_4Pt_9Al_{13}$ measured at 0.1 T. H/M above 100 K follows the Curie-Weiss law (see the solid line in the figure. 4). The μ_{eff} and the θ_P derived by the Curie-Weiss fitting are 3.53 μ_B/Ce and -91.4 K, respectively. Since the derived μ_{eff} is close to the theoretical one for the free Pr^{3+} ion (3.58 μ_B/Ce), we can consider that the Pr ions in $Pr_4Pt_9Al_{13}$ are trivalent. The negative θ_P indicates that the antiferromagnetic interaction acts between the Pr magnetic moments. The magnetic susceptibility M/H shows a cusp at $T_N = 2.6$ K as

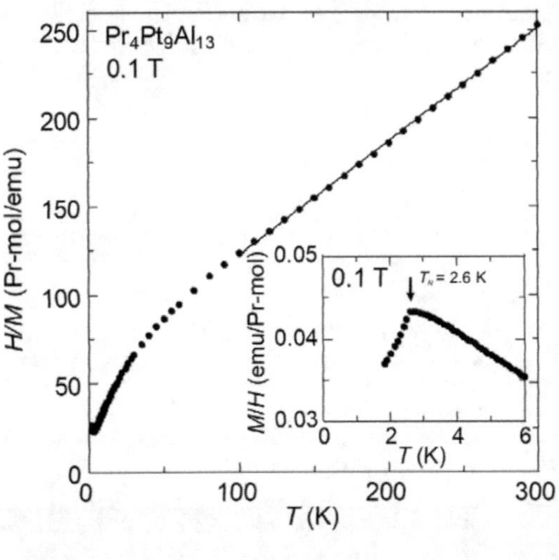

Figure 4. Temperature dependence of the inverse magnetic susceptibility H/M of $Pr_4Pt_9Al_{13}$ measured at 0.1 T. The inset repserents the low-temperature part of M/H.

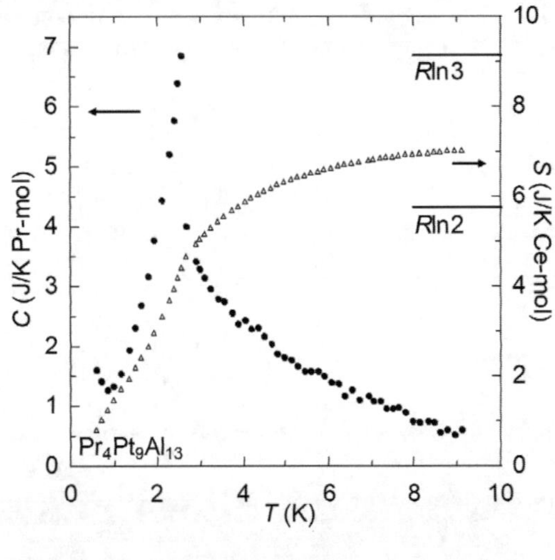

Figure 5. Temperature dependence of the specific heat C and the total entropy S of $Pr_4Pt_9Al_{13}$.

shown by the arrow in the inset of the figure. 4. The cusp of *M*/*H* and the negative θ_P indicate the occurrence of antiferromagnetic transition at T_N.

The electrical resistivity ρ of $Pr_4Pt_9Al_{13}$ (not shown) decreases with decreasing temperature and shows an increase below T_N. The increase below T_N is probably due to the formation of a superzone gap accompanied by the antiferromagnetic transition.

Figure 5 represents the temperature dependence of the specific heat *C* of $Pr_4Pt_9Al_{13}$. The λ-type anomaly at T_N indicates that the second-order antiferromagnetic transition occurs at T_N. The total entropy *S* calculated by the same method for $Ce_4Pt_9Al_{13}$ is also shown in the figure. 5. The *S* value at T_N is 78% of *R*ln2 and exceeds *R*ln2 above 3.8 K. The nine-fold ground multiplet of Pr^{3+} splits into nine singlets under the crystalline electric field with orthorhombic or monoclinic symmetry. The abovementioned *S* value suggests that the first excited and/or the second excited singlets are located at the vicinity of the ground singlet. We therefore consider that the pseudo doublet ground state is responsible for the antiferromagnetic transition. The increase in *S* below 0.8 K is probably attributed to the nuclear specific heat of Pr^{3+} ions.

4. Conclusion

In this study, polycrystalline samples of new orthorhombic compounds $R_4Pt_9Al_{13}$ (R = Ce, La) have been synthesized and their physical properties down to 0.4 K have been examined.

Magnetic-susceptibility measurement has revealed that the Ce ions in $Ce_4Pt_9Al_{13}$ are trivalent. $Ce_4Pt_9Al_{13}$ is a ferromagnetic or ferrimagnetic Kondo-lattice compound with the transition temperature at $T_C = 0.88$ K. No superconductivity has been detected above 0.4 K.

We have found that $Pr_4Pt_9Al_{13}$ exists as the isomorphous compound of $Ce_4Pt_9Al_{13}$. Magnetic-susceptibility measurement has revealed that the Pr ions in $Pr_4Pt_9Al_{13}$ are trivalent. The pseudo doublet ground state of Pr^{3+} is responsible for the antiferromagnetic transition at 2.6 K.

Acknowledgment

This work was partly supported by JPJS KAKENHI Grant Number JP15H05882 (J-Physics).

References

[1] Aoki D, Ishida K and Flouquet J 2019 *J. Phys. Soc. Jpn.* **88** 022001
[2] Morozova Yu, Gribanov A, Murashova E, Dunaev S, Grytsiv A, Rogl P, Giester G and Kaczorowski D 2018 *J. Alloys Compd.* **767** 496

Low-temperature physical properties of a new orthorhombic compound $Ce_2Ir_3Sb_4$

T Komoda[1], E Matsuoka[1], H Sugawara[1], T Sakurai[2] and H Ohta[3]

[1]Department of Physics, Graduate School of Science, Kobe University, Kobe, Japan
[2]Research Facility Center for Science and Technology, Kobe University, Kobe, Japan
[3]Molecular Photoscience Research Center, Kobe University, Kobe, Japan

E-mail: matsuoka@crystal.kobe-u.ac.jp

Abstract. We have synthesized polycrystalline samples of a new orthorhombic compound $Ce_2Ir_3Sb_4$ and investigated their physical properties by measuring the magnetization, the electrical resistivity, and the specific heat down to 0.4 K. From the results of electrical resistivity measurements, it was found that $Ce_2Ir_3Sb_4$ is a Kondo lattice compound. We found that $Ce_2Ir_3Sb_4$ does not show any phase transition down to 0.4 K. The specific heat peak at 1.3 K is not due to a phase transition but due to the Schottky specific heat caused by the crystalline electric field splitting of the 4*f* level of Ce ions.

1. Introduction

Compounds containing the rare-earth element Ce exhibit various magnetic ground state such as magnetic ordering and heavy fermion state due to the strong interaction between the 4*f* electrons and the conduction electrons. It is well known that the competition between the Ruderman-Kittel-Kasuya-Yosida interaction and the Kondo effect plays a key role to determine the magnetic ground state [1]. Recent interest in these compounds has been focused on anomalous physical properties such as unconventional superconductivity appearing in the vicinity of a quantum critical point where the ground state changes from a magnetically ordered state to a heavy fermion state. One of the possible methods to search for new compounds with quantum criticality and concomitant anomalous physical properties is to search for compounds that do not show magnetic ordering down to low temperatures. As the compound without magnetic ordering, we have focused on the newly synthesized orthorhombic compounds $RE_2Ir_3Sb_4$ (RE = La, Ce, Pr, Nd) (space groupe *Pnma*, D_{2h}^{16}, No.62) [2]. The crystal structure of $RE_2Ir_3Sb_4$ is shown in the figure 1. It has been reported that $Ce_2Ir_3Sb_4$ has the trivalent Ce ions and does not show magnetic ordering down to 3 K [2]. In this study, we prepared polycrystalline samples of $Ce_2Ir_3Sb_4$ and investigated their physical properties by measuring the magnetization, the electrical resistivity, and the specific heat down to 0.4 K.

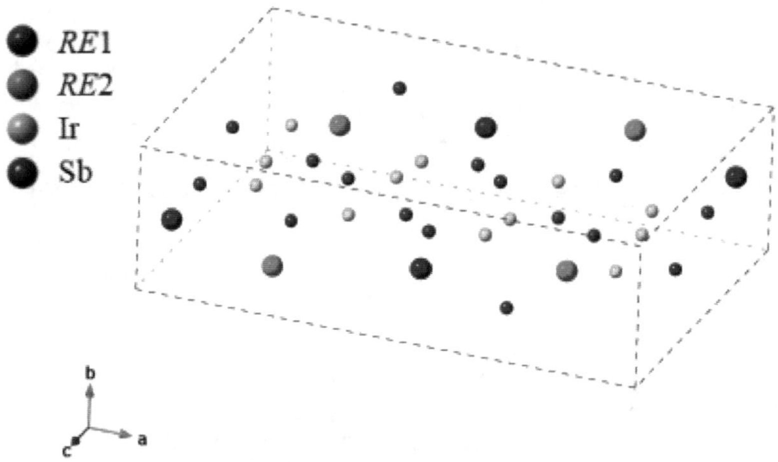

Figure 1. The crystal structure of $Ce_2Ir_3Sb_4$. The dotted lines represent a chemical unit cell.

2. Experimental Methods

Since the direct arc melting of the constituent elements is not possible due to the high vapor pressure of Sb, polycrystalline samples were prepared by the following procedure. First, Ce and Ir with the molar ratio 2:3 were melted using a tetra arc furnace to prepare the intermediate alloy Ce_2Ir_3. Next, powdered Ce_2Ir_3 and Sb with the molar ratio 1:4 were mixed in the N_2 atmosphere. The mixture was sealed in an evacuated quartz tube and sintered at 950 °C for 62 hours in a muffle furnace. The sintered samples used for the magnetization and the resistivity measurements are fragile and crumbles easily. We therefor melted the sintered sample by using a tetra-arc furnace. The melted sample was then sealed in an evacuated quartz tube and annealed at 950 °C for 156 hours using a muffle furnace. The annealed sample was used for the specific-heat measurement. Powder X-ray diffraction measurements revealed that almost all of the Bragg peaks can be indexed on the basis of the reported crystal structure. Small peaks of unidentified impurity phases were also observed in both samples. The lattice parameters were calculated to be a = 16.182, b = 4.564, c = 10.964 Å for the sintered sample and a = 16.182, b = 4.563, c = 10.968 Å for the annealed sample. These values are in good agreement with the reported values within the experimental accuracy. The magnetization M of the sintered sample was measured using a SQUID magnetometer (Quantum Design, MPMS) at temperature range of $1.8 \leqq T \leqq 300$ K and magnetic field range of $0 \leqq \mu_0 H \leqq 5.0$ T. The electrical resistivity ρ of the sintered sample was measured by with a DC four-terminal method using a ^{3}He cryostat at temperature range of $0.4 \leqq T \leqq$ 287 K. The specific heat C of the annealed sample was measured with the thermal relaxation method using a ^{3}He cryostat at a temperature range of $0.4 \leqq T \leqq 10$ K.

3. Results and Discussion

Figure 2 shows the temperature dependence of the inverse magnetic susceptibility H/M of $Ce_2Ir_3Sb_4$. The H/M obeys the modified Curie-Weiss law $H/M = \{C_{curie}/(T - \theta_P) + \chi_0\}^{-1}$, where C_{curie}, θ_P, and χ_0 are the Curie constant, the paramagnetic Curie temperature, and the sum of the Pauli paramagnetic contribution and the core diamagnetic one, respectively. The red line represents the fitted curve. The estimated parameters are C_{curie} = 0.7828 emu K/Ce-mol θ_P = -24.4 K, and $\chi_0 = -1.79\times10^{-4}$ emu/Ce-mol. From this C_{curie} value, the effective magnetic moment μ_{eff} was calculated to be 2.50 μ_B/Ce. Since this value is close to the theoretical value of 2.54 μ_B/Ce for Ce^{3+}, it was found that the Ce ions in $Ce_2Ir_3Sb_4$

is in the trivalent state as reported in the ref. [2]. The negative θ_P indicates that the magnetic interaction between Ce magnetic moments is antiferromagnetic. The magnetic susceptibility *M*/*H* did not show any anomaly above 1.8 K as shown in the inset of the figure 2.

Figure 3 shows the magnetic-field dependence of the magnetization *M* at 1.8 K of $Ce_2Ir_3Sb_4$. The *M* increases with increasing magnetic field and shows no anomalies. Such a field dependence as well as the low-temperature *M*/*H* behavior indicate that the magnetic state of $Ce_2Ir_3Sb_4$ is paramagnetic down to 1.8 K.

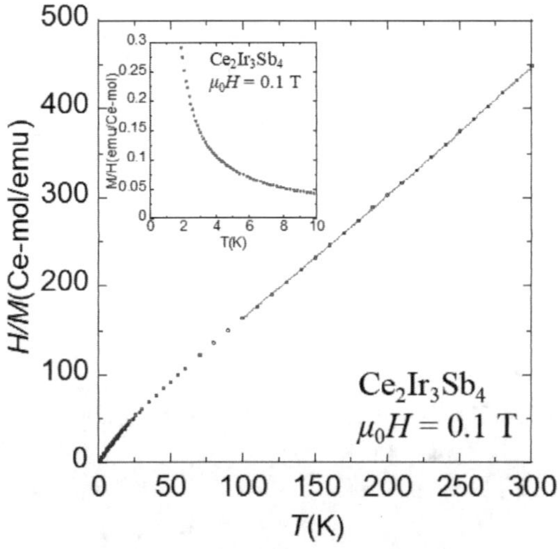

Figure 2. Temperature dependence of the inverse magnetic susceptibility *H*/*M* of $Ce_2Ir_3Sb_4$ measured at 0.1 T. The inset shows the low-temperature part of the magnetic susceptibility *M*/*H*.

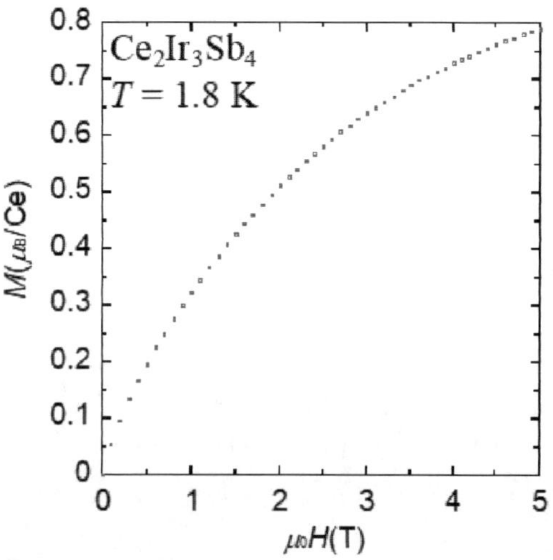

Figure 3. Magnetic-field dependence of the magnetization *M* of $Ce_2Ir_3Sb_4$ measured at 1.8 K.

Figure 4 shows the temperature dependence of the electrical resistivity of $Ce_2Ir_3Sb_4$. The normalized electrical resistivity $\rho(T)/\rho(287\ K)$ is shown in the figure 4 since it is difficult to deduce sample dimension due to the porosity of the sintered sample. The electrical resistivity decreases with decreasing temperature, followed by a minimum at 25 K and a broad maximum at 2.1 K. Below 2.1 K, the ρ decreases with decreasing temperature. In addition, the temperature dependence of the electrical resistivity is proportional to -ln*T* in the range of 10.5 to 5 K, indicating that $Ce_2Ir_3Sb_4$ is the Kondo lattice compound. Since the aforementioned *M*/*H* and the specific heat mentioned later do not show any anomalies at 2.1 K, it is expected that the decrease in the resistivity below 2.1 K can be ascribable to the development of a coherent state of the Kondo lattice.

Figure 5 shows the temperature dependence of the specific heat *C*. The *C* increases as the temperature decreases from 10 to 1.3 K. Below the peak temperature at 1.3 K, the *C* decreases rapidly at lower temperatures. We now consider that the specific heat behavior can be explained as the Schottky specific heat caused by the crystalline electric field splitting of the 4*f* level of Ce^{3+}. In fact, the calculated magnetic specific heat under the assumption that the first and the second excited doublets are located at 3 and 100 K, respectively, can reproduce the overall temperature dependence of the experimental *C* as shown by the blue line in the figure 5.

Journal of Physics: Conference Series **2164** (2022) 012039 doi:10.1088/1742-6596/2164/1/012039

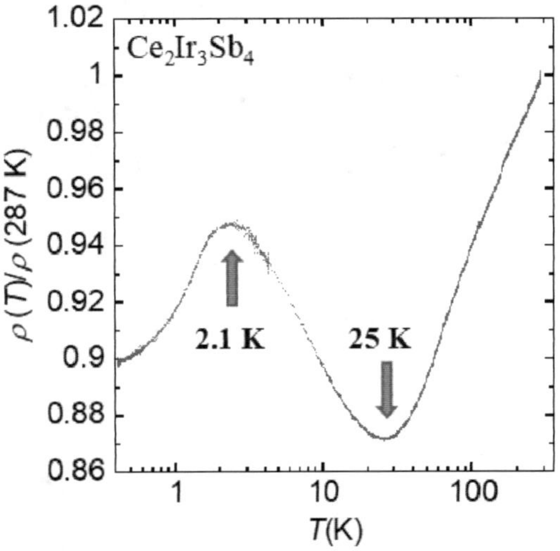

Figure 4. Temperature dependence of the normalized electrical resistivity $\rho(T)/\rho(287$ K) of $Ce_2Ir_3Sb_4$.

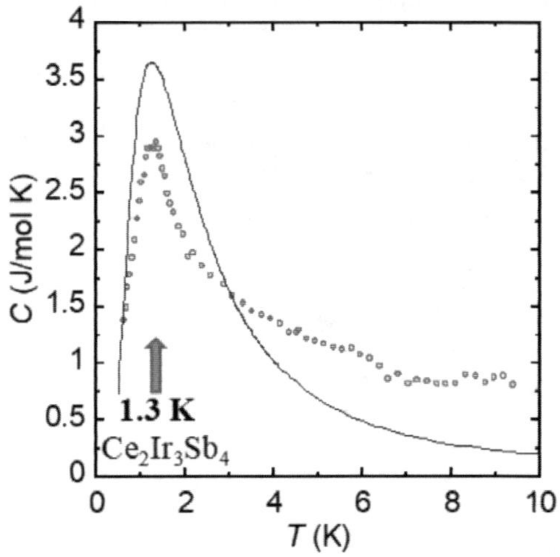

Figure 5. Temperature dependence of the specific heat *C*. The blue line is the calculated magnetic specific heat with the first and the second excited doublets at 3 and 100 K, respectively.

4. Conclusion

In this study, polycrystalline samples of $Ce_2Ir_3Sb_4$ were prepared and their magnetic and transport properties were examined by the magnetization, the electrical resistivity, and the specific heat measurements. From the results of the electrical resistivity measurement, $Ce_2Ir_3Sb_4$ was found to be the Kondo lattice compound. The temperature dependence of the specific heat below 10 K can be explained as the Schottky specific heat caused by the crystalline electric field splitting of the 4*f* level of Ce^{3+}.

Acknowledgment
This work was supported partly by JSPS KAKENHI Grant Number 20K03861.

References

[1] Doniach S 1977 *Physica B+C* **91** 231
[2] Konrad S, Wilfried H, Ute Ch. R, Rolf-Dieter H, and Rainer P 2011 *Z. Naturforsch.B* **66** 777

Electrical Resistivity Measurements of Antiferromagnetic Compound Ce_3TiSb_5 under Pressure

Masahiro Shinozaki[1*], Gaku Motoyama[1], Shijo Nishigori[2], Akira Yamaguchi[3], Yu Yamane[3], Tetsuya Mutou[1], Kenji Fujiwara[1], Masahiro Manago[1], Kiyotaka Miyoshi[1], and Akihiko Sumiyama[3]

[1] Department of Material Science, Shimane University, Shimane 690-8504, Japan
[2] ICSR, Shimane University, Shimane 690-8504, Japan
[3] Department of Material Science, University of Hyogo, Hyogo 678-1297, Japan

*n20d102@matsu.shimane-u.ac.jp

Abstract. We performed electrical resistivity measurements of Ce_3TiSb_5 under pressure. From the pressure dependences of antiferromagnetic ordering temperature T_N and lower anomalous temperature T^*, temperature–pressure phase diagram of Ce_3TiSb_5 up to 2.3 GPa has been constructed. T^* rapidly decreases with increasing pressure and disappears around 1 GPa. T_N increases with increasing pressure, however a hump structure of electrical resistivity below T_N becomes small. Hence, some change for the magnetic ordered state is expected to occur in higher pressure region.

1. Introduction

Odd-parity multipole ordering has attracted great attention owing to various unique properties such as cross-correlation phenomena [1]. An antiferromagnetic (AFM) compound with broken local inversion symmetry at magnetic ion site is an ideal system for studying odd-parity multipole ordering. On the UNi_4B, which has a toroidal ordered state with a vortex-like magnetic structure, some cross-correlation responses and an unconventional multipole ordered state predicted by theoretical study have been experimentally verified [2-4]. Recently, we reported that Ce_3TiBi_5 also shows magnetoelectric (ME) effect which is considered to be originated from magnetic ferrotoroidal order [5, 6]. There are many family ternary compounds expressed as R_3TX_5 (R: rare-earth element, T: transition metal, X: p-block element). Hence, it is expected to be an existence of the odd-parity multipole ordering other than Ce_3TiBi_5.

On the other hand, CeCoSi, in which there is no local inversion symmetry at the Ce site, has the complex magnetic temperature–pressure (T–P) phase diagram and the possibility of the antiferroquadrupolar ordering at T_0 has been extensively studied [7-9]. The phase transition temperature of T_0 shows strong pressure dependence. Note that the broad kink anomaly at T_0 appeared at low pressure ranges changes a broad hump anomaly at high pressure ranges [7].

Ce_3TiSb_5 is a hexagonal system with space group $P6_3/mcm$ (No. 193) and has identical crystal structure as Ce_3TiBi_5 [10]. Ce_3TiSb_5 also shows AFM transition, although the transition temperature of T_N is 5.2 K and the direction of ordered Ce magnetic moments is presumed to differ from Ce_3TiBi_5 from the anisotropy of the magnetic susceptibility [11]. There is a possibility of odd-parity multipole ordering in Ce_3TiSb_5 due to the expected magnetic structure. Meanwhile, Ce_3TiSb_5 has not been reported about physical properties under pressure. Pressure dependence of T_N is one of the keys to investigate the magnetic phase realizing in this system. To discuss about cross-correlation phenomena, it is necessary to elucidate the detail of the ordered state. In the present work, we show the results of electrical resistivity measurements of Ce_3TiSb_5 under pressure and *T–P* phase diagram up to 2.3 GPa in this system.

2. Experimental

Single-crystalline samples were grown by a Sn-flux method, as described in the literature by M. Matin *et al.* [11]. Electrical resistivity was measured by a standard four-terminal method using indenter-type pressure cell. Daphne oil 7373 was used as the pressure transmitting media.

3. Results

Figure 1 shows temperature dependence of electrical resistivity (ρ) of Ce_3TiBi_5 and Ce_3TiSb_5 at ambient pressure. An electric current was applied parallel to the *c*-axis. Both $\rho(T)$ curves of Ce_3TiBi_5 and Ce_3TiSb_5 exhibit a double hump structure: one is around 100 K and the other is at lower temperature (Ce_3TiBi_5: ~7 K, Ce_3TiSb_5: ~9 K). The double hump structure is considered to originate in Kondo effect in the excited CEF states. $\rho(T)$ of Ce_3TiBi_5 exhibits a clear kink at AFM transition temperature T_N = 5.0 K, then $\rho(T)$ only decreases with decreasing temperature. Note, however, that the derivative resistivity with temperature ($d\rho/dT$) shows a broad peak about 3 K. In contrast, $\rho(T)$ of Ce_3TiSb_5 exhibits a minimum at T_N = 5.2 K, then a hump structure appears at just below T_N. This anomaly reminds us the charge-density-wave (CDW) or spin-density-wave (SDW) transition with Fermi surface reconstruction. This is consistent with the result of neutron scattering experiment [12]. The difference of the $\rho(T)$ between Ce_3TiBi_5 and Ce_3TiSb_5 is considered to be due to the magnetic anisotropy, but microscopic studies as NMR and neutron diffraction measurements are needed in the future. We can also notice there is another feature of $\rho(T)$ of Ce_3TiSb_5 which is a small anomaly at T^* = 2.5 K. Ce_3TiSb_5 often contains Sn as an impurity phase at the crystal growth by a Sn-flux method, then the sudden decrease at about 3.5 K is often observed [11]. However, the anomaly at T^* is also observed with applying higher magnetic field than the critical field of Sn. Therefore, this anomaly is considered to be an intrinsic behavior for Ce_3TiSb_5. T^* is close to the temperature at which one of the two kinds of the magnetic scattering intensities on the neutron powder diffraction measurements increases [12].

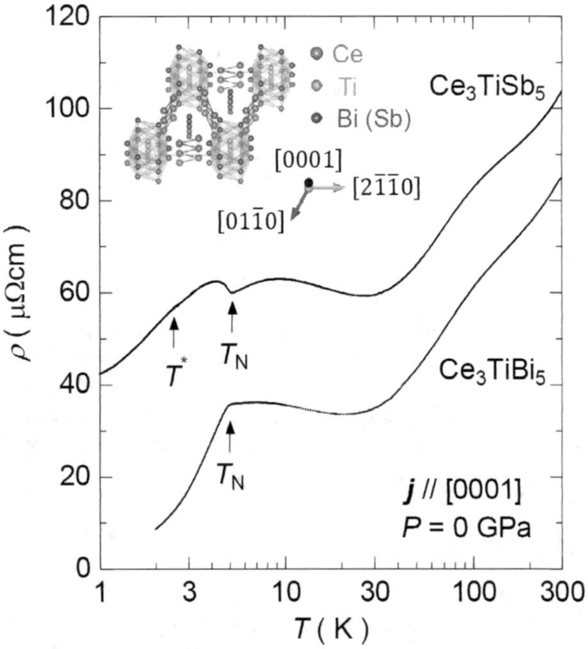

Figure 1. Temperature dependences of electrical resistivity of Ce_3TiX_5 (X = Bi, Sb) at ambient pressure. Crystal structure of Ce_3TiX_5 is drawn in the inset.

Next, we show temperature dependence of ρ of Ce_3TiSb_5 at various pressure in Fig. 2. An electric current was applied parallel to the *c*-axis because of the very thin need-shape single crystal. The double hump structure caused by Kondo effect is enhanced with increasing pressure. Similar behaviors are seen in $\rho(T)$ of Ce_3TiBi_5 [13]. On the $\rho(T)$ of the Ce_3TiBi_5, the higher hump of the double hump structure clearly changes to a peak anomaly above 6 GPa. On the other hand, $\rho(T)$ of the Ce_3TiSb_5 at

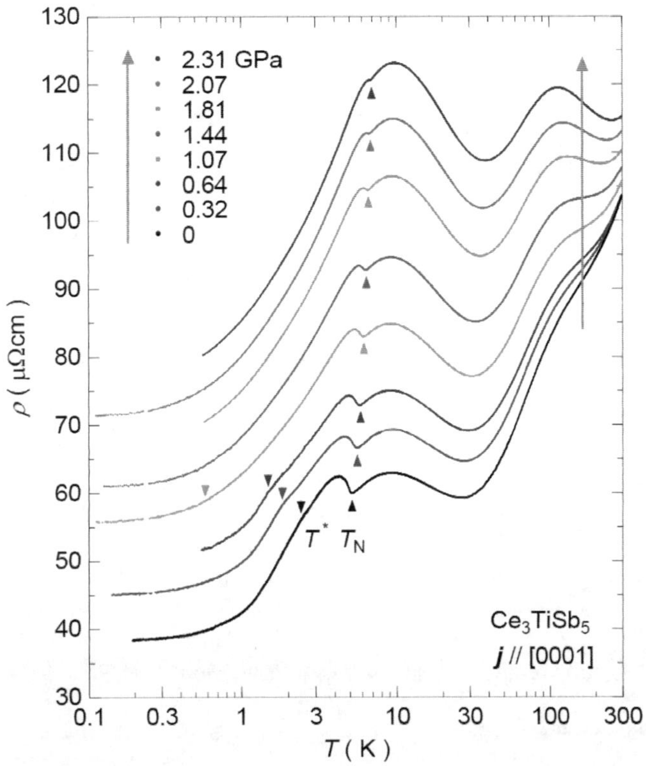

Figure 2. Temperature dependences of electrical resistivity of Ce_3TiSb_5 under several pressure.

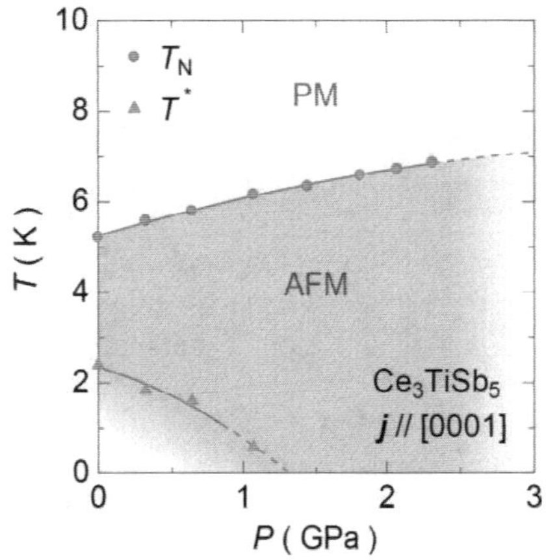

Figure 3. Temperature–pressure phase diagram of Ce_3TiSb_5.

1.81 GPa clearly exhibits a peak anomaly at the higher hump of the double hump structure. These differences are considered to be due to the chemical pressure effect replacing Bi with Sb.

Down triangle symbols indicate the anomaly at T^*. T^* rapidly decreases with increasing pressure and disappears around 1 GPa. From a perspective of the anomaly appearing temperature at ambient pressure, the anomaly at T^* is consistent with the ordered phase observed by neutron scattering measurement below 3 K. However, the decrease of the lattice constant along the a-axis in the ordered phase is observed at ambient pressure [12]. It seems to contradict the decrease of T^* under pressure. In the previous report of magnetic properties of Ce_3TiSb_5, there is no clear anomaly at T^* in the temperature dependence of the magnetic susceptibility and only small broad anomaly below 2 K on the temperature dependence of the specific heat was observed.

Upward triangle symbols indicate the AFM ordering at T_N. T_N increases with increasing pressure and reaches 6.7 K at 2.31 GPa. The hump anomaly just below T_N becomes smaller with pressure application despite the increase in T_N. This hump structure is expected to disappear around 3 GPa by further pressurization. This behaviour implies the possibility that some change occurs in AFM ordered state. In the case of that the hump anomaly just below T_N changes to an anomaly with only decreasing, the order parameter of AFM ordered state might be changed by pressure.

Finally, we show the T–P phase diagram of Ce_3TiSb_5 in Fig. 3. The T–P phase diagram has been constructed from the results of Fig. 2. T_N monotonically increases with increasing pressure at least up to 2.31 GPa. The increase of T_N under pressure is more gradual than that of Ce_3TiBi_5. The double hump structure on Ce_3TiSb_5 caused by Kondo effect is enhanced by the application of lower pressure than Ce_3TiBi_5. Due to the substitution of Bi with Sb, it is expected that quantum critical point of Ce_3TiSb_5 emerges lower pressure region than Ce_3TiBi_5. The existence of the high-pressure phase and the details of the low-temperature phase are still unknown, but the existence of these phases is interesting because the changes in the order parameter affect the multipole ordered state and the ME effect derived from it.

4. Summary

We performed electrical resistivity measurements of Ce_3TiSb_5 under pressure. T–P phase diagram of Ce_3TiSb_5 up to 2.3 GPa has been constructed. The lower anomaly at T^* is found to rapidly decrease with increasing pressure and disappears around 1 GPa. T_N is still increasing in this pressure range. In contrast, the hump structure of ρ below T_N becomes small. Thus, we expect some change of AFM ordered state with further pressurization.

Acknowledgment
This work was supported by JSPS KAKENHI Grant No. 21K03447. One of the starting materials of 4N purity Sb was provided by NIHON SEIKO Co., Ltd.

References

[1] S. Hayami, M. Yatsushiro, Y. Yanagi, and H. Kusunose, Phys. Rev. B **98**, 165110 (2018).
[2] H. Saito, K. Uenishi, N. Miura, C. Tabata, H. Hidaka, T. Yanagisawa, and H. Amitsuka, J. Phys. Soc. Jpn. **87**, 033702 (2018).
[3] T. Yanagisawa, H. Matsumori, H. Saito, H. Hidaka, H. Amitsuka, S. Nakamura, S. Awaji, D. I. Gorbunov, S. Zherlitsyn, J. Wosnitza, K. Uhlířová, M. Vališka, and V. Sechovský, Phys. Rev. Lett. **126**, 157201 (2021).
[4] C. Tabata, H. Sagayama, H. Saito, H. Nakao, and H. Amitsuka, J. Phys. Soc. Jpn. **90**, 064601 (2021).
[5] M. Shinozaki, G. Motoyama, M. Tsubouchi, M. Sezaki, J. Gouchi, S. Nishigori, T. Mutou, A. Yamaguchi, K. Fujiwara, K. Miyoshi, and Y. Uwatoko, J. Phys. Soc. Jpn. **89**, 033703 (2020).
[6] M. Shinozaki, G. Motoyama, T. Mutou, S. Nishigori, A. Yamaguchi, K. Fujiwara, K. Miyoshi, and A. Sumiyama, JPS Conf. Proc. **30**, 011189 (2020).
[7] E. Lengyel, M. Nicklas, N. Caroca-Canales, and C. Geibel, Phys. Rev. B **88**, 155137 (2013).
[8] H. Tanida, K. Mitsumoto, Y. Muro, T. Fukuhara, Y. Kawamura, A. Kondo, K. Kindo, Y. Matsumoto, T. Namiki, T. Kuwai, and T. Matsumura, J. Phys. Soc. Jpn. **88**, 054716 (2019).
[9] M. Manago, H. Kotegawa, H. Tou, H. Harima, and H. Tanida, J. Phys. Soc. Jpn. **90**, 023702 (2021).
[10] S. H. D. Moore, L. Deakin, M. J. Ferguson, and A. Mar, Chem. Mater. **14**, 4867–4873 (2002).
[11] M. Matin, R. Kulkarni, A. Thamizhavel, S. K. Dhar, A. Provino, and P. Manfrinetti, J. Phys. Condens. Matter **29**, 145601 (2017).
[12] C. Ritter, A. K. Pathak, R. Filippone, A. Provino, S. K. Dhar, and P. Manfrinetti, J. Phys. Condens. Matter **33**, 245801 (2021).
[13] M. Tsubouchi, G. Motoyama, J. Gouchi, K. Miyoshi, S. Nishigori, K. Fujiwara, T. Mutou, and Y. Uwatoko, JPS Conf. Proc. **30**, 011102 (2020).

Possible routes for the synthesis of nanowires of intermetallic compounds: The case of $CeIn_3$

M. H. Carvalho[1], G. S. Freitas[1], J. C. Souza[1], R. B. Campanelli[1], H. B. Pizzi[1], S. G. Mercena[1], M.V. Puydinger dos Santos[1], F. Béron[1], P. F. S. Rosa[2], K. R. Pirota[1], P. G. Pagliuso[1]

[1]Instituto de Física "Gleb Wataghin", UNICAMP, Campinas-SP, 13083-857, Brazil.
[2]Los Alamos National Laboratory, Los Alamos, New Mexico 87545, USA
E-mail: helenacc@unicamp.br

Abstract. In this work, we investigated the role of different parameters in the synthesis of intermetallic nanowires of $CeIn_3$ by the metallic-flux nanonucleation (MFNN) method such as template pore diameter, crystallization temperature, heat treatment temperature, and synthesis time. Depending on the growing parameters, we obtained $CeIn_3$ nanowires (d ~ 350 nm) or $CeAlO_3$ nanotubes. For the nanowires, we observed a suppression of the $CeIn_3$ antiferromagnetic transition from the bulk T_N ~ 10 K to the nanowire system T_N ~ 3 K, which may be associated with the dimensionality affecting the interplay between magnetic exchange interactions, crystalline electrical field, and Kondo effects. We assume that the $CeAlO_3$ nanotubes may result from a reaction with the alumina template and consequent rare-earth oxidation. Our work shows that even it is a great challenge to find the correct growth path of a particular intermetallic compound, the MFNN method can be a promising route to obtain rare-earth based nanowires.

1. Introduction

Correlated intermetallic electron systems exhibit emergent physical phenomena, such as unconventional superconductivity, heavy fermion behavior, complex magnetic ordering, non-Fermi-Liquid behavior, and quantum criticality [1]. These phenomena emerge from the collective behavior of interacting electrons, which is strongly dependent on the system dimensionality [2]. The study of emergent phenomena in reduced dimension systems has been previously reported in different systems [3–6]. Exploring the Fe_3Ga_4 system, Moura *et al.* reported a complex change in the phase diagram as a function of the dimensionality [3]. While bulk Fe_3Ga_4 shows a transition from the ferromagnetic (FM) to antiferromagnetic (AFM) state, Fe_3Ga_4 nanowires with diameter ($d \approx 250$ nm) obtained by metallic-flux nanonucleation (MFNN) method [7] exhibit a transition from FM to ferrimagnetic or coexistence of FM and AFM. Previous studies also reported by the synthesis through the MFNN technique and further macroscopic characterization of β-Ga nanowires with a diameter of 140 nm and length around 3.8 μm [4]. In this case, it was possible to demonstrate the stabilization of a weakly coupled type II superconductor with Tc ≈ 6.2 K. Cruz *et al.* reported the synthesis of oxide-shell-protected Mn_5Si_3 nanowires with a diameter between 85 and 850 nm by the MFNN method [5]. Electrical characterization showed that nanowires are metallic at low temperatures. Rosa *et al.* performed a study on the dimensionality effect on the physical properties of the intermetallic compound $GdIn_3$ [6]. Single crystals and nanowires ($d \approx 200$ nm and length $l \approx 30$ μm) of $GdIn_3$ were also grown by the MFNN method. Magnetic susceptibility and specific heat measurements showed a drastic suppression of the AFM order of the bulk system (T_N^{bulk} = 45 K) for the T_N^{nano} = 3.8 K nanowire system. Since Gd^{3+} is an S ion (L = 0), such reduction was associated with changes in the Ruderman–Kittel–Kasuya–Yoshida (RKKY) magnetic interaction.

Due to the flexibility of the RIn_3 compound, which allows chemical substitutions in different crystallographic sites, this system represents an excellent opportunity to study the interplay between crystalline electrical field (CEF), Kondo effects, magnetic interactions, and dimensionality [2]. In particular, $CeIn_3$ crystallizes in the simple cubic structure $AuCu_3$ (space group *Pm-3m*) but has a variety of interesting properties [8]. At ambient pressure, $CeIn_3$ orders antiferromagnetically with T_N ~ 10 K. At pressures of P ~ 25 kbar, the system becomes superconducting with a T_c^{max} = 0.2 K. This unconventional superconductivity appears around a quantum critical point associated with the destruction of the magnetically ordered state at $T = 0$ K [9]. Here, dimensionality might be another ingredient to understanding the quantum criticality in heavy fermion systems.

In this work, we report the results of growing $CeIn_3$ nanowires using the MFNN method. We systematically investigated the influence of the pore size and template crystallization temperature, heat treatment temperature, and synthesis time on the nanowire growth. In addition, we present the morphological and magnetic characterizations of $CeIn_3$ nanowires and $CeAlO_3$ nanotubes. Our work shows that the MFNN method is a possible but challenging route to obtain rare-earth based nanowires.

2. Experimental

Several attempts to grow $CeIn_3$ were performed using the metallic flux nanonucleation (MFNN) method. [7]. This method is a combination of the conventional metal flux growth technique with the addition of a nanometric template for nanowire growth. In this study, we used an Al_2O_3 template. For the growth of $CeIn_3$ nanowires, a stoichiometry relation of 1 Ce:10 In was used with Al_2O_3 membranes with pore sizes equal or smaller than (155 ± 25) nm. The crucible was covered by quartz wool and sealed inside an evacuated quartz tube. We explored the influence of the pore size, membrane crystallization temperature (T_{cryst} = 900 or 1150 °C), maximum heat treatment temperature (T_{max} = 850, 950, 1100, 1150, 1200 °C) and synthesis time (2, 8, 12, 24, 48 h) on the nanowire growth. After the heat treatment, the excess In flux was centrifuged. Both the membrane and the $CeIn_3$ crystals were mechanically removed from the crucible.

The morphology and energy dispersion X-ray spectrometry (EDS) analysis of the $CeAlO_3$ nanotubes were investigated at an FEI Nova NanoLab 200 Dual Beam system (FIB/SEM). The images of the nanowires and their EDS composition mappings were performed at an FEI Inspect F50 microscope. X-ray diffraction measurements of nanotubes and templates were performed at the Brazilian Synchrotron Light Laboratory (LNLS). Magnetization measurements were performed using a commercial magnetometer.

3. Results and discussion

3.1. Morphology and Composition.

We observed nanowires wherein T_{cryst} = 900 °C and T_{max} = 1100 °C. Figure 1(a) shows the Scanning Electron Microscopy (SEM) image of an isolated nanowire with a diameter of (346 ± 8) nm. As shown in Figures 1 (b) and (c), EDS maps confirm that both Ce and In are present at In Kα (Figure 1(b)) and for Ce Lα (Figure 1(c)) energies, which indicates the formation of $CeIn_3$ nanowires.

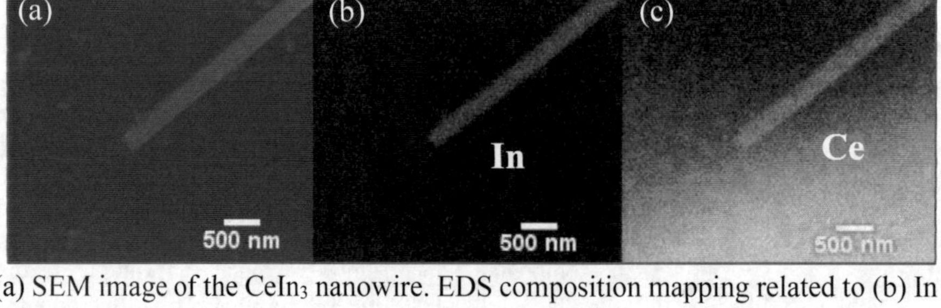

Figure 1. (a) SEM image of the $CeIn_3$ nanowire. EDS composition mapping related to (b) In $K\alpha$ and (c) Ce $L\alpha$ energies.

Several growths were performed with a mean pore diameter equal to or smaller than (155 ± 25) nm. For each attempt, we varied the T_{max} (850, 950, 1100, 1150, 1200 °C) or the synthesis time (2, 8, 24, 48 h). Regardless of the chosen parameters, all attempts performed for these templates yielded a presence of nanotubes, as evidenced by the SEM image shown in Figures 2 (a,b). The exception was a growth performed at T_{max} = 850 °C with T_{cryst} = 900 °C, for which In nanowires were obtained inside the pores (not shown here). Figure 2 (c) shows the point EDS spectrum of a nanotube. We observe the presence of only Ce, Al, and O, which is not consistent with the targeted $CeIn_3$ phase. Figure 2d shows the X-ray diffraction patterns of an alumina template with nanotubes as well as an empty Al_2O_3 membrane. The orange squares and blue triangles correspond to the indexation from the Inorganic Crystal Structure Database (ICSD no. 72558 and 171679) for the $CeAlO_3$ and In phases, respectively. Both XRD patterns shown in Fig 2d are consistent with the alumina phase. In the case of the template with nanotubes, we also identified peaks consistent with the In and with the tetragonal phase of $CeAlO_3$ (space group: *P4/mmm)* with lattice parameters $a = b = 3.767$ Å and $c = 3.797$ Å [10]. Therefore, XRD measurements and EDS analyses indicate that the nanotubes present the $CeAlO_3$ phase.

The formation of $CeAlO_3$ nanotubes is possibly associated with the chemical composition of the Al_2O_3 templates. Previous results suggest that Al_2O_3 templates with a "honeycomb" geometry, i.e., hexagonally ordered circular pores, have a duplex structure of pore walls in terms of chemical composition, an acid-anion contaminated outer oxide layer next to the pores with a relatively pure inner wall oxide [11,12]. In this perspective, the formation of nanotubes may occur due to the reactivity of the Ce^{3+} ions with the relatively pure oxide wall. Exploring further routes by changing the chemical composition of the template could be a promising path to be explored.

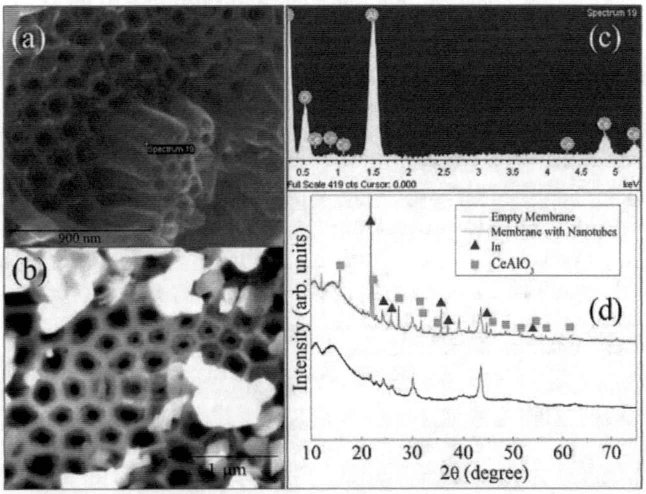

Figure 2. SEM image in (a, b) and EDS spectrum in (c) for the nanotubes of $CeAlO_3$; (d) XRD measurements for an empty Al_2O_3 template and for template with nanotubes.

3.2. *Magnetic characterization.*

We investigated the dimensionality role in the magnetic properties of the $CeIn_3$ bulk and nanowire system. Figure 3 (a) shows the temperature dependence of magnetic susceptibility χ(T), for 1 kOe. Both curves were fitted to the Curie-Weiss law plus a temperature-independent term, $\chi(T) = \chi_0 + C/(T - \theta_{CW})$. Table 1 shows the fitted parameters. The effective moment (μ_{eff}) for the Ce^{3+} ions, extracted from the Curie–Weiss constant (C), is in agreement with the theoretical value μ_{eff} =2.54 μ_B [8]. Besides that, we observe a suppression of the antiferromagnetic ordering temperature from T_N^{bulk} = 10 K (black arrow) to T_N^{nano} = 3 K (red arrow). This decrease is possibly associated with changes in the interplay between the RKKY exchange interaction, CEF, and Kondo effects.

Table 1 Fitted parameters for both bulk and nanowire $CeIn_3$ systems of a Curie–Weiss law plus a temperature-independent term, $\chi(T) = \chi_0 + C/(T - \theta_{CW})$.

$CeIn_3$	θ_{CW} (K)	χ_0(emu/mol.Oe)	μ_{eff} (μ_B)
Bulk	-55	7.4 x 10^{-4}	2.6
Nanowire	-57	-8.4 x10^{-4}	2.1

Figure 3b shows the dependence of susceptibility as a function of temperature measured at 10 kOe for $CeAlO_3$ nanotubes grown at different temperatures and synthesis time. The curves show no phase transition, exhibiting a paramagnetic behavior throughout the studied temperature range.

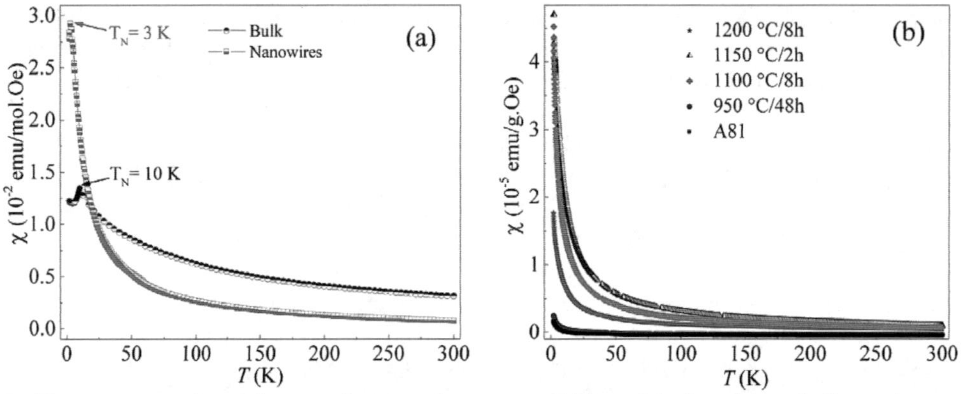

Figure 3. Temperature dependence of magnetic susceptibility: (a) for $CeIn_3$ bulk and nanowire, evidencing the Néel temperature; (b) for the Al_2O_3 template (A81) and $CeAlO_3$ nanotubes grown at different synthesis temperatures.

4. Conclusions

In this work, we explored the routes for synthesizing intermetallic nanowires of $CeIn_3$ via the MFNN method. $CeIn_3$ nanowires were successfully obtained in only one batch with an average diameter of (346 ± 8) nm. Suppression of the AFM transition from ($T_N^{bulk} = 10$ K) to ($T_N^{nano} = 3$ K) was observed in the nanowires. This reduction is possibly associated with the dimensionality affecting the interplay between the RKKY exchange interaction, CEF, and Kondo effects. For several other attempts, the presence of $CeAlO_3$ nanotubes was observed. The MFNN method is a promising but challenging alternative for the fabrication of nanowires of intermetallic systems.

Acknowledgments

This work was supported by the FAPESP (grants No 2020/10580-8, 2020/12283-0, 2018/11364-7, 2017/10581-1, 2012/04870-7), CAPES and CNPq Brazilian agencies. The authors would like to acknowledge the LNNano, CCSNano-UNICAMP, and LNLS.Work at Los Alamos National Laboratory (LANL) was supported by the Los Alamos Laboratory Directed Research and Development program through project 20190076ER.

References

[1] Paschen S and Si Q 2021 *Nat. Rev. Phys.* **3** 9–26
[2] Shishido H, Shibauchi T, Yasu K, Kato T, Kontani H, Terashima T and Matsuda Y 2010 *Science.* **327** 980–3
[3] Moura K O, Oliveira L A S, Rosa P F S, Jesus C B R, Saleta M E, Granado E, Béron F, Pagliuso P G and Pirota K R 2016 *Sci. Rep.* **6** 1–9
[4] Moura K O, Pirota K R, Béron F, Jesus C B R, Rosa P F S, Tobia D, Pagliuso P G and De Lima O F 2017 *Sci.Rep.* **7** 3–9
[5] Cruz A dos S E d., Puydinger dos Santos M V., Campanelli R B, Pagliuso P G, Bettini J, Pirota K R and Béron F 2021 *Nanoscale Adv.* **3** 3251–9
[6] Rosa P F S, De Oliveira L A S, De Jesus C B R, Moura K O, Adriano C, Iwamoto W, Garitezi T M, Granado E, Saleta M E, Pirota K R and Pagliuso P G 2014 *Solid State Commun.* **191** 14–8
[7] K.R. Pirota, F. Beron, L.A.S. Oliveira, K.O. Moura, M. Knobel P G P, T.M. Garifezi, C.B.R. de Jesus, C. Rettori, C. Adriano, P.F.S. Rosa, R R Urbano and W.A. Iwamoto, L.C. Arzuza P G C 2014 P BR patent 10 2014 019794 0
[8] Berry N, Bittar E M, Capan C, Pagliuso P G and Fisk Z 2010 *Phys. Rev. B Condens. Matter Mater. Phys.* **81** 1–9
[9] Moll P J W, Helm T, Zhang S S, Batista C D, Harrison N, McDonald R D, Winter L E, Ramshaw B J, Chan M K, Balakirev F F, Batlogg B, Bauer E D and Ronning F 2017 *Quantum Mater.* **2**
[10] Tanaka M, Shishido T, Horiuchi H, Toyota N, Shindo D and Fukuda T 1993 *J. Alloys Compd.* **192** 87–89
[11] Thompson G E and Wood G C 1981 *Nature* **290** 230–2
[12] Le Coz F, Arurault L and Datas L 2010 *Mater. Charact.* **61** 283–8

Magnetic-field-tuned valence susceptibility in $CeCo_{1-x}Fe_xSi$ revealed by magnetostriction

V F Correa[1], I Curci[1], J G Sereni[1], N Caroca Canales[2], C Geibel[2]

[1] Centro Atómico Bariloche (CNEA) and Instituto Balseiro (U. N. Cuyo), 8400 Bariloche, Río Negro, Argentina

[2] Max-Planck-Institut für Chemische Physik fester Stoffe, D-01187 Dresde, Germany

E-mail: victor.correa@cab.cnea.gov.ar

Abstract. We report magnetoelastic studies in the $CeCo_{1-x}Fe_xSi$ alloys. Iron doping has a profound effect on the low temperature linear magnetostriction. At $T = 2$ K, the strength of the magnetostriction peaks at $x = 0.23$ where it reaches a value as large as $\frac{\Delta L}{L} = 3\times10^{-3}$ in an applied magnetic field $B = 16$ T. This Fe concentration corresponds to the critical one x_c above which long-range antiferromagnetic order is no longer observed. The progressive increment of the hibridization between the Ce $4f$ orbital and the conduction band, as the magnetic order vanishes, gives rise to a sizeable valence susceptibility that can be finely tuned by the magnetic field explaining the large magnetostrictive effect around x_c. At higher x, the magnetostriction decreases, in agreement with a weaker valence susceptibility resulting from a stronger hibridization.

1. Introduction

Magnetostriction (MS) has demonstrated to be an outstanding tool to study magnetic correlations due to the strain dependence of the exchange couplings [1]. Here we use MS as a probe of a different property: the onset of a slight valence change in a Cerium alloy. $CeCo_{1-x}Fe_xSi$ orders antiferromagnetically ($T_N = 8.8$ K) in the stoichiometric limit ($x = 0$) [2]. Iron doping is detrimental to the magnetic order, suppressing it at the critical concentration $x_c \approx 0.23$ (inset of figure 1) [3]. Beyond this doping level, the progressive hibridization between the Ce^{3+} $4f$ electron and the conduction band electrons gives rise firstly to a moderate heavy fermion to end up in an intermediate valence regime in the other stoichiometric limit ($x = 1$) [4].

2. Results

The linear MS is strongly dependent on iron content [5, 6]. This is summarized in figure 1, where the linear MS of polycrystalline samples for four different Fe concentrations at low temperature ($T \approx 2$ K) is shown. It evolves from a moderate strength at $x = 0.15$ ($\frac{\Delta L}{L}$ (16 T) $= 4\times10^{-4}$) to a very large strength at x_c ($\frac{\Delta L}{L}$ (16 T) $= 3\times10^{-3}$). Further doping reduces the effect again.

The unit-cell volume V_u shows an interesting evolution with the Fe content as well [3]. It remains unchanged up to $x \approx 0.15$. Beyond this doping, V_u monotonously decreases reaching $\sim$ 1% contraction at $x = 1$. This is indicative of an incipient volume reduction of the Ce ions: a partial delocalization of the $4f$ electron in the Ce^{3+} ion with the corresponding increase in the effective valence towards a Ce^{4+} character. X-ray absorption spectroscopy experiments confirm this trend as they show a 10% valence difference between CeCoSi and CeFeSi [7]. These results suggest that the large MS observed at x_c is related to an important valence susceptibility tuned by the magnetic field B.

The Ce $4f$ orbital, sited well below the Fermi energy ϵ_F in CeCoSi, approaches ϵ_F as x increases. At x_c, it is basically leveled with ϵ_F and it hibridizates with the conduction band giving rise to a very narrow f-band. As a consequence the effective valence of the Ce ions increase. A moderate magnetic field of the order of the f-bandwidth splits this band moving the spin-up component below ϵ_F. Thus, a mostly localized 3+ valence state is recovered. Accordingly, the lattice expands bringing about the large MS observed around x_c. This magnetic field-tuned valence susceptibility lessens at higher x due to the ever-growing hibridization. Thus, the MS effect is reduced.

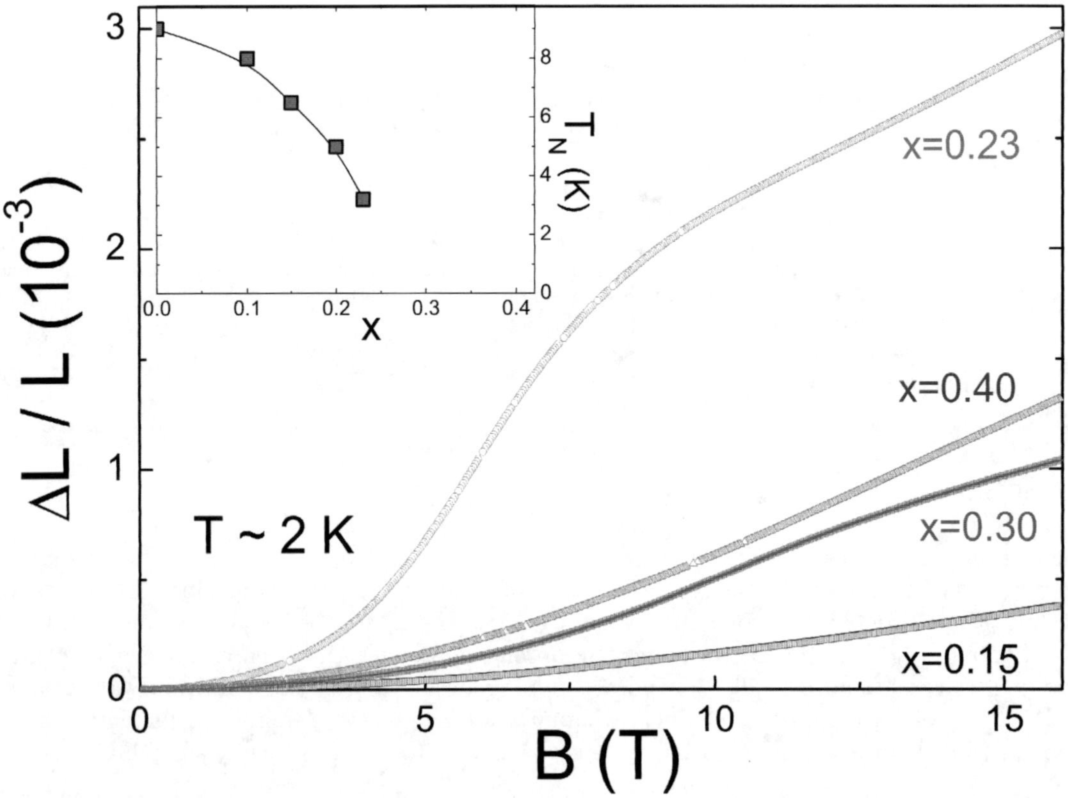

Figure 1. Magnetostriction for different Fe doping x at $T \approx 2$ K. Inset: Néel temperature T_N dependence on x.

The MS displays a step-like shape around x_c (see figure 1). This step is the hallmark that the f-band full split has been accomplished. This is better seen in figure 2 where the MS as well as its derivative (i.e. the magnetostriction coefficient $\lambda = \frac{1}{L}\frac{\partial L}{\partial B}$) are plotted. The top and bottom panels correspond to $x = 0.23$ and $x = 0.3$, respectively. A clear peak at $B_m \approx 6$ T is observed in λ for $x = 0.23$. Though more diluted and wider, the peak still survives for $x = 0.3$ but is shifted to higher field ($B_m \approx 10$ T). This is nothing but the consequence of a stronger hibridization at $x = 0.3$: a larger field is needed to split a broader band.

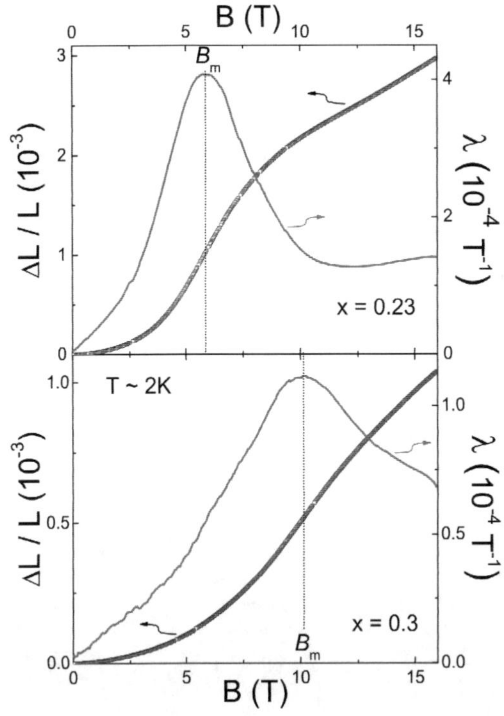

Figure 2. Magnetostriction and magnetostriction coefficient $\lambda = \frac{1}{L}\frac{\partial L}{\partial B}$ at $T \approx 2$ K for $x = 0.23$ (top panel) and $x = 0.3$ (bottom panel).

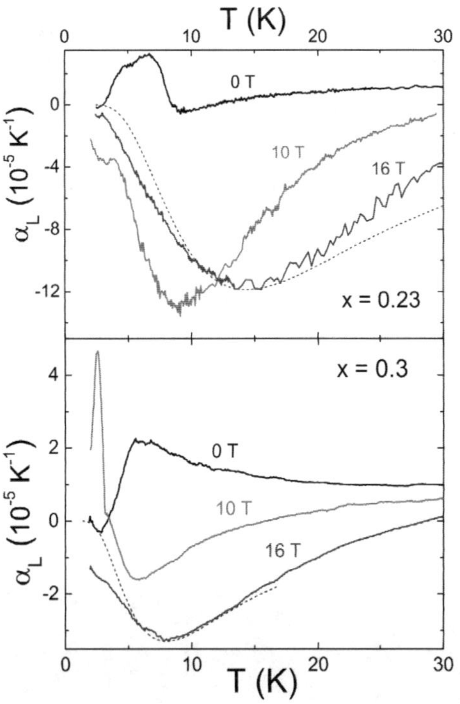

Figure 3. Thermal-expansion coefficient at different fields for $x = 0.23$ (top panel) and $x = 0.3$ (bottom panel). Dotted curves are fits to the 16 T curves using a two-level model.

At high enough fields above B_m the spin-down and spin-up split bands should resemble a two-level system. Thus, a Schottky-like behavior is expected in the thermal-expansion coefficient. Figure 3 shows the temperature dependence of the linear thermal-expansion coefficient $\alpha_L = \frac{1}{L}\frac{\partial L}{\partial T}$ at different magnetic fields. Again, the top and bottom panels correspond to $x = 0.23$ and $x = 0.3$, respectively. The zero field α_L, positive in almost the whole field range, shows a broad peak at low-T probably associated with residual short-range magnetic correlations. This feature is washed out as the field is increased. In turn, α_L displays a broad negative minimum that, for a given field, is shifted to higher-T for $x = 0.23$ relative to $x = 0.3$, in agreement with a larger valence susceptibility for the former concentration. At $B = 16$ T, and assuming the thermal expansion to be proportional to the specific heat, α_L can be reasonably fitted with a two-level model (dotted curves in figure 3) in both cases.

3. Conclusions

A considerable magnetostructural coupling is present in the $CeCo_{1-x}Fe_xSi$ intermetallic alloys. The large magnetostriction observed when the magnetic order is suppressed seems to be caused by a significant valence susceptibility at the onset of a valence instability. As the iron content goes beyond the critical concentration $x_c \approx 0.23$, the increasing hibridization between the f-orbitals and the conduction band weakens the magnetostructural coupling making the lattice less susceptible to the magnetic field.

Acknowledgments
V F C and J G S are members of CONICET, Argentina. Work partially supported by ANPCyT PICT2019-02396 and SIIP-UNCuyo 06/C559.

References
[1] Betancourth D, Correa V F, Facio J I, Fernández J, Vildosola V, Lora-Serrano R, Cadogan J R, Aligia A A, Cornaglia P S and García D J 2019 *Phys. Rev.* B **99** 134406
[2] Chevalier B and Matar S F 2004 *Phys. Rev.* B **70** 174408
[3] Sereni J G, Gómez Berisso M, Betancourth D, Correa V F, Caroca Canales N and Geibel C 2014 *Phys. Rev.* B **89** 035107
[4] Welter R, Venturini G and Malaman B 1992 *J. Alloys Compd.* **189** 49
[5] Correa V F, Betancourth D, Sereni J G, Caroca Canales N and Geibel C 2016 *J. Phys.: Condens. Matter* **28** 346003
[6] Correa V F, Villagrán Asiares A G, Betancourth D, Encina S, Pedrazzini P, Cornaglia P S, García D J, Sereni J G, Maiorov B, Caroca Canales N and Geibel C 2019 *Phys. Rev.* B **100** 184409
[7] Isnard O, Miraglia S, Welter R and Malaman B 1999 *J. Synchrotron Rad.* **6** 701

Effect of 3*p*- and 5*d*-electron doping on the Kondo Semiconductor $CeFe_2Al_{10}$

Rajesh Tripathi[1,2], D T Adroja[1,3], M R Lees[4], V K Anand[5,6], A Sundaresan[2], S Langridge[1], A Bhattacharyya[7], Y Muro[8], K Hayashi[9], T Takabatake[9]

[1]ISIS Facility, STFC, Rutherford Appleton Laboratory, Chilton, Oxon OX11 0QX, United Kingdom
[2]Jawaharlal Nehru Centre for Advanced Scientific Research, Jakkur, Bangalore 560064, India
[3]Highly Correlated Matter Research Group, Physics Department, University of Johannesburg, P.O. Box 524, Auckland Park 2006, South Africa
[4]Department of Physics, University of Warwick, Coventry CV4 7AL, UK
[5]Helmholtz-Zentrum Berlin für Materialien und Energie GmbH, Hahn-Meitner Platz 1, D-14109 Berlin, Germany
[6]University of Petroleum and Energy Studies, Dehradun, Uttarakhand, 248007, India
[7]Department of Physics, Ramakrishna Mission Vivekananda Educational and Research Institute, Belur Math, Howrah 711202, West Bengal, India
[8]Liberal Arts and Sciences, Faculty of Engineering, Toyama Prefectural University, Imizu 939-0398, Japan
[9] Graduate School of Advanced Science and Engineering, Hiroshima University, Higashi-Hiroshima 739-8530, Japan

E-mail: rajeshtripathi@jncasr.ac.in, devashibhai.adroja@stfc.ac.uk

Abstract. We examined the effect of 3*p*- and 5*d*-electron doping on the Kondo semiconductor $CeFe_2Al_{10}$ by means of the electrical resistivity (ρ), magnetic susceptibility (χ), and specific heat (C) measurements. The results show that in the 3*p*-electron-doped system $CeFe_2Al_{10-y}Si_y$, the semiconducting behavior is suppressed for $y = 0.05$, and the system adopts a metallic ground state with an increase in the density of states at the Fermi level. The Si substitution leads to a large decrease in the paramagnetic Weiss temperature θ_P indicating a reduction in *c*-*f* hybridization strength, however the Si does not induce magnetic order up to $y = 0.3$ down to 2 K. The systematic changes in $\rho(T)$, $\chi(T)$, and $C(T)$ are similar to those for 5*d*-electron doped system $CeFe_{2-x}Ir_xAl_{10}$, although, Ir substitution induces a bulk antiferromagnetic transition below 3.1 K in $CeFe_{1.7}Ir_{0.3}Al_{10}$. These changes can be explained by the collapse of the hybridization gap due to the suppression of the *c*-*f* hybridization effect. Our results further confirm that the collapse of the spin/charge gap by an excess electron doping is one of the universal features of the Kondo semiconductors CeT_2Al_{10} (T = Fe, Ru, and Os).

1. Introduction

The Kondo semiconductors CeT_2Al_{10} (T = Fe, Ru, and Os) have been studied intensively as the compounds with T = Ru and Os exhibit antiferromagnetic (AFM) phase transitions at unusually high temperature (T_N = 27.3 and 28.7 K, with a very low ordered state moment $\mu \simeq 0.40$ and 0.29 μ_B/Ce for T = Ru and Os, respectively) despite the Ce-Ce distances of more than 5Å in the cagelike orthorhombic crystal structure [1]. Surprisingly, these T_N values are much higher than those in the isostructural gadolinium compounds GdT_2Al$_{10}$ (T_N = 17.5 K for T = Ru and 18 K for T = Os with $\mu \simeq 7\mu_B$) [2]. Optical conductivity measurements on CeT_2Al$_{10}$ (T = Ru and Os) have revealed that a charge-density-wave-like instability develops along the b axis at temperatures slightly higher than T_N which induces unconventional AFM ordering [3]. Furthermore, CeT_2Al$_{10}$ with T = Fe, exhibits semiconducting behavior with nonmagnetic ground state down to 45 mK [4]. Indeed, the temperature dependent resistivity obeys the thermal activation-type form $\rho(T) \propto \exp(\Delta/k_BT)$, for T <20 K, exhibiting a semiconducting behavior. The temperature dependence of the magnetic susceptibility suggests an intermediate valence behavior for $CeFe_2Al_{10}$ with a strong magnetic anisotropy owing to the strong c-f hybridization [4]. The optical conductivity measurements on $CeFe_2Al_{10}$ at low temperatures, have revealed an optical transition across the hybridization gap between the conduction band and the localized $4f$ states, namely c–f hybridization, appears at 55 meV [5]. Due to the highest degree of c-f hybridization, $CeFe_2Al_{10}$ is expected to have the largest real gap or pseudogap feature among the CeT_2Al$_{10}$ (T = Fe, Ru, Os) materials [5].

In the present study, we have observed marked changes in the thermal and transport properties of $CeFe_2Al_{10}$ induced by $3p$ (Si for Al)- and $5d$ (Ir for Fe)-electron doping. The low-temperature energy gap, which is proposed to be a consequence of strong c-f hybridization is suppressed by a small amount of Si/Ir substitution, and the system adopts a metallic ground state with an increase in the density of states at the Fermi level. With increasing Ir concentration x in $CeFe_{2-x}Ir_xAl_{10}$, the system undergoes long-range AFM order below 3.1 K for $x = 0.3$. On the contrary, Si substitution does not induce magnetic order up to $y = 0.3$ down to 2 K. The emergence of AFM order in $CeFe_{1.7}Ir_{0.3}Al_{10}$ was attributed to the combined effect of weakening of c-f exchange interaction and the lattice contraction along the b axis [6].

2. Experimental details

Polycrystalline samples of $CeFe_2Al_{10-y}Si_y$ ($y = 0.1, 0.2, 0.3$), and $CeFe_{2-x}Ir_xAl_{10}$ ($x = 0, 0.08, 0.16, 0.3$) were prepared by standard arc melting of stoichiometric quantities of high purity elements. The obtained ingots were then annealed for a week at 1000 °C. Powder x-ray diffraction confirmed the $YbFe_2Al_{10}$-type orthorhombic structure (space group $Cmcm$)[Fig. 1(a)]. The crystal structures were refined by the Rietveld method and the lattice parameters as a function of the x and y are shown in Fig. 1(a) and Fig. 1(b), respectively. All the samples were single phase except for the sample $y = 0.3$, containing a small amount of $CeAl_{0.9}Si_{1.1}$, which orders ferromagnetically with T_C = 11 K [7]. The the lattice parameters and hence unit cell volume of $CeFe_2Al_{10-y}Si_y$ decreases with increasing y, while the lattice parameters exhibit anomalous changes for $CeFe_{2-x}Ir_xAl_{10}$ such that the unit cell volume remains almost unchanged. The $\rho(T)$ measurements were performed using a standard four-probe method using the resistivity option of a Quantum Design Physical Property Measurement System (PPMS). The $C(T)$ was measured by the relaxation method in a PPMS. The $\chi(T)$ was measured by using a commercial SQUID magnetometer in a magnetic field B = 0.1 T.

3. Results and Discussion

Figure 2(a) shows the temperature dependence of electrical resistivity $\rho(T)$ for $CeFe_2Al_{10-y}Si_y$ (y = 0, 0.1, 0.2, and 0.3) down to 2 K. The ρ for $y = 0$ increases with decreasing T, with a broad hump near T_{coh} = 75 K and a sharp upturn below T = 20 K. The low-temperature behavior is well described by the activation law, $\rho(T) = \rho_0 \exp(\Delta/2k_BT)$, presenting a typical Kondo semiconducting behavior. The $-\ln T$ behavior above T_{coh} can be ascribed to Kondo scattering on the crystal field excited states. For $y \geq 0.1$, the Kondo semiconducting behavior below T = 20 K abruptly disappears and $\rho(T)$ becomes metallic at low temperature. A similar change in $\rho(T)$ is also seen with x for $CeFe_{2-x}Ir_xAl_{10}$ [Fig. 2(b)]. Therefore, the metallization in $CeFe_2Al_{10}$ seems to be favored by both $3p$ and $5d$ electron doping. In Si-doped $CeFe_2Al_{10-y}Si_y$,

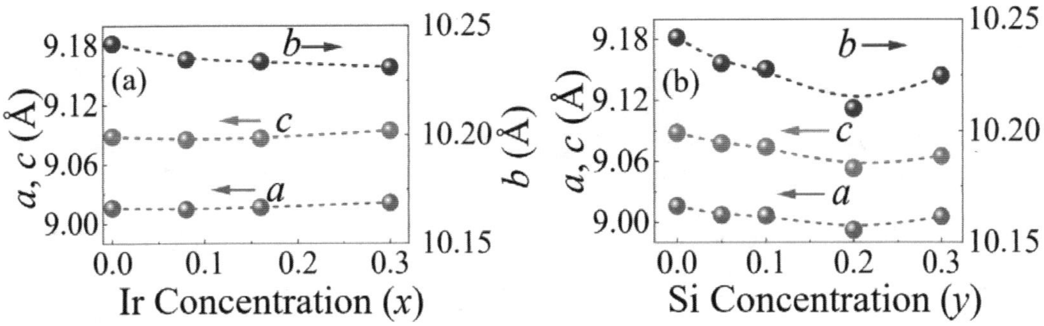

Figure 1. (a)The variation of the lattice parameters as a function of Ir concentration x for $CeFe_{2-x}Ir_xAl_{10}$ (b) The variation of the lattice parameters as a function of Si concentration y for $CeFe_2Al_{10-y}Si_y$.

the low-temperature maximum in ρ at $T_0 \sim 30$ K for $y = 0.1, 0.2, 0.3$ can be attributed to the onset of Kondo coherence to the crystal field ground state and the high temperature $-\ln T$ is the result of the Kondo scattering from the excited CEF states of Ce^{3+}. A similar feature in $\rho(T)$ is also seen for $CeOs_2Al_{10}$ where a metallic ground state has been realized with both electron (Ir/Rh) and hole (Re)-doping [12].

Figures 2(c) and 2(d) shows the temperature dependence of magnetic susceptibility χ for $CeFe_2Al_{10-y}Si_y$ and $CeFe_{2-x}Ir_xAl_{10}$. The low temperature χ increases with increasing y, however, no signature of bulk magnetic ordering could be seen down to 2 K up to $y = 0.3$. The inverse susceptibility (χ^{-1} vs T) data for $CeFe_2Al_{10-y}Si_y$ (not shown here) follow the Curie–Weiss like behavior at $T > 100$ K. A least squares fit to the $\chi(T)$ data above 100 K with the modified Curie–Weiss law yielded the effective magnetic moment $\mu_{\rm eff}$ = 2.32(4), 2.25(1), 2.20(3) and 2.13(1) μ_B/Ce, and paramagnetic Weiss temperatures Θ_P of –377(6), –346(3), –147(1), and –112(3) K for the samples with y = 0, 0.1, 0.2, and 0.3, respectively. The estimated $\mu_{\rm eff}$ are slightly smaller than 2.54 μ_B, expected for a free Ce^{3+} ion. This discrepancy may be due to crystal-field effects. A large and negative value of Θ_P is a common feature in Ce-based compounds with strong $c-f$ hybridization [8]. The variation of $|\Theta_P|$ with y for $CeFe_2Al_{10-y}Si_y$ suggests that the Si substitution suppresses the c–f hybridization and leads to the localization of the Ce-4f electrons.

On the other hand, Ir substitution for Fe changes the system more strongly from the intermediate valence regime to a localized regime owing to the suppression of c-f hybridization. In fact, $\chi(T)$ of $CeFe_{2-x}Ir_xAl_{10}$ for $x = 0.3$ exhibits a sharp peak at $T = 3.1$ K [shown by an arrow in Fig. 2(d)], indicating a phase transition to an AFM ordered state, which was confirmed through a muon spin rotation study [6]. The previous studies on CeT_2Al_{10} (T = Ru, Os), report a gap related to the charge- and spin-density waves along the b axis, in addition to a hybridization gap along all three axes [9]. It is also found that the application of uniaxial pressure along the b axis strongly enhanced the T_N of CeT_2Al_{10} (T = Ru, Os) [10]. In $CeFe_{1.7}Ir_{0.3}Al_{10}$ we have observed an anomalous change in the lattice parameters, in particular a contraction along the b axis, which may be taken as an effective uniaxial pressure in the system. This in addition to the weakening of the c-f hybridization gives rise to magnetic ordering in contrast to Si doping. Here it is important to note that the 3p (Si)- and 4d (Rh)-electron doping in $CeRu_2Al_{10}$ play an equivalent role in both the formation of hybridization gap and the unusual AFM ordering [11]. The observed long-range ordered magnetic ground state with Ir substitution rather than the paramagnetic ground state seen with Si substitution could, therefore, be attributed to the combined effect of b axis behavior and weakens strength of the c-f hybridization.

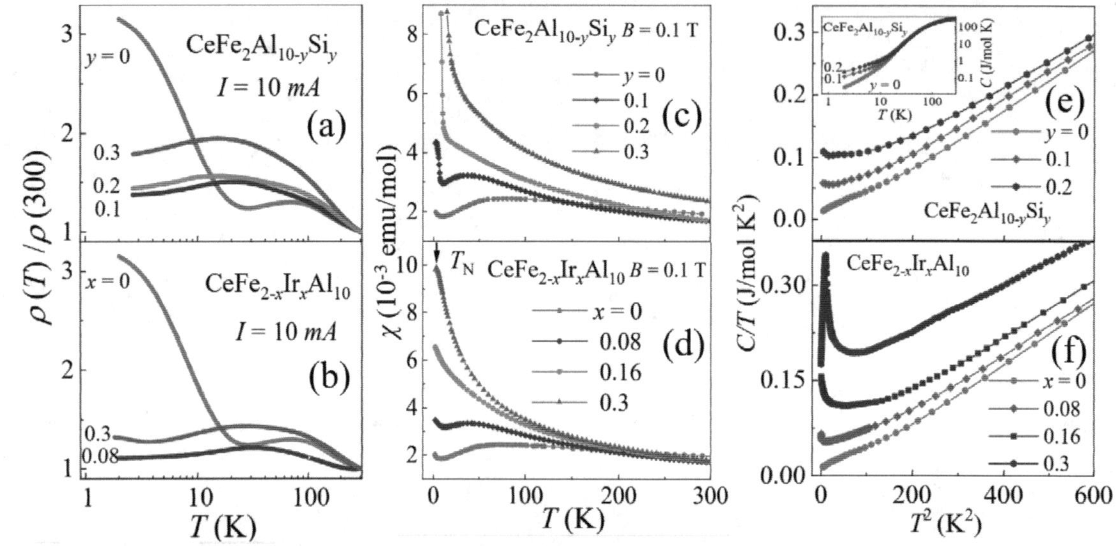

Figure 2. (a) T dependence of ρ for polycrystalline samples of $CeFe_2Al_{10-y}Si_y$. (b) $\rho(T)$ of $CeFe_{2-x}Ir_xAl_{10}$. (c) $\chi(T)$ for polycrystalline samples of $CeFe_2Al_{10-y}Si_y$. (d) $\chi(T)$ of $CeFe_{2-x}Ir_xAl_{10}$. (e) C/T versus T^2 for $CeFe_2Al_{10-y}Si_y$. Inset: Heat capacity $C(T)$ for polycrystalline samples of $CeFe_2Al_{10-y}Si_y$ on a log-log plot. (f) C/T versus T^2 for $CeFe_{2-x}Ir_xAl_{10}$.

The double logarithmic plot of heat capacity C versus T for $CeFe_2Al_{10-y}Si_y$ ($y = 0$, 0.1, and 0.2) are shown in the inset of Fig. 2(e). At $T < 10$ K, C increases with increasing y without any sign of magnetic ordering, reminiscent of heavy-fermion-like behavior. On the other hand, for $x = 0.3$ in $CeFe_{2-x}Ir_xAl_{10}$, a λ-type anomaly is observed at $T \sim 3$ K further indicating the onset of long-range AFM ordering. Figures 2(e) and 2(f) show the variation of C/T with T^2 for $CeFe_2Al_{10-y}Si_y$ and $CeFe_{2-x}Ir_xAl_{10}$, respectively. For $y = 0$, C/T (or γ) at 2 K is almost zero, suggesting a very small or zero density of states at the Fermi level. However, for $y \geq 0$, γ is significantly enhanced and reaches 0.11 J/mole K^2 for $y = 0.2$. Similar behavior is also seen for $CeFe_{2-x}Ir_xAl_{10}$ [Fig. 2(f)]. In other T-site substituted systems e.g. $Ce(Fe_{1-x}Rh_x)_2Al_{10}$, $Ce(Ru_{1-x}Rh_x)_2Al_{10}$ or $Ce(Os_{1-x}Ir_x)_2Al_{10}$ [12], the γ value increases rapidly with x, which is attributed to a rapid collapse of a hybridization gap (or spin gap) and the appearance of conduction electrons with high effective mass at the Fermi level. We, therefore, anticipate that the enhancement of γ in the heat capacity of $CeFe_2Al_{10-y}Si_y$ and $CeFe_{2-x}Ir_xAl_{10}$ is associated with the appearance of a finite density of states close to the Fermi level. Support for this scenario comes from the behavior of the low temperature electrical resistivity. Thus, the above characteristic doping effect is common in all the CeT_2Al_{10} systems (T= Fe, Ru, and Os) with both $3p$ and $4d/5d$-electron doping.

4. Conclusions

We have investigated the magnetic, thermal, and electronic transport properties of Si- and Ir-substituted Kondo semiconducting compound $CeFe_2Al_{10}$ using $\rho(T)$, $\chi(T)$, and $C(T)$ measurements. In $CeFe_2Al_{10-y}Si_y$, the localized nature of the Ce ion is much enhanced by Si doping. However, no bulk magnetic ordering is observed down to 2 K even up to $y = 0.3$. On the other hand, Ir substitution induces a bulk AFM transition below $T_N = 3.1$ K in $CeFe_{1.7}Ir_{0.3}Al_{10}$. The observed long-range ordered magnetic ground state with Ir substitution rather than the paramagnetic ground state seen with Si substitution is attributed to the combined effect of changes in the b axis and a weakening of the c-f hybridization. Our results further suggests that the electron doping achieved by substituting Fe with Ir (or Al with Si), weakens the c-f hybridization and destroys the Kondo semiconducting character and the spin/charge gapped state at low temperatures.

Acknowledgements

DTA would like to thank the Royal Society of London for the Newton Advanced fellowship funding between UK and China, and International Exchange funding between UK and Japan. DTA also thanks EPSRC UK for funding (grant ref. EP/W00562X/1). RT thanks the Indian Nanomission for a post-doctoral fellowship. Y. M and T. T acknowledge the financial supports from JSPS, grant numbers JP26400363, JP15K05180, JP16H01076, and JP21K03473.

References

[1] T. Nishioka, et al., J. Phys.Soc. Jpn. 78, 123705(2009); H. Kato, et al., J. Phys. Soc. Jpn. 80, 073701 (2011).
[2] Y. Muro, et al., J. Phys. Soc. Jpn. 80, SA021 (2011).
[3] S. I. Kimura, et al., Phys. Rev. Lett. 106, 056404 (2011).
[4] Y. Muro, et al., J. Korean Phys. Soc. 63, 508 (2013); Y. Muro, et al., J. Phys. Soc. Jpn. 78, 083707 (2009); H. Tanida, et al., J. Phys. Soc. Jpn. 83, 084708 (2014).
[5] S. Kimura1, et al., J. Phys. Soc. Jpn. 80, 033702 (2011); J. Kawabata, et al., Phys. Rev. B 92, 201113(R) (2015); D . T. Adroja, et al., Solid State Phenom. 257, 11 (2017).
[6] R.Tripathi, et al., Phys. Rev. B. (accepted)(2021).
[7] K. Hayashi, et al., J. Phys. Soc. Jpn. 85, 034714 (2016).
[8] J. G. Sereni, et al., Phys. Rev. B 75, 024432 (2007).
[9] S.-I. Kimura, et al., Phys. Rev. Lett. 106, 056404 (2011); Phys. Rev. B 91, 241120(R) (2015).
[10] K. Hayashi, et al., Phys. Rev. B 96, 245130 (2017).
[11] K. Hayashi, et al., J. Phys. Soc. Jpn. 85, 034714 (2016).
[12] H. Tanida, et al., Phys. Rev. B 92, 235154 (2015); H. Tanida, et al., Phys. Rev. B 92, 235154 (2015); J. Kawabata, et al., Phys. Rev. B 89, 094404 (2014).

Electronic structure of a Kondo lattice system $CeCuAs_2$

Sawani Datta[1], Ram Prakash Pandeya[1], Arka Bikash Dey[2], Christoph Schlueter[2] Andrei Gloskovskii[2], Thiago Peixoto[2], Arumugam Thamizhavel[1], Kalobaran Maiti[1,*]

[1]Tata Institute of Fundamental Research, Mumbai, India
[2]DESY, Petra III, Hamburg, Germany

*E-mail for corresponding author : kbmaiti@tifr.res.in

Abstract. We study the electronic properties of a Ce-based Kondo material, $CeCuAs_2$ employing high-resolution hard *x*-ray photoemission spectroscopy. The measurements were done with different photon energies to probe the surface-bulk differences of the electronic structure. Experimental results show significant difference in the hybridization physics for the surface and bulk electronic structures indicating different Ce valency at the surface compared to the bulk. Surface termination appears to play an important role in the correlation physics of this system. In addition, the experimental spectra show loss features due to plasmon excitations. These results bring out complexity of this novel Kondo lattice system that does not show magnetic order down to the lowest temperature studied and have significantly different surface-bulk properties.

1. Introduction

The rare-earth-based systems are one of the most intriguing topics in condensed matter physics due to their unusual properties arising from strong electron correlation [1]. These materials show several exotic phenomena like Kondo behavior [1-3], unconventional superconductivity [4], the fluctuating valency [5-6], *etc.* arising from the hybridization between the localized $4f$ electronic states and the valence states. Among the rare earth series, Ce has some unique properties, as the radial distribution of the $4f$ electron is relatively more extended for Ce leading to a relatively larger itineracy of the $4f$ electrons, proximity of the occupied $4f$ band to the Fermi level and the presence of one $4f$ electron [7]. Ce has predominantly magnetic trivalent ($4f^1 5d6s^2$) and non-magnetic tetravalent ($4f^0 5d^2 6s^2$) electronic configurations. The mixing of these two configurations in the ground state is associated to the fluctuating valency in Ce compounds. Many Ce-based Kondo lattice systems possesses uncompensated Ce moments leading to magnetically ordered ground states. $CeCuAs_2$ is a unique system which does not show magnetic order down to the lowest temperature studied and hence is a good platform to study the Kondo physics [8].

In this paper, we have report the electronic structure of $CeCuAs_2$ studied by hard *x*-ray photoemission spectroscopy. $CeCuAs_2$ forms in $ZrCuSi_2$ type layered tetragonal structure. The unit cell forms by stacking of the As(2)-CeAs(1)-Cu-As(1)Ce-As(2) layer. Here, As(1) and As(2) are the two non-equivalent As atoms with different Wycoff positions. This crystal has a space group *P4/nmm* (129), with the lattice constants, a = 4.045 Å and c = 10.111 Å. The bulk properties studies of this material show several interesting phenomena. For example, electrical transport exhibits logarithmic increment of resistivity, ρ with the decrease in temperature, T [8]. Below 5 K, the temperature dependence of the resistivity changes to $\rho \propto T^{-0.6}$ [8]. This is different from the behaviour of other members of $RCuAs_2$ (here R = rare earth element) family [9]. In magnetic measurements, no long-range magnetic order has been observed from room temperature (RT) down to 45 mK. The effective magnetic moment in the paramagnetic phase is found to be 2.33 μ_B/Ce and 2.50 μ_B/Ce at low and high temperature regimes, respectively for the magnetitic field applied perpendicular to c axis. Such reduced effective magnetic moment from the moment of the trivalent Ce atom (2.54 μ_B/Ce) indicates compensation of the moment due to Kondo coupling. The other possibility could be an over valency greater than (+3) in this material, fluctuating valency case, etc. The hard x-ray photoemission spectroscopy study reported earlier suggests a pseudogap at the Fermi level [10]. We prepared high quality single crystals of this material and found interesting features that may be corroborated to the exoticity of this material.

2. Experimental details:

We have prepared high-quality single crystals of $CeCuAS_2$ using flux method. To check the elemental composition, we have performed energy dispersive analysis of x-rays and found close to 1:1:2 ratio of the elements. By this observation we can rule out the presence of an easily formable Cu_2As phase in this material. Good crystallinity of the sample is confirmed by the bright spots in the Laue diffraction pattern shown in the Fig. 1.

The hard x-ray photoemission spectroscopy (HAXPES) was carried out using 2500, 6000 and 8000 eV photon energies at 45 K and 120 K. The experiments were performed at the P22 beamline of Petra III, Desy, Germany. The energy resolution was about 150 meV at 6000 eV photon energy. The sample was cleaved in the ultrahigh vacuum chamber before the measurements.

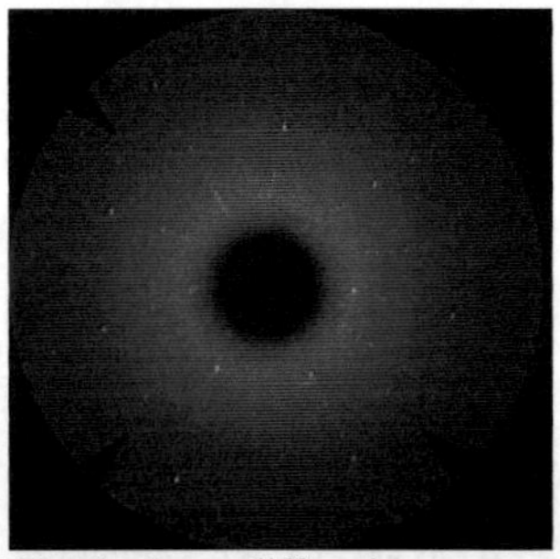

Fig. 1. Laue diffraction pattern of $CeCuA_2$

3. Results and discussions

In Fig.2, we have shown the Cu $2p$ core level spectra of $CeCuAs_2$ measured at 45 K using 6 keV photon energy. The experimental data is shown by the black open circles. We have fitted the curve to identify each peak separately, the overall simulated envelop is shown by the red line superimposed over the experimental data. There are two strong, intense, peak structures in the spectrum representing the spin-orbit split features, Cu $2p_{1/2}$and Cu $2p_{3/2}$. Binding energies of these features are 952.8 eV and 933 eV; the spin-orbit coupling is about 19.8 eV. These lineshape of the features is asymmetric with the tail in the higher binding energy side. This type of tail appears due to the excitation of the valence electron across the Fermi level along with the core level excitations and hence, this is an indication of the metallic ground state of the material [11]. In addition, we observe a broad hump lying about 15.8 eV higher binding energy relative to the main peaks. Similar features have been observed for other core level excitations. Thus, these features can be attributed to the loss feature due to excitation of the collective modes; here the energy of the loss feature suggests plasmon excitations. The single peak structure of Cu $2p$ core level peaks suggests (+1) valency of Cu in this material. This result is consistent with the earlier reported HAXPES study [10].

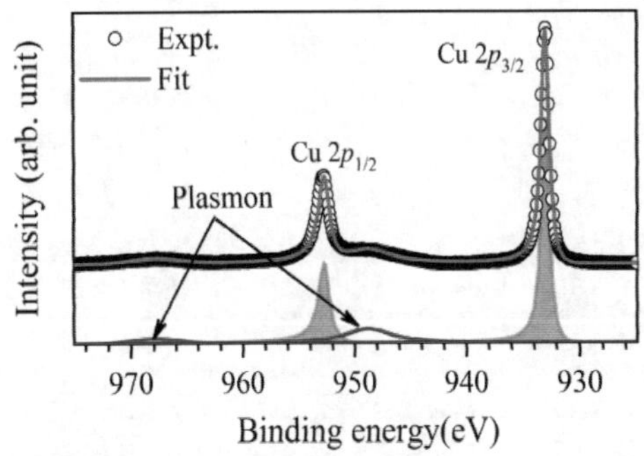

Fig. 2. Cu $2p$ core level spectrum (open circle) and the fit curve (line). Peaks in the lower panel are the constituent peaks.

In Fig.3, we show the Ce 3d core level spectra of $CeCuAs_2$ measured at 45 K using 6 keV and 2.5 keV photon energies. The experimental data are shown by the black open circles. There are many distinct features in each of these spectra. The intensity of the peaks depends on the excitation photon energies. Since the photoemission escape depth strongly depends on the kinetic energy of the photoelectrons, these results suggest significant surface-bulk differences in the electronic structure. To examine these spectra in more details, we have simulated the experimental spectra with a set of asymmetric Voigt functions. The fit results are shown by the red line superimposed over the experimental data in the figure.

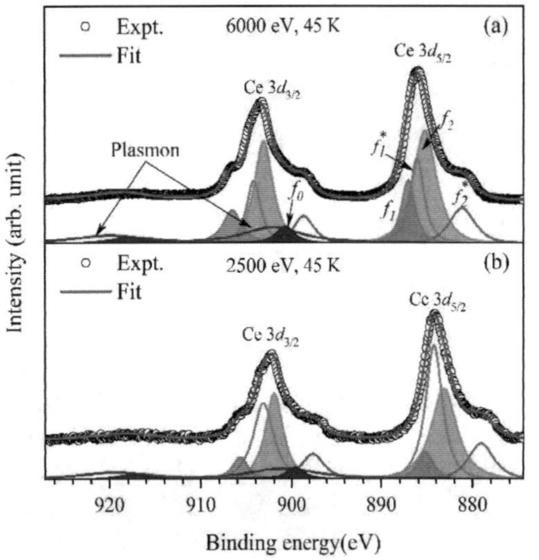

Fig. 3. Ce 3d core level spectra at (a) 6 keV and (b) 2.5 keV photon energy and 45 K temperature. The symbols are the experimental data and the line are simulated data. The shaded peaks are the bulk feature and the same color-outlined peaks are the surface feature.

The Ce 3d spectra exhibit three-peak structures corresponding to different final states of photo-excitations. These final states are often termed as poorly screened f_1 feature (final state electronic configuration $3d^9 4f^1$) and well-screened f_2 feature where a valence electron hopped to the 4f level of the photoemission site to screen the core hole, the final state electronic configuration will be $3d^9 4f^{\,2}$). In addition, these materials show f_0 feature having the final state electronic configuration $3d^9 4f^0$, which corresponds to a tetravalent like Ce site and often show growth at low temperatures. This feature is considered as a signature of Kondo effect [12]. In Kondo lattice system, an increment of the f_0 feature intensity is observed with the lowering of the sample temperature due to the formation of the Kondo singlet [13]. In this material, the position of these f_0, f_1 and f_2 features associate to Ce-3$d_{5/2}$ features are respectively 899.6 eV, 885.2 eV and 883.2 eV. In Fig.3, the f_0, f_1 and f_2 features are shown by blue, grey and green shaded region respectively. Their spin orbit split peaks appear at about 18.8 eV higher binding energy side of each above-mentioned features respectively.

In the experimental data, we find a second set of the f_1 and f_2 peaks. The features for 3$d_{5/2}$ excitations appear at 884.3 eV and 879.1 eV. In Fig.3, they are shown by the grey and green line and marked as f_1^* and f_2^* respectively. The intensity of these peaks becomes stronger at 2.5 keV data compared to 6 keV data. Since the technique is more bulk sensitive at 6 keV compared to 2.5 keV case, these features are attributed to the surface electronic structure. Evidently the unstarred ones are the bulk peaks and appear at higher binding energies relative to the surface peaks. This suggests that the effective Madelung potential at the bulk Ce sites is higher than that at the surface Ce sites. Moreover, we observe that the intensity of the bulk f_2 feature is larger than bulk f_1 peak which indicates larger hybridization in the bulk electronic structure that leads to more efficient screening of the core holes. The surface features show the opposite scenario reflecting the more local behavior of the surface Ce [6]. Ce 3d spectra also show plasmon features as found in Cu 2p spectrum.

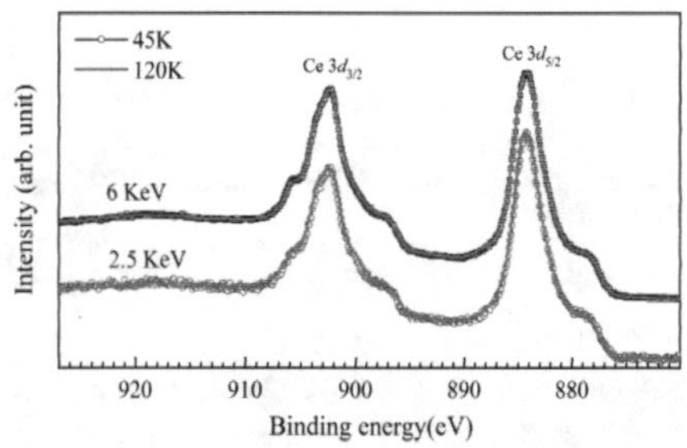

Fig. 4. Temperature evolution of Ce 3d spectra. Top and the bottom panels are the spectra in 6 keV and 2.5 keV photon energy at 45 K (black Line + symbol) and 120 K (red line) temperature.

In Fig.4., we show the temperature variation of Ce 3*d* spectra. We do not observe any change in any of the 2.5 keV and 6 keV data. This is unusual for a Kondo lattice system hosting f_0 feature. Considering the fact that Ce moments are not fully compensated in the magnetization results, this observation suggests that the experimental temperature is much higher than the temperature regime required to observe changes in f_0 feature. Further studies are required to understand this scenario.

4. Conclusion

In summery we have studied the electronic properties of $CeCuAs_2$ with high-resolution hard *x*-ray photoemission spectroscopy. Core level spectra show signature of plasmon loss features and Cu is found to be monovalent in this material. We find the signature of strong Ce 4*f*-valence state hybridization and significant differences in the surface and bulk electronic structure. Though the bulk measurements show Kondo behavior, we have not observed temperature dependence of the Kondo feature down to 45 K studied, which is the lowest temperature achievable in the experimental setup. We hope these results will help to initiate further studies in this direction.

References

[1] Stewart G. R. 1984 *Rev. Mod. Phys.* **56**, 755; Steglich F. 1991 *J. Magn. Magn. Mater.* **100** 186;
[2] Patil S. *et. al.* 2010 *J. Phys.: Condens. Matter* **22** 255602; *ibid.* 2008 *Phys. Rev. B* **77**, 094443;
[3] Maiti K. *et al.* 2011 *J. Phys.: Conf. Ser.* **273**, 012042; Patil S. *et al.* 2011 *J. Phys.: Condens. Matter* **23** 495601;
[4] Watanabe S. and Miyake K. 2011 *J. Phys.: Condens. Matter* **23,** 094217;
[5] Sereni J. G. *et. al.*1989 *Modern Physics Letters B* **03** 16;
[6] Laubschat C. *et.al.* 1992 *Surface Science* **269/270**, 605-609;
[7] Patil S. *et. al.* 2010 *Phys. Rev. B*, **82**, 104428;
[8] Sengupta K. *et. al.* 2004 *Phys. Rev. B* **70**, 064406;
[9] Sengupta K. *et al.* 2004 *Physica B* **348,** 465;
[10] Chainani A. *et al.* 2014 *Phys. Rev. B* **89**, 235117;
[11] Wertheim C.K., Hufner S. 1975 *Phys. Rev. Lett.* **34**, 53;
[12] Gunnarsson O. and Schonhammer K. 1983 *Phys. Rev. B* **28**, 4315;
[13] Gunnarsson O. and Schonhammer K. 1985 *Phys. Rev. B* **31**, 4815.

Modeling of Co impurity in Cu host using DFT+ED

M Tchaplianka, A B Shick and J Kolorenč

Institute of Physics, Czech Academy of Sciences, Na Slovance 2, 182 21 Prague, Czech Republic

E-mail: tchaplianka@fzu.cz

Abstract. We investigate the electronic structure of a cobalt atom in a copper host using the density functional theory and the exact diagonalization of an Anderson impurity model. The spectral functions and spin moments at the impurity are calculated and are found to be close the ones calculated by DFT+QMC [1].

1. Introduction

The first-principles study of magnetic atoms embedded in non-magnetic metals is valuable for understanding quantum many-body systems, since it can provide a theoretical explanation for phenomena such as the Kondo effect in material-specific settings. One promising method for solving such a system is a combination of the density functional theory (DFT) and the exact diagonalization (ED) of the single impurity Anderson model (DFT+ED). Within the ED method, the hybridization between the correlated d orbitals and the host is approximated by discretized "bath" of the conduction electrons. Some of the advantages of using the ED over the quantum Monte Carlo (QMC) with the continuous bath are that it can work at very low temperatures, and can include the spin-orbit coupling (SOC) and the crystal field (CF) simultaneously.

We examine the electronic structure of $CoCu_{15}$, a material approximating a single cobalt impurity in a copper bulk. The solution is performed in a system with no SOC and close to room temperature so that it can be directly compared to the results obtained with DFT+QMC [1]. This work follows an earlier study comparing the two methods, where the DFT+ED calculations were performed with a smaller discretized "bath", at low temperature and with SOC [2].

2. Methods

The material under study is modeled as $CoCu_{15}$, a $2 \times 2 \times 2$ *fcc* supercell of Cu with a single atom substituted with a Co atom. The Co impurity atom has 10 electrons in its $3d$ shell. The cubic symmetry of the crystal makes the $3d$ shell diagonal in the basis of cubic harmonics: $m \in \{xy, xz, yz, x^2-y^2, 3z^2-r^2\}$. The crystal field splits the orbitals into two blocks, the triply-degenerate t_{2g} (for xy, xz and yz) and the doubly-degenerate e_g (for x^2-y^2 and $3z^2-r^2$), where the e_g block has a higher energy.

The dynamics of the Co $3d$ electrons are described using the Anderson impurity model, in a basis where each spin-orbital $d_{m\sigma}$ is coupled to bath states $b_{mk\sigma}$ [3]. The impurity Hamiltonian is

$$H_{\text{imp}} = \sum_{m\sigma} \left(\epsilon_d + h^{\text{CF}}_{mm}\right) n^d_{m\sigma} + \sum_{mk\sigma} \epsilon_{mk} n^b_{mk\sigma} + \sum_{mk\sigma} \left(V_{mk} d^\dagger_{m\sigma} b_{mk\sigma} + \text{h.c.}\right) + U_{\text{Coul}}, \tag{1}$$

Table 1. Values of the "bath" parameters ϵ_{mk} and V_{mk} obtained by fitting the hybridization function obtained from DFT, for models III (left) and IV (right).

m	k	ϵ_{mk} (eV)	V_{mk} (eV)
t_{2g}	1	−2.062	1.150
	2	0.316	0.407
	3	3.655	1.508
e_g	1	−2.337	1.077
	2	0.022	0.146
	3	1.395	0.790

m	k	ϵ_{mk} (eV)	V_{mk} (eV)
t_{2g}	1	−2.320	1.172
	2	−0.053	0.136
	3	0.432	0.476
	4	3.955	1.511
e_g	1	−2.693	1.104
	2	−0.132	0.206
	3	0.252	0.280
	4	2.306	0.830

where $n^b_{mk\sigma} = b^\dagger_{mk\sigma} b_{mk\sigma}$, $n^d_{m\sigma} = d^\dagger_{m\sigma} d_{m\sigma}$ and $U_{\text{Coul}} = \sum_{abcd\sigma\sigma'} d^\dagger_{a\sigma} d^\dagger_{b\sigma'} d_{c\sigma'} d_{d\sigma}/2$. The terms in eq. (1) represent, in order, the energy of the impurity orbitals (split into an average d-shell energy ϵ_d and a traceless crystal-field potential h^{CF}_{mm} that is diagonal in the basis of the cubic harmonics), the bath energy levels coupled to each spin-orbital in the impurity, the hybridization between the bath and the impurity electrons, and the spherically symmetric Coulomb interaction of the impurity electrons.

The continuous bath-impurity hybridization strength for an orbital m, $[V_m(\epsilon)]^2 = -\text{Im}\left[\Delta_m(\epsilon)\right]/\pi = -\text{Im}\left[G^{-1}_m(\epsilon)\right]/\pi$ [3], is obtained from the local Green's function of the impurity $G_m(\epsilon)$. In the space of Matsubara frequencies, the local Green's function G_m can be expressed in terms of the corresponding density of states (DOS) $\rho_m(\epsilon)$,

$$G_m(i\omega) = \int_{-\infty}^{D} \mathrm{d}\epsilon \frac{\rho_m(\epsilon)}{i\omega_n - \epsilon - \epsilon_d - h^{\text{CF}}_{mm}} = \frac{1}{i\omega_n - \epsilon_d - h^{\text{CF}}_{mm} - \Delta_m(i\omega_n)}, \tag{2}$$

where $\rho_m(\epsilon)$ is calculated in the local density approximation using the full-potential linearized augmented plane wave method (FLAPW) [4, 5]. Note that in order to achieve the correct analytic behavior of $G(i\omega)$, the DOS has to be properly normalized, so that $\int_{-\infty}^{D} \rho_m(\epsilon) = 1$ for the chosen upper cutoff D. The crystal field of each block was determined as the first moment of the DOS. The crystal field energies were found to be $h^{\text{CF}}_{t_{2g}} = 0.039$ eV and $h^{\text{CF}}_{e_g} = -0.059$ eV.

The Matsubara frequency hybridization function of the discrete Anderson model in eq. (1) is $\tilde{\Delta}_m(i\omega_n) = \sum_{k=1}^{K} V^2_{km}/(i\omega_n - \epsilon_{km})$. The parameters ϵ_{mk} and V_{mk} were obtained by fitting this expression to the continuous hybridization function appearing in Eq. (2) by minimizing the residual function $f_m(\{\epsilon_{km}, V_{km}\}) = \sum_{n=1}^{N_\omega} \frac{1}{\sqrt{\omega_n}} \left|\tilde{\Delta}_m(i\omega_n) - \Delta_m(i\omega_n)\right|^2$ [6], with $N_\omega = 637$. The parameter K defines the number of "bath" orbitals included in the discrete model: 30 spin-orbitals for $K = 3$ (model III), and 40 spin-orbitals for $K = 4$ (model IV).

The interaction term in Eq. (1) was parametrized with the Slater integrals $F_0 = 4.0$ eV, $F_2 = 7.75$ eV and $F_4 = 4.85$ eV, which correspond to an intra-orbital repulsion $U = 4.0$ eV and to an exchange interaction $J = 0.9$ eV, the characteristic values for 3d metals [7]. In order to correct for the double counting of the Hartree-like contribution in the interaction term of eq. (1), the parameter μ needs to be subtracted from the energy of each spin-orbital. This parameter defines the value of ϵ_d in eq. (1) in the ED calculations, and was set to the three values used in ref. [1], namely 26 eV, 27 eV and 28 eV.

The exact diagonalization was performed using the Lanczos method in a reduced many-body Hilbert space, which includes only states with a limited number of excitations (in the

Table 2. The total occupation $\langle N\rangle$, impurity occupation $\langle n_d\rangle$ and impurity spin as functions of μ, calculated by the exact diagonalization and compared to the QMC calculations from ref. [1].

	μ (eV)	$\langle N\rangle$	$\langle n_d\rangle$	S	$Z_{t_{2g}}$	Z_{e_g}
	26	21.93	7.47	1.13	0.06	0.25
DFT+ED, III	27	22.31	7.77	0.99	0.10	0.23
	28	23.67	8.17	0.79	0.31	0.34
	26	28.02	7.46	1.14	0.12	0.29
DFT+ED, IV	27	28.71	7.77	0.99	0.30	0.23
	28	29.92	8.20	0.79	0.43	0.40
	26		7.51	1.02	0.38	0.39
QMC [1]	27		7.78	0.92	0.42	0.47
	28		8.06	0.82	0.48	0.56

form of occupied bath orbitals above Fermi energy and holes in the bath orbitals below Fermi energy) [6, 8]. This reduction significantly lowers the computational cost of diagonalization. The excitation cutoff, which is the maximal number of excitations allowed in each state, was set to 5 for model III and 4 for model IV. Those cutoffs are the largest values allowed by memory constraints.

3. Results

The bath parameters shown in table 1 were used to create six discrete Anderson models – three μ values of 26 eV, 27 eV and 28 eV for each of the models III and IV. For each model, several observables were measured as grand canonical averages – the total number of electrons, the occupation of the impurity d shell, and the spin moment in the d shell. The observables have been measured at room temperature, $\beta = 1/k_{\mathrm{B}}T = 40\ \mathrm{eV}^{-1}$. The spin moment S was found using the expectation value $\langle S^2\rangle = S\,(S+1)$. The results are shown in table 2. For every value of μ, the observables of models III and IV are in good agreement with each other, which demonstrates convergence with respect to the bath-size parameter K. The observables of all models are in good agreement with the corresponding DFT+QMC results as well.

The self-energy $\Sigma\,(\omega)$ was computed from the Green's function of the Anderson impurity model as $\Sigma\,(\omega) = \omega - H^{(0)}_{\mathrm{imp}} - G^{-1}_{\mathrm{imp}}\,(\omega)$ where $H^{(0)}_{\mathrm{imp}}$ is the impurity Hamiltonian without the Coulomb term [8]. It was used to calculate the quasi-particle weight of each orbital, $Z_m = \left(1 - \mathrm{Re}\left[\frac{\partial \Sigma_m(\omega)}{\partial\omega}\right]\right)^{-1}$, which was used in ref. [1] to estimate the Kondo temperature.

The results are shown in table 2 together with the results from ref. [1]. For the t_{2g} states, our quasi-particle weights behave qualitatively similarly to the ones calculated by QMC, increasing with $\langle n_d\rangle$. For the e_g states, the quasi-particle weight for $\mu = 27$ eV breaks the pattern, as it is slightly lower than for $\mu = 26$ eV.

The spectral function of each discrete model was calculated from the impurity's Green's function. The spectra are depicted in figure 1. For each value of μ, the spectra of model III and IV are largely in agreement with each other. However, the DOS for $\mu = 27$ eV does not match the corresponding plot from ref. [1] very well. The discrepancy between the spectra and quasi-particle weights found by ED and QMC is an artifact of the relatively sparse discretization of the impurity model used in the ED method. Clearly, the energy-resolved quantities are much more susceptible to the discretization errors than averaged quantities like occupations and spin moments discussed above.

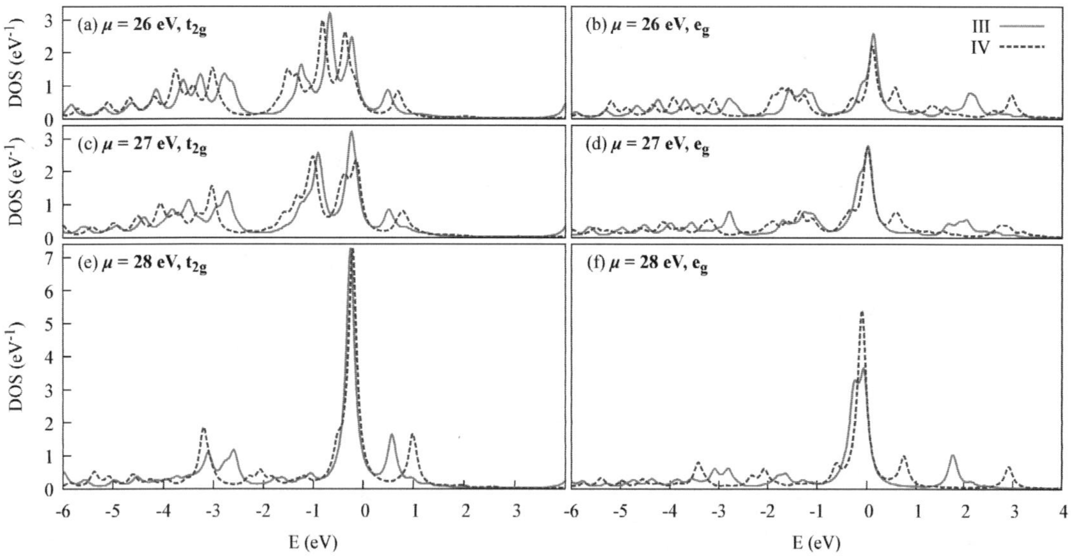

Figure 1. Density of states of the t_{2g} and e_g orbitals for model III and IV, for μ values of 26 eV, 27 eV and 28 eV. Lorenzian broadening with FWHM of 0.2 eV was applied.

4. Summary

We present DFT+ED calculations performed on a model of a Co impurity in a Cu bulk. The material is described with an Anderson impurity model, parameters of which have been found by DFT in the local density approximation. Two bath sizes and three double-counting parameters have been used in our study. The reported impurity occupations and spin moments are found to be essentially converged with respect to the size of the bath and to agree with the DFT+QMC calculations [1]. The convergence of the spectral functions (and related quantities like the quasi-particle weights) is not as good. Still, the qualitative behavior of the quasi-particle weights is mostly in agreement with DFT+QMC as well. Our research indicates that DFT+ED is a useful method for studying the electronic structure of systems with magnetic impurities.

Acknowledgments

Financial support was provided by Operational Programme Research, Development and Education financed by European Structural and Investment Funds and the Czech Ministry of Education, Youth and Sports (Project No. SOLID21 - CZ.02.1.01/0.0/0.0/16_019/0000760), and by the Czech Science Foundation (GACR) under the grant No. 21-09766S.

Bibliography

[1] Surer B, Troyer M, Werner P, Wehling T O, Läuchli A M, Wilhelm A and Lichtenstein A I 2012 *Phys. Rev. B* **85** 085114
[2] Kozub A L, Kolorenč J and Shick A B 2016 *Preprint* arXiv:1601.06022
[3] Gunnarsson O and Schönhammer K 1983 *Phys. Rev. B* **28** 4315
[4] Wimmer E, Krakauer H, Weinert M and Freeman A 1981 *Phys. Rev. B* **24** 864
[5] Mannstadt W and Freeman A J 1997 *Phys. Rev. B* **55** 13298
[6] Mozara R, Valentyuk M, Krivenko I, Şaşıoğlu E, Kolorenč J and Lichtenstein A 2018 *Phys. Rev. B* **97** 085133
[7] Şaşıoğlu E, Friedrich C and Blügel S 2011 *Phys. Rev. B* **83**(12) 121101
[8] Kolorenč J, Shick A B and Lichtenstein A I 2015 *Phys. Rev. B* **92**(8) 085125

Nanoprobe study of the electric field driven insulator-to-metal transition in $GaMo_4S_8$

H. Koussir[1], I. Lefebvre[1], M. Berthe[1], Y. Chernukha[1], J.Tranchant[2], B. Corraze[2], E. Janod[2], L. Cario[2], B. Grandidier[1] and P. Diener[1]

[1] Univ. Lille, CNRS, Centrale Lille, Univ. Polytechnique Hauts-de-France, Junia-ISEN, UMR 8520 - IEMN, F-59000 Lille, France

[2] Institut des Matériaux Jean Rouxel (IMN), Université de Nantes, CNRS, 2 Rue de la Houssinière, BP32229, 44322 Nantes Cedex 3, France

E-mail: pascale.diener@junia.com

Abstract. The resistive switching observed under electric pulses in Mott materials has a high potential for micro and nanoelectronics. Here we report on the study of the resistive switching observed at the surface of single crystals of the canonical Mott semiconductor $GaMo_4S_8$. The study is made using a multiprobe setup with 4 nanopositionable tips under the supervision of a high resolution scanning electron microscop. We find a resistivity of $38\,\Omega.cm$ by four-point probe measurements, in agreement with the literature. The volatile insulator to metal transition is studied with a two probes configuration for interelectrode distances varying between 4 and 200 microns. Finite element simulations are performed to determine the spatial distribution of the electric field prior to the transition. Our results are in agreement with i) an intrinsic voltage threshold of 60 mV independent of the interelectrode distance ii) a maximum electric field close to the electrodes and iii) a threshold electric field of 0.2 kV/cm.

1. Introduction

Mott materials exhibit an amazingly rich physics, with interesting perspectives for the micro/nanoelectronics industry [1, 2, 3]. This paper focuses on the electric field driven, volatile, insulator to metal transition observed in canonical narrow gap (0.1 - 0.5 eV) Mott insulators [4]. A voltage pulse induces an electronic avalanche specific to Mott insulators that allows the out of equilibrium system to explore a metastable metallic state [5]. The specific dynamics of the system is used to implement integrated artificial neurons [6]. In this context, the study of the transition at the nanoscaleis of high importance for the downscaling of future Mottronic devices using this transition.

2. Results

The $GaMo_4S_8$ single crystals are obtained using the method described in [5]. They are cleaved and glued with a conductive epoxy to a gold plated, p-doped silicon substrate. We checked that the sample substrate interface is sufficiently resistive in order to have a negligible contribution of the substrate to all measured sample resistances. The electrical properties have been obtained using the Nanoprobe, which consist of four Scanning Tunneling Microscop (STM) tips in an ultra high vacuum chamber, monitored under the supervision of a high resolution scanning electron microscop (SEM). The sharp tungsten tips are prepared using an electrochemical etching, and annealed under vacuum to ensure a metallic apex. To check the quality of the tip-sample contact, current voltage measurements between each tip and the grounded substrate are performed and only tips showing a linear characteristic, associated with an ohmic contact, were considered. The sample resistance R_S is measured with the four-point probe method to avoid additional contact resistances, and with the sample substrate disconnected from the ground. Figure 1a shows a

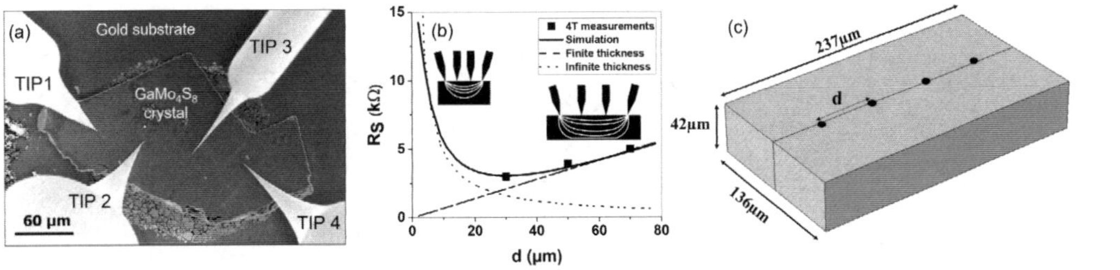

Figure 1. a) SEM picture of the $GaMo_4S_8$ single crystal S1 and the tips in the four-point probe configuration. b) Sample resistance R_S as a function of the interelectrode distance d. The experimental data are compared to simulations and to the expected analytic dependence at small d (*Infinite thickness*) and large d(*Finite thickness*). The insert shows a picture of the lines of current. c) Sample geometry used for COMSOL simulations.

typical SEM picture of a single crystal and the four STM tips. Figure 1b exhibits the variation of R_S in sample S1 for several interelectrode distances d from 30 to 70 μm. The resistance is expected to increase linearly with interelectrode distance for large values of d, and to decrease such as $R \propto 1/d$ for small d values [9]. These two limiting behaviour are pictured in Figure 1b: the first corresponds to the finite thickness case and the second corresponds to the semi infinite thickness case. Our data are expected to be in between the two regimes because the thickness is of the order of d.

Finite element simulations with COMSOL Multiphysics have been performed to fit the data, with the sample resistivity as a free parameter. Figure 1c show the sample geometry used for the simulations. The best fit, shown in Fig. 1b is obtained using a resistivity of 38 Ωcm. This value is in excellent agreement with the resistivity of 40 Ωcm reported previously on single crystals of the same batch [5]. We have also compared four probes measurements $R_{4P} = R_S$ and two probes measurements $R_{2P} = R_S + 2R_c$ at the same position to estimate the contact resistance R_c at tip-sample interfaces. We find a reproducible $R_c \simeq 300 k\Omega$. This value is well above the typical values ($< 1k\Omega$) obtained with the same setup on equivalent narrow gap semiconductors. Such a high contact resistance may be due to the presence of interface states specific to strongly correlated electronic systems, as discussed in [10].

We now focus on the electrical response of the sample at higher voltage using a two probes configuration and the sample floating. Figure 2a shows a typical current voltage curve measured at the surface of a $GaMo_4S_8$ single crystal. Transitions appear as a saturation of the current at threshold voltages $V^* = \pm 12,0\,V$. V^* has been measured for two samples, S2 and S3, and for different interelectrode distances. As the measurements are performed with a two probes configuration, the voltage V_S across the sample differs from V^*, due to the potential drop at the contacts. Figure 2b gives the equivalent circuit of the measured resistance between the two electrodes, $R = R_S + 2R_C$, and indicates the measured voltage threshold V^* and the voltage threshold inside the sample V_S^*. V_S^* is calculated from the voltage divider relationship $V_S^* = V^*.R_S/R$. R is determined from a linear fit in the low field, ohmic regime, as shown in the insert of Figure 1a. R_S is determined from the simulation of the resistance in S2 and S3 following the procedure used for sample S1, with the same resistivity value.

Figure 2 (c-d-e) show R, V^* and V_S^* as a function of the interelectrode distance d. In the two probes configuration, the relatively high total resistance, above $600\,k\Omega$ is mainly due to the contact resistances. The measured threshold voltage weakly increases with d for distances approximately below $50\,\mu m$, ranging from $12\,V$ to $16\,V$ and becomes independent of d above this distance. In another hand, the sample threshold voltage is $V_S^* \simeq 60 \pm 20\,mV$ with no clear dependence as a function of d.

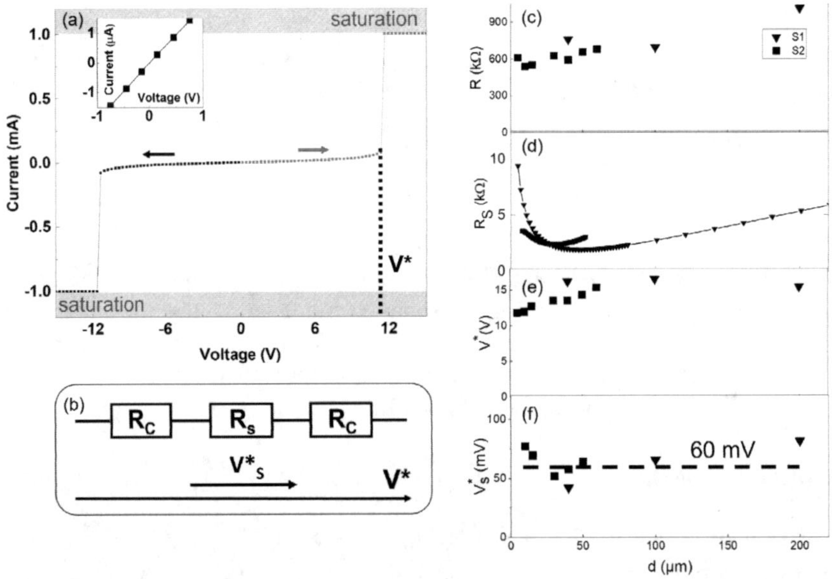

Figure 2. a) Typical current-voltage curve. V^* is the threshold voltage. Arrows indicates the voltage sweep directions. Grey areas indicate the values for which the measured current saturates. Insert: low voltage fit to determine the resistance R. b) Equivalent circuit in the two tip configuration with the contact resistances R_C and sample resistance R_S. V^* is reached for a voltage V_S^*. c,d,e,f) R, R_S, V^* and V_S^* as a function of the interelectrode distance d.

The determination of the electric field threshold E^* from V_S^* depends on the electrodes geometry. According to the literature, the insulator to metal transition occurs as soon as one sample area reaches the threshold electric field [11]. The transition then propagates from near to far up to the creation of a metallic path between the electrodes. E^* then corresponds to the maximal value of the electric field seen locally by the sample for an applied voltage $V = V_S^*$.

Figure 3 presents the simulation of the distribution of the electrical field in the sample, for an applied voltage of $60\,mV$ corresponding to V_S^*. As shown in figure 3a, the electrodes are metal contact defined as disks of radius $r = 1\,\mu m$ at the surface, separated by a distance d. An adaptative mesh as been used to resolve the field distribution. Figure 3b exhibits a typical profile of the electric field along the direction of the applied electric field, at the surface and at $500\,nm$ below. Three characteristic electric fields are labeled E_D, E_E and E_M in the figure. E_D corresponds to a divergence of the electric field at the so called triple point, because it is the junction of the sample, electrode and vacuum. E_E is located below the electrode, at the contact center, and E_M is in the middle of the interelectrode distance.

One could attribute the threshold value E^* to the electric field found at the diverging point E_D. However, we believe that E_D is not the relevant electric field, principally because our experiments show a reproducible value for the voltage threshold, independent of the probe-sample contact. In contrast E_D is expected to depends strongly on the contact geometry with

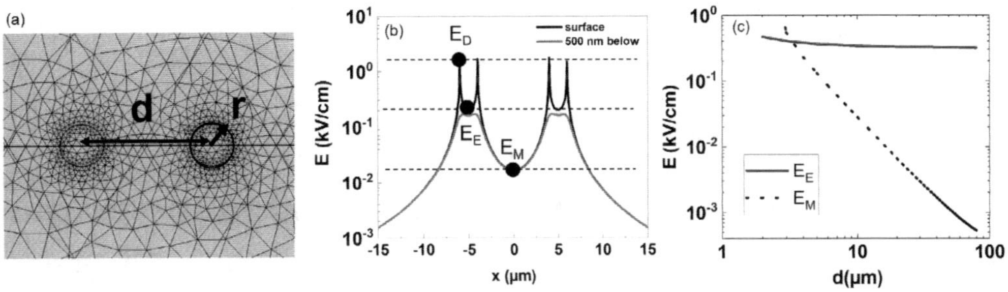

Figure 3. Comsol simulations of the electric field map in the sample for an applied voltage of $60\,mV$ a) Electrode geometry and adaptive mesh used for the simulations. d is the interelectrode distance and $r = 1\,\mu m$ is the radius of the metallic disks corresponding to the contacts. b) Electric field profile at the surface and 500 nm below. c) E_E and E_M as a function of d.

minimal values for obtuse angle contacts [12]. We rather attribute the threshold electric field to E_E, which is the maximum electric field point as soon as the divergences are excluded. Indeed, E_E corresponds to a plateau as shown figure 3b and its value varies slowly when going below the sample surface. E_E is also independent of the interelectrode distance, as shown in Figure 3c. This is compatible with the literature, because the electric field threshold is expected to be an intensive parameter. According to this analysis, we find $E^* = E_E \simeq 0.2\,kV/cm$, which compares well with the value of approximately $0.1\,kV/cm$ at ambient temperature deduced from [5]. In complement, figure 3c shows a change of regime below $d < 3\mu m$, where an increase of the electric field is expected when the electric field E_M between the electrode becomes preponderant.

3. Conclusion

The resistive switching probed at the surface of $GaMo_4S_8$ has been rationalized using Comsol simulations. The values obtained for the resistivity and the electric field threshold compare well with the literature. This study provides a framework for analysis of the resistive switching as a function of the electrodes geometry.

4. Acknowledgments

This work was supported by the RENATECH network, I-SITE ULNE (R-20-004), Hauts-de-France region and the French National Research Agency (ANR-21-CE24-0001-01).

5. References

[1] Takagi H and Hwang H Y 2010 *Science* **327** 1601–02
[2] Basov D N, Averitt R D and Hsieh D 2017 *Nat. Mater.* **16** 1077–88
[3] Wang Y, Kang K M, Kim M, Lee H S, Waser R, Wouters D, Dittmann R, Yang J J and Park H H 2019 *Materials Today* **28** 63–80
[4] Janod E. *et al. Adv. Func. Mater.* 2015 **25** 6287-305
[5] Diener P *et al.* 2018 *Phys. Rev. Lett.* **121** 016601
[6] Adda C *et al. J. Appl. Phys.* 2018 **124** 152124
[7] Dubost V, Cren T, Vaju C, Cario L, Corraze B, Janod E, Debontridder F. and Roditchev D 2013 *Nano Lett.* **13** 3648-53
[8] Durand C, Capiod P, Berthe M, Nys JP, Krzeminski C, Stiévenard D, Delerue C and Grandidier B 2014 *Nano Lett.* **14** 5636–40
[9] Miccoli I, Edler F, Pfnür H and Tegenkamp C 2015*J. Phys.: Condens. Matter* **27** 223201
[10] Oka T and Nagaosa N 2005 *Phys. Rev. Lett.* **95** 266403
[11] Stoliar P, Cario L, Janod E, Corraze B, Guillot-Deudon C, Salmon-Bourmand S, Guiot V, Tranchant J, and Rozenberg M 2013 *Adv. Mater.* **25** 3222-6
[12] Bayer C, Baer E, Waltrich U, Malipaard D and Schletz A 2015 *IEEE Trans. Dielectr. Electr. Insul.* **22** 257-65

Memory and rejuvenation effect on the cluster-glass phase of 1-D spin-chain compound $Ca_3Co_{2-x}Bi_xO_6$

Gajendra Singh Bisht, D. Pal

Department of Physics, Indian Institute of Technology Guwahati, Guwahati, 781039, Assam, India

E-mail: dpal@iitg.ac.in

Abstract. We studied the structural, static, and dynamic properties of $Ca_3Co_{2-x}Bi_xO_6$ (x = 0 and 0.2) by means of powder X-ray diffraction and magnetization measurements. We reveal the rejuvenation phenomena in magnetic relaxation under cyclic temperature change and quenching below T_{c2}. The reproduced step in the memory curve confirms the striking feature of rejuvenation effect in these samples. The observed memory and rejuvenation effects are discussed in the framework of hierarchical model of glassy phase.

1. Introduction

Study of aging, memory, and rejuvenation effects in a glassy system has attracted enormous attraction in the last few decades [1]. The glassy feature in a system is highly frustration related which arises due to competing magnetic interaction. $Ca_3Co_2O_6$ is one of the geometrically frustrated compounds which belongs to the family of Quasi 1-D spin-chain compound of structure $A_3BB'O_6$ [2]. It crystallizes in the K_4CdCl_6 (rhombohedral) type structure with space group $R\bar{3}c$. The crystal structure of $Ca_3Co_2O_6$ consists of an infinite number of chains of alternating face-sharing CoO_6 octahedral (Co1) and CoO_6 trigonal prisms (Co2), which are running along the c-axis. It possesses an Ising-like character with spin parallel to the chains, while the intra-chain and inter-chain coupling are ferromagnetic and antiferromagnetic, respectively, which give rise to geometric frustration and unusual magnetic behavior [3, 4].

The interest in relaxation phenomena revived by the effect of temperature changes, which revealed non-trivial memory and rejuvenation effects in the magnetic system [5, 6]. In this work, we carried out a comprehensive study of the structural and magnetic properties of $Ca_3Co_{2-x}Bi_xO_6$. The magnetic memory and rejuvenation effects in the system has been demonstrated by the magnetic relaxation measurements.

2. Experimental Details

Polycrystalline samples of $Ca_3Co_{2-x}Bi_xO_6$ (x = 0 and 0.2) were prepared by standard solid-state reaction method. Powder X-ray diffraction measurements were performed on all the samples at room temperature using a Rigaku X-ray diffractometer (XRD) (TTRAX III) with Cu-Kα (λ = 1.5406 Å) radiation from scattering angle (2θ) 10° to 80°. The magnetization data were taken using a vibrating sample magnetometer (VSM) probe installed in a quantum design physical property measurement system (PPMS).

3. Results

X-ray diffraction pattern of $Ca_3Co_{2-x}Bi_xO_6$ (x = 0 and 0.2) samples were taken at room temperature. Rietveld refinement was carried out by considering that the substituted Bi atom replaces the Co from the $6a$ site. Fig. 1 presents the graphical output of the Rietveld refinement

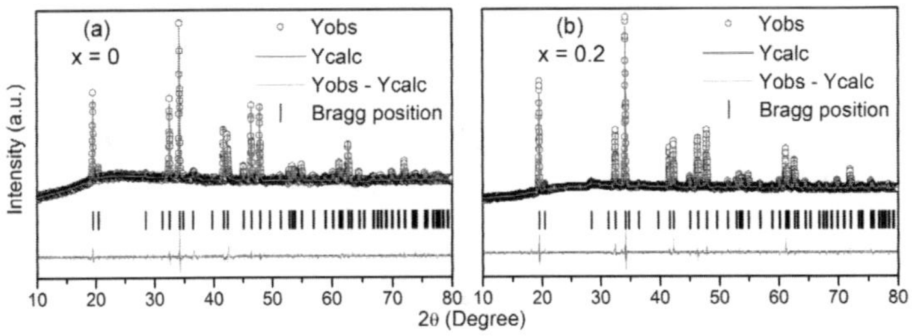

Figure 1: Rietveld refined XRD pattern of $Ca_3Co_{2-x}Bi_xO_6$ (x = 0 and 0.2) samples.

performed to the XRD pattern of x = 0 and 0.2 samples. All the diffraction peaks were indexed within the rhombohedral structure that belongs to the space group $R\bar{3}c$. The lattice parameters a and c obtained from the refinement increases for x = 0.2 sample. It is consistent with the fact that the Bi^{3+} (1.24 Å) ion has a larger ionic radius compared to Co^{3+} (0.61 Å) ion. The structure visualization and bond-length analysis have been done using VESTA software. The intra-chain bond-length (Co1-Co2) increases from 2.596 Å for the parent compound to 2.598 Å for x = 0.2 sample. In the same way, the inter-chain bond-length (Co2-Co2) increases from 5.520 Å for the parent compound to 5.523 Å for x = 0.2 sample. According to these results, we could expect an interplay of reduced dimensionality and bond randomness effect introduced by Bi-substitution, due to which the exchange interaction between magnetic ions may get modified.

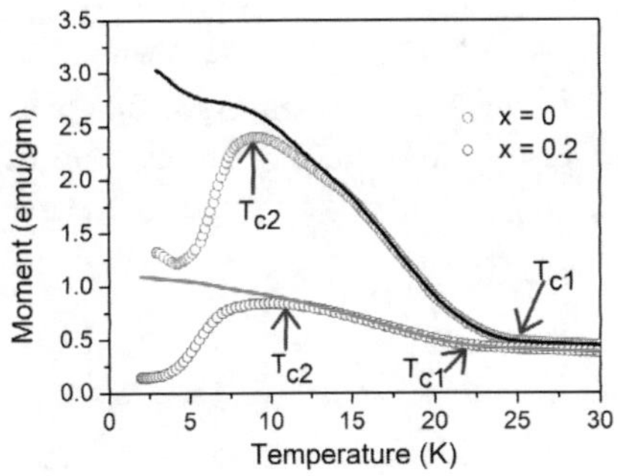

Figure 2: M(T) curves measured under ZFC condition (open-circle) in an applied field of 1000 Oe for $Ca_3Co_{2-x}Bi_xO_6$ (x = 0 and 0.2) samples. The solid line represents the FC curves under the same applied field.

Fig. 2 depicts the temperature dependence of magnetization for $Ca_3Co_{2-x}Bi_xO_6$ (x = 0 and 0.2) samples measured under zero-field cooled (ZFC) and field cooled (FC) conditions in an applied magnetic field of 1000 Oe. For the parent compound, the magnetization starts rising below 25 K, and the ZFC-FC curves bifurcate at temperature 16 K. Below this temperature, the ZFC curves attain a maximum, while the FC magnetization increases monotonously with decreasing temperature. The divergence between M(T) ZFC-FC curves followed by a cusp in the ZFC curve are usually characteristics of glassy behavior. Similar to the parent compound, the magnetization for x = 0.2 sample starts rising below 22 K. The curve changes its behavior for low-temperature, i.e., the sharp peak in the ZFC curves for x = 0 sample broadens for x = 0.2 sample and shifts towards higher temperature. In the earlier study for the parent compound, the sudden increase in the magnetization below 25 K (T_{c1}) is a transition from paramagnetic to partially disordered antiferromagnetic state [3], and the cusp at 9.1 K (T_{c2}) has a place on spin-cluster-glass [3]. The transition temperature T_{c1} and T_{c2} were determined from dM/dT

vs. T plot. It is found that with Bi-substitution, the T_{c1} decreases from 25 K for the parent compound to 22 K for x = 0.2 sample. On the other hand, the T_{c2} increases from 9.1 K for the parent compound to 11.2 K for x = 0.2 sample, which indicates the increment in glassy feature due to Bi-substitution. We speculate that the modification in the relative strength of intra-chain and inter-chain interactions is the origin of the feature due to the random occupation of the 6*a* site by Bi atoms. Furthermore, the susceptibility data in the high-temperature region obey the Curie-Weiss law $\chi_{mol} = \frac{C_{mol}}{T-\theta}$, where C_{mol} is molar curie constant, and θ is Curie temperature. As expected from the non-magnetic ion substitution, the μ_{eff} decreases from 5.6 μ_B for the parent compound to 5.03 μ_B for x = 0.2 sample. On the other hand, the θ remains fixed (31 K) for both samples.

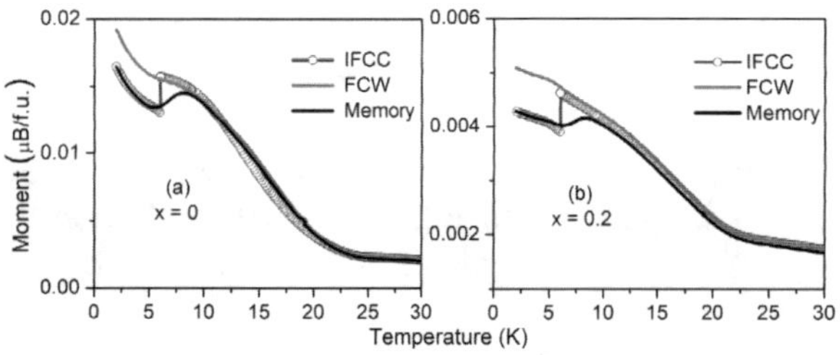

Figure 3: Memory effect experiment: Interrupted cooling magnetization (IFCC), warming cycle (FCW), and continuous warming memory curves for $Ca_3Co_{2-x}Bi_xO_6$ (x = 0 and 0.2) samples.

In order to demonstrate the memory in the system, the samples were cooled down from room temperature to 6 K in an applied field of 100 Oe, and the magnetization was recorded. The cooling process was interrupted at T = 6 K, and the magnetic field was switched off for 5400 sec. In this duration, the samples were allowed to relax. The same magnetic field (100 Oe) was re-applied immediately after the relaxation, and the cooling was resumed. The obtained magnetization curve is designated as interrupted field cooled cooling (IFCC) curve, and it produces a step-like feature at 6 K (Fig. 3). After reaching 2 K, the samples were heated under the same field without interruption, and M(T) was recorded up to 30 K, which is designated as memory curve (Fig. 3). FC curve in the same field (FCW) without any interruption was measured for reference (Fig. 3). Interestingly, the obtained memory curve at interrupting point (T = 6 K) attempts to follow the path traced during IFCC process, which is a clear signature of the magnetic memory in the system.

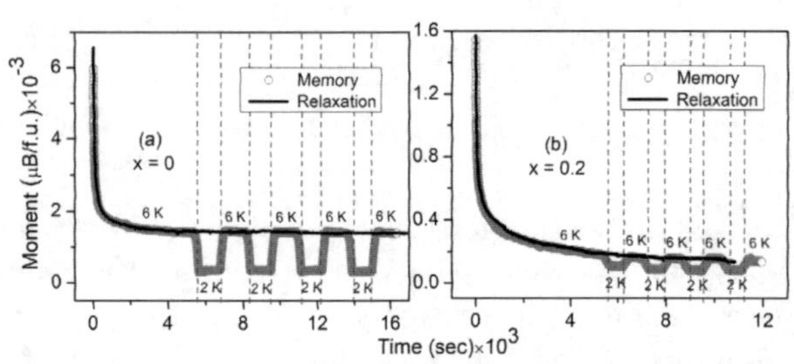

Figure 4: Memory curves generated by following the decreasing temperature cycle protocol (open circle). Solid-line represents the 6 K relaxation curve for $Ca_3Co_{2-x}Bi_xO_6$ (x = 0 and 0.2) samples.

To understand the memory effect in details, we perform the magnetic relaxation measurement M(t) under decreasing temperature cycle. In this protocol, the samples were cooled down from room temperature to 6 K in a field of 100 Oe. At 6 K, the M(t) was recorded for 5400 sec after switching off the field. To generate the memory curve, we have repeatedly followed a two-step process after finishing the relaxation. In this process, the samples were cooled down to 2 K from 6 K in zero field, and M(t) was measured for t_m (t_m=900 sec for x = 0 sample and 450 sec for x = 0.2 sample) followed by warming of samples back to 6 K in zero field and re-recording the M(t) for t_m. Fig. 4 shows the memory curves generated by following the decreasing temperature cycle protocol. It is observed that the state of the samples before cooling is recovered when the sample is cycled back to the initial temperature (6 K). This is a straightforward demonstration of the rejuvenation of magnetic relaxation and memory effect in a cluster-glass system, where the sample remembers its previous state even after experiencing a large change in magnetization. The 6 K relaxation curve under the same field (100 Oe) was measured for reference (Fig. 4).

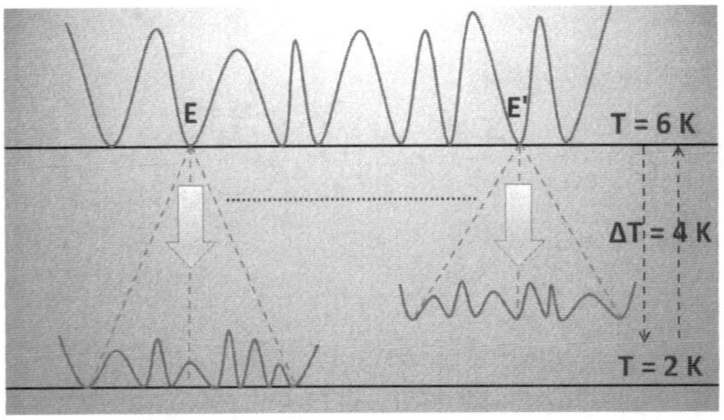

Figure 5: Splitting of two free-energy landscape E and E$'$ according to hierarchical model of glassy phase under decreasing temperature cycle.

The observed memory and rejuvenation effect in this system can be discussed in the framework of hierarchical model of glassy phase [7]. In the hierarchical model, the glassy phase is a multi-valleyed free energy landscape at temperatures below T_{c2}. These free energy valleys are metastable states which split into new sub valleys on decreasing the temperature and get merged on increasing the temperature. Fig. 5 shows the schematic representation of free-energy landscape at T = 2 and 6 K under decreasing temperature cycle. For the decreasing temperature cycle (T-ΔT), each metastable state splits into a set of sub valleys while cooling the samples to 2 K from 6 K. In the present case, the ΔT = 4 K is large enough that the energy barrier separating the metastable states become too high, and the system cannot overcome these barriers during the time t_m. Thus, the relaxation occurs only within the sub valleys of each set. As the temperature is restored to 6 K, the sub valleys and barriers merge back to the original free-energy landscape, and the relaxation at 6 K is not disturbed by the intermediate relaxations at 2 K.

Conclusion

We prepared polycrystalline samples of $Ca_3Co_{2-x}Bi_xO_6$ (x = 0 and 0.2) by standard solid-state reaction method. Magnetic properties were found to be sensitive to Bi-substitution, due to which the T_{c1} decreases while, the T_{c2} increases for x = 0.2 sample. Our study throws fresh light on the memory and rejuvenation effects in this system. A detailed analysis via relaxation measurements demonstrates the evolution of the system through a number of metastable states and memory effects. It is to be mentioned that a comprehensive study which involves dc magnetization and magnetic relaxation measurements for intermediate Bi-composition will strengthen the current observations.

References

[1] Du J, Zhang B, Zheng R K and Zhang X X 2007 *Phys. Rev. B* **75** 014415
[2] Kamiya Y and Batista C D 2012 *Phys. Rev. Lett.* **109** 067204
[3] Bisht G S and Pal D 2021 *J. Magn. Magn. Mater.* **537** 168141
[4] Hardy V, Flahaut D, Lees M R and Petrenko O A 2004 *Phys. Rev. B* **70** 214439
[5] Bag P, Baral P R and Nath R 2018 *Phys. Rev. B* **98** 144436
[6] Khan N, Mandal P and Prabhakaran D 2014 *Phys. Rev. B* **90** 024421
[7] Lefloch F, Hammann J, Ocio M and Vincent E 1992 *Europhysics Letters* **18** 647-652

Coherent, time-correlated tunneling of density wave electrons

John H Miller, Jr * and Martha Y Suárez-Villagrán

Department of Physics and Texas Center for Superconductivity, University of Houston, Houston, Texas 77204-5005 USA

* Corresponding author, email: jhmiller@uh.edu

Abstract. A growing body of evidence reveals that charge density wave (CDW) transport is a high-temperature cooperative quantum phenomenon. According to the time-correlated soliton tunneling (ST) model, quantum solitons, or electron-phonon correlates within the CDW condensate, act much like electrons tunneling through a Coulomb-blockade tunnel junction. Pair creation of charged fluidic soliton droplets is prevented by their electrostatic energy below a Coulomb-blockade threshold electric field. Above threshold, the quantum fluid flows in a periodic fashion, via a hybrid between Zener-like and coherent Josephson-like tunneling. We summarize the time-correlated ST model and compare model simulations with experiment. The ST model shows excellent agreement with coherent voltage oscillations, and with CDW current-voltage characteristics. Finally, we discuss implications for physics and potential applications.

1. Introduction

Charge density waves (CDWs) exhibit correlated flow of electrons at the highest known temperatures of any macroscopic electron condensate [1-4]. CDW dynamical behavior has been observed above room temperature [5], in some cases even above the boiling point of water [6-8]. In linear chain compounds, the CDW condensate modulates the charge along each of N parallel chains, $\rho_i(x,t) = \rho_i^0(x,t) + \rho_i^1 \cos[2k_F x - \phi_i(x,t)]$, where kinks in ϕ_i carry charge and can transport electric current. Spin density waves are similar and can be viewed as two out-of-phase CDWs for the spin-up and -down sub-bands.

Considerable evidence shows that CDWs do not classically "slide," but instead transport electric current as a fundamentally quantum process. In the classical picture, the CDW is expected to slide when the applied electric field tilts the washboard pinning potential enough for classical depinning. Just below this classical depinning field, the CDW should displace by at least a quarter wavelength, equivalent to a phase displacement of $\pi/2$, or 90°. The restoring force near threshold, moreover, should become vanishingly small, leading to a divergent dielectric response [3], as shown in figure 1. These classical predictions are refuted by: 1) NMR experiments [9] showing only 2° CDW phase displacement just below threshold; and flat 2) dielectric (figure 1 and ref. [3]), and 3) harmonic mixing [10] responses vs. bias field below threshold. These experiments reveal that the CDW remains near the bottom of the pinning potential well, and that the threshold field for nonlinear transport is often much smaller than the classical depinning field.

The Coulomb blockade effect in single electron tunneling shows how a threshold voltage or field can arise from a purely quantum process [11]. A similar Coulomb blockade mechanism was found to yield a threshold field for charged soliton pair creation [12]. This model, based on the (1+1)-D massive Schwinger model [13], was extended to a description of time-correlated soliton tunneling (ST) in a 1-D

CDW model [14]. Oscillations of period $h/2e$, in CDW conductance vs. magnetic flux of TaS_3 rings, provided clear evidence for the quantum nature of CDW transport [15,16]. This Aharonov-Bohm-like behavior, first reported in 2009 [15], motivated further development of the time-correlated ST model. Later results, reported in 2012 [16], included telegraphic switching effects suggesting transitions between macroscopically distinct states. Additional evidence for quantum behavior includes linear ac admittance and mixing experiments showing agreement with photon-assisted tunneling theory [3,10].

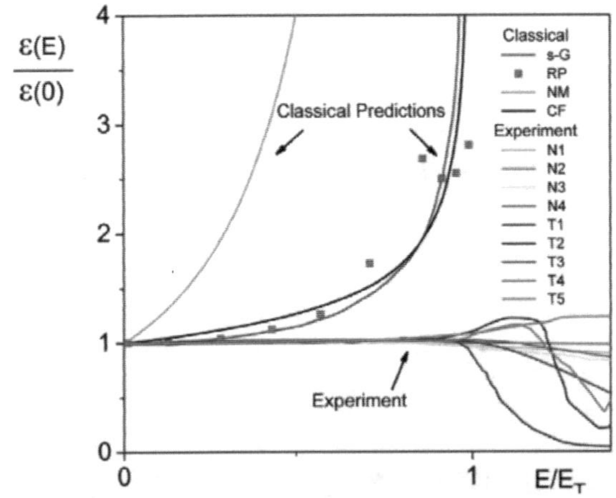

Figure 1. Dielectric response vs. bias field [3], showing classical predictions vs experiment. Classical models include classical sine-Gordon (s-G); random pinning (RP); renormalization group (NM); and incommensurate harmonic chain (CF) models. See [3] for details.

2. Time-correlated soliton tunneling model

We treat CDW transport as the periodic flow of a quantum fluid of electron–phonon correlates, or quantum solitons, within the condensate [2-4]. Pair creation of charged soliton droplets by an applied field E is prevented by their electrostatic energy below a Coulomb blockade threshold field, $E_T = E^*/2$, where E^* is the internal field produced by a soliton-antisoliton pair. This threshold corresponds to a "vacuum angle" $\theta = 2\pi E/E^* = \pi$, and is often much smaller than the classical depinning field. In a pinned CDW, the potential energy of the ith chain is $u[\phi_i] = 2u_0[1 - \cos[\phi_i(x)]] + u_E[\theta - \phi_i(x)]^2$. The quadratic term is the electrostatic energy resulting from the applied field and internal fields created by kinks at the boundaries. Above threshold, $\theta(t)$ is related to the evolving displacement charge $Q(t)$ by $\theta(t) = 2\pi Q(t)/Q_0$, where $Q_0 = 2eN$ is the charge of a fluidic soliton domain wall, an aggregate of many single-chain soliton dislocations.

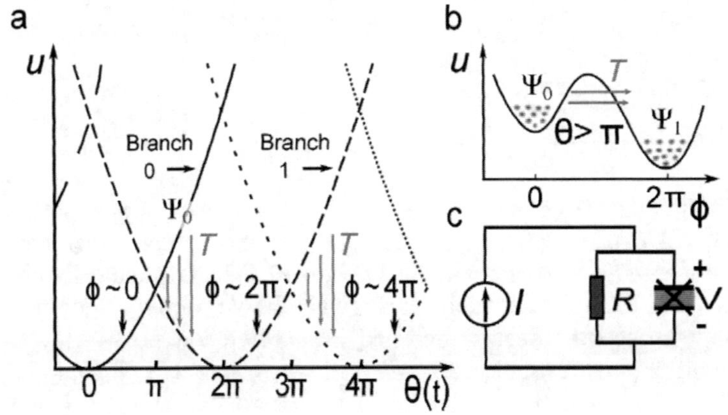

Figure 2. (a) Potential energy vs θ for $\phi \sim 2\pi n$. **(b)** $u(\phi)$ when $\theta = 2\pi E/E^* > \pi$ as the phases ϕ_i tunnel coherently into the next well. **(c)** Time-correlated ST model [2] .

Figure 2 (a) plots u vs θ when the energy is minimized for $\phi \sim 2\pi n$ when $u_E \ll u_0$. The phases ϕ_i tunnel coherently into the next well via a matrix element T [figure 2(b)] as each charging energy branch,

$\sim (\theta - 2\pi n)^2$, crosses the next at the instability points $\theta = 2\pi(n + 1/2)$. The time-correlated soliton tunneling model [2-4] includes a shunt resistance R, representing normal, uncondensed electrons, in parallel with a capacitive tunnel junction depicting soliton tunneling [figure 2(c)]. Advancing the phases by $2\pi n$ creates multiple pairs of fluidic domain walls that are driven outward by the applied field. As in time-correlated single electron tunneling, the voltage is proportional to net displacement charge, $V = (Q_0/2\pi C)[\theta - 2\pi n]$. If the expectation value $\langle\phi\rangle$ among N parallel chains advances by a fraction or non-integer multiple of $2\pi n$, the voltage and normal current are $V = (Q_0/2\pi C)[\theta - \langle\phi\rangle]$ and $I_n = V/R$, respectively. Using $I_{CDW} = I - I_n = (Q_0/2\pi)(d\theta/dt)$, and defining $\omega = 2\pi I/Q_0$ and $\tau = RC$, yield the time evolution equation, $d\theta/dt = \omega - [\theta - \langle\phi\rangle]/\tau$. We compute $\langle\phi\rangle$ by solving the Schrödinger equation:

$$i\hbar\, d\Psi_{n,n+1}/dt = U_0\Psi_{n,n+1} + T\Psi_{n+1,n}. \tag{1}$$

This describes Josephson-like coupling between successive branches via the tunneling matrix element T, which has a Zener-like force dependence. Here Ψ_n depicts the macrostate amplitude for the system to be on branch n [figure 2(a)]. We interpret Ψ_n to be classically robust order parameters, whose magnitudes grow and diminish when the system evolves between successive branches [2-4].

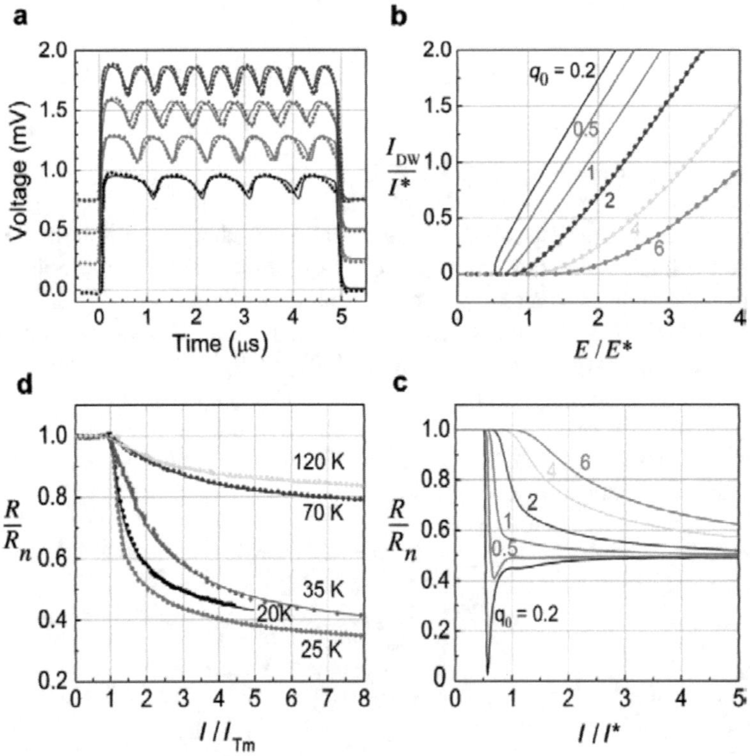

Figure 3. **(a)** Theoretical (solid lines) vs experimental (dashed lines [17]) voltage oscillations of an $NbSe_3$ crystal at 52 K for current pulse amplitudes (bottom to top, offset by 0, 0.25, 0.5, and 0.75 mV for clarity): 9:90, 10:89, 11:49, and 11:88 μA. **(b)** Simulated density wave (DW) current vs field for several values of $q_0 \sim E_0/E_T$. Dotted lines: Bardeen's modified Zener function [18] , $I_{DW} \propto [E - E_T]\exp[E_0/E]$. **(c)** Simulated differential resistance, $R = dV/dI$, where $I^* = E^*\ell/R_n$ and R_n is the normal resistance at zero bias. **(d)** Theoretical (solid lines) vs experimental (dotted lines) normalized dV/dI vs. I for $NbSe_3$ [2,3].

The quantum fluid thus flows in drip-like fashion as microscopic entities tunnel coherently from one charging energy macrostate to the next. Using this model, we have performed simulations of coherent voltage oscillations, narrow-band noise, and current–voltage characteristics [2-4]. Some examples are shown in figure 3. Despite the model's simplicity, we find unsurpassed agreement between theory and experiment. Work is in progress to use the ST model to study possible collective quantum CDW transport up to 474 K [8].

3. Concluding remarks

CDW transport is emerging as an extraordinary collective quantum phenomenon, at the highest known temperatures for any Earth-bound system. Revealing the underlying quantum mechanisms will likely constitute one of the most transformative breakthroughs in physics, with implications extending well beyond condensed matter. Practical implications include possible forms of quantum information processing, which do not require cooling with a dilution refrigerator. These include quantum reservoir computing, suggested by rapid CDW learning phenomena, in which the CDW would act as a physical reservoir between input and output neural network layers [4]. Related quantum processes include time-correlated tunneling of flux vortices in layered superconductors, important for applications of high-T_c superconductivity [19]. Such advances will be enabled by understanding of the quantum transport mechanisms involving topological deformations in CDWs and related systems.

4. Acknowledgments

The authors gratefully acknowledge support from the State of Texas through the Texas Center for Superconductivity at the University of Houston.

5. References

[1] Monceau P 2012 *Adv. Phys.* **61** 325
[2] Miller J H Jr, Wijesinghe A I, Tang, Z and Guloy A M 2012 *Phys. Rev. Lett.* **108** 036404
[3] Miller J H, Wijesinghe A I, Tang Z and Guloy A M 2013 *Phys. Rev.* B **87** 115127
[4] Miller J H Jr and Suárez-Villagrán M Y 2021 *Appl. Phys. Lett.* **118** 184002
[5] Sinchenko A A, Lejay P and Monceau P 2012 *Phys. Rev.* B **85** 241104
[6] Zybtsev S G, Pokrovskii V Y, Nasretdinova V F and Zaitsev-Zotov S V 2009 *Appl. Phys. Lett.* **94** 152112
[7] Zybtsev S G, Pokrovskii V Y, Nasretdinova V F and Zaitsev-Zotov S V 2012 *Physica B: Cond. Matt.* **407** 1696
[8] Zybtsev S G, Pokrovskii V Y, Nasretdinova V F, Zaitsev-Zotov S V, Zupanič E, van Midden M A and Pai W W 2021 *J. Alloys and Compounds* **854** 157098
[9] Ross J H, Wang Z and Slichter C P 1986 *Phys. Rev. Lett.* **56** 663
[10] Miller J H, Thorne R E, Lyons W G, Tucker J R and Bardeen J 1985 *Phys. Rev.* B **31** 5229
[11] Averin D V and Likharev K K 1986 *J. Low Temp. Phys.* **62** 345
[12] Krive I V and Rozhavsky A S 1985 *Solid State Commun.* **55** 691
[13] Coleman S 1976 *Ann. Phys.* **101** 239
[14] Miller J H, Ordóñez C and Prodan E 2000 *Phys. Rev. Lett.* **84** 1555
[15] Tsubota M, Inagaki K and Tanda S 2009 *Physica B: Cond. Matt.* **404** 416
[16] Tsubota M, Inagaki K, Matsuura T and Tanda S 2012 *EPL* (*Europhys. Lett.*) **97** 57011
[17] Jones T C, Wu X, Simpson C R, Clayhold J A and McCarten J P 2000 *Phys. Rev.* B **61** 10066
[18] Bardeen J 1980 *Phys. Rev. Lett.* **45** 1978
[19] Miller J H and Villagrán M Y S 2017 *Cond. Matter* **2** 21

Antiphase oscillations in the time-resolved spin structure factor of a photoexcited Mott insulator on a square lattice

Takami Tohyama[1], Kazuya Shinjo[1], and Kenji Tsutsui[2]

[1] Department of Applied Physics, Tokyo University of Science, Tokyo 125-8585, Japan
[2] Synchrotron Radiation Research Center, National Institutes for Quantum and Radiological Science and Technology, Hyogo 679-5148, Japan

E-mail: tohyama@rs.tus.ac.jp

Abstract.
We investigate momentum-dependent transient spin dynamics after photoirradiation in a two-dimensional Mott insulator described by a half-filled Hubbard model on a square lattice by using a numerically exact-diagonalization technique. We find a temporal oscillation in the static spin structure factor of a $\sqrt{18} \times \sqrt{18}$ periodic lattice. The oscillation exhibits an antiphase behavior between, for example, two orthogonal momentum directions parallel and perpendicular to the electric field of a pump pulse. The origin of the antiphase oscillation is attributed to a photoexcited wave function that has overlap with eigenstates producing the B_{1g} bimagnon Raman intensity. Observing such antiphase oscillations will be a big challenge for time-resolved resonant elastic x-ray scattering experiments.

1. Introduction

The recent development of time-resolved resonant inelastic x-ray scattering (trRIXS) and time-resolved resonant elastic scattering (trREXS) tuning incident x rays for the L edge of transition metals opens a new avenue for probing magnetic properties, from which one can investigate novel photoinduced nonequilibrium phenomena in the wide range of momentum and energy spaces [1, 2]. Motivated by the development, we theoretically investigate momentum-dependent spin dynamics that evolves after pumping within a femtosecond timescale in the antiferromagnetic Mott insulator on a square lattice. Using a numerically exact-diagonalization technique for a half-filled Hubbard model with a $\sqrt{18} \times \sqrt{18}$-site lattice, we find novel momentum-dependent transient spin dynamics. In particular, we demonstrate characteristic temporal oscillations for the static spin structure factor, showing an antiphase behavior for two orthogonal directions in the momentum space, which are parallel and perpendicular to the electric field of a pump pulse. Analyzing a time-dependent wave function after pumping, we find that their oscillation period in time is determined by bimagnon excitation in the Mott insulator [3]. This theoretical prediction will be confirmed for Mott insulating cuprates and iridates once trREXS is ready for a femtosecond timescale.

This paper is organized as follows. The half-filled Hubbard model, time-dependent wave function, and static spin structure factor are introduced in Sec. 2. Time-resolved static structure

Journal of Physics: Conference Series **2164** (2022) 012049 doi:10.1088/1742-6596/2164/1/012049

factors after photoirradiation by a pump pulse is calculated in Sec. 3. Finally, a summary is given in Sec. 4.

2. Model and method

In order to describe Mott insulating state, we take a single-band Hubbard model on a square lattice at half filling given by

$$\hat{H}_0 = -t_{\rm h} \sum_{i\delta\sigma} \hat{c}^{\dagger}_{i\sigma} \hat{c}_{i+\delta\sigma} + U \sum_{i} \hat{n}_{i\uparrow} \hat{n}_{i\downarrow}, \tag{1}$$

where $\hat{c}^{\dagger}_{i\sigma}$ is the creation operator of an electron with spin σ at site i, number operator $\hat{n}_{i\sigma} = \hat{c}^{\dagger}_{i\sigma}\hat{c}_{i\sigma}$, $i+\delta$ represents four nearest-neighbor sites around site i and $t_{\rm h}$, and U are the nearest-neighbor hopping and on-site Coulomb interaction, respectively.

We incorporate an external spatially homogeneous electric field applied along the x direction via the Peierls substitution in the hopping terms, $\hat{c}^{\dagger}_{i\sigma}\hat{c}_{i+\delta\sigma} \rightarrow e^{-iA_x(t)\delta_x}\hat{c}^{\dagger}_{i\sigma}\hat{c}_{i+\delta\sigma}$, leading to the time-dependent Hamiltonian $\hat{H}(t)$. Here, δ_x is the x component of δ and $A_x(t)$ is the x component of vector potential given by

$$A_x(t) = A_0 e^{-(t-t_0)^2/(2t_{\rm d}^2)} \cos[\omega_{\rm p}(t-t_0)], \tag{2}$$

where a Gaussian-like envelope centered at t_0 has a temporal width $t_{\rm d}$ and a central frequency $\omega_{\rm p}$. Hereafter we use $t_{\rm h} = 1$ as the energy unit and $1/t_{\rm h}$ as the time unit.

The time-dependent wave function $|\psi(t)\rangle$ is described by the time-dependent Schrödinger equation. When a time step Δt between two wave functions is as small as $\Delta t = 0.01$ used in our calculations, the solution of the Schrödinger equation is given by

$$|\psi(t+\Delta t)\rangle = e^{-i\hat{H}(t)\Delta t}|\psi(t)\rangle = \sum_{n} e^{-i\epsilon_n(t)\Delta t}|\phi_n(t)\rangle\langle\phi_n(t)|\psi(t)\rangle, \tag{3}$$

where $\epsilon_n(t)$ and $|\phi_n(t)\rangle$ are eigenvalue and eigenvector of $\hat{H}(t)$. One of techniques for calculating (3) based on a exact-diagonalization scheme is the use of a Taylor expansion [4]:

$$|\psi(t+\Delta t)\rangle = \sum_{n=0}^{\infty} \frac{(-i\hat{H}(t)\Delta t)^n}{n!}|\psi(t)\rangle \equiv \sum_{n=0}^{\infty} |\varphi_n(t)\rangle. \tag{4}$$

Each term is iteratively obtained by $|\varphi_{n+1}(t)\rangle = -i\hat{H}(t)\Delta t/(n+1)|\varphi_n(t)\rangle$ starting from $|\varphi_0\rangle = |\psi(t)\rangle$. The summation can be truncated if the norm $\langle\varphi_n(t)|\varphi_n(t)\rangle$ is smaller than a critical value such as 10^{-14}.

Since $|\psi(t)\rangle$ can be expanded by the eigenstates $|\varphi_n\rangle$ of $\hat{H}_0$, the weight of each eigenstate in a given $|\psi(t)\rangle$ can be measured by a spectral representation such as

$$W(\omega;t) = \sum_{n} |\langle\varphi_n|\psi(t)\rangle|^2 \, \delta\left(\omega - \epsilon_n + \epsilon_0\right), \tag{5}$$

where ϵ_n $(n = 0, 1, 2, \cdots)$ is the engenenergy of $\hat{H}_0$.

The time-resolved static spin structure factor is defined as $S\left(\mathbf{q};t\right) = \langle\psi\left(t\right)|\, \hat{S}^z_{\mathbf{q}}\hat{S}^z_{-\mathbf{q}}\,|\psi\left(t\right)\rangle$,where $\hat{S}^z_{\mathbf{q}} = \sum_i e^{-i\mathbf{q}\cdot\mathbf{R}_i}\hat{S}^z_i$ with $\hat{S}^z_i$ being the z component of spin operator at site i with position vector $\mathbf{R}_i$.

We consider $\hat{H}(t)$ for a $\sqrt{18}\times\sqrt{18}$ lattice with periodic boundary conditions. We take $U = 10$, which is larger than the band width $8t_{\rm h}$. For the pump pulse, we set $A_0 = 0.5$, $t_d = 0.5$, $t_0 = 0$, and $\omega_{\rm p} = 4$, 10, and 20. Since the lower and upper bounds of optical absorption spectrum are around $\omega = 6$ and $\omega = 17$ for the $\sqrt{18}\times\sqrt{18}$ Hubbard lattice [5], the energy absorbed into the system is small for $\omega_{\rm p} = 4$ and 20 due to an nonresonance condition.

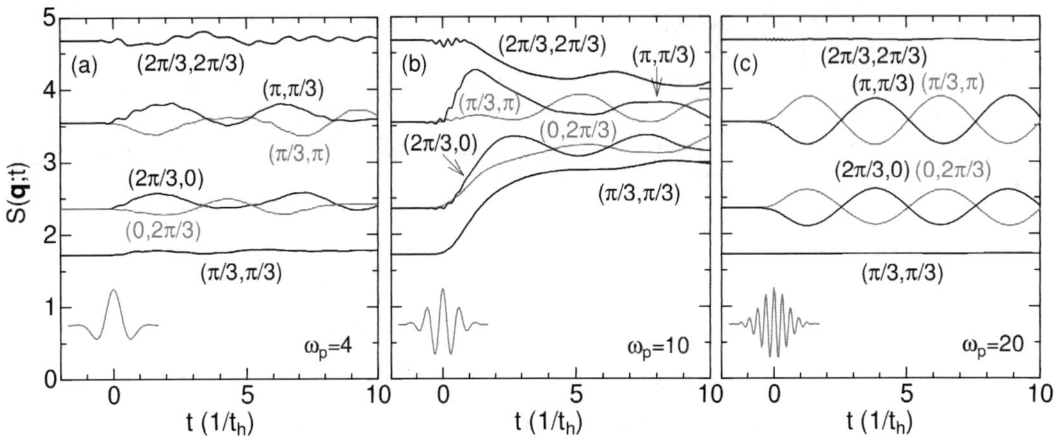

Figure 1. Time-resolved static spin structure factor $S(\mathbf{q};t)$ for the half-filled $\sqrt{18} \times \sqrt{18}$ Hubbard lattice with $t_\mathrm{h} = 1$ and $U = 10$. (a) $\omega_\mathrm{p} = 4$, (b) $\omega_\mathrm{p} = 10$, and (c) $\omega_\mathrm{p} = 20$. The color of solid lines corresponds to the color of the $\mathbf{q}$ value denoted in the panel. The profile of a pump pulse is shown at the bottom left-hand side in each panel, whose amplitude is of arbitrary unit.

3. Results

Figure 1 shows $S(\mathbf{q};t)$ for several values of $\mathbf{q}$ together with the profile of a pump pulse. The values of $S(\mathbf{q};t)$ change under the pump pulse and show oscillating behaviors with t after turning off the pump pulse. For $\omega_\mathrm{p} = 4$ and 20 shown in Fig. 1(a) and 1(c), respectively, time-averaged values are almost the same as the values before pumping. This is due to an off-resonance condition of ω_p. On the other hand, significant changes occur for $\omega_\mathrm{p} = 10$: $S(\mathbf{q};t)$ for small $\mathbf{q}$ such as $(\pi/3, \pi/3)$ enhances, while that for large $\mathbf{q}$ such as $(2\pi/3, 2\pi/3)$ decreases. This is because resonant photoirradiation destroys short-ranged antiferromagnetic correlation, leading to the suppression of spin correlation for large $\mathbf{q}$ but the enhancement for small $\mathbf{q}$.

We find in Fig. 1 an interesting oscillating behavior of $S(\mathbf{q};t)$, where two momenta equivalent under the reflection with respect to the (1,1) axis $[(2\pi/3,0)/(0,2\pi/3)$ and $(\pi,\pi/3)/(\pi/3,\pi)]$ exhibit out-of-phase time dependence. To understand the origin of the antiphase oscillation, we focus on $S(\mathbf{q};t)$ at $\mathbf{q} = (2\pi/3,0)$ and $(0,2\pi/3)$. During pumping centered at $t = 0$ $S(\mathbf{q} = (2\pi/3,0);t)$ $[S(\mathbf{q} = (2\pi/3,0);t)]$ decreases [increases] for $\omega_\mathrm{p} = 20$ but vise versa for $\omega_\mathrm{p} = 10$ and 4. This ω_p dependence is understood if one regards the pump pulse as an approximate periodic pulse and applies the Floquet theory for the Hubbard model as discussed in [3].

Following this initial change inducing the antiphase condition, an oscillation with the period ~ 5 emerges after turning off a pump pulse. At $\omega_\mathrm{p} = 20$, this period produces a large peak at $\omega \sim 1$ in the power spectrum. Therefore, we need to clarify the origin of the energy $\omega \sim 1.25$ (not shown here). For this propose, it is reasonable to decompose the photoinduced wave function after pump pulse into the eigenstates of time-independent Hamiltonian $\hat{H}_0$. Figure 2 shows $W(\omega;t)$ at $t = 2$, where the pump pulse is turned off. For simplicity, we neglect the $\omega = 0$ component and focus on finite energies. We find a weight at $\omega = 1.25$ for $\omega_\mathrm{p} = 20$ as shown in Fig. 2(c). Analyzing eiegenstates of a subspace with zero total momentum of $\hat{H}_0$, we realize that the eigenstate at $\omega = 1.25$ has $B_{1\mathrm{g}}$ symmetry. This means that the oscillation is a Rabi type between the ground state and the $B_{1\mathrm{g}}$ eigenstate, leading the antiphase behavior [3]. We note that the $B_{1\mathrm{g}}$ eigenstate is a Raman active state and gives the highest spectral weight in a bimagnon Raman process [6]. With decreasing ω_p, higher-energy weights appear above the $B_{1\mathrm{g}}$ state, and their weights become relatively large at $\omega_\mathrm{p} = 4$ [see Fig. 2(a)]. Their symmetry is

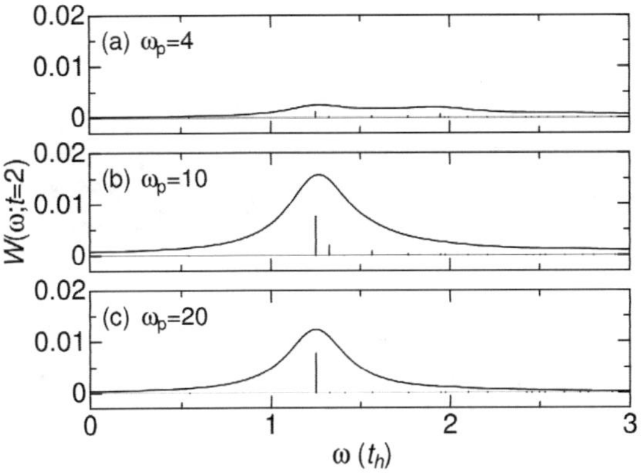

Figure 2. The spectral weight of eigenststes in $|\psi(t)\rangle$ at $t = 2$ measured by $W(\omega; t = 2)$ for the half-filled $\sqrt{18} \times \sqrt{18}$ Hubbard lattice with $t_{\rm h} = 1$ and $U = 10$. (a) $\omega_{\rm p} = 4$, (b) $\omega_{\rm p} = 10$, and (c) $\omega_{\rm p} = 20$. The weight at $\omega = 0$ is removed from each panel.

predominantly A_{1g} not contributing to the antiphase oscillations.

The oscillation period estimated form our calculation is $5/t_{\rm h}$, which is roughly 30 fs. Recent trREXS has reported the time resolution of 150 fs [2]. Therefore, the resolution should be improved to detect the antiphase oscillation predicted by our calculation. We believe that recent rapid development of trREXS/trRIXS will enable this improvement in the near future.

4. Summary

We have investigated the momentum-dependent transient spin dynamics in the Mott insulator on a square lattice after photoirradiation. Using a exact diagonalization technique for a half-filled Hubbard model on the square lattice, we have found that after turning off a pump pulse, the static spin structure factor temporally oscillates. Examining the components of time-dependent wave function, we have confirmed that the oscillation comes from a Rabi-type one between the ground state and the Raman-active B_{1g} state [3]. We have also found that an antiphase behavior in the oscillations between originally equivalent two momenta, for example, two orthogonal momentum directions parallel and perpendicular to the electric field of a pump pulse. We hope that trREXS will be able to detect these predicted oscillations in the near future.

Acknowledgments

This work was supported by CREST (Grant No. JPMJCR1661), the Japan Science and Technology Agency, by the Japan Society for the Promotion of Science, KAKENHI (Grants No. 19H01829, No. JP19H05825, No. 17K14148, No. 21H03455) from Ministry of Education, Culture, Sports, Science, and Technology (MEXT), Japan, by JST PRESTO (Grant No. JPMJPR2013), by MEXT HPCI Strategic Programs for Innovative Research (SPIRE; hp200071), and by QST President's Strategic Grant (QST Advanced Study Laboratory). Part of the numerical calculation was carried out using HOKUSAI at RIKEN, the supercomputer system at the information initiative center, Hokkaido University, the facilities of the Supercomputer Center at Institute for Solid State Physics, University of Tokyo, and the supercomputing facilities in QST.

References

[1] Dean M P M *et al.* 2016 *Nat. Mater.* **15** 601
[2] Mazzone D G *et al.* 2021 *Proc. Natl. Acad. Sci. USA* **118** e2103696118
[3] Tsutsui K, Shinjo K and Tohyama T 2021 *Phys. Rev. Lett.* **126** 127404
[4] Shirakawa T, Miyakoshi S and Yunoki S 2020 *Phys. Rev.* B **101** 174307
[5] Tohyama T, Inoue Y, Tsutsui K and Maekawa S 2005 *Phys. Rev.* B **72** 045113
[6] Tohyama T, Onodera H, Tsutsui K and Maekawa S 2002 *Phys. Rev. Lett.* **89** 257405, and references therein.

Photoinduced Phase Transition in Two-Band model on Penrose Tiling

Ken Inayoshi, Yuta Murakami, and Akihisa Koga
Department of Physics, Tokyo Institute of Technology, Meguro, Tokyo 152-8551, Japan
E-mail: k-inayoshi@stat.phys.titech.ac.jp

Abstract. We study the effects of the photo irradiation on the band insulating state in the two-band Hubbard model on the Penrose tiling. Examining the time- and site-dependent physical quantities, we find that the excitionic state is dynamically induced with site-dependent order parameters. It is also clarified that, in the excitonic state induced by the photo irradiation, local oscillatory behavior appears in the electron number as well as in the order parameter, which should be characteristic of the quasiperiodic lattice.

1. Introduction

Quasicrystal is an interesting system, which is characterized by a long-range order with no translational symmetry [1]. In the field of quasicrystals, the discovery of an Al-Si-Ru semiconducting approximant [2] stimulates theoretical and experimental investigations of the electron correlation effects on the semiconducting quasicrystals. In our previous works [3, 4], we have studied the excitonic insulating phase [5] on the Penrose tiling [6] to clarify the role of the quasiperiodic structure in the equilibrium and nonequilibrium states. One of the interesting features is that local charge fluctuations are induced by the photo irradiation in the BEC-type excitonic insulating state. Then, a question arises: are similar charge fluctuations induced even in the photo-excited band insulator on the quasiperiodic lattice? It is instructive to treat the band insulating state on the Penrose tiling to study nonequilibrium behavior after the photo irradiation.

2. Model and Method

We consider the two-band Hubbard model [7, 8], whose Hamiltonian is given as

$$\hat{H}_0 = -J \sum_{\langle i,j\rangle\sigma} (\hat{c}^\dagger_{i\sigma}\hat{c}_{j\sigma} - \hat{f}^\dagger_{i\sigma}\hat{f}_{j\sigma}) + \frac{D}{2}\sum_{i\sigma}(\hat{n}_{ci\sigma} - \hat{n}_{fi\sigma}) - \mu\sum_{i\sigma}(\hat{n}_{fi\sigma} + \hat{n}_{ci\sigma})$$
$$+ U\sum_i(\hat{n}_{ci\uparrow}\hat{n}_{ci\downarrow} + \hat{n}_{fi\uparrow}\hat{n}_{fi\downarrow}) + V\sum_{i\sigma\sigma'}\hat{n}_{ci\sigma}\hat{n}_{fi\sigma'}, \tag{1}$$

where $\hat{a}^\dagger_{i\sigma}$ creates an electron at site i with spin $\sigma \in \{\uparrow, \downarrow\}$ in the $a(= c, f)$-band, and $\hat{n}_{ai\sigma} = \hat{a}^\dagger_{i\sigma}\hat{a}_{i\sigma}$. J is the hopping amplitude between the nearest neighbor sites, D is the energy difference between two bands, and μ is the chemical potential. U (> 0) is the intraband onsite interaction and V (> 0) is the interband onsite interaction. We consider here the Penrose tiling

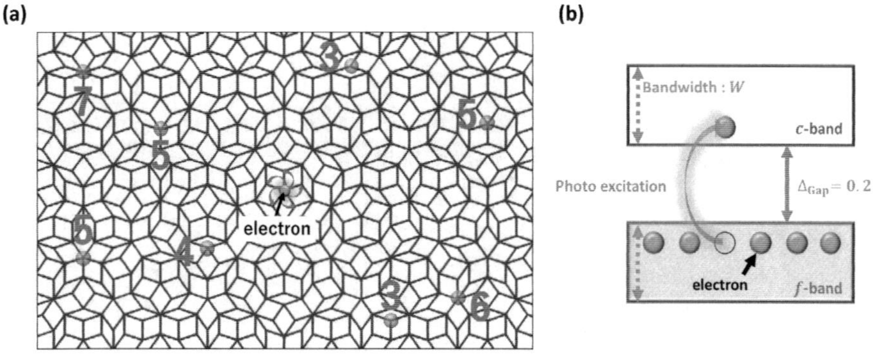

Figure 1. (a) Penrose tiling. The integer for each site represents its coordination number. (b) Schematic picture of the photo excitation of the band insulating state.

as a simple quasiperiodic lattice, as shown in Fig. 1(a). Electronic properties on the Penrose tiling have been discussed in the tight-binding [9, 10, 11, 12, 13], single-band Hubbard [14, 15], and Anderson lattice models [16]. One of the important features of the Penrose tiling is that the number of nearest neighbor sites (coordination number) α takes 3 to 7 [see Fig. 1(a)]. In this study, we focus on the α dependence of the physical quantities to discuss the effects of the quasiperiodic tiling. In the following, our discussions are restricted to the half-filled paramagnetic system, where $\mu = U/2 + V$ and the spin indices are omitted.

To study the photo irradiation in the system, we introduce the Hamiltonian for the dipole transition as

$$H'(t) = F(t)\sum_i (\hat{c}_i^\dagger \hat{f}_i + \text{h.c.}), \tag{2}$$

where $|F(t)|$ is the time dependent amplitude of the external field. Here, we employ the time-dependent real-space mean-field (MF) approximation. This method has an advantage in treating the large system, which allows us to discuss the effect of the quasiperiodic structure. The electron number of the a-band and the local order parameter at the ith site are defined as $n_{ai}(t) = \langle\psi(t)|\hat{n}_{ai}|\psi(t)\rangle$, and $\Delta_i(t) = \langle\psi(t)|\hat{c}_i^\dagger \hat{f}_i|\psi(t)\rangle$. Here $|\psi(t)\rangle = T_t \exp\left[-i\int_0^t \hat{H}_{\rm MF}^{\rm total}(t')dt'\right]|\psi(0)\rangle$, where T_t is the time-ordering operator and $|\psi(0)\rangle$ is the ground state of $\hat{H}_{\rm MF}^{\rm total}(t=0)$. The explicit form of the MF Hamiltonian is given as,

$$\begin{aligned}\hat{H}_{\rm MF}^{\rm total}(t) &= -J\sum_{\langle i,j\rangle}(\hat{c}_i^\dagger \hat{c}_j - \hat{f}_i^\dagger \hat{f}_j) + \frac{D}{2}\sum_i(\hat{n}_{ci} - \hat{n}_{fi}) - \mu\sum_i(\hat{n}_{fi} + \hat{n}_{ci}) \\ &+ U\sum_i(n_{ci}(t)\hat{n}_{ci} + n_{fi}(t)\hat{n}_{fi}) + 2V\sum_i(n_{fi}(t)\hat{n}_{ci} + n_{ci}(t)\hat{n}_{fi}) \\ &- \sum_i\Big[\{V\Delta_i(t) - F(t)\}\,\hat{f}_i^\dagger \hat{c}_i + \text{h.c.}\Big].\end{aligned} \tag{3}$$

We define the single-particle density matrix $\rho_{ia,jb}(t) = \langle\psi(t)|\hat{b}_j^\dagger \hat{a}_i|\psi(t)\rangle$. The time evolution of density matrix is given by $i\partial_t \boldsymbol{\rho}(t) = \left[\boldsymbol{H}^{\rm MF}(t), \boldsymbol{\rho}(t)\right]$. Here, $\boldsymbol{H}^{\rm MF}(t)$ is the matrix representation of $\hat{H}_{\rm MF}^{\rm total}(t)$. Since $\boldsymbol{H}^{\rm MF}(t)$ depends on $\boldsymbol{\rho}(t)$, one can numerically solve this differential equation.

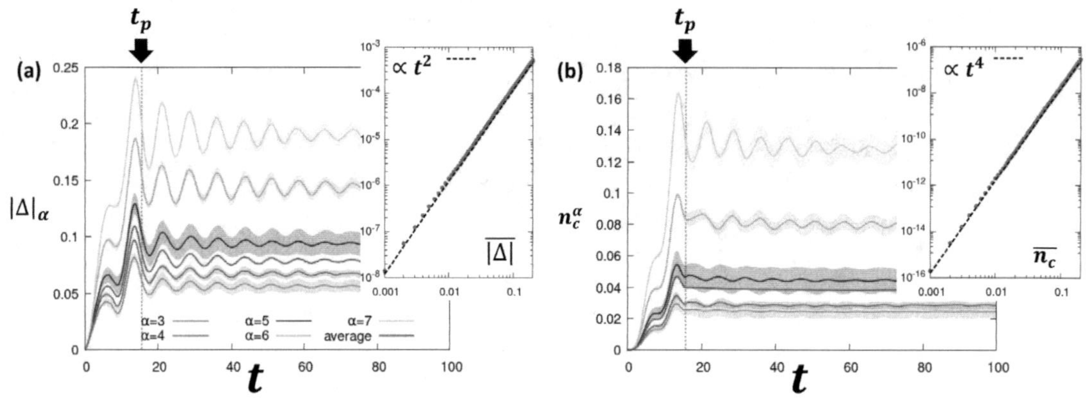

Figure 2. Time evolution of (a) $|\Delta|_\alpha$ and (b) n_c^α for each α after the single-cycle pulse is injected with $F_0 = -0.07$. We also plot the bulk averaged values, $\overline{|\Delta|}$ and $\overline{n_c}$. Note that we define those averaged values in the *bulk* region which excludes edge sites. Therefore, $\overline{n_c}$ is not conserved (slightly oscillates) although the total number of conduction electrons is conserved at $t > t_p$. Right panels show the log-log plots of $\overline{|\Delta|}$ and $\overline{n_c}$.

We note that $\partial_t n_{ci}(t) = -iJ\{\sum_m \rho_{ic,mc}(t) - \sum_m \rho_{mc,ic}(t)\}$ if $F(t) = 0$, where m runs the nearest neighbor sites for site i. Therefore, $\sum_i \partial_t n_{ci}(t) = 0$. This implies that the total number of conduction electrons is conserved when $F(t) = 0$ (for example, after the photo irradiation). However, this does not necessarily imply the absence of local charge fluctuations. In fact, we have found local charge oscillation triggered by the photo irradiation in the two-band system in the BEC regime [4]. In this study, we examine local physical quantities to clarify whether or not such charge fluctuations are induced by the photoinduced phase transition in the band insulating state.

In the following, we take J as the unit of the energy. For simplicity, we set the parameters as $U = D = 4$. We treat the Penrose tiling with the total sites $N = 11006$ under the open boundary condition. It has been clarified that in the case of $V < V_c$ (~ 4.3), the order parameter is finite and the excitonic insulating state is realized. On the other hand, when $V > V_c$, the large interband interaction stabilizes the band insulating state with the energy gap $\Delta_{\rm Gap} = 2V - W$, where W ($= 8.46$). In this study, to discuss the effects of the photo irradiation on the band insulating state, we set the interband interaction as $V = 4.33$, where $\Delta_{\rm Gap} = 0.2$, see Fig. 1(b). Then, we apply the single-cycle pulse field to the two-band system, which is given as $F(t) = F_0 \sin \omega t\, \theta(t)\theta(t_p - t)$ with the magnitude of the external field $|F_0|$, the frequency ω, the Heaviside step function $\theta(t)$, and the light irradiation time $t_p = 2\pi/\omega$ [7]. Setting $\omega = 2\Delta_{\rm Gap}$, we study the coordination number dependent quantities $\Delta_\alpha(t) = \sum_{i \text{ with } Z_i=\alpha} |\Delta_i(t)|/N'_\alpha$ and $n_c^\alpha(t) = \sum_{i \text{ with } Z_i=\alpha} n_{ci}(t)/N'_\alpha$ to discuss the photoinduced phase transition in the two-band system on the quasiperiodic tiling. Here, N'_α is the number of the lattice sites with $Z_i = \alpha$ ($= 3, \cdots, 7$) in the bulk region.

3. Results

When $t < 0$, no field is applied and the band insulating state with $\Delta_{\rm Gap} = 0.2$ is realized with $\Delta_i = 0$ and $n_{ci} = 0$. Switching on the photo irradiation at $t = 0$, local order parameter $|\Delta_i|$ and its bulk average $\overline{|\Delta|}$ is induced, as shown in Fig. 2(a). We also find in Fig. 2(b) that the bulk averaged value $\overline{n_c}$ becomes finite, implying that some conduction electrons are excited. Our MF calculations clearly show $\overline{|\Delta|} \sim t^2$ and $\overline{n_c} \sim t^4$, which means that the photoinduced phase transition occurs at $t = 0$. We also find that $\Delta_\alpha(t)$ and $n_c^\alpha(t)$ are well classified by the coordination number, as seen in Figs. 2(a) and (b). We wish to note that when $t > t_p$, the total number of conduction electrons is constant, while $n_c^\alpha(t)$ shows oscillatory behavior. This implies that local charge fluctuations are induced even in the band insulating state. This is similar to the behavior in the BEC-type excitonic state in the two-band model on the Penrose tiling [4]. Therefore, the charge fluctuations are characteristic nonequilibrium phenomena peculiar to the strong coupling regime of quasicrystalline excitonic state.

4. Summary

To summarize, we have studied the effects of the photo irradiation on the band insulating state in the two-band Hubbard model on the Penrose tiling by means of the time- and site-dependent MF approximation. It is found that the photoinduced phase transition to the excitonic state occurs in the quasiperiodic lattice as in normal crystal [17, 18, 19]. In the photo-induced state, the interband interaction is so large that the electron-hole pairs are tightly coupled and these are almost localized around each lattice site. Therefore, physical quantities are well classified by the coordination number. In general, each site in a quasiperiodic tiling is topologically inequivalent. It is expected that this unusual charge oscillation should appear in the realistic quasicrystals and approximants, which will be discussed in the future.

Acknowledgments

K.I. acknowledges the financial support from Advanced Human Resource Development Fellowship for Doctoral Students, Tokyo Institute of Technology. This work was supported by a Grant-in-Aid for Scientific Research from JSPS, KAKENHI Grant Nos. JP20K14412, JP20H05265, JP21H05017 (Y.M.), JP21H01025, JP19H05821, JP18K04678, JP17K05536 (A.K.), JST CREST Grant No. JPMJCR1901 (Y.M.).

References

[1] Shechtman D, Blech I, Gratias D and Cahn J W 1984 *Phys. Rev. Lett.* **53**(20) 1951–1953
[2] Iwasaki Y, Kitahara K and Kimura K 2019 *Phys. Rev. Materials* **3**(6) 061601
[3] Inayoshi K, Murakami Y and Koga A 2020 *J. Phys. Soc. Jpn.* **89** 064002
[4] Inayoshi K, Murakami Y and Koga A 2021 (*Preprint* `arXiv:2107.05304`)
[5] Jérome D, Rice T M and Kohn W 1967 *Phys. Rev.* **158**(2) 462–475
[6] Penrose R 1974 *Bull. Inst. Maths. Appl.* **10** 266
[7] Tanaka Y, Daira M and Yonemitsu K 2018 *Phys. Rev. B* **97**(11) 115105
[8] Tanaka Y and Yonemitsu K 2020 *Phys. Rev. B* **102**(7) 075118
[9] Kohmoto M and Sutherland B 1986 *Phys. Rev. B* **34**(6) 3849–3853
[10] Arai M, Tokihiro T, Fujiwara T and Kohmoto M 1988 *Phys. Rev. B* **38**(3) 1621–1626
[11] Kraus Y E, Lahini Y, Ringel Z, Verbin M and Zilberberg O 2012 *Phys. Rev. Lett.* **109**(10) 106402
[12] Kraus Y E and Zilberberg O 2012 *Phys. Rev. Lett.* **109**(11) 116404
[13] Fan J and Huang H 2021 *Frontiers of Physics* **17** 13203
[14] Takemori N and Koga A 2015 *J. Phys. Soc. Jpn.* **84** 023701
[15] Koga A and Tsunetsugu H 2017 *Phys. Rev. B* **96**(21) 214402
[16] Takemura S, Takemori N and Koga A 2015 *Phys. Rev. B* **91**(16) 165114
[17] Östreich T and Schönhammer K 1993 *Z. Phys. B* **91** 189–197
[18] Perfetto E, Sangalli D, Marini A and Stefanucci G 2019 *Phys. Rev. Materials* **3**(12) 124601
[19] Murakami Y, Schüler M, Takayoshi S and Werner P 2020 *Phys. Rev. B* **101**(3) 035203

Out of equilibrium charge transport in molecular electronic devices

Andrea Nava, Domenico Giuliano, Luca Lepori and Marco Rossi

Dipartimento di Fisica, Università della Calabria, Arcavacata di Rende I-87036, Cosenza, Italy
INFN - Gruppo collegato di Cosenza, Arcavacata di Rende I-87036, Cosenza, Italy

E-mail: andrea.nava@fis.unical.it

Abstract. Using the Lindblad equation approach, we study the nonequilibrium stationary state of a benzene ring connected to two reservoirs in the large bias regime, a prototype of a generic molecular electronic device. We show the emergence of an optimal working point (corresponding to a change in the monotonicity of the stationary current, as a function of the applied bias) and its robustness against chemical potential and bond disorder.

1. Introduction

Molecular electronics is a promising theoretical and experimental research field focusing on the possibility to implement single molecules in quantum circuit in order to realize molecular switches, logic gate, and junctions. It is, therefore, of the utmost importance to provide models describing molecules attached to two, or more, macroscopic contacts gates and the currents across as a function of the voltage bias or of the temperature gradient between the contacts.

In this manuscript, we model the charge transport mechanisms in such systems using the master equation formalism, already implemented for one-dimensional chains [1, 2, 3]. To be specific, we resort to the Lindblad equation (LE) approach to open quantum systems. This approach allows to derive the effective dynamics of the system by integrating over the reservoir degrees of freedom under the Markovian approximation, consisting in neglecting memory effects of the reservoirs. In the following, we first describe the general method and then apply it to a simple paradigmatic example, the benzene ring.

2. The model

The starting point is the Hückel Hamiltonian for the π-electrons of a generic N-atom molecule,

$$H = - \sum_{j \neq k = 1}^{N} t_{j,k} c_j^\dagger c_k + \sum_{j=0}^{N} \epsilon_j c_j^\dagger c_j \quad . \tag{1}$$

In Eq.[1], $c_j^\dagger, c_j$ are single-electron creation and annihilation operators at molecular site j, satisfying the canonical anticommutation relations $\{c_j, c_k^\dagger\} = \delta_{j,k}$, $t_{j,k}$ is the single-electron hopping strength between the j-th and the k-th atoms and ϵ_j is the on-site energy. For the sake of simplicity, we neglect the interaction between electrons, which can be easily implemented following Ref.[3].

We are interested in studying the behavior of the molecule when it is connected to two or more external reservoirs through some of its atoms. The LE consists of a first order differential equation for the time evolution of the system density matrix $\rho(t)$, given by

$$\dot{\rho}(t) = -i[H, \rho(t)] + \sum_k (L_k \rho(t) L_k^\dagger - \frac{1}{2}\{L_k^\dagger L_k, \rho(t)\}) \quad . \tag{2}$$

The first term at the right-hand side of Eq.[2] is the Liouvillian that describes the unitary evolution determined by the Hamiltonian, H. The second term, the Lindbladian, includes dissipation and decoherence on the system dynamics introduced by the reservoirs. It depends on the so-called "jump" operators L_k, which are determined by the coupling between the system and the reservoirs.

In the following, we consider reservoirs that locally inject fermions to, or extract fermions from, a generic site j of the molecule, at given and fixed rates. We describe the injecting and extracting reservoirs at site j in terms of the Lindblad operators $L_{in,j}$ and $L_{out,j}$, given by $L_{in,j} = \sqrt{\Gamma_j} c_j^\dagger$, $L_{out,j} = \sqrt{\gamma_j} c_j$, with Γ_j and γ_j being the coupling strengths respectively determining the creation and the annihilation of a fermion at site j. Once we determine $\rho(t)$ by solving Eq.[2], we compute the (time dependent) expectation value of any observable O using $\langle O(t) \rangle = \mathrm{Tr}[O\rho(t)]$. For the sake of our analysis, we compute the currents flowing from the reservoirs into the site j, $I_{in,j}(t)$, or from site j to the reservoir, $I_{out,j}(t)$. These are given by $I_{in,j}(t) = \Gamma_i(1 - \langle n_j(t) \rangle)$ and $I_{out,j}(t) = \gamma_i \langle n_j(t) \rangle$, with $n_j(t) = c_j^\dagger c_j$ the number operator. The net current exchanged at time t between the reservoirs and the site j is given by $I_j(t) = I_{in,j}(t) - I_{out,j}(t)$.

Using Eq.[2], we compute the current across a small Benzene ring connected to two reservoirs. The system asymptotically evolves to a non-equilibrium steady state (NESS), which we determine from the condition $\dot{\rho} = 0$. In such a state we compute the stationary current in order to find the optimal working point (OWP), i.e. the change in the monotonicity of the stationary current as a function of the applied bias, for a molecular electronic device.

3. Benzene ring

The Hückel Hamiltonian for the Benzene ring is given by [4]

$$H_B = \begin{pmatrix} \epsilon_1 & -t_{12} & 0 & 0 & 0 & -t_{16} \\ -t_{21} & \epsilon_2 & -t_{23} & 0 & 0 & 0 \\ 0 & -t_{32} & \epsilon_3 & -t_{34} & 0 & 0 \\ 0 & 0 & -t_{43} & \epsilon_4 & -t_{45} & 0 \\ 0 & 0 & 0 & -t_{54} & \epsilon_5 & -t_{56} \\ -t_{61} & 0 & 0 & 0 & -t_{65} & \epsilon_6 \end{pmatrix} , \tag{3}$$

whith real hopping amplitude, $t_{ij} = t_{j,i}^*$, between consecutive atoms of the ring. In the following we set $t_{ij} = t = 2eV$ and $\epsilon_j = 0$ except for $j = 5$, where a local gate potential, ϵ_g, is applied (values are taken from Ref.[4]). We connect the Benzene molecule to two reservoirs, one located on the site 1 and another located on the site L with $L = 2$ for the ortho-configuration, $L = 3$ for the meta-configuration and $L = 4$ for the para-configuration (see Fig.1 for a pictorial representation of the setup).

As a paradigmatic regime, we consider the large bias limit, in which one of the reservoirs, the one coupled to the site 1, acts as an electron "source", by only injecting electrons in the system ($\gamma_1 = 0$), and the other, the one coupled to the site L, acts as an electron "drain", by only absorbing electrons from the system ($\Gamma_L = 0$). As a result, electrons entering at the site 1 must travel all the way down to the site L, before leaving the system. Accordingly, the boundary dynamics is determined only by the coupling strengths, $\Gamma_1 = \gamma_L \equiv \Gamma$, while the bulk

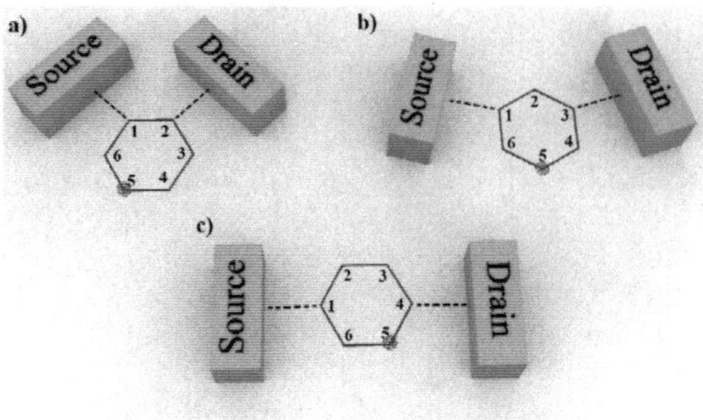

Figure 1. A Benzene molecule connected to a source and drain reservoir in the a) ortho, b) meta and c) para configurations. Blue dot represents the gate voltage ϵ_g.

dynamics only depends on the hopping strengths t and the gate voltage ϵ_g. In Fig.[2a] we plot the stationary current across the benzene ring, $I = I_{in,1} = -I_{out,L}$, against the coupling strength to the two reservoirs for the para, ortho and meta configurations, comparing it with the case of the one-dimensional chain discussed in Ref.[3]. Two interesting features emerge: first, the OWP, already observed for a linear chain [3], is present also in the benzene ring; second, the para and ortho configurations exhibit higher values of the current compared to the one-dimensional chain. This suggests that the introduction of molecular electronic devices within a quantum network can, in principle, improve its performance. In Fig.[2b] we compute the current across the ring as a function of the gate voltage ϵ_g, at fixed applied bias. Interestingly, the stationary current of the three possible configurations show opposite behaviors as a function of ϵ_g, decreasing in the para configuration, increasing in the meta configuration and keeping almost constant in the ortho configuration.

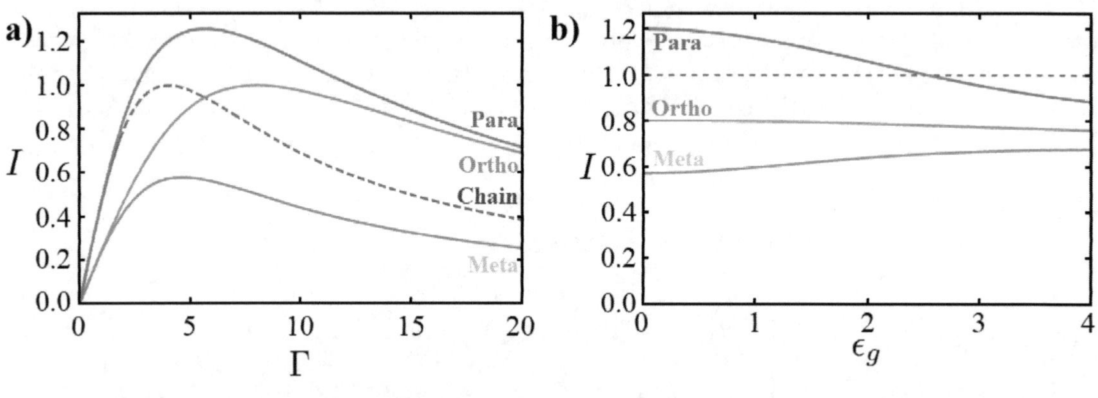

Figure 2. Panel a) current across a benzene molecule in terms of the coupling strength $\Gamma = \Gamma_1 = \gamma_L$ in the large bias regime for the three configurations. Dashed blue curve corresponds to the current across a linear chain of 6 sites. We set $t = 2eV$ and $\epsilon_g = 0$. The maximum of the curve is the OWP [3].
Panel b) current across a benzene molecule in terms of the gate voltage ϵ_g in the large bias regime. Dashed blue curve corresponds to the current across a chain of 6 sites (no gate voltage applied). We set $t = 2eV$ and $\Gamma_1 = \gamma_L = 4$ corresponding to the OWP for the linear chain.

An interesting test concerns the robustness of the OWP against disorder. To introduce disorder in the benzene ring we can, e.g., randomize the on-site energy ϵ and/or the bond electron hopping strength t. Disorder is, in general, expected to substantially affect the transport properties of the system, especially in lower dimensions. In the following, we compare the current across the benzene ring, in the para, meta and ortho configurations, with the current across a one

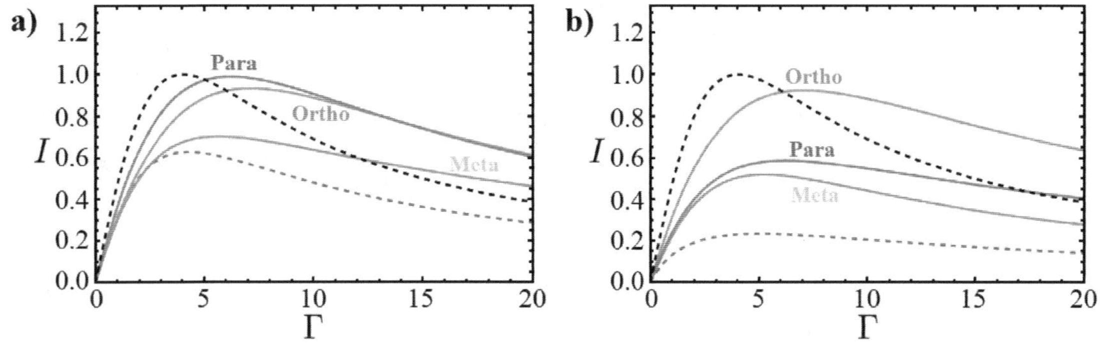

Figure 3. Current across a disordered benzene molecule in terms of the coupling strength $\Gamma = \Gamma_1 = \gamma_L$ in the large bias regime for the three configurations. Dashed black curve corresponds to the current across a clean chain of 6 sites (like in Fig.[2]), dashed blue curve to the current across a disordered chain of 6 sites. We set $t = 2eV$, $\epsilon_g = 0$ and a) strong "chemical potential disorder", $\sigma_\epsilon = 1.2$, $\sigma_t = 0$, b) strong "bond disorder", $\sigma_\epsilon = 0$, $\sigma_t = 1.2$.

dimensional chain [3]. Technically, we realize this by setting [5] $\epsilon_{\rm eff} \to \epsilon_j = \epsilon + \delta\epsilon_j$ and $t_{\rm eff} \to t_{i,j} = t + \delta t_{i,j}$ with $i < j = 1, ..., L$. The $\{\delta\epsilon_j\}$ are independent random variables described by a probability distribution $P_\epsilon[\{\delta\epsilon_j\}] = \prod_{j=1}^{\ell} p_\epsilon(\delta\epsilon_j)$. Specifically, we choose $p_\epsilon(\delta\epsilon)$ to be the probability distribution for the $\{\delta\epsilon\}$, with average $\overline{\delta\epsilon} = \int dx\, x p_\epsilon(x) = 0$, and with variance $\sigma_\epsilon^2 = \int dx\, x^2 p_\epsilon(x)$. We use the uniform probability distributions given by

$$p_\epsilon(\delta\epsilon) = \begin{cases} \frac{1}{2\sqrt{3}\sigma_\epsilon} \ , \text{ for } -\sqrt{3}\sigma_\epsilon \leq \delta\epsilon \leq \sqrt{3}\sigma_\epsilon \\ 0 \ , \text{ otherwise} \end{cases} , \tag{4}$$

with equivalent formulas holding for the hopping strength fluctuations $\delta t_{i,j}$. Having assumed the probability distribution in Eq.[4], we can estimate the disorder-averaged current distribution at given values of σ_ϵ and σ_t. In Fig.[3], we show the behavior of the stationary current, as a function of applied bias, in the presence of a strong chemical potential disorder (panel a) or a strong bond disorder (panel b). Interestingly, while the one-dimensional chain is strongly affected by disorder with a remarkable reduction of the current, the benzene ring is much more robust. Indeed, comparing panel a) and b), we observe that the para configuration is robust againts chemical potential disorder but is affected by the bond one, the meta configuration current is enhanced by chemical disorder and the ortho configuration is quite unaffected by the disorder. These effects are expected to be a consequence of the interferences between the two branches of the ring.

4. Conclusions

Using the LE framework, we have discussed the main features of the NESS arising in a small disordered molecular electronic device connected to two reservoirs in the large bias limit. The preliminary results shown in the manuscript suggest that the LE framework can be succesfully extended to non-linear systems, like molecular electronic devices, small DNA chains or junctions of quantum wires in the presence of two or more reservoirs in order to investigate their transport properties under the effect of a large voltage bias [6, 7, 8, 9, 10]. Furthermore, the role of disorder and/or interaction can be discussed and, if a multi-site Lindblad bath is considered [11], it is also possible to introduce a temperature bias to investigate the thermal transport across such devices and, for example, the breakdown of the Wiedemann-Franz law [12, 13].

Acknowledgments
A. N. was financially supported by POR Calabria FESR-FSE 2014/2020 - Linea B) Azione 10.5.12, grant no. A.5.1. D. G., L. L. and M. R. acknowledge financial support from Italy's MIUR PRIN projects TOP-SPIN (Grant No. PRIN 20177SL7HC).

References

[1] Benenti G, Casati G, Prosen T and Rossini D 2009 *Europhysics Letters* **85** 37001
[2] Benenti G, Casati G, Prosen T, Rossini D and Žnidarič M 2009 *Phys. Rev.* B **80** 035110
[3] Nava A, Rossi M and Giuliano D 2021 *Phys. Rev.* B **103** 115139
[4] Vahedi J and Sartipi Z 2015 *Molecular Physics* **113** 1422
[5] Nava A, Giuliano R, Campagnano G and Giuliano D 2017 *Phys. Rev.* B **95** 155449
[6] Goyer F, Ernzerhof M and Zhuang M 2007 *The Journal of Chemical Physics* **14** 144104
[7] Lambropoulos K and and Simserides C 2019 *Symmetry* **11** 968
[8] Nava A, Giuliano R, Campagnano G and Giuliano D 2016 *Phys. Rev.* B **94** 205125
[9] Giuliano D, Lepori L and Nava A 2020 *Phys. Rev.* B **101** 195140
[10] Giuliano D, Nava A and Sodano P 2020 *Nuclear Physics* B **960** 115192
[11] Guimarães P H, Landi G T, de Oliveira M J 2016 *Phys. Rev.* E **94** 032139
[12] Buccheri F, Nava A, Egger R, Sodano P, Giuliano D 2021, arXiv:2108.04156
[13] Giuliano D, Nava A, Egger R, Sodano P, Buccheri F 2021 arXiv:2108.04149

Journal of Physics: Conference Series **2164** (2022) 012052 doi:10.1088/1742-6596/2164/1/012052

Inelastic neutron scattering study of crystalline electric field excitations in a cubic compound $PrMgNi_4$

Y. Kusanose[1], T. Onimaru[1], Y. Yamane[1,*], K. Umeo[2], T. Takabatake[1], T. Guidi[3], D. Le[3] and D. T. Adroja[3,4]

[1] Department of Quantum Matter, Graduate School of Advanced Science and Engineering, Hiroshima University, Higashi-Hiroshima, Hiroshima 739-8530, Japan

[2]Integrated Experimental Support/Research Division, N-BARD, Hiroshima University, Higashi-Hiroshima, Hiroshima 739-8526, Japan

[3]ISIS Facility, Rutherford Appleton Laboratory, Chilton, Didcot, Oxon, OX11 0QX, United Kingdom

[4]Highly Correlated Matter Research Group, Physics Department, University of Johannesburg, Auckland Park 2006, South Africa

E-mail: `d206038@hiroshima-u.ac.jp`

Abstract. We performed inelastic neutron scattering experiments on $PrMgNi_4$ with the cubic $MgSnCu_4$-type structure. The magnetic excitations were observed at 1.1, 2.5, 5.9, 10.7, and 11.7 meV. Since the energy of the excitations is constant for $|\boldsymbol{Q}| < 5$ Å^{-1}, they are ascribed to the crystalline electric field (CEF) excitations of the Pr^{3+} ions. We adopted CEF parameters of $W = -3.3$ K and $x = 0.80$ for the cubic T_d point group to reproduce the strong excitations at 1.1, 10.7, and 11.7 meV. The calculated CEF level scheme reveals the Γ_3 doublet ground state with the quadrupolar degrees of freedom and the excited states of Γ_4 triplet (1.16 meV), Γ_1 singlet (2.78 meV), and Γ_5 triplet (11.6 meV). This scheme, however, does not explain the observed two excitations at 2.5 and 5.9 meV. These additional excitations may arise from splitting of the doublet by symmetry lowering due to excess Mg atoms occupying the Pr sites. This splitting is probably responsible for the absence of the long-range quadrupole order in $PrMgNi_4$.

1. Introduction

Cubic praseodymium-based intermetallic compounds with nonmagnetic Γ_3 doublet crystalline electric field (CEF) ground states of $4f^2$ electrons have received much interest. They show a variety of phenomena such as multipole order, heavy fermion superconductivity and non-Fermi liquid state originating from quadrupolar degrees of freedom in the Γ_3 doublet through strong exchange interaction with conduction electrons (c–f hybridization). For example, $PrPb_3$ exhibits an antiferroquadrupole order at $T_Q = 0.4$ K [1], below which the electric quadrupoles are modulated sinusoidally [2]. In contrast, $PrAg_2In$ [3] and $PrMg_3$ [4] exhibit no quadrupole order in spite of the Γ_3 doublet ground states. The absence of the quadrupole order may be ascribed to the Kondo effect due to strong c–f hybridization, because the specific heat divided

*Present address: Department of Material Science, Graduate School of Science, University of Hyogo, Kamigori, Hyogo 678-1297, Japan

by temperature, C/T, is strongly enhanced on cooling below 1 K. Another possibility is splitting of the Γ_3 doublet by symmetry lowering of the Pr site due to the atomic disorder inherent in their Heusler-type structures.

Recently, we have proposed that a cubic compound $PrMgNi_4$ possesses the Γ_3 doublet ground state [5]. $PrMgNi_4$ crystallizes in the cubic $MgSnCu_4$-type structure with the space group of $F\bar{4}3m$ [6]. A single-crystal X-ray structural analysis revealed that the excess Mg atoms occupy 4.5% of the Pr sites [5]. The Schottky-type anomaly in $C(T)$ and van-Vleck paramagnetic behavior of the magnetic susceptibility suggest that the CEF ground state is the Γ_3 doublet. Though no long-range quadrupole order was observed down to 0.1 K, both $C(T)$ and the electrical resistivity $\rho(T)$ exhibit shoulder-like anomalies at around 0.7 K. Since the shoulder of $C(T)$ becomes more pronounced when the amount of excess Mg atoms is decreased [7], the absence of the long-range quadrupole order may be attributed to symmetry lowering of the Pr site owing to the excess Mg atoms substituting for the Pr atoms. In addition, the shoulders of $C(T)$ and $\rho(T)$ at around 0.7 K are robust against the magnetic field up to 5 T, indicating development of intersite correlation of the quadrupoles in the ground state doublet [8]. In the present work, we performed inelastic neutron scattering experiments to confirm the Γ_3 doublet ground state by measuring the CEF excitations of a Pr^{3+} ion.

2. Experiments

The inelastic neutron scattering experiments were performed using the time-of-flight neutron spectrometer MARI installed at ISIS facility in Rutherford Appleton Laboratory. The temperature was changed from 5 to 120 K using a closed cycle refrigerator. The incident neutron energy was selected at 11.8 and 30.0 meV. Powdered samples of $PrMgNi_4$ of 4.60 g and the nonmagnetic reference compound $LaMgNi_4$ of 1.54 g each were obtained by crushing single-crystalline samples grown by the Mg self-flux method [5].

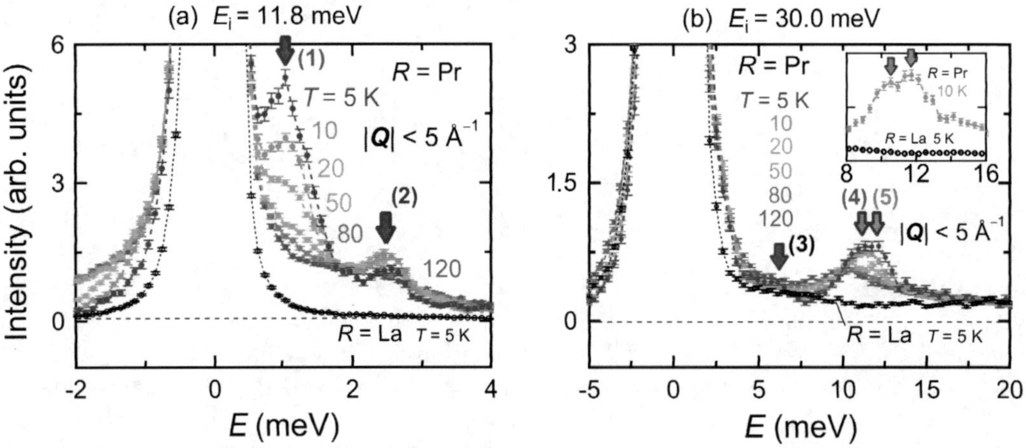

Figure 1. Inelastic neutron scattering spectra of $PrMgNi_4$ for $5 \leq T \leq 120$ K with incident neutron energy of (a) $E_i = 11.8$ and (b) 30.0 meV. Magnetic excitations were observed at 1.1 meV labeled (1), and (2) 2.5, (3) 5.9, (4) 10.7 and (5) 11.7 meV. Black circles are the data for the nonmagnetic counterpart $LaMgNi_4$ measured at 5 K. The inset of (b) shows the data for $8 \leq E \leq 16$ meV.

3. Results and Discussion

Figure 1 shows the inelastic neutron scattering spectra of $PrMgNi_4$ from 5 to 120 K with the incident neutron energy of (a) E_i = 11.8 and (b) 30.0 meV. The spectra are evaluated by integrating the neutron scattering intensity of $I(\boldsymbol{Q}, E)$ data for $|\boldsymbol{Q}| < 5$ Å^{-1}, where $\boldsymbol{Q}$ is the wave-vector transfer and E the energy transfer. As a reference for the phonon contribution, the spectra of the nonmagnetic counterpart $LaMgNi_4$ are shown with the black circles. In the spectra with E_i = 11.8 meV shown in Fig. 1(a), magnetic excitations were observed at E = 1.1 and 2.5 meV. In Fig. 1(b), a shoulder appears at around 5.9 meV and excitations at 10.7 and 11.7 meV overlap. Since the energy of these excitations is nearly constant for $|\boldsymbol{Q}| < 5$ Å^{-1}, they are ascribed to the CEF excitations of the Pr ions.

Focussing on the temperature variations of the CEF excitations, the intensity of the excitation peak at 1.1 meV becomes half with increasing temperatures from 5 to 20 K, indicating a transition from a low-lying state to an excited state. On the other hand, the intensity of the excitation peak at 2.5 meV increases from 5 to 20 K but decreases above 50 K. The temperature variation of the shoulder at 5.9 meV is small. The two peaks at 10.7 and 11.7 meV merge with increasing temperatures to 20 K. The shift of the peaks to the lower energy is probably ascribed to additional phonon contributions.

With the CEF effect of the cubic T_d point group, the nine-fold multiplet for the Pr^{3+} ion with $J = 4$ is split into the Γ_1 singlet, Γ_3 doublet, and Γ_4 and Γ_5 triplets. The scattering intensity due to the CEF excitations is estimated from the matrix elements of $|\langle\Gamma_i|J_z|\Gamma_j\rangle|^2$ [9]. Non-zero values of the matrix elements indicate the excitations between the CEF multiplets. The matrix elements of $|\langle\Gamma_i|J_z|\Gamma_j\rangle|^2$ are estimated as follows; 6.67 for Γ_1 and Γ_4, 9.33 for Γ_3 and Γ_4, 4.0 for Γ_3 and Γ_5, and 3.50 for Γ_4 and Γ_5. Thereby, the magnetic excitations between the CEF levels are rigidly restricted as shown with the arrows in the inset of Fig. 2(a). A cubic Hamiltonian matrix is diagonalized to give the expectation values of the CEF energy as a function of two CEF parameters of x and W [10]. Here, x is the ratio of fourth- to sixth-order terms and W the

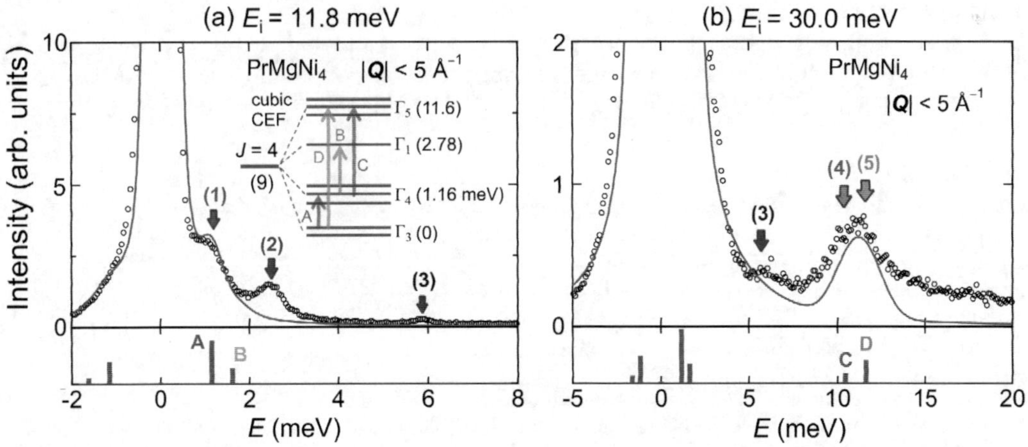

Figure 2. Observed and simulated inelastic neutron spectra of $PrMgNi_4$ at 20 K with the incident neutron energy of (a) E_i = 11.8 and (b) 30.0 meV. The (red) solid lines are simulations with cubic CEF parameters of $W = -3.3$ K and $x = 0.80$. The bars indicate the relative intensity of the magnetic excitations between the CEF levels as follows; A : $\Gamma_3 \rightarrow \Gamma_4$, B : $\Gamma_4 \rightarrow \Gamma_1$, C : $\Gamma_4 \rightarrow \Gamma_5$, D : $\Gamma_3 \rightarrow \Gamma_5$. The inset shows the CEF level scheme of the Pr ion with the above CEF parameters. The arrows indicate the allowed excitations between the CEF levels. The observed excitations labeled (2) and (3) are not assigned by the simulation with the above CEF parameters.

overall energy scale. Referring to the CEF parameters of $W = -2.0$ K and $x = 0.77$ proposed to reproduce the temperature variation of the magnetic susceptibility and the Schottky specific heat [5], we adopt modified parameters of $W = -3.3$ K and $x = 0.80$ to reproduce the inelastic neutron scattering spectra. The calculations are shown with the (red) solid lines in Figs. 2(a) and (b). The excitation peaks at 1.1, 10.7, and 11.7 meV are reproduced. The above CEF parameters give the CEF level scheme as the Γ_3 doublet ground state and excited states of Γ_4 triplet (1.16 meV), Γ_1 singlet (2.78 meV), and Γ_5 triplet (11.6 meV) as shown in the inset.

On the other hand, the observed excitations at 2.5 and 5.9 meV are not assigned by the simulation with the above CEF parameters. These additional excitations are probably attributed to the symmetry lowering of the Pr sites due to the excess Mg atoms substituting for the Pr atoms as proposed by the single-crystal X-ray structural analysis [5]. To reproduce the additional excitations, it is necessary to include other CEF parameters, e.g., for a trigonal point group. Thereby, the absence of the long-range quadrupole order in $PrMgNi_4$ is attributed to the splitting of the Γ_3 doublet by the symmetry lowering of the Pr atom.

4. Summary

To determine the CEF level scheme of the Pr^{3+} ions in the cubic compound $PrMgNi_4$, inelastic neutron scattering experiments were performed. We observed CEF excitations at $E = 1.1$, 2.5, 5.9, 10.7, and 11.7 meV. The CEF parameters of $W = -3.3$ K and $x = 0.80$ for the cubic T_d point group were chosen to reproduce the excitations at 1.1, 10.7, and 11.7 meV, indicating that the CEF ground state is the Γ_3 doublet. The obtained CEF level scheme, however, does not explain the excitations at 2.5 and 5.9 meV. The additional peaks are probably attributed to symmetry lowering of the Pr sites due to excess Mg atoms substituted for the Pr atoms. Thereby, symmetry lowering of the Pr sites due to the atomic disorder splits the Γ_3 doublet. This splitting of the Γ_3 doublet is responsible for the absence of the long-range quadrupole order in $PrMgNi_4$.

Acknowledgments

We acknowledge support from the Center for Emergent Condensed-Matter Physics (ECMP), Hiroshima University. This work was financially supported by Grants in Aid from MEXT/JSPS of Japan, Grants Nos. JP15H05886 (J-Physics) and JP18H01182.

References

[1] Morin P, Schmitt D and du Tremolet de Lacheisserie E 1982 *J. Magn. Magn. Mater.* **30** 257
[2] Onimaru T, Sakakibara T, Aso N, Yoshizawa H, Suzuki H S and Takeuchi T 2005 *Phys. Rev. Lett.* **94** 197201
[3] Yatskar A, Beyermann W P, Movshovich R and Canfield P C 1996 *Phys. Rev. Lett.* **77** 3637
[4] Tanida H, Suzuki H S, Takagi S, Onodera H and Tanigaki K 2006 *J. Phys. Soc. Jpn.* **75** 073705
[5] Kusanose Y, Onimaru T, Park G B, Yamane Y, Umeo K, Takabatake T, Kawata N and Mizuta T 2019 *J. Phys. Soc. Jpn.* **88** 083703
[6] Kadir K, Noréus D and Yamashita I 2002 *J. Alloys Compd.* **345** 140
[7] Kusanose Y and Onimaru T 2021 private communication
[8] Kusanose Y, Onimaru T, Park G B, Yamane Y, Umeo K, Takabatake T, Kawata N and Mizuta T 2020 *JPS Conf. Proc.* **30** 011160
[9] Birgeneau R J 1972 *J. Phys. Chem. Solids* **33** 59
[10] Lea K R, Leask M J M and Wolf W P 1962 *J. Phys. Chem. Solids* **23** 1381

Magnetic structure of an antiferromagnet $NdRh_2Zn_{20}$ investigated by powder neutron diffraction

R. Yamamoto[1], **Y. Shimura**[1], **K. Umeo**[2], **T. Takabatake**[1], **F. Damay**[3], **J.-M. Mignot**[3], **and T. Onimaru**[1]

[1]Department of Quantum Matter, Graduate School of Advanced Science and Engineering, Hiroshima University, Higashi-Hiroshima, Hiroshima 739-8530, Japan
[2]Department of Low Temperature Experiment, Integrated Experimental Support/Research Division, N-BARD, Hiroshima University, Higashi-Hiroshima, Hiroshima 739-8526, Japan
[3]Laboratoire Léon Brillouin, CEA-Saclay, 91191 Gif sur Yvette, France

E-mail: d200355@hiroshima-u.ac.jp

Abstract. Powder neutron diffraction measurements were carried out to investigate the magnetic structure of the antiferromagnet $NdRh_2Zn_{20}$. The magnetic reflections observed for $T \leq 1.0$ K are indexed by the propagation vector of $\boldsymbol{k}$ = (1/2, 1/2, 1/2). From the analysis of the intensity of magnetic reflections, a magnetic structure described by the Γ_6 representation is proposed. The magnetic moments are parallel to the $[11\bar{2}]$ or $[\bar{1}10]$ direction and stacked with a sequence of up-up-down-down along the [111] direction.

1. Introduction

Rare-earth-based intermetallic compounds RTr_2X_{20} (R: rare-earth element, Tr: transition metal, and X = Al, Zn, and Cd) crystallize in the cubic $CeCr_2Al_{20}$-type structure with the space group of $Fd\bar{3}m$ (# 227, O_h^7) [1]. This series of compounds display a variety of strongly correlated electronic phenomena such as multipole orders and superconductivity in $PrTr_2Zn_{20}$ (Tr = Rh and Ir) and $PrTr_2Al_{20}$ (Tr = V and Ti) [2, 3, 4, 5, 6, 7], non-Fermi liquid behaviors in $PrTr_2Zn_{20}$ (Tr = Rh and Ir) [8, 9, 10], and heavy fermion states in $YbTr_2Zn_{20}$ (Tr = Fe, Ru, Os, Co, Rh, and Ir) [11, 12, 13]. Up till now, antiferromagnetic (AFM) structures of this series of compounds were only reported for RCo_2Zn_{20} (R = Gd and Tb) [14, 15]. The antiferro-type quadrupole structure of $PrIr_2Zn_{20}$ was determined by observing magnetic-field-induced magnetic dipoles [16]. It is notable that the ordering structures are commonly characterized by a propagation vector $\boldsymbol{k}$ = (1/2, 1/2, 1/2).

Nd-based compounds $NdTr_2X_{20}$ exhibit ferromagnetic or antiferromagnetic transitions below 2 K [17, 18, 19, 20, 21, 22, 23]. The crystalline electric field (CEF) ground states of the Nd ion are mostly the Γ_6 doublets. Interestingly, emergence of the two-channel Kondo effect was proposed with a seven orbital impurity Anderson model [24]. Among them, $NdRh_2Zn_{20}$ undergoes an antiferromagnetic transition at T_N = 0.94 K [23]. The magnetic specific heat C_m shows a pronounced peak at T_N, and the peak shifts to lower temperature as magnetic field is increased to 1.5 T. The magnetic entropy S_m is 60% of $R\ln 2$ at T_N and reaches $R\ln 2$ on heating to 5 K. This reduction of S_m at T_N indicates the presence of magnetic fluctuations above T_N. In this paper, we report a magnetic structure of $NdRh_2Zn_{20}$ studied by powder neutron diffraction measurements. The determination of the magnetic structure of $NdRh_2Zn_{20}$ gives a hint to understand a variety of phenomena in RTr_2Zn_{20}.

2. Experimental procedures

Polycrystalline samples of $NdRh_2Zn_{20}$ were synthesized by the melt-growth method. High purity elements of Nd (99.9%), Rh (99.9%), and Zn (99.9999%) were loaded in the a boron nitride crucible and sealed into a quartz ampule under 1/3 atmospheric pressure of Ar. The sealed ampule was heated to 1050°C, and then cooled down to 300°C. The samples were characterized by powder X-ray diffraction measurements. Neutron diffraction experiments for powdered sample were carried out at the G4-1 diffractometer installed at the Orphée reactor in Laboratoire Léon Brillouin. The incident neutron wavelength was 2.426 Å. The temperature was changed from the lowest temperature of 0.046 K to 1.5 K by using a ^{3}He-^{4}He dilution refrigerator. The Rietveld analysis was performed with a program of the FULLPROF suite to refine the crystal and magnetic structures [25].

3. Results and Discussion

Figure 1(a) shows the neutron powder diffraction pattern of $NdRh_2Zn_{20}$ at T = 1.5 K > T_N. Nuclear peaks are indexed with the cubic $CeCr_2Al_{20}$-type structure [1]. The solid line is a fit with the Rietveld analysis. As shown in Fig. 1(b), additional peaks appear in the diffraction pattern at 0.046 K < T_N. The difference in the intensity, $I(T = 0.046\text{ K}) - I(T = 1.5\text{ K})$, is plotted in Fig. 1(c). The peaks are indexed by the propagation vector $\boldsymbol{k}$ = (1/2, 1/2, 1/2). This development of the superlattice reflections is the evidence of

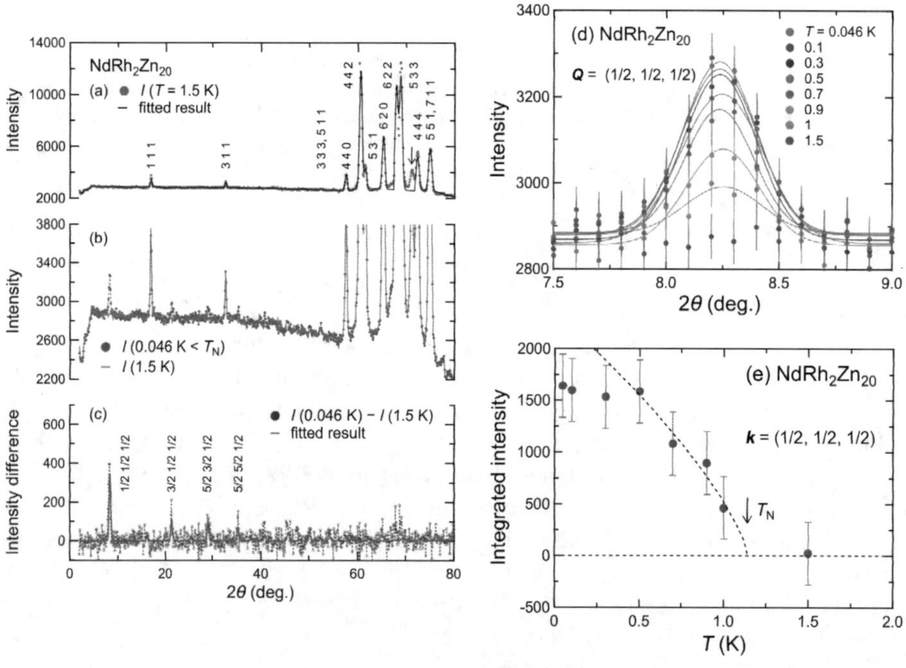

Figure 1. (a) Neutron powder diffraction patterns of $NdRh_2Zn_{20}$ at T = 1.5 K > T_N. (a) Green circles are data and the black solid line is a fit to the nuclear peaks by the Rietveld analysis. The peak at 71.2° (small arrow) comes from an extrinsic cause. (b) Diffraction patterns at 0.046 K (red circles) and 1.5 K (green solid line), which are below and above T_N. (c) Difference of the intensities at 0.046 K and 1.5 K. Magnetic reflections are visible as sharp peaks. The red solid line is a fit to the magnetic peaks. See the main text for detail. (d) Profiles of the peaks at $\boldsymbol{Q}$ = (1/2, 1/2, 1/2) for $0.046 \leq T \leq 1.5$ K. The profiles below 1 K are fitted with a Gaussian function. (e) Temperature dependence of the integrated intensities of the peak at $\boldsymbol{Q}$ = (1/2, 1/2, 1/2). The arrow indicates T_N = 1.14 K determined by a fit to the data for $0.5 \leq T \leq 1.0$ K with a power law expression of $(1 - T/T_N)^{2\beta}$ for β = 0.36.

the antiferromagnetic order.

Fig. 1(d) shows profiles of the peak at $\boldsymbol{Q}$ = (1/2, 1/2, 1/2) from 0.046 to 1.5 K. The peaks for $T \leq 1.0$ K are fitted with a Gaussian function. The temperature dependence of the integrated intensity is plotted in Fig. 1(e). The integrated intensity increases on cooling below 1.0 K and tends to saturate at $T < 0.5$ K. The temperature variation of the intensity is fitted with a power law expression $(1 - T/T_N)^{2\beta}$ from 0.5 to 1.0 K. Here, the critical exponent β was fixed to 0.36 according to the 3D Heisenberg model [26]. The transition temperature was thereby deduced to be 1.14(6) K. This value is slightly higher than T_N = 0.94 K determined from the peak top of the specific heat data [23].

Table 1. Irreducible representations (IR) and basis vectors (BV) for the space group of $Fd\bar{3}m$ (Origin choice: 2) with the propagation vector $\boldsymbol{k}$ = (1/2, 1/2, 1/2). The decomposition of the magnetic representation for the Nd site (0.125, 0.125, 0.125). The Nd atoms of the nonprimitive basis are defined according to 1: (0.125, 0.125, 0.125) and 2: (0.875, 0.875, 0.875). The imaginary components of the BV are zero.

IR	BV	Atom	BV components		
			$m_{\parallel a}$	$m_{\parallel b}$	$m_{\parallel c}$
Γ_2	ψ_1	1	1	1	1
		2	1	1	1
Γ_3	ψ_2	1	1	1	1
		2	−1	−1	−1
Γ_5	ψ_3	1	0.866	−0.866	0
		2	−0.866	0.866	0
	ψ_4	1	0.5	0.5	−1
		2	−0.5	−0.5	1
Γ_6	ψ_5	1	0.5	0.5	−1
		2	0.5	0.5	−1
	ψ_6	1	−0.866	0.866	0
		2	−0.866	0.866	0

The irreducible representations are calculated for the Wyckoff position 8*a* in the space group $Fd\bar{3}m$ for the Nd site. According to the representation analysis with a program SARAh [27], there are four possible irreducible representations as listed in Table 1. The propagation vector and the irreducible representations in Table 1 are labeled following the Kovalev notation [28]. The imaginary components for all the irreducible representations are zero.

The strong intensity at (1/2, 1/2, 1/2) rules out the Γ_2 and Γ_3 representations, since they constrain the moments along the [111] direction. The best agreement between the experimental data and the calculation is achieved by a magnetic structure with the Γ_6 representation. Therefore, the magnetic moments are parallel to the $[11\bar{2}]$ or $[\bar{1}10]$ direction. Here, the magnetic moments are assumed to align along the $[11\bar{2}]$ direction based on the basis vector ψ_5 for Γ_6. The solid red line in Fig. 1(c) is a fit with the Riedveld analysis. The reliability factors are R_{wp} = 4.59, R_B = 8.71, R_F = 5.42, and χ^2 = 29.4. The amplitude of the magnetic moments is 1.39(2) μ_B/Nd. This value agrees with that of 1.33 μ_B/Nd expected for the Γ_6 double ground state. We analyze the data based on another model in which the ordered magnetic moments parallel to the $[\bar{1}10]$ direction. This model also reproduces the magnetic reflections. Because the magnetic structure is constructed by a linear combination of the two basis vectors of ψ_5 and ψ_6 of Γ_6 as shown in Table 1, the direction of the magnetic moments could not be distinguished from the present powder diffraction data. To determine the direction of the magnetic moments, it is necessary to perform neutron diffraction experiments with a single-crystalline sample.

It is noteworthy that the propagation vector of $\boldsymbol{k}$ = (1/2, 1/2, 1/2) is common among those for the isostructural compounds $R\text{Co}_2\text{Zn}_{20}$ (R = Gd and Tb) and $\text{PrIr}_2\text{Zn}_{20}$ [14, 15, 16]. The magnetic structure for $GdCo_2Zn_{20}$ is the same as that for $NdRh_2Zn_{20}$, where the alignment of the magnetic moments is described by the Γ_6 representation [14], while the magnetic structure of $TbCo_2Zn_{20}$ is characterized by the Γ_5 representation.

4. Conclusions

Powder neutron diffraction measurements of an antiferromagnet $NdRh_2Zn_{20}$ with T_N = 0.94 K were performed from 0.046 to 1.5 K. Magnetic reflections due to the antiferromagnetic order with the propagation vector $\boldsymbol{k}$ = (1/2, 1/2, 1/2) were observed at $T \leq 1.0$ K. We propose the AFM structure described by the Γ_6 representation, where the magnetic moments lying in the (111) plane are stacked with a sequence of up-up-down-down along the [111] direction.

5. Acknowledgments

The author would like to thank D. Ueta and K. Iwasa for helpful discussions. We acknowledge support from the Center for Emergent Condensed-Matter Physics (ECMP), Hiroshima University. This work was financially supported by Grants in Aid from MEXT/JSPS of Japan, Grants No. JP18H01182.

References

[1] Nasch T, Jeitschko W and Rodewald U C 1997 *Z. Naturforsch. B* **52** 1023
[2] Onimaru T, Matsumoto K T, Inoue Y F, Umeo K, Saiga Y, Matsushita Y, Tamura R, Nishimoto K, Isao I, Suzuki T and Takabatake T 2010 *J. Phys. Soc. Jpn.* **79** 033704
[3] Onimaru T, Matsumoto K T, Inoue Y F, Umeo K, Sakakibara T, Karaki Y, Kubota M and Takabatake T 2011 *Phys. Rev. Lett.* **106** 177001
[4] Sakai A and Nakatsuji S 2011 *J. Phys. Soc. Jpn.* **80** 063701
[5] Sakai A, Kuga K and Nakatsuji S 2012 *J. Phys. Soc. Jpn.* **81** 083702
[6] Tsujimoto M, Matsumoto Y, Tomita T, Sakai A and Nakatsuji S 2014 *Phys. Rev. Lett.* **113** 267001
[7] Onimaru T and Kusunose H 2016 *J. Phys. Soc. Jpn.* **85** 082002
[8] Onimaru T, Izawa K, Matsumoto K T, Yoshida T, Machida Y, Ikeura T, Wakiya K, Umeo K, Kittaka S, Araki K, Sakakibara T and Takabatake T 2016 *Phys. Rev. B* **94** 075134
[9] Yoshida T, Machida Y, Izawa K, Shimada Y, Nagasawa N, Onimaru T, Takabatake T, Gourgout A, Pourret A, Knebel G and Brison J P 2017 *J. Phys. Soc. Jpn.* **86** 044711
[10] Tsuruta A and Miyake K 2015 *J. Phys. Soc. Jpn.* **84** 114714
[11] Torikachvili M S, Jia S, Mun E D, Hannahs S T, Black R C, Neils W K, Martien D, Bud'ko S L and Canfield P C 2007 *Proc. Natl. Acad. Sci. U.S.A.* **104** 9960
[12] Saiga Y, Matsubayashi K, Fujiwara T, Kosaka M, Katano S, Hedo M, Matsumoto T and Uwatoko Y 2008 *J. Phys. Soc. Jpn.* **77** 053710
[13] Honda F, Taga Y, Hirose Y, Yoshiuchi S, Tomooka Y, Ohya M, Sakaguchi J, Takeuchi T, Settai R, Shimura Y, Sakakibara T, Sheikin I, Tanaka T, Kubo Y and Ōnuki Y 2014 *J. Phys. Soc. Jpn.* **83** 044703
[14] Mardegan J R L, Francoual S, Fabbris G, Veiga L S I, Strempfer J, Haskel D, Ribeiro R A, Avila M A and Giles C 2016 *Phys. Rev. B* **93** 024421
[15] Tian W, Christianson A D, Zarestky J L, Jia S, Bud'ko S L, Canfield P C, Piccoli P M B and Schultz A J 2010 *Phys. Rev. B* **81** 144409
[16] Iwasa K, Matsumoto K T, Onimaru T, Takabatake T, Mignot J M and Gukasov A 2017 *Phys. Rev. B* **95** 155106
[17] Isikawa Y, Ejiri J, Mizushima T and Kuwai T 2013 *J. Phys. Soc. Jpn.* **82** 123708
[18] Namiki T, Murata Y, Baba K, Shimojo K and KNishimura 2014 *JPS Conf. Proc.* **3** 011055
[19] Wakiya K, Matsumoto K T, Onimaru T, Umeo K and Takabatake T 2015 *Phys. Procedia* **75** 511
[20] Namiki T, Nosaka K, Tsuchida K, Lei Q, Kanamori R and Nishimura K 2016 *J. Phys.: Conf. Ser.* **983** 012017
[21] Namiki T, Lei Q, Isikawa Y and Nishimura K 2016 *J. Phys. Soc. Jpn.* **85** 073706
[22] Yamane Y, Yamada R J, Onimaru T, Uenishi K, Wakiya K, Matsumoto K T, Umeo K and Takabatake T 2017 *J. Phys. Soc. Jpn.* **86** 054708
[23] Yamamoto R, Onimaru T, Yamada R J, Yamane Y, Shimura Y, Umeo K and Takabatake T 2019 *J. Phys. Soc. Jpn.* **88** 044703
[24] Hotta T 2017 *J. Phys. Soc. Jpn.* **86** 083704
[25] Rodriguez-Carvajal J 1993 *Physica B* **192** 55
[26] Ashcroft N W and Mermin N D 1976 *Solid State Physics* (Saunders College, New York)
[27] Wills A S 2003 *Physica B* **276–278**

[28] Kovalev O V 1993 *Representations of the Crystallographic Space Groups* 2nd ed (Gordon and Breach Science, Yverdon, Switzerland.)

Crystal-field excitations and quadrupolar fluctuations of 4f-electron systems studied by polarized light scattering

M Ye[1,*], H H Kung[2], P F S Rosa[3], E Rosenberg[4], J Kim[5], X H Xu[1], E D Bauer[3], Z Fisk[6], I R Fisher[7], S W Cheong[1] and G Blumberg[1,†]

[1] Department of Physics and Astronomy, Rutgers University, Piscataway, NJ 08854, USA
[2] Stewart Blusson Quantum Matter Institute, University of British Columbia, V6T 1Z4 Vancouver, Canada
[3] Los Alamos National Laboratory, Los Alamos, NM 87545, USA
[4] Department of Physics, University of Washington, Seattle, WA 98105, USA
[5] Korea Atomic Energy Research Institute, 34057 Daejeon, Republic of Korea
[6] Department of Physics and Astronomy, University of California, Irvine, CA 92697, USA
[7] Department of Applied Physics, Stanford University, CA 94305, USA

E-mail: * `mye@physics.rutgers.edu`, † `girsh@physics.rutgers.edu`

Abstract. We present Raman-scattering results for three materials, CeB_6, $TbInO_3$, and $YbRu_2Ge_2$, to illustrate the essential aspects of crystal-field (CF) excitations and quadrupolar fluctuations of 4f-electron systems. For CF excitations, we illustrate how the 4f orbits are split by spin-orbit coupling and CF potential by presenting spectra for inter- and intra-multiplet excitations over a large energy range. We discuss identification of the CF ground state and establishment of low-energy CF level scheme from the symmetry and energy of measured CF excitations. In addition, we demonstrate that the CF linewidth is a sensitive probe of electron correlation by virtue of self-energy effect. For quadrupolar fluctuations, we discuss both ferroquadrupolar (FQ) and antiferroquadrupolar (AFQ) cases. Long-wavelength quadrupolar fluctuations of the same symmetry as the FQ order parameter persists well above the transition temperature, from which the strength of electronic intersite quadrupolar interaction can be evaluated. The tendency towards AFQ ordering induces ferromagnetic correlation between neighboring 4f-ion sites, leading to long-wavelength magnetic fluctuations.

1. Introduction

Because of the heaviness of f elements and the compactness of f orbits, the energy scale of crystal-field (CF) effect is much weaker than that of spin-orbit coupling (SOC) for 4f-electron systems. As a result, CF interaction plays an important role in the low-temperature physics of these materials [1, 2, 3, 4]. In particular, if the CF ground state possesses orbital degeneracy, it supports multipole moments, with quadrupole moments being the most common one. Although the field of CF analysis is well established with a long history, predictive understanding of structure-property relationships is still in its infancy for f-electron systems, and new questions continuously emerge with advances in material synthesis and characterization.

For investigation of the new questions related to CF and quadrupolar physics of 4f-electron systems, polarization-resolved Raman spectroscopy is a well suited experimental technique.

Besides supreme spectral resolution, this method offers the ability to disentangle the long-wavelength excitation spectra into individual symmetry channels. By controlling the polarization of the incident and scattered light, we can identify the symmetry of different CF levels and separate quadrupolar excitations from other signals.

Here we first examine the CF excitations of CeB_6 and $TbInO_3$ to illustrate the relationship between CF linewidth and electron correlation effects. We show that Kondo effect in heavy-fermion metal CeB_6 leads to consistent temperature dependence of CF linewidth and electric resistivity data from Ref. [5]. Ferroelectric insulator $TbInO_3$ is a spin-liquid (SL) candidate, but the symmetry of the proposed SL ground state remains unclear [4, 6]. We demonstrate that the ground state of Tb^{3+} ions is a non-Kramers doublet, which suggests a triangular-lattice spin liquid. We further relate the enhanced linewidth of its CF modes to the magnetic fluctuations near the SL ground state. Then, we examine the quasi-elastic fluctuations of CeB_6 and $YbRu_2Ge_2$. CeB_6 has an antiferroquadrupolar (AFQ) phase below 3.2 K [7]. Above the transition temperature, the tendency towards AFQ ordering induces magnetic fluctuations. The corresponding Raman susceptibility, and magnetic susceptibility data from Ref. [8] show consistent temperature dependence. $YbRu_2Ge_2$ has a ferroquadrupolar (FQ) phase below 10 K [9]. We present the quadrupolar fluctuations, and evaluate the electronic intersite quadrupolar coupling by analyzing the corresponding Raman susceptibility whose temperature dependence is consistent with that of quadrupole-strain susceptibility derived from elastoresistivity data of Ref. [9].

2. Crystal-field excitations

CeB_6 has a simple $4f^1$ Ce^{3+} electronic configuration; the $4f$ orbits of Ce^{3+} ions are split into sixfold $^2F_{5/2}$ and eightfold $^2F_{7/2}$ multiplets. The electronic configuration for $TbInO_3$ is $4f^8$; the lowest-lying ^7F term of Tb^{3+} ions is split into 7 multiplets, with J=6 having the lowest energy and J=0 having the highest energy. In Fig. 1 we show the inter-multiplet excitations for CeB_6 and $TbInO_3$. It is clear that SOC becomes larger with atomic number: for Ce^{3+} ion (Z=58) the energy difference between $^2F_{5/2}$ and $^2F_{7/2}$ multiplets is around 0.3 eV; for Tb^{3+} ion (Z=65) the energy separation between $^7F_6$ and $^7F_0$ is around 0.7 eV.

The CF potential leads to fine structure within each multiplet, and the CF level scheme of the lowest-energy multiplet is most relevant to the low-temperature physics of the system. For CeB_6, the $^2F_{5/2}$ multiplet is split into Γ_8 ground state and Γ_7 excited state, and one intra-multiplet CF excitation within $^2F_{5/2}$ multiplet is observed [Fig. 1(a)] (As the linewidth of this excitation is around 4 meV, the previously proposed 2.5 meV splitting of the Γ_8 ground state [10] does not contradict our data). For $TbInO_3$, the $^7F_6$ multiplet is split into 5 singlets and 4 doublets, resulting in a total of 9 CF transitions within $^7F_6$ multiplet [Fig. 1(c)]. Multiple scattering geometries need to be employed to identify the symmetry of the CF transitions and in turn that of the CF states; only a doublet ground state is consistent with the experimental results [11].

Although only the energy and symmetry of the CF excitations are required for determining the low-energy CF level scheme, the spectral linewidth of these excitations provides further information relevant to scattering and fluctuation effects. In CeB_6, the CF linewidth is observed to increase on cooling below the temperature at which the resistivity data from Ref. [5] show local minimum [Fig. 1(a) Inset]. Such behavior is a manifestation of Kondo effect, which shortens the lifetime of the excited CF states [12]. In $TbInO_3$, the enhanced linewidth of the CF transitions within the $^7F_6$ multiplet is associated with the the magnetic fluctuations near the spin-liquid ground state: the dynamics of the correlated yet nonordered magnetic moments manifests itself through the width of CF excitations.

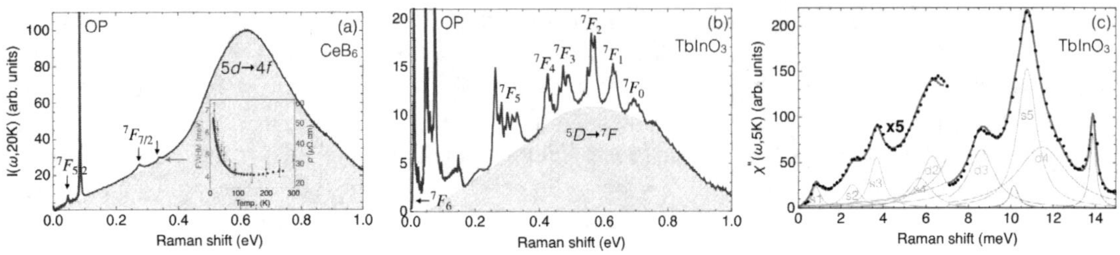

Figure 1. The inter- and intra-multiplet excitations for CeB_6 (a) and $TbInO_3$ (b-c). The Raman features are superposed on a broad photoluminescence continuum (PL, shaded with blue). For metallic CeB_6 the PL originates from the transition from broad $5d$ band bottom to narrow $4f$ band; for insulating $TbInO_3$ the PL results from the transition from ^5D term to ^7F term. The strong, sharp modes at low energy are optical phonons (OP). The inset of (a) compares the linewidth of the 0.34 eV CF excitation, and the electric resistivity taken from Ref. [5]. (c) The intra-multiplet CF transitions within the $^7F_6$ multiplet for $TbInO_3$. The Raman data are represented by black dots. The red lines represent the fits by Lorentzian lineshapes. The orange lines show doublet-to-singlet CF modes labeled by the final singlet state, and the green lines show doublet-to-doublet CF modes; the phonon modes are shown in purple. The Raman data and fitting curves below 7 meV are multiplied by a factor of 5 for clarity.

3. Quadrupolar fluctuations

In Fig. 2 we show the quasi-elastic Raman response for CeB_6 and $YbRu_2Ge_2$, which can be well modelled by a Drude lineshape. The corresponding Raman susceptibility is obtained by virtue of Kramers-Kronig relation. Due to the short wavevector of photons compared to the size of Brillouin zone, inelastic light scattering can only probe quadrupolar excitations at the Brillouin zone center. Thus for quadrupolar ordering at finite wavevector, including the case of AFQ order, the quadrupolar fluctuations are not observed in the corresponding symmetry channel. However, on cooling towards the AFQ transition temperature, long-wavelength fluctuations in the magnetic channel are observed for CeB_6 [Fig. 2(a)], suggesting development of ferromagnetic (FM) correlations. The FM correlations are induced by the tendency towards AFQ ordering: as the orbits at neighboring Ce sites are quasi-degenerate, the intra-atomic Hund's rule favors the spins to be aligned[12].

For $YbRu_2Ge_2$, the FQ phase is a realization of electronic nematic states since the electronic properties spontaneously break the fourfold rotational symmetry of the tetragonal crystal. Above the FQ transition temperature, fluctuations in the quadrupolar channel are enhanced on cooling [Fig. 2(c)]. A Curie-Weiss fit to the corresponding Raman susceptibility renders a Weiss temperature of -2±3 K [Fig. 2(d)], indicating that the susceptibility essentially follows Curie law, and the electronic intersite quadrupolar coupling is weak. It is the relatively strong coupling between the quadrupole moments and lattice strain fields, analogous to cooperative Jahn-Teller effect, that is essential for the FQ transition at finite temperature [13].

4. Conclusion

In summary, we use three $4f$-electron systems, CeB_6, $TbInO_3$, and $YbRu_2Ge_2$, as examples to show the key features and underlying physics of CF excitations and quadrupolar fluctuations. The symmetry and energy of low-energy CF transitions are crucial for establishing the CF level scheme of the lowest-lying multiplet; the linewidth of CF excitations provides information on electron correlation. At low temperature, the charge quadrupoles of $4f$ electrons tend to order in a uniform arrangement (FQ order) or a staggered arrangement (AFQ order). The quadrupolar

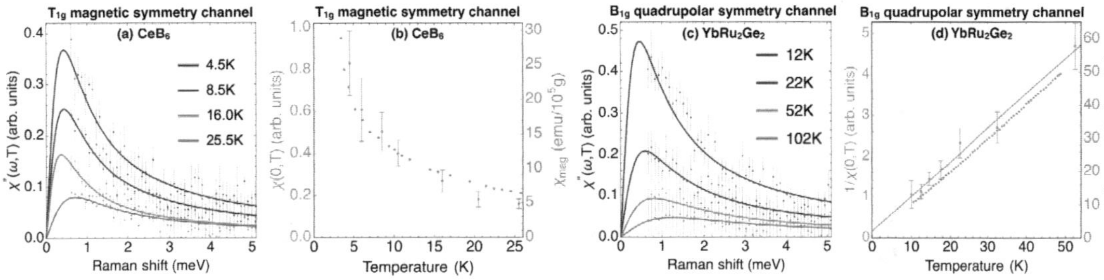

Figure 2. The quasi-elastic Raman response and the corresponding Raman susceptibility for CeB_6 and $YbRu_2Ge_2$. (a) Fluctuations in the magnetic symmetry channel for CeB_6. (b) The Raman susceptibility derived from the Raman response shown in (a) [left axis], compared with the magnetic susceptibility taken from Ref.[8] [right axis]. (c) Fluctuations in the quadrupolar symmetry channel for $YbRu_2Ge_2$. (d) The inverse Raman susceptibility derived from the Raman response shown in (c) [left axis], compared with the inverse quadrupole-strain susceptibility derived from elastoresistivity of Ref.[9] [right axis]. For (a) and (c), the curves represent Drude fits; in (d), the line represents Curie-Weiss fit.

fluctuations above the FQ transition temperature can be directly probed by inelastic light scattering, and the corresponding susceptibility provides the strength of the electronic intersite quadrupolar interactions. The AFQ fluctuations at finite wavevector are not directly accessible to Raman scattering, but the induced long-wavelength magnetic fluctuations can be probed.

Acknowledgments

The spectroscopic work was supported by NSF DMR-1709161 and DMR-2105001. The sample preparation for CeB_6 was supported by DOE Office of Basic Energy Sciences, Division of Materials Science and Engineering. The sample preparation for $TbInO_3$ was supported by DOE DEFG02-07ER46382. The sample preparation for $YbRu_2Ge_2$ was supported by Gordon and Betty Moore Foundation GBMF9068.

References

[1] Jensen J and Mackintosh A R 1991 *Rare Earth Magnetism: Structures and Excitations* (*The International Series of Monographs on Physics* no 81) (Oxford: Clarendon Press)

[2] Fulde P 2012 *Electron Correlations in Molecules and Solids* 3rd ed (*Springer Series in Solid-State Sciences* no 100) (Berlin: Springer)

[3] Willers T, Strigari F, Hu Z, Sessi V, Brookes N B, Bauer E D, Sarrao J L, Thompson J D, Tanaka A, Wirth S, Tjeng L H and Severing A 2015 *Proc. Natl. Acad. Sci. U.S.A* **112** 2384

[4] Clark L, Sala G, Maharaj D D, Stone M B, Knight K S, Telling M T F, Wang X, Xu X, Kim J, Li Y, Cheong S W and Gaulin B D 2019 *Nat.* Phys. **15** 262

[5] Takase A, Kojima K, Komatsubara T and Kasuya T 1980 *Solid State Commun.* **36** 461

[6] Kim M G, Winn B, Chi S, Savici A T, Rodriguez-Rivera J A, Chen W C, Xu X, Li Y, Kim J W, Cheong S W and Kiryukhin V 2019 *Phys. Rev.* B **100** 024405

[7] Cameron A S, Friemel G and Inosov D S 2016 *Rep. Prog. Phys.* **79** 066502

[8] Kawakami M, Kunii S, Komatsubara T and Kasuya T 1980 *Solid State Commun.* **36** 435

[9] Rosenberg E W, Chu J H, Ruff J P C, Hristov A T and Fisher I R 2019 *Proc. Natl. Acad. Sci. U.S.A* **116** 7232

[10] Zirngiebl E, Hillebrands B, Blumenröder S, Güntherodt G, Loewenhaupt M, Carpenter J M, Winzer K and Fisk Z 1984 *Phys. Rev.* B **30** 4052

[11] Ye M, Xu X, Wang X, Kim J, Cheong S W and Blumberg G 2021 *Phys. Rev.* B **104** 085102

[12] Ye M, Kung H H, Rosa P F S, Bauer E D, Fisk Z and Blumberg G 2019 *Phys. Rev.* Mater. **3** 065003

[13] Ye M, Rosenberg E W, Fisher I R and Blumberg G 2019 *Phys. Rev.* B **99** 235104

$CeNi_4Cr$ compound and its thermodynamic properties

Mariana Zapotoková, Marián Reiffers

Faculty of Sciences, University of Prešov, 17. novembra 1, SK 080 78 Presov, Slovakia

mariana.zapotokova@unipo.sk

Abstract. $CeNi_5$ as a mother compound is well known as a Stoner enhanced paramagnet characterized by the spin fluctuation contribution on its transport properties. Previous work on $CeNi_4Cr$ compound showed typical features of mixed valence systems with indication of tendency to heavy fermion behavior [1]. The theoretically predicted phase transition into antiferromagnetic order wasn´t observed down to 2 K. In this work we present the effect of spin fluctuations of the $CeNi_4Cr$ compound focused on its thermodynamic properties

1. Introduction

In previous works, Jain et al. [1] have studied the effect of partial substitution of Ni by Cr in the $CeNi_5$ intermetallic compound to search materials for hydrogen storage. The mother compound $CeNi_5$ crystallizes in the hexagonal $CaCu_5$ type, and it results to be a Stoner enhanced paramagnet characterized by a spin fluctuation (SF) contribution on its transport properties [2]. Murugan et al. [3] have revealed using first-principles density functional calculations that $CeNi_4Cr$ prefers the antiferromagnetic state. In our previous work [4] we have found that $CeNi_4Cr$ shows typical features of mixed valence systems with indication of tendency to heavy fermion behavior. The lack of a strong evidence of theoretically predicted phase transition into antiferromagnetic order may be due to Kondo screening of the magnetic moment. In order to study instability of the compound $CeNi_4Cr$ we have provided further investigation and analysis of heat capacity measurements with possible spin fluctuations contribution.

The measurements were carried out on a piece of the pollycrystalline $CeNi_4Cr$ compound prepared by induction melting of the constituent element under argon atmosphere. The $CeNi_4Cr$ compound crystallizes in the hexagonal $CaCu_5$-type structure, space group P6/mmm. The lattice constants are a = 4.931Å and c = 4.037Å. Heat capacity measurements were performed in Kosice, Slovakia by a Physical Property Measurement System (PPMS) commercial device (Quantum Design) in the temperature range 2-300 K using the two-τ model of the relaxation method in applied magnetic field up to 9 T.

2. Heat capacity

The temperature dependence of heat capacity of $CeNi_4Cr$ doesn´t show any sign of transition into magnetically ordered state in contradiction with predictions calculated by [3] – see Fig. 1. Moreover no influence of magnetic field on heat capacity dependence is evident [Fig.1].
In order to study spin fluctuation effect on heat capacity we used the formula (1)

$$C_p(T) = \gamma T + A^* T^3 ln\left(T/T_{sf}\right) + \frac{24{,}9416 + 0{,}05313 x_D^2 + 9{,}85.10^{-4} x_D^4 + 4{,}8.10^{-7} x_D^6}{1 + 0.0521 x_D^2 + 8{,}71.10^{-4} x_D^4 + 4{,}56.10^{-6} x_D^6 + 2.10^{-8} x_D^8} \quad (1)$$

Journal of Physics: Conference Series **2164** (2022) 012055 doi:10.1088/1742-6596/2164/1/012055

where first term corresponds to electronic contribution with Sommerfeld coefficient γ, second term corresponds to spin fluctuation contribution with coefficient A* and characteristic spin fluctuation temperature T_{sf} [5]. The third term is an expression of phonon contribution to heat capacity with $x_D = \theta_D/T$, θ_D is the Debye temperature, used by numerical analysis [6] and very good fits experimental data. The example of such fitting for compound $CeNi_4Cr$ is shown in Fig. 2. It is visible that fitting with included spin fluctuation term sets better (pink curve) than without it (dark blue – represents the curve with fixed parameters obtained from pink curve).

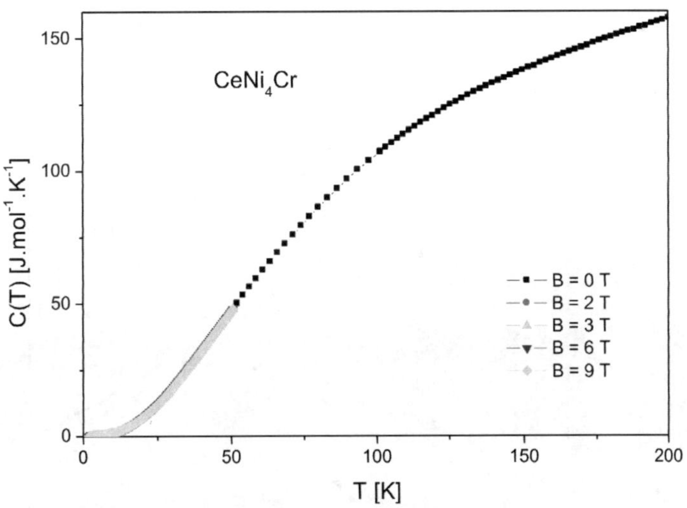

Fig. 1. Heat capacity dependences of $CeNi_4Cr$ compound in different applied magnetic field.

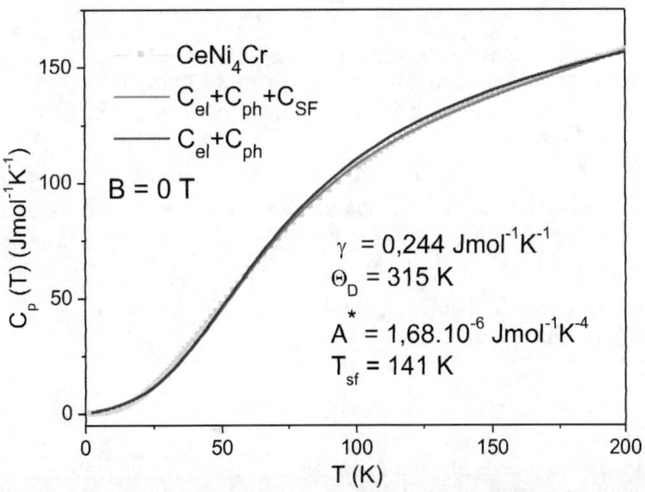

Fig. 2. Temperature dependence of heat capacity of $CeNi_4Cr$ compound, pink curve represents the fitting function (1) with the free parameters γ, θ_D, A*, T_{sf}, dark blue curve represents the fitting function with SF contribution with the fixed values γ, θ_D obtained from pink curve.

Magnetic contribution to heat capacity was obtained by extraction of heat capacity dependence of isostructural $LaNi_5$ crystallizing in CaCu5 type of structure, space group P_6/mmm. We analysed the magnetic contribution by means of Schottky effect using formula (2)

$$C_{mag}(T)/T = A^*T^2 ln(T/T_{sf}) + \frac{R}{T^3}\left[\frac{\sum_{i=0}^{n-1}\Delta_i^2 e^{-\Delta_i/T}}{\sum_{i=0}^{n-1} e^{-\Delta_i/T}} - \left(\frac{\sum_{i=0}^{n-1}\Delta_i e^{-\Delta_i/T}}{\sum_{i=0}^{n-1} e^{-\Delta_i/T}}\right)^2\right] \quad (2)$$

where R is gas constant, n denotes to the number of energy levels and Δ0=0. In the hexagonal symmetry the crystal field splits the Ce^{3+} ($J = 5/2$) multiplet into 3 dublets ($n=3$). Magnetic contributions to heat capacity of $CeNi_4Cr$ is shown in Fig. 3 (symbols). Fitting of Schottky using (2) without term of $A^*T^2ln(T/T_{sf})$ corresponding to SF contribution is represented by blue curve. The cyan curve represents fitting of Schottky contribution using (2). One can see that the fitting including spin fluctuation contributions fits better than separate Schottky effect.

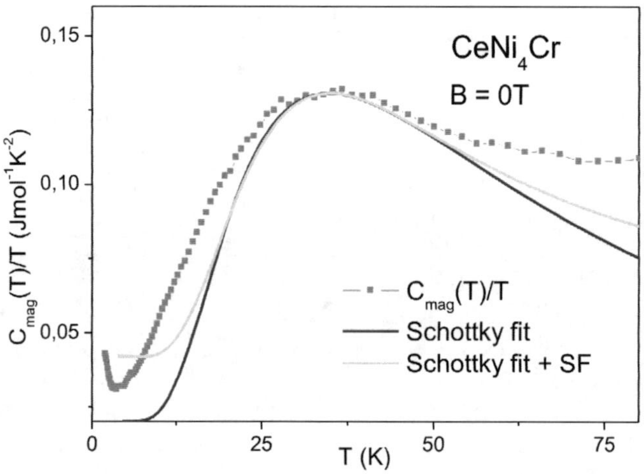

Fig. 3. Magnetic contribution to heat capacity as a function of temperature of $CeNi_4Cr$ (pink symbols). Cyan curve represents fitting using (2) with SF term, blue curve represents fitting using standard formula by means of (2) without SF term.

The obtained values of energy levels of CEF are in good agreement with theoretically expected values in [7] and with the values of similar compound $CeNi_4Si$ [8]. The coefficients A* and T_{sf} determined from magnetic contribution to heat capacity are in good agreement with the values determined from the total heat capacity. On the basis of [5] and classification of spin fluctuators therein we assigned $CeNi_4Cr$ compound to type 3 of spin fluctuators.

3. Conclusions

The temperature dependence of heat capacity of $CeNi_4Cr$ confirms no sign of transition into magnetically ordered state in agreement with [4]. Analysis C vs. T using formula with electronic, phonon and spin fluctuation contributions shows influence of spin fluctuation on the heat capacity dependence. Further analysing of magnetic contribution in means of Schottky effect and inclusion of spin fluctuation term has confirmed that spin fluctuation effect is not negligible. The level scheme of crystal electric field splitting is in good agreement with the scheme obtained in $CeNi_4Si$ [8].

References

[1] R. K. Jain, A. Jain, S. Agarwal, N.P. Lalla, V. Ganesan, D.M. Phase, I.P. Jain., J. Alloys Comp. 430 (2007) 165.
[2] M. Reiffers, M.D. Mea, E. Bauer, G. Pristaš, J. Magn. Magn. Mater. 272-276 (2004) 605-607Another reference
[3] P. Murugan, M.S. Bahramy, Y. Kawazoe, Phys. Rev. B 77 (2008) 064401
[4] T.Toliński et al., J. Magn. Magn. Mater. 321 (2009) 1121–1124
[5] K. Ikeda et al. , J. Magn. Magn. Mater. 100 (1991) 292-321
[6] P. Svoboda, P. Javorský et al. , Phys. Rev. B 63 (2001) 212408,
[7] S. Nasu, N.H. Neumann, N. Marzouk, R.S. Craig, W.E. Wallace, J. Phys. Chem. Solids 52 (1971) 2779-2783
[8] M. Falkowski, A. Kowalczyk, T. Toliński, G. Chełkowska, Materials Science Poland 25 (2007).

Signature of CEF-phonon coupling in Kondo lattice system $CeCuGa_3$

V K Anand[1,2], A Fraile[3] D T Adroja[1,4], Shivani Sharma[1], Rajesh Tripathi[1,5], C Ritter[6], C de la Fuente[7], P K Biswas[1], V Garcia Sakai[1], A del Moral[7] and A M Strydom[4]

[1]ISIS Facility, Rutherford Appleton Laboratory, Chilton, Didcot, Oxon, OX11 0QX, United Kingdom
[2]Department of Physics, University of Petroleum and Energy Studies, Dehradun, Uttarakhand, 248007, India
[3]Nuclear Futures Institute, Bangor University, Bangor, LL57 1UT, UK
[4]Highly Correlated Matter Research Group, Physics Department, University of Johannesburg, P.O. Box 524, Auckland Park 2006, South Africa
[5]Jawaharlal Nehru Centre for Advanced Scientific Research, Jakkur, Bangalore 560064, India
[6]Institut Laue-Langevin, Boite Postale 156, 38042 Grenoble Cedex, France
[7]Departamento de Fisica de Materia Condensada, Facultad de Ciencias, Universidad de Zaragoza, E-50009 Zaragoza, Spain

E-mail: vivekkranand@gmail.com, devashibhai.adroja@stfc.ac.uk

Abstract. We present the results of inelastic neutron scattering (INS) measurements on Kondo lattice heavy fermion $CeCuGa_3$. The low-temperature magnetic susceptibility exhibits an anomaly near 2.6 K associated with an antiferromagnetic transition which is further confirmed by the heat capacity measurement. The analysis of magnetic heat capacity $C_{\rm mag}(T)$ based on the crystal electric field (CEF) model reveals an overall CEF splitting of $\sim$ 21 meV. The INS data reveal two strong magnetic excitations near 4.5 and 6.9 meV, which, however failed to reproduce the observed $C_{\rm mag}(T)$. We therefore analyze the INS data by a model based on magneto-elastic (i.e. CEF-phonon) coupling which suggests that the excitations observed near 4.5 and 6.9 meV originate from the CEF-phonon coupling as observed in $CeCuAl_3$ and $CeAuAl_3$.

1. Introduction

Crystal electric field (CEF) effect plays an important role in dictating the physical properties of rare-earth intermetallic compounds. Among the rare-earth intermetallics, Ce-compounds are well known to present diverse physical properties [1-3]. The electronic ground state of a Ce-compound is determined by competing RKKY, Kondo and crystal electric field (CEF) interactions. Ce^{3+} ($4f^1$) with $J = 5/2$ has six-fold degenerate ground state multiplet which when subject to CEF environment (lower than a cubic point symmetry) splits into three doublets. As specified by Kramer's degeneracy theorem, being a Kramers ion, for Ce^{3+} one would expect only two CEF excitations from the ground state in the paramagnetic state for a low point symmetry. In this regard recent observations of three CEF excitations in some of the tetragonal Ce-compounds such as $CeCuAl_3$ [4] and $CeAuAl_3$ [5] are quite striking, and provide evidence for strong CEF-phonon coupling (magneto-elastic coupling) in these compounds.

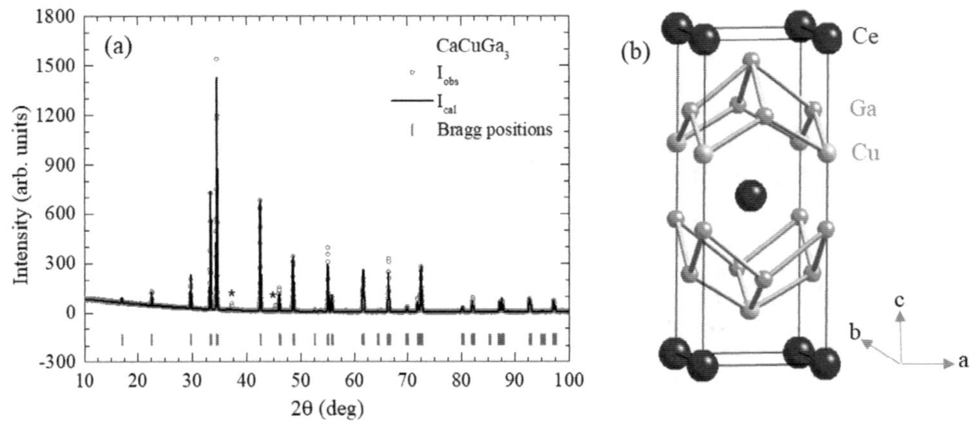

Figure 1. (a) Room temperature x-ray powder diffraction pattern of $CeCuGa_3$ along with the structural refinement profile for noncentrosymmetric $BaNiSn_3$-type tetragonal structure (space group $I4mm$). The short vertical bars indicate the Bragg peak positions. Asterisks mark the unidentified impurity peaks. (b) The unit cell representing the $BaNiSn_3$-type noncentrosymmetric tetragonal (space group $I4mm$) crystal structure of $CeCuGa_3$.

Here we present another tetragonal compound, namely $CeCuGa_3$, which shows signature of CEF-phonon coupling. Earlier investigations on Kondo lattice heavy fermion $CeCuGa_3$ [6-14] have shown a sample dependent ground state. Depending on Cu/Ga stoichiometry as well as crystal structure, different groups have reported different magnetic states, such as paramagnetic behavior down to 0.4 K [7], antiferromagnetic (AFM) ordering below 1.9 K [8, 10] and ferromagnetic ordering below 4 K [13]. It was suggested that it is the degree of Cu-Ga disorder that plays a key role in adapting the crystal structure, whether $ThCr_2Si_2$-type tetragonal structure ($I4/mmm$) or $BaNiSn_3$-type tetragonal structure ($I4mm$), and controls the ground state properties of $CeCuGa_3$.

2. Experimental

A polycrystalline sample of $CeCuGa_3$ was prepared by standard arc-melting on water-cooled copper hearth under inert atmosphere. In order to homogenize, the arc-melted button was flipped and re-melted several times, and annealed at 900° C for a week. The quality of sample was checked by room temperature x-ray diffraction (XRD) on powdered sample. Magnetic susceptibility χ vs temperature T was measured by a commercial superconducting quantum interference device (SQUID) magnetometer (MPMS, Quantum Design). The heat capacity $C_p(T)$ was measured by relaxation method using a physical properties measurements system (PPMS, Quantum Design). The inelastic neutron scattering measurement was performed on the MARI time of flight spectrometer at the ISIS Facility, UK. The powdered sample was wrapped in thin Al-foils and mounted inside a thin-walled cylindrical Al-can which was then cooled down to 4.7 K in a top-loading closed cycle refrigerator with He-exchange gas. The INS data were collected at 4.7 K and 100 K using neutrons with incident energies $E_i = 15$ and 40 meV.

3. Results and Discussion

Crystallography and the sample quality were checked by room temperature x-ray powder diffraction. The powder XRD pattern is shown in Fig. 1(a). The XRD data were refined by Rietveld refinement using the software FullProf. The structural Rietveld refinement profile fit for the noncentrosymmetric $BaNiSn_3$-type tetragonal structure (space group $I4mm$) is shown

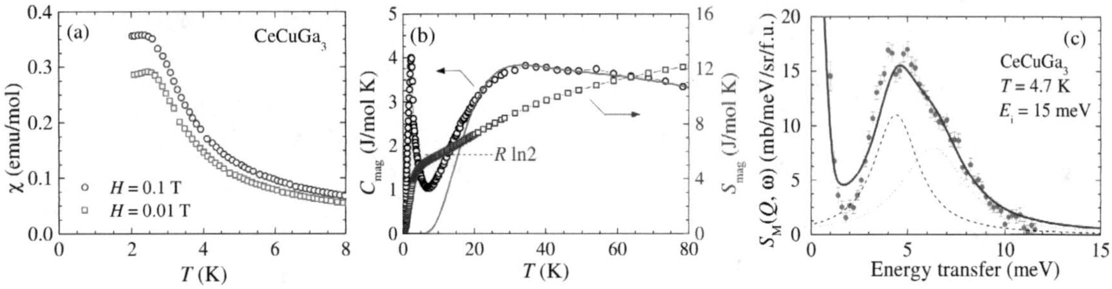

Figure 2. (a) The low temperature magnetic susceptibility χ versus temperature T of $CeCuGa_3$ measured in applied fields $H = 0.01$ and 0.1 T. (b) Magnetic contributions to heat capacity $C_{\rm mag}(T)$ and magnetic entropy $S_{\rm mag}(T)$. The solid red curve represents the Schottky contribution to $C_{\rm mag}$ due to crystal electric field (CEF). (c) Q-integrated ($0 \le Q \le 2.5$ Å^{-1}) inelastic magnetic scattering intensity $S_{\rm M}(Q,\omega)$ versus energy transfer E measured with $E_i =$ 15 meV at 4.7 K. The solid blue curve represents the fit of the INS data based on magneto-elastic CEF-phonon coupling model. The dashed and dotted curves represent the two excitations.

in Fig. 1(a). An almost single phase nature of sample is inferred from the XRD data and fit. The XRD also shows the presence of a tiny amount of unidentified impurity in the sample, the impurity peaks are marked with asterisks in Fig. 1(a). The refinement yielded the lattice parameters $a = 4.2533(1)$ Å and $c = 10.4497(2)$ Å which are in very good agreement with the literature values [10]. In $BaNiSn_3$-type tetragonal structure the atoms Ce, Cu, Ga1 and Ga2 occupy the Wyckoff positions 2a $(0,0,0)$, 2a $(0,0,z_{\rm Cu})$, 2a $(0,0,z_{\rm Ga1})$, and 4b $(0,0,z_{\rm Ga2})$, respectively. The values of the z-coordinates obtained from the refinement are $z_{\rm Cu} = 0.597(2)$, $z_{\rm Ga1} = 0.373(2)$, and $z_{\rm Ga2} = 0.259(2)$. The $BaNiSn_3$-type noncentrosymmetric tetragonal structure of $CeCuGa_3$ is shown in Fig. 1(b). As can be seen from Fig. 1(b), this structure lacks an inversion symmetry along the c-axis.

Figure 2(a) shows the low-T $\chi(T)$ data measured in magnetic fields $H = 0.01$ and 0.1 T. At $H = 0.01$ T $\chi(T)$ exhibits an anomaly near 2.6 K which shifts to 2.5 K as H is increased to 0.1 T, suggesting an antiferromagnetic transition. Above 100 K, the $\chi(T)$ follows a modified Curie-Weiss behavior $\chi = \chi_0 + C/(T - \theta_{\rm p})$ with $\chi_0 = 4.8 \times 10^{-4}$ emu/mol, effective moment $\mu_{\rm eff} = 2.37\,\mu_{\rm B}$ and Weiss temperature $\theta_p = -18(2)$ K [15]. The value of $\mu_{\rm eff}$ is close to the theoretical value of $2.54\,\mu_{\rm B}$/Ce expected for Ce^{3+} ions. An AFM transition is also inferred from the heat capacity data which shows a sharp λ-type anomaly near 2.3 K. The magnetic contribution to heat capacity $C_{\rm mag}(T)$ obtained after subtracting the phonon contribution (equivalent to heat capacity of nonmagnetic reference $LaCuGa_3$ [15]) is shown in Fig. 2(b) along with the magnetic entropy $S_{\rm mag}(T)$. The AFM ordering is further confirmed by muon spin relaxation and neutron powder diffraction (NPD) measurements [15]. The NPD reveals an incommensurate magnetic structure described by wave vector $\mathbf{k} = (0.148, 0.148, 0)$ corresponding to a longitudinal spin density wave with ordered moments in the ab-plane [15]. In addition to λ-type anomaly, $C_{\rm mag}(T)$ also presents a broad Schottky-type anomaly associated with crystal electric field. We analyzed the $C_{\rm mag}(T)$ by a three-level CEF scheme [16, 17] [see Fig. 2(b)] which suggests that with an oveall splitting of 240 K, the first excited doublet lies 70 K above the ground state. The splitting energy of 70 K obtained so is consistent with $S_{\rm mag}(T)$ which attains a value of $R\ln 4$ near 66 K [Fig. 2(b)]. The CEF level scheme deduced so is consistent with the one obtained from the analysis of single crystal $\chi(T)$ by Joshi *et al.* [13].

Figure 2(c) shows the Q-integrated ($0 \le Q \le 2.5$ Å^{-1}) one-dimensional energy cut of magnetic contribution $S_M(Q,\omega)$ to INS response of $CeCuGa_3$ at 4.7 K. A relatively broad CEF excitation consisting of two closely situated magnetic excitations is seen. We analyzed the INS data by a model based on crystal field, the CEF Hamiltonian for Ce^{3+} ions in tetragonal symmetry is given by $H_{\rm CEF} = B_2^0 O_2^0 + B_4^0 O_4^0 + B_4^4 O_4^4$, where B_n^m represent the CEF parameters and O_n^m the Stevens operators [18]. Our analysis of INS data by this model suggests two closely situated CEF excitations at 4.5 and 6.9 meV. The overall splitting of 6.9 meV in this model is much lower than the value 20.7 meV (240 K) inferred above from the analysis of $C_{\rm mag}(T)$. Joshi *et al.* [13] also suggested an overall splitting of 19.6 meV (228 K) from the analysis of $\chi(T)$. Further, the CEF parameters obtained fail to explain the $\chi(T)$ and $C_{\rm mag}(T)$ data. This inconsistency suggests that the 4.5 and 6.9 meV excitations do not originate only from single-ion CEF transition, and very likely results from the splitting of a CEF level due to CEF-phonon coupling. We therefore analyzed the INS data by a model based on magneto-elastic (CEF-phonon) coupling model, i.e., by including the magneto-elastic as well as CEF-phonon terms to the Hamiltonian, such that $H_{\rm tot} = H_{\rm CEF} + H_{\rm CEF-ph} + H_{\rm MEL}$, with $H_{\rm CEF-ph} = \hbar\omega_0(a_u^+ a_u + 1/2)$ and $H_{\rm MEL} = M^\gamma(a_u + a_u^+)O_u$ [4, 15]. The analysis of INS data by this model suggests three CEF excitations at 4.6, 7.2 and 28.2 meV. The calculated intensity of 28.2 meV excitation is very weak, which also agrees with our 40 meV INS data [15]. The fit of INS data by this model is shown by solid blue curve in Fig. 2(c) along with the contributions from individual excitations at 4.6 and 7.2 meV. The CEF parameters obtained from the CEF-phonon model also explain very well the anisotropic behaviour of single crystal $\chi(T)$, and $C_{\rm mag}(T)$ data [15].

4. Conclusion

We have investigated the crystal field states of $CeCuGa_3$ using inelastic neutron scattering and heat capacity measurements. The analysis of INS data by a model based on CEF suggests two excitations near 4.5 and 6.9 meV. On the other hand, $C_{\rm mag}(T)$ suggests an overall CEF splitting of 20.7 meV. We therefore analyzed the INS data by a model based on magneto-elastic (CEF-phonon) coupling which suggests that the 4.5 and 6.9 meV excitations originate from CEF-phonon coupling similar to the homologue compounds $CeCuAl_3$ and $CeAuAl_3$.

References

[1] Stewart G R 1984 *Rev. Mod. Phys.* **56** 755; Stewart G R 2001 Rev. Mod. Phys. **73** 797; Stewart G R 2006 Rev. Mod. Phys. **78**, (2006) 743.
[2] Löhneysen H v, Rosch A, Vojta M and Woelfle P 2007 *Rev. Mod. Phys.* **79** 1015
[3] Pfleiderer C 2009 *Rev. Mod. Phys.* **81** 1551
[4] Adroja D T, Moral A del, Fuente C de la, Fraile A, Goremyhkin E A, Taylor J W, Hillier A D and Fernandez-Alonso F 2012 *Phys. Rev. Lett.* **108** 216402
[5] Čermák P, Schneidewind A, Liu B, Koza M M, Franz C, Schönmann R, Sobolev O and Pfleiderer C 2019 *Proc. Natl. Acad. Sci.* **116** 6695
[6] Sampathkumaran E V and Das I 1992 *Solid State Commun.* **81** 901
[7] Mentink S A M, Bos N M, van Rossum B J, Nieuwenhuys G J, Mydosh J A and Buschow K H J, J. Appl.Phys. **73**, 6625 (1993).
[8] Martin J M, Paul D M, Lees M R, Werner D and Bauer E, J. Magn. Magn. Mater. **159**, 223 (1996).
[9] Aoyama S, Ido H, Nishioka T, Kontani M 1996 *Czechoslov. J. Phys.* **46** 2069
[10] Martin J M, Lees M R, Paul D M, Dai P, Ritter C and Bi Y J 1998 *Phys. Rev. B* **57** 7419
[11] Kontani M, Motoyama G, Nishioka T and Murase K 1999 *Physica B* **259**, 24
[12] Oe K, Kawamura Y, Nishioka T, Kato H, Matsumura M and Kodama K 2010 *J. Phys.: Conf. Ser.* **200** 012147
[13] Joshi D A, Burger P, Adelmann P, Ernst D, Wolf T, Sparta K, Roth G, Grube K, Meingast C and Löhneysen H v 2012 *Phys. Rev. B* **86** 035144
[14] Przybylski P, Pikul A P, Kaczorowski D, Wiśniewski P 2014 *J. Phys. Chem. Solids* **75** 1284
[15] Anand V K *et al.* (to be published)
[16] Layek S, Anand V K and Hossain Z 2009 *J. Magn. Magn. Mater.* **321** 3447
[17] Anand V K, Adroja D T, Britz D, Strydom A M, Taylor J W and Kockelmann W 2016 *Phys. Rev. B* **94** 014440
[18] Stevens K W H 1952 *Proc. Phys. Soc. London A* **65** 209

DFT+U study on the multipolar nematic state of nonmagnetic FeSe

Takemi Yamada and Takami Tohyama

Department of Applied Physics, Faculty of Science, Tokyo University of Science, Tokyo, Japan

E-mail: t-yamada@rs.tus.ac.jp

Abstract. In this study, we investigate the electronic structure of FeSe in the normal and nematic states based on the DFT+U method in combination with the multipole analyses, and discuss their microscopic properties and origin. In the normal state, the topological change of the Fermi surface occurs before the nematic transition with increasing the on-site Coulomb interaction U. The resulting nematic ground state is a multipolar state having both hexadecapoles in the E-representation and multipoles in the B_2-representation on each Fe site, where the appearance of the E-type multipoles induces non-trivial spontaneous orbital hybridizations between d_{xy}, $d_{x^2-y^2}$ and d_{xz}/d_{yz} orbitals. Such the spontaneous orbital mixings due to E-type multipoles is inherent in the FeSe and related systems, and is expected to play an important role in the formation of the nematic states in other iron-based superconductors.

1. Introduction

The iron-based superconductor FeSe [1] has been known to exhibit the so-called "nematic state", in which the rotational symmetry is broken while keeping the translational symmetry, below the structural phase transition temperature T_S=90 K [2]. The splitting of the d_{xz}/d_{yz} degenerate orbitals in the electronic bands measured in the angle-resolved photoemission spectroscopy (ARPES) experiment is much larger than that expected from the lattice distortion, suggesting that the nematic state is of electronic origin [3]. In addition, unlike other iron-based superconducting materials, there is no magnetic ordering, while the orbital fluctuations are enhanced towards T_S, suggesting the possibility of orbital-dependent correlation effects [4] and BCS-BEC crossover [5], which has gathered much attention.

Recent intensive search for low-energy states of FeSe in the ARPES experiments [6, 7, 8] has revealed the detailed band structures, which are essentially compensation metals involving d_{xz}, d_{yz}, and d_{xy} orbitals. The Fermi surfaces (FSs) of the normal state, consisting of two hole FSs (h-FSs) around the Γ point and two electron FSs (e-FSs) around the M point, are much smaller than that of the density functional theory (DFT) calculation [7]. On the other hand, for the nematic state, the band splitting due to the d_{xz}-d_{yz} orbital ordering are observed and results in a single hole and electron surface [8]. Furthermore, a one-side gap opening has been observed, where the gap opens in the M-Σ_y direction while remains unchanged in the M-Σ_x direction [8].

More recently, it has been reported that symmetry broken nematic states can be obtained from the first-principles calculations by adjusting the initial charge distribution, raising a lot of interest [9]. One of them called as the γ nematic state is particularly noteworthy in that it reproduces not only the d_{xz}-d_{yz} band splitting but also the one-side gap opening behavior, which seems to correspond well to experiment. However, such symmetry-breaking DFT calculations are few and far between, and it is important to confirm them by other reliable methods.

In this study, we aim to clarify the normal and nematic states of nonmagnetic FeSe by performing DFT+U calculations based on the all-electron full-potential code, WIEN2k [10]. In addition to the detailed analysis of the changes in the normal state up to the nematic state, we have evaluated the order parameters in the nematic state without any assumptions from the multipole moment calculations to elucidate the mechanism of the nematic state [11].

2. Method and Formulation

In the present DFT+U calculation, we adopt the self interaction correction, corresponding to the fully localized atomic limit, and determine the charge density self-consistently by solving the Kohn-Sham equation with the DFT+U potential $v_U(\boldsymbol{r})$ which is explicitly given by,

$$v_U(\boldsymbol{r}) = \sum_{\nu}\sum_{mm'}\sum_{\sigma\sigma'} v_{mm'}^{\sigma\sigma',\nu} |\nu m\sigma\rangle \langle \nu m'\sigma'| , \tag{1}$$

$$v_{mm'}^{\sigma\sigma',\nu} = \delta_{mm'}\delta_{\sigma\sigma'}\frac{U}{2}(1-n_m^{-\sigma,\nu}) + (1-\delta_{mm'}\delta_{\sigma\sigma'})(-Un_{mm'}^{\sigma\sigma',\nu}), \tag{2}$$

where $|\nu m\sigma\rangle$ denotes the local basis set with Fe atom ν=Fe1,Fe2 and orbital m and spin σ, and $v_{mm'}^{\sigma\sigma',\nu}$ acts as an effective local potential including the Coulomb interaction U in the muffin-tin radius $R_{\rm MT}^{\nu}$, the occupation number $n_m^{\sigma,\nu}$ and the density matrix $n_{mm'}^{\sigma\sigma',\nu}$. Here we restrict the magnetic moment to be zero, keeping in mind that FeSe is nonmagnetic and drop the spin index σ, yielding the relation: $n_{mm'}^{\uparrow,\nu}$=$n_{mm'}^{\downarrow,\nu}$=$n_{mm'}^{\nu}$ and n_{mm}^{ν}=$n_m^{\nu}/2$. In self-consistent field (SCF) calculation, we use the experimental lattice parameters and $\boldsymbol{k}$-mesh of $38^2\times26$ and $R_{\rm MT}^{\rm Fe(Se)}=2.25$ (2.14) bohr with $R_{\rm MT}K_{\rm max}$=8-9.

In order to obtain the symmetry broken nematic states, first, we perform SCF calculations with a structure in which Fe coordinates are shifted by $\boldsymbol{\delta}=\delta(1,1,0)$ and δ=$O(10^{-2})$. Next, by using the previous converged charge density as the new initial charge density for the undistorted structure with $\boldsymbol{\delta}=\boldsymbol{0}$, we can obtain the symmetry broken nematic states as done in Ref. [9]. The technical details are presented in the Supplemental material of Ref. [11].

3. Results

First we investigate the change of the normal states by plotting the band energy $\varepsilon_{\boldsymbol{k}n}$ at Γ, Z (near the h-FSs) and M points (near the e-FSs) as shown in Figs.1 (a), (b) and (c), respectively, where the Fermi energy $\varepsilon_{\rm F}$=0 eV. When U=0 eV, we have obtained $\varepsilon_{\boldsymbol{k}n}>0$ at Γ and Z points as shown in Figs.1 (a) and (b), yielding the three h-FSs consisting of the 28th, 29th, and 30th bands as shown in our previous study [11] and the DFT calculation of FeSe [7].

With increasing U, the 29th and 30th bands from the degenerated d_{xz}, d_{yz} orbitals increase at Γ point, while the 28th band, mainly from the d_{xy} orbital, decreases, and a Lifshitz transition occurs at $U\simeq2$ eV (=$U_{\rm LT}$), where the most inner hole FS consisting of the 28th band disappears. Similar behavior is obtained at Z point, where the 28th (30th, 31st) band from the d_{xy} (d_{xz}, d_{yz}) orbital decreases (increases), but the 29th band from the Se-p_z orbital also decreases and falls below $\varepsilon_{\rm F}$ for a second Lifshitz transition as shown in Figs.1 (b). On the other hand, at M points, e-FSs consisting of the 31st and 32nd bands with d_{xz}, d_{yz} orbitals approach $\varepsilon_{\rm F}$ monotonically with increasing U and does not undergo any topological change in contrast to the h-FSs as shown in Figs.1 (c), where $\varepsilon_{\boldsymbol{k}n}$ of A point is almost the same as that of M point.

Next, the partial density-of-states (pDOS) $\rho_m(\varepsilon)$ for U=0 eV and 3 eV are plotted in Fig.1 (f) and (g), respectively, indicating that the low-energy states are dominated by d_{xy} and d_{xz}/d_{yz} orbitals. As shown in Fig.1 (f) for U=0 eV, $\rho_{xy}(\varepsilon)$ ($\rho_{xz/yz}(\varepsilon)$) has a distinct shoulder (a clear peak) structure around +0.1 eV (−0.3 eV) originating from the bands in the Γ-Z (M-A) direction,

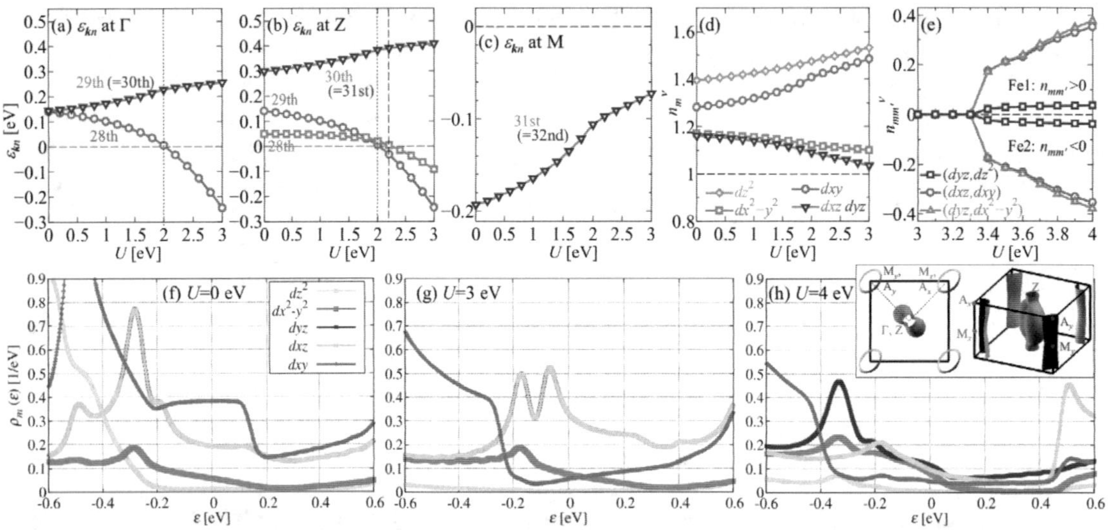

Figure 1. (a)-(e) U-dependence of (a)-(c) the band energy $\varepsilon_{\boldsymbol{k}n}$ at (a) Γ, (b) Z and (c) M, where the vertical lines in (a) and (b) are drawn as a guide. (d) The occupied electron number n_m^ν in the normal states and (e) the off-diagonal density matrix $n_{mm'}^\nu$ in the nematic states. (f)-(h) The pDOS for (f) U=0 eV, (g) 3 eV and (h) 4 eV. The inset of (h) shows the FSs for U=4 eV.

and $\rho_{xy}(0)$ ($\rho_{xz/yz}(0)$) has the largest (second largest) value of $\rho_{xy}(0) \sim 0.4$ eV^{-1} ($\rho_{xz/yz}(0) \sim 0.2$ eV^{-1}). With increasing U, $\rho_{xy}(\varepsilon)$ near ε_F decreases and its shoulder structure drops below ε_F for $U > U_{LT}$, while $\rho_{xz/yz}(\varepsilon)$ near ε_F increases and its peak approaches ε_F. For U=3 eV, the shoulder of $\rho_{xy}(\varepsilon)$ reaches around -0.24 eV as shown in Fig.1 (g), and $\rho_{xz/yz}(\varepsilon)$ becomes a two-peak structure keeping the degeneracy, where the shallower peak at -0.07 eV and the deeper peak at -0.17 eV are due to a flatband-like degeneracy of the M-A direction and a slight dip towards low-energy side of the same bands of the M-G and A-Z directions, respectively.

These orbital-dependent changes are caused by the occupied electron number n_m^ν, where n_{xy} and n_{z^2} are larger than $n_{xz/yz}$ and $n_{x^2-y^2}$, and the former (latter) increase (decrease) with increasing U as shown in Fig.1 (d), giving rise to a strong downward energy level shift for the d_{xy} orbital as compared to the $d_{xz/yz}$ orbital. As a result, the Lifshitz transition to the two h-FSs and e-FSs accompanying with the dominant orbital switching from the d_{xy} orbital to the d_{xz}/d_{yz} orbital can be commonly realized in the FeSe and related system as long as $n_{xy} > n_{xz/yz}$.

Under such the Lifshitz transitioned normal states for $U > U_{LT}$, we have found that further increase of U stabilizes the symmetry broken nematic states above U=3.4 eV, as already shown in our previous study [11], where a multipolar nematic state becomes ground state with the hexadecapoles ($Q_{4\alpha,x}$, $Q_{4\alpha,y}$, $Q_{4\beta,x}$, $Q_{4\beta,y}$) and multipoles (O_{xy}, $Q_{4\beta,z}$) corresponding to the E and B_2 irreducible representation of the D_{2d} point group at each Fe site (see table in Ref. [11]).

Here we focus on the E-type multipoles dominating over the ordered states in order to clarify their roles for the nematic states, since they are not well understood compared to the B_2-type multipoles inducing the d_{xz}-d_{yz} orbital splitting. We have observed spontaneous orbital hybridizations above the nematic transition U=3.4 eV from the density matrix $n_{mm'}^\nu$ for the E-type multipoles as a function of U as shown in Fig.1 (e), where two dominant components of ($d_{yz}, d_{x^2-y^2}$) and (d_{xz}, d_{xz}) have the same sign and almost the same value at each Fe site and it is also clear that the sign of $n_{mm'}^{\mathrm{Fe1}}$ and $n_{mm'}^{\mathrm{Fe2}}$ is reversed, indicating an antiferro multipole order. The corresponding E-type multipoles are found to be the quadrupole O_{yz} and hexadecapole $Q_{4\beta,x}$ by taking x, y in the Fe-Fe direction in contrast to the previous study [11], and their expression are given with the density matrix $n^{\nu}_{mm'}$ which are explicitly written by,

$$O_{yz} = \sqrt{\frac{6}{7}}\left(n^{\nu}_{1,2} - n^{\nu}_{3,4} - \frac{1}{\sqrt{3}}n^{\nu}_{3,5}\right),\ Q_{4\beta,x} = \frac{3}{\sqrt{14}}\left(\frac{4}{3}n^{\nu}_{1,2} + n^{\nu}_{3,4} - \frac{1}{\sqrt{3}}n^{\nu}_{3,5}\right), \quad (3)$$

where 1:d_{xy} 2:d_{xz} 3:d_{yz} 4:$d_{x^2-y^2}$ 5:d_{z^2}. From Eq. (3), it is clear that the hexadecapole $Q_{4\beta,x}$ (quadrupole O_{yz}) becomes large (small) values, due to the combination (cancellation) of the density matrices $n^{\nu}_{1,2}$ and $n^{\nu}_{3,4}$. It is worth noting that the present orbital hybridization via the density matrices has the intrinsic property of inducing the higher-order multipoles.

Finally, the pDOS and FSs of the nematic state for U=4 eV are shown in Fig.1 (h). The degenerated pDOS $\rho_{xz/yz}(\varepsilon)$ for U=3 eV [see Fig.1 (g)] is clearly separated, with the peak of $\rho_{yz}(\varepsilon)$ ($\rho_{xz}(\varepsilon)$) splitting to lower (higher) energy due to the B_2-multipole order [11]. In addition to this d_{xz}-d_{yz} splitting, we have obtained a remarkable suppression of $\rho_m(\varepsilon)$ near $\varepsilon_{\rm F}$ involving $\rho_{xy}(\varepsilon)$ and $\rho_{x^2-y^2}(\varepsilon)$, which is responsible for the energy gain of the nematic state by forming a partial gap-like structure around 0.2 eV$\sim$0.4 eV due to the E-type multipole mixing. The corresponding FSs shown in the inset of Fig.1 (h) are consistent with the experiment in terms of the single h-FS (Γ) and e-FS (M) with the asymmetric shape along $\rm M_x$-Γ-$\rm M_y$ direction [8]. However, the obtained size of the FSs does not agree with the experiment, and it is an important issue to develop the theory to incorporate more advanced many-body correlation effects.

4. Summary

In summary, we have investigated the normal and nematic states of FeSe using the DFT+U method together with multipole analyses. We have found that the normal state undergoes a change in electronic state associated with the Lifshitz transition, and the nematic ground state is a coupled multipole state with the spontaneous hybridization due to the E-type hexadecapole. The obtained results are consistent with the ARPES [8] and the previous DFT [9], indicating that the multipolar nematic state is a universally stabile solution in the FeSe system.

Acknowledgments

This work was supported by the "Quantum Liquid Crystals" No. JP19H05825 KAKENHI on Innovative Areas from JSPS of Japan.

References

[1] Hsu F C, Luo J Y, Yeh K W, Chen T K, Huang T W, Wu P M, Lee Y C, Huang Y L, Chu Y Y, Yan D C and Wu M K 2008 *Proc. Natl. Acad. Sci. U.S.A.* **105** 14262
[2] Shibauchi T, Hanaguri T and Matsuda Y 2020 *J. Phys. Soc. Jpn.* **89** 102002
[3] Shimojima T, Suzuki Y, Sonobe T, Nakamura A, M Sakano J O, Yoshioka K, Kuwata-Gonokami M, Ono K, Kumigashira H, Böhmer A E, Hardy F, Wolf T, Meingast C, v Löhneysen H, Ikeda H and Ishizaka K 2014 *Phys. Rev. B* **90** 121111(R)
[4] Lanată N, Strand H U R, Giovannetti G, Hellsing B, de' Medici L, and Capone M 2013 *Phys. Rev. B* **87** 045122
[5] Kasahara S, Sato Y, Licciardello S, Ćulo M, Arsenijevič S, Ottenbros T, Tominaga T, Böker J, Eremin I, Shibauchi T, Wosnitza J, Hussey N E and Matsuda Y 2020 *Phys. Rev. Lett.* **124** 107001
[6] Suzuki Y, Shimojima T, Sonobe T, Nakamura A, Sakano M, Tsuji H, Omachi J, Yoshioka K, Kuwata-Gonokami M, Watashige T, Kobayashi R, Kasahara S, Shibauchi T, Matsuda Y, Yamakawa Y, Kontani H and Ishizaka K 2015 *Phys. Rev. B* **92** 205117
[7] Watson M D, Kim T K, Haghighirad A A, Davies N R, McCollam A, Narayanan A, Blake S F, Chen Y L, Ghannadzadeh S, Schofield A J, Hoesch M, Meingast C, Wolf T and Coldea A I 2015 *Phys. Rev. B* **91** 155106
[8] Yi M, Pfau H, Zhang Y, He Y, Wu H, Chen T, Ye Z R, Hashimoto M, Yu R, Si Q, Lee D H, Dai P C, Shen Z X, Lu D H and Birgeneau R J 2019 *Phys. Rev. X* **9** 041049
[9] Long X, Zhang S, Wang F and Liu Z 2020 *npj Quantm Mater.* **5** 50
[10] Blaha P, Schwarz K, Tran F, Laskowski R, Madsen G K H and Marks L D 2020 *J. Chem. Phys.* **152** 074101
[11] Yamada T and Tohyama T 2021 *arXiv:2107.07244* Accepted in PRB

Charge and crystal field electronic states in $CeRh_2Si_2$ and Ba_2IrO_4

R. J. RADWANSKI[1,2], **D. M. NALECZ**[1] **and Z. ROPKA**[2]
[1]Institute of Physics, Pedagogical University, 30-084 Krakow, Poland
[2]Center of Solid State Physics, Snt Filip 5, 31-150 Krakow, Poland
E-mail: `sfradwan@cyf-kr.edu.pl`

Abstract. We point out the scientific importance of the increasing evidence for the existence of the discrete low-energy meV-energy states in compounds containing atoms with incomplete 4f, 5f, 3d shells and recently in iridates (5d shell). We point out that the realized charge state of the 3d/4f/5f/4d/5d atoms/ions in oxides is very close to the formal one obtained within the ionic model. In Ba_2IrO_4 the charge states are very close to Ba^{2+}, Ir^{4+} and O^{2-}. In $CeRh_2Si_2$, which is intermetallic and considered as a Kondo-lattice antiferromagnet, by detailed analysis of temperature dependence of the specific heat we have proved the realization of the trivalent charge state of practically **all** cerium ions. In the magnetic state, below T_N of 36 K, the Kramers doublet becomes split and there appears a spin gap growing to 6 meV at T= 0 K.

1. Introduction

Description of magnetic and electronic properties of compounds containing $3d$-/$4f$-/$5f$-open-shell atoms has been very vital part of the condensed-matter physics in last fifty years. The interest has been associated with such topics as Mott insulators, NiO [1], $FeBr_2$ [2] or $LaCoO_3$ [3], and heavy-fermion anomalous phenomena occuring in intermetallics with Ce, Yb and uranium (UPd_2Al_3 [4]) atoms. These anomalous properties are very different from classical behavior observed in, e.g. $ErNi_5$ [5], Ho_2Co_{17} [6], $Nd_2Fe_{14}B$ [7] intermetallics. $CeRh_2Si_2$ is considered as a Kondo-lattice antiferromagnet with T_N of 36 K [8]. Recently iridates with open $5d$ shell like Sr_2IrO_4 [9] and Ba_2IrO_4 [10], enter to the experimental and theoretical field.
The aim of the present contribution is a problem of formal and realized charge states in compounds containing open-shell atoms/ions.

2. Formal and real charge state in Ba_2IrO_4

Each theoretical study practically starts with an assumption/a concept about the charge state of the open-shell atoms constituting a considered compound on basis of experimental facts about its electronic and magnetic properties. In literature there is discussion about the role of $3d$-/$4f$-/$5f$-/$4d$-/$5d$− (we sometimes will call all these atoms/ions as transition-metal atoms/ions) open-shell electrons - should they be described as itinerant or localized? We are strongly localized. In fact, such question should be formulated rather as the localized or the itinerant/band model is a better starting point for the physically adequate theoretical description of electronic and magnetic properties, remembering also about different temperature regions.

Our description of Ba_2IrO_4 starts with the ionic model. It demands the tetravalent Ir charge state. Such starting point, quite obvious to experimentalists, is in sharp contrast to results of a quite recent publication [11] in Phys. Rev. B in which authors have claimed that in Ba_2IrO_4 the valency of Ba is 1.5+ (in contrast to the expected charge state of 2+) and iridium of 2.32+ only (instead of the expected 4+). We are strongly localized, i.e. we think that a number of valence electrons in the open electronic shell should be close to the integer number. Authors of the recalled paper have strongly stated that accepting the charge state in oxides on basis of the ionic model, called by these authors as "naive", is completely erroneous as the real charge state is largely different. Authors of the recalled paper, in the developed $d-p$ model, have assumed substantial oxygen hybridization and very strange barium valence as $Ba^{1.5+}$. In a consequence they rejected "the idealistic formal ionic model" accepting $Ba^{1.5+}$ and $Ir^{2.32+}$ ionic states.

The publication of so exotic theoretical results in Phys. Rev. B proves the scientific necessity of the formulation of reasonable and physically adequate Quantum Atomistic Solid State theory (QUASST) [1, 12]. QUASST points the substantial preservation of the ionic integrity and importance of the local effects, like the crystal field (CEF) and the spin-orbit (s-o) coupling. An experimental hallmark of QUASST is observation, apart of the large 5-100 eV multiplet electronic structure, of the quasi-atomic discrete electronic structure in the meV-energy scale. QUASST underlines that the strong electron correlations are predominantly of the on-site atomistic/ionic origin. We claim that in Ba_2IrO_4 the charge distribution is very close to the ionic, Ba^{2+}, O^{2-} and Ir^{4+}. The multiple electronic structure observed in Ba_2IrO_4 at neighborhood of 0.6 eV, originating from large intra-atomic spin-orbit coupling, is a finger print of the tetravalent iridium charge state.

3. Formal and real charge state in intermetallics: $CeRh_2Si_2$

Determination of the charge state of all atoms in intermetallic is rather problematic in comparison to oxides/fluorines. However, one of us (RR) already in 1984 has applied, working at the University of Amsterdam in the group of Prof. J. J. M. Franse, the crystal-field theory to METALLIC Ho_2Co_{17} [6]. By successful description of its electronic and magnetic properties we have proved the existence of the Ho^{3+} ions in this intermetallic system. Later low-energy crystal field states, energies and eigenfunctions, associated with Er^{3+} and Nd^{3+} ions have been described in conventional $ErNi_5$ [5] and $Nd_2Fe_{14}B$ [7] intermetallics. An our description of normal-state electronic and magnetic properties of UPd_2Al_3 [4], above the heavy-fermion and superconducting states (above, say 3 K) with U^{3+} ions was later confirmed by experimental discovery of CEF states in inelastic-neutron scattering experiments [13]. The U^{3+} ion has the Kramers doublet ground state. Then in year of 1992 one of us (RR) stated that the heavy-fermion phenomena are largely associated with a difficulty in removing the Kramers degeneracy in trivalent Ce, Yb and U intermetallics (abstract at the 1st SCES-92 Conf. Sendai-92: R. J. Radwanski, "Formation of the heavy-fermion state - an explanation in a model traditionally called localized" poster 8P-92 SCES-1992, Sendai-92).
It is worth to remind that it was time (1980-2010) that such concepts like ionic states in intermetallics, CEF in heavy-fermion intermetallics, s-o coupling in $3d$ oxides, discrete states, below, say, 10 meV, in $3d$ oxides) have been treated as scientific heresies.
We underline the existence of the crystal-field quasi-atomic states in these metallic systems having in mind, apart of subsequent rejections at SCES Conf., an administration decision of the Polish Highest Scientific Committee (BCK-V-P-1262/02, 2003) that "applying the crystal-field theory to compounds conducting current is erroneous" [14]). The obvious nonsense, but such nonsense postponed the real professor scientific carrier by 10 years - it documents that the existence of crystal-field ionic states and discrete low-energy electronic structure in intermetallics was not obvious as it is at present. This scientific situation has radically changed after the

discovery in the group of Prof. Frank Steglich, of the ionic Yb^{3+} state and its the Kramers-doublet ground state in the profound heavy fermion $YbRh_2Si_2$ [15]. The electron spin resonance has been observed at 1.5 K, i.e. deeply in the Kondo/heavy-fermion state owing to its Kondo temperature T_K of 25 K.

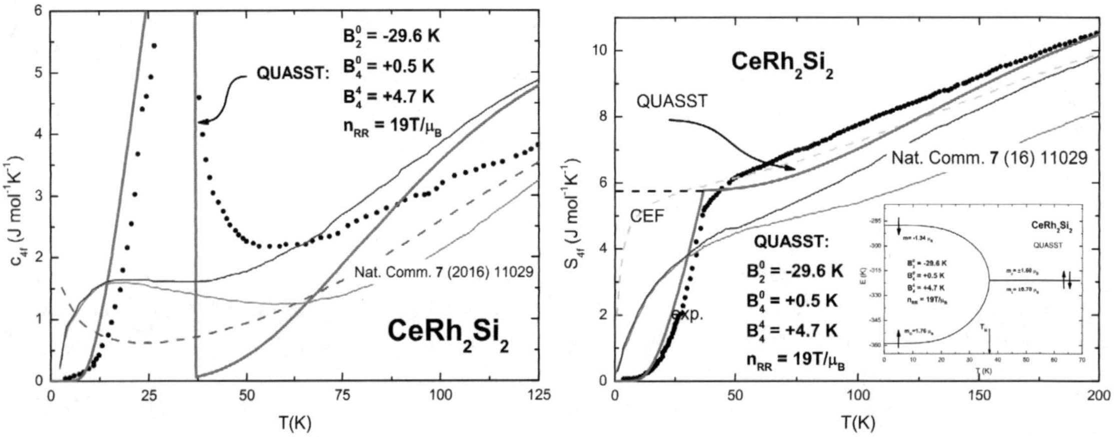

Figure 1. The calculated, within the CEF-based QUASST approach, temperature dependence of the $4f$ electron (magnetic+CEF) contribution, originating from localized crystal-field states of the Ce^{3+} ion, to the specific heat (left) and to the entropy (right) of $CeRh_2Si_2$ (thick solid red line with a peak at T_N at 36 K) together with experimental data (black dots), reported by Patil *et al.* [16] and with their three theoretical calculations (thin lines).
Coloured thin lines show the curves calculated by Patil-16 [16] for different values of the CEF splitting, of the Kondo temperature T_K, and of the width Γ_i of the first excited CEF level. Patil-16 results have been calculated with parameters: T_K= 3.0 meV, Γ_1 = 5.1 meV, Δ_1 =48 meV and Δ_2 =62 meV - solid magenda line; 3.0, 5.1, 30 and 52 meV - solid blue line, and 0.3, 15, 30 and 52 meV - dashed green line. Colours are the same as in the original paper of Patil-16. In the inset the calculated splitting of the Kramers doublet below T_N and the opening of the spin gap is shown.

$CeRh_2Si_2$, isostructural to $YbRh_2Si_2$, has been considered as a Kondo-lattice antiferromagnet with T_N of 36.5 K [16]. XAS [8], ARPES [16] and RIXS [17] experiments revealed two CEF states at 30/48 and 52/62 meV, respectively, expected for the $4f^1$ configuration in the tetragonal symmetry confirming earlier CEF findings on basis of classical magnetic measurements [18, 19]. Apart of ARPES experiment Patil *et al.* [16] [later denoted as Patil-16] have tried to describe the temperature dependence of the specific heat. For the theoretical description of $4f$ contribution c_{4f}(T) they have used two literature models both taking into account the hybridization and Kondo interactions. Despite of taking very different parameters they did not get a satisfactory description below 150 K. Moreover, their description of c_{4f}(T) and the involved entropy S_{4f}(T) below Neel temperature was completely wrong, as one can see on Fig. 1. Our crystal-field based QUASST description [20] very well describes c_{4f}(T) and S_{4f}(T) in the whole temperature range, from zero up to measured 300 K, together with very good description below T_N. A λ-type peak at 36 K is also nicely reproduced.
Good description below T_N of 36 K (=3 meV) confirms that CEF states in $CeRh_2Si_2$ are remarkably thin (in the energy scale their width is lower than 0.3 meV) and that practically **all**

Ce atoms/ions contribute to the experimental specific heat. An obtained value of the molecular-field coefficient of 19 T/μ_B indicates that each Ce ion experiences the molecular field of 30 T. This value is close to value of 26 T at which the antiferromagnetic structure is broken by external magnetic field [21]. Our model predicts, in contrary to models used in Ref. [16], opening of a spin gap in the antiferromagnetic state, see inset in Fig. 1. Its value grows to 6 meV at T= 0 K. We think that appearance of such gap should be experimentally verified.

In conclusion, we have established the charge state of iridium in Ba_2IrO_4 as Ir^{4+} and of cerium in $CeRh_2Si_2$ as Ce^{3+}. Our results are in agreement with realization of different integer valences in rhenium [22] and osmium [23, 24] oxides. Good description below T_N of 36 K (=3 meV) confirms that CEF states in $CeRh_2Si_2$ are remarkably thin (in the energy scale their width is lower than 0.3 meV) and that practically **all** Ce atoms/ions contribute to the experimental low-temperature specific heat. We conclude that hybridization/mixed valence phenomena are very weak in $CeRh_2Si_2$.

References

[1] Radwanski R J and Ropka Z 2000 *Acta Phys. Pol.* A **97** 963; 2006 *Acta Physica* **1** 26; [arXiv:0606604].
[2] Ropka Z, Michalski R and Radwanski R J 2001 *Phys. Rev.* B **63** 172404
[3] Ropka Z and Radwanski R J 2003 *Phys. Rev.* B **67** 172401
[4] Radwanski R J, Nalecz D M and Ropka Z 2016 *Acta Phys. Pol.* A **130** 545
[5] Radwanski R J, Kim-Ngan N H, Kayzel F E, Franse J J M, Gignoux D, Schmitt D and Zhang F Y 1992 *J. Phys.: Condens. Matter* **4** 8853
[6] Franse J J M, de Boer F R, Frings P H, Gersdorf R, Menovsky A A, Muller F A, Radwanski R J and Sinnema S 1985 *Phys. Rev.* B **31** 4347
[7] Radwanski R J and Franse J J M 1989 *J. Magn. Magn. Mater.* **80** 14
[8] Willers T, Adroja D, Rainford B, Hu Z, Hollmann N, Korner P, Chin Y -Y, Schmitz D, Hsieh H, Lin H -J, Chen C T, Bauer E D, Sarrao J L, McClellan K J, Byler D, Geibel C, Steglich F, Aoki H, Lejay P, Tanaka A, Tjeng L H and Severing A 2012 *Phys. Rev.* B **85** 035117
[9] Bertinshaw J, Kim Y K, Khaliullin G and Kim B J, *Annu. Rev. Condens. Matter Phys.* 2019 **10** 315 Bertinshaw J, Kim J K, Porras J, Ueda K, Sung N H, Efimenko A, Bombardi A, Jungho Kim, Keimer B and Kim B J 2020 *Phys. Rev.* B **101** 094428
[10] Moser S, Moreschini L, Ebrahimi A, Dalla Piazza B, Isobe M, Okabe H, Akimitsu J, Mazurenko V V, Kim K S, Bostwick A, Rotenberg E, Chang J, Ronnow H M and Grioni M 2014 *New J. Phys.* **16** 013008
[11] Rosciszewski K and Oles A M 2016 *Phys. Rev.* B **93** 085106
[12] Radwanski R J, Michalski R and Ropka Z 2000 *Acta Phys. Pol.* B **31** 3079; Radwanski R J, Michalski R and Ropka Z 2000 *Physica* B **276-278** 803
[13] Krimmel A, Loidl A, Eccleston R, Geibel C and Steglich F 1996 *J. Phys.: Condens. Matter* **8** 1677
[14] Radwanski R J 2008 *Acta Physica* **12-13** 1; 2008 **16-17** 1
[15] Sichelschmidt J, Ivanshin V A, Ferstl J, Geibel C and Steglich F 2003 *Phys. Rev. Lett.* **91** 156401
[16] Patil S, Generalov A, Guttler M, Kushwaha P, Chikina A, Kummer K, Rodel T C, Santander-Syro A F, Caroca-Canales N, Geibel C, Danzenbacher S, Kucherenko Y, Laubchat C, Allen J W and Vyalikh D V 2016 *Nat. Commun.* **7** 11029
[17] Amorese A, Caroca-Canales N, Seiro S, Krellner C, Ghiringhelli G, Brokes N B, Vyalikh D V, Geibel C and Kummer K 2018 *Phys. Rev.* B **97** 245130
[18] Knafo W, Aoki D, Vignolles D, Vignolle B, Klein Y, Jaudet C, Villaume A, Proust C and Flouquet J 2010 *Phys. Rev.* B **81** 094403
[19] Settai R, Misawa A, Araki S, Kosaki M, Sugiyama K, Takeuchi T, Kindo K, Haga Y, Yamamoto E and Onuki Y 1997 *J. Phys. Soc. Jpn* **66** 2260
[20] Radwanski R J, Z. Ropka and D. M. Nalecz 2019 *Mater. Chem. Phys.* **232** 301
[21] Abe H, Kitazawa H, Suzuki H, Kido G and Matsumoto T 1998 *Physica* B **246-247** 141; Abe H, Kitazawa H, Suzuki H, Kido G and Matsumoto T 1998 *J. Magn. Magn. Mater.* **177-181** 479
[22] Hirai D and Hiroi Z 2021 *J. Phys.: Condens. Matter* **33** 135603
[23] Erickson A S, Misra S, Miller G J, Gupta R R, Schlessinger Z, Harrison W A, Kim J M and Fisher I R 2007 *Phys. Rev. Lett.* **99** 016404
[24] Paramekanti A, at this SCES-20 Conference

Enhanced Spin Orbit Interaction in Actinides

Xiao Yuan[1], P.S. Riseborough[1], Eleonir Calegari[2], Sergio Magalhaes[3]

[1] Physics Dept., Temple University, Philadelphia, Pa 19122, USA
[2] Departamento de Física, UFSM, 97105-900, Santa Maria, RS, Brazil
[3] Instituto de Fisica, UFRGS, 91501-970 Porto Alegre, RS, Brazil

E-mail: `prisebor@temple.edu`

Abstract. We investigate the effect of many-body interactions on the spin-orbit coupling of anisotropic metals. We use the Underscreened Anderson Lattice Model that was proposed to describe actinide compounds. The Coulomb interactions induce off-diagonal correlations that enhance the components of the spin-orbit coupling. Modest values of the Coulomb interaction U can significantly enhance the spin-orbit coupling and effect the electronic spectrum. The enhancements are most pronounced for systems that are on the verge of magnetic instabilities. The enhancement is anisotropic in crystals with non-cubic symmetries and can lead to giant magnetic anisotropies in paramagnetic states.

1. Introduction

Actinide materials exhibit strong electronic correlations and spin-orbit interactions. The 5f wave functions are spatially extended and often exhibit giant magnetic anisotropies, such as in the cubic ferromagnetic uranium monochalcogenides [1], tetragonal URu_2Si_2 [2, 3] and ordered phase of α-plutonium [4, 5]. The giant magnetic anisotropies persist in non-magnetic states. It has been proposed that the anisotropies are due to the enhanced spin-orbit coupling strengths caused by electron correlations [6]. The Underscreened Kondo Model was formulated to describe the properties of actinide materials, such as the uranium monochalcogenides where the resistivity shows logarithmic temperature variations over several decades of temperature [7] but order ferromagnetically at temperatures of the order of 100 K, and have ordered magnetic moments of the order of a Bohr magneton [1]. This has been described in terms of the underscreened Kondo Model [8, 9] in which the Kondo effect only partially screens the local f moment, and the residual local moments order at low temperatures. In the simplest version of the underscreened Anderson Lattice Model, the atomic f orbitals are degenerate with the result that only a single combination of the f orbitals hybridizes and participates in the Kondo screening. The remaining f bands do not hybridize and host the formation of the residual low temperature local moments. Here we shall consider the effect of spin-orbit interaction in the underscreened Anderson Lattice Model, and find that the itinerant character of the model and the Coulomb interaction leads to an enhanced spin-orbit coupling which is anisotropic in non-cubic materials [10].

2. The Underscreened Anderson Lattice Model

We consider a simple orthorhombic crystal, in which a set of A atoms are located at the $(0, 0, 0)$ sites and the second set of B atoms are located at $(\frac{a_x}{2}, \frac{a_y}{2}, \frac{a_z}{2})$ sites. The A atoms host the itinerant conduction states, whereas the B sites host f states that are subjected to spin-orbit coupling and strong electron-electron interactions. We consider the conduction band as non-degenerate and the f orbitals as triply-degenerate. The Hamiltonian can be represented as the sum of three terms,

$$\hat{H} = \hat{H}_A + \hat{H}_B + \hat{H}_{AB} \tag{1}$$

where $\hat{H}_A$ describes the Hamiltonian for the conduction band and is given by

$$\hat{H}_A = \sum_{\underline{k},\sigma} \epsilon_a(\underline{k})\, c^{\dagger}_{\underline{k},\sigma}\, c_{\underline{k},\sigma} \tag{2}$$

in which $c^{\dagger}_{\underline{k},\sigma}$ and $c_{\underline{k},\sigma}$, respectively create and annihilate an electron with spin σ in the conduction band with Bloch wave vector $\underline{k}$. The Hamiltonian for the electrons in the orbitals of the B atoms are described by the sum of a tight-binding Hamiltonian $\hat{H}_0$, the spin-orbit interaction, $\hat{H}_{SO}$ and $\hat{H}_{int}$ which describes the Coulomb interactions

$$\hat{H}_B = \hat{H}_0 + \hat{H}_{SO} + \hat{H}_{int} \tag{3}$$

The tight-binding Hamiltonian for the localized orbitals, $\hat{H}_0$, is given by

$$\hat{H}_0 = \sum_{\underline{k},\beta,\sigma} \epsilon_\beta(\underline{k})\, f^{\dagger}_{\underline{k},\beta,\sigma}\, f_{\underline{k},\beta,\sigma} \tag{4}$$

where $\beta = x, y, z$, and where $f^{\dagger}_{\underline{k},\beta,\sigma}$ and $f_{\underline{k},\beta,\sigma}$, respectively create or annihilate an electron of spin σ, $\sigma = \pm 1$, and Bloch wave-vector $\underline{k}$ in the β-th orbital on the B atom. In the cubic harmonic Bloch representation, the spin-orbit interaction has the form

$$\begin{aligned}\hat{H}_{SO} &= \frac{\lambda_z}{2} \sum_{\underline{k},\sigma} i\,\sigma \left[f^{\dagger}_{\underline{k},y,\sigma} f_{\underline{k},x,\sigma} - f^{\dagger}_{\underline{k},x,\sigma} f_{\underline{k},y,\sigma} \right] \\ &- \frac{\lambda_y}{2} \sum_{\underline{k},\sigma} \sigma \left[f^{\dagger}_{\underline{k},z,\sigma} f_{\underline{k},x,\overline{\sigma}} + f^{\dagger}_{\underline{k},x,\overline{\sigma}} f_{\underline{k},z,\sigma} \right] \\ &+ \frac{\lambda_x}{2} \sum_{\underline{k},\sigma} i \left[f^{\dagger}_{\underline{k},z,\sigma} f_{\underline{k},y,\overline{\sigma}} - f^{\dagger}_{\underline{k},y,\overline{\sigma}} f_{\underline{k},z,\sigma} \right]\end{aligned} \tag{5}$$

where we have introduced different coupling strengths in anticipation of the result that the Coulomb enhancement will break the atomic symmetry. The interaction between the f electrons is denoted as $\hat{H}_{int}$. The interaction is composed of the on-site Coulomb interaction U and the Hund's rule interaction J between the electrons in an atomic shell. In principle, the atomic interactions U and J should depend on the pair of orbital quantum numbers m and m'. We shall neglect this dependence but instead require that the interaction be invariant under both spin and orbital rotations [11]. In the tetragonal basis, the interaction has the form

$$\begin{aligned}\hat{H}_{int} &= \frac{1}{2} \sum_{i,\alpha,\beta,\sigma} U\, \hat{n}_{i,\alpha,\sigma}\, \hat{n}_{i,\beta,\overline{\sigma}} \\ &+ \frac{1}{2} \sum_{i,\alpha\neq\beta,\sigma} (U - J)\, \hat{n}_{i,\alpha,\sigma}\, \hat{n}_{i,\beta,\sigma} \\ &- \frac{1}{2} \sum_{i,\alpha\neq\beta,\sigma} J\, f^{\dagger}_{i,\alpha,\sigma}\, f_{i,\alpha,\overline{\sigma}}\, f^{\dagger}_{i,\beta,\overline{\sigma}}\, f_{i,\beta,\sigma} \\ &+ \frac{1}{2} \sum_{i,\alpha,\sigma} J\, \hat{n}_{i,\alpha,\sigma}\, \hat{n}_{i,\alpha,\overline{\sigma}}\end{aligned} \tag{6}$$

where the sums over α and β run through x, y, z. Thus, the strength of the intra-orbital Coulomb interaction, (i.e. $\alpha = \beta$) is given by $U' = U + J$ and, therefore, is distinct from the inter-orbital interaction U. The hybridization term that couples the A and B orbitals on nearest-neighbor atoms has the form

$$\hat{H}_{AB} = \sum_{\underline{k},\beta,\sigma} i \; V_\beta(\underline{k}) \left[c^\dagger_{\underline{k},\sigma} f_{\underline{k},\beta,\sigma} - f^\dagger_{\underline{k},\beta,\sigma} c_{\underline{k},\sigma} \right] \tag{7}$$

where the k-dependent hybridization matrix elements are given by

$$V_\beta(\underline{k}) = 8 \, V_{AB} \frac{a_\beta}{\sqrt{\sum_\gamma a_\gamma^2}} \sin \frac{k_\beta a_\beta}{2} \prod_{\alpha \neq \beta} \left(\cos \frac{k_\alpha a_\alpha}{2} \right) \tag{8}$$

in which V_{AB} is the Slater-Koster parameter $t^\sigma_{A,B}(R)$ where R is the distance between the atoms. The hybridization is primarily responsible for the width of the 5f band.

3. Results and Discussion

The electrons in the hybridized bands are subjected to Coulomb and spin-orbit scattering which interfere constructively. In the mean-field approximation, the induced Coulomb correlations have the same symmetries as the components of the spin-orbit interaction and produce an anisotropic enhancement

$$\begin{aligned} - i \, \sigma \frac{\lambda_z}{2} &= - i \, \sigma \frac{\lambda_0}{2} - U < f^\dagger_{i,y,\sigma} f_{i,x,\sigma} > \\ - \sigma \frac{\lambda_y}{2} &= - \sigma \frac{\lambda_0}{2} - U < f^\dagger_{i,z,\sigma} f_{i,x,\bar{\sigma}} > \\ - i \frac{\lambda_x}{2} &= - i \frac{\lambda_0}{2} - U < f^\dagger_{i,z,\sigma} f_{i,y,\bar{\sigma}} > \end{aligned} \tag{9}$$

The off-diagonal spin-flip correlations are obtained from the respective Green's functions

$$\begin{aligned} G_{z,\sigma,x,\bar{\sigma}}(\omega, \underline{k}) &= - \sigma \left[\frac{\lambda_y}{2} \left((\omega - \epsilon_y)(\omega - \epsilon_a) - V_y^2 \right) - \frac{\lambda_x \lambda_z}{4} (\omega - \epsilon_a) \right] / D_4(\omega, k) \\ &+ i \, \sigma \frac{\lambda_x}{2} \frac{V_x \, V_y}{D_4(\omega, \underline{k})} \end{aligned} \tag{10}$$

and

$$\begin{aligned} G_{y,\bar{\sigma},z,\sigma}(\omega, \underline{k}) &= - i \left[\frac{\lambda_x}{2} \left((\omega - \epsilon_x)(\omega - \epsilon_a) - V_x^2 \right) - \frac{\lambda_y \lambda_z}{4} (\omega - \epsilon_a) \right] / D_4(\omega, k) \\ &+ \sigma \frac{\lambda_y}{2} \frac{V_x \, V_y}{D_4(\omega, \underline{k})} \end{aligned} \tag{11}$$

where $D_4(\omega, \underline{k})$ is the secular determinant which determines the degenerate quasiparticle dispersion relations expected for a time-reversal invariant centro-symmetric material. The enhancement of the α-component of the spin-orbit interaction contains terms which are proportional $\lambda_\alpha V_x V_y$ that are expected to vanish for a centro-symmetric system. Similar terms also appear in the expression for the directionally dependent magnetic susceptibility which, if non-zero, would produce an anisotropy along the (1,1,1) direction for a cubic system. We note that, for compounds with strong hybridization and spin-orbit coupling, the formation of a magnetic moment can induce a slight structural distortion that stabilizes the $(1, 1, 1)$ direction as an easy-axis. This is in accord with recent observations on ferromagnetic US [12].

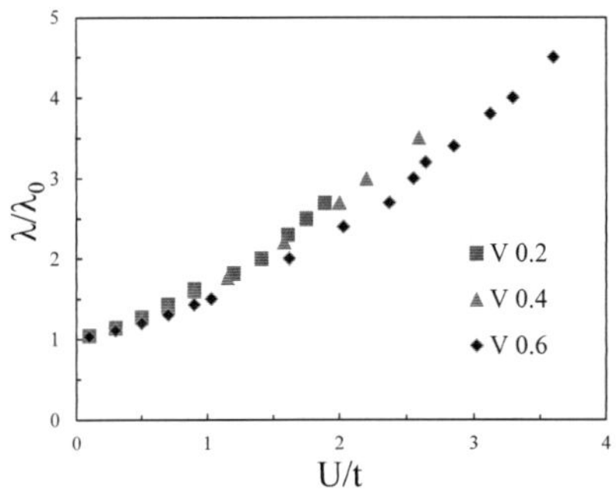

Figure 1. The ratio of the enhanced spin orbit coupling to the atomic value, as a function of U/t for various values of the hybridization V/t.

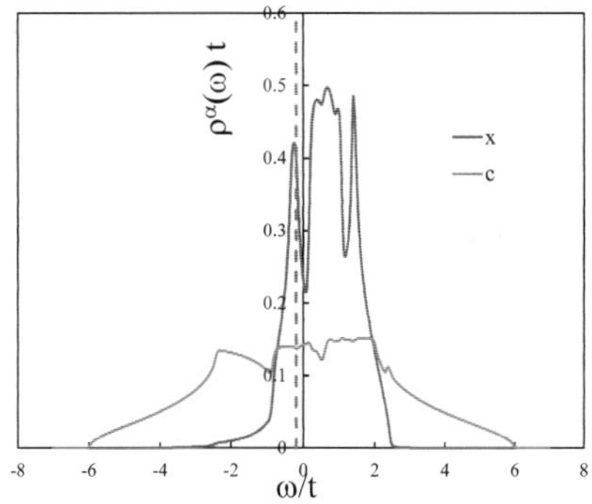

Figure 2. The partial conduction band and f sub-band density of states for $V/t = 0.15$ and $U/t \sim 1$. The Fermi-energy is indicated by the dashed vertical line.

For cubic compounds, the spin-orbit interaction is isotropic. The enhancement of the spin orbit coupling λ/λ_0 in the paramagnetic state is shown as a function of U for various values of the hybridization in fig(1). The conduction band width has been set at $12t$, the f tight-binding parameters have been set at 0.3 t, the atomic spin-orbit coupling is $\lambda_0 = 0.24t$ and $J = U/6$. It is seen that the enhancement increases with U and is maximum at the critical value above which the system becomes magnetic. The critical value of U increases when V is increased. The largest enhancements are found at the quantum critical point, and the are maximal for large V. The single particle f and conduction and f band partial densities of states are shown in fig(2) for $U/t \approx 1$. It is seen that the enhanced spin-orbit coupling produces pseudo-gaps in the f density of states at energies associated with avoided band crossings. The enhancement of the spin-obit coupling is associated with the increased stability that occurs when the chemical potential is inside a pseudo-gap. Hence, the enhanced spin-orbit coupling is expected to be manifest in the density of states as probed in ARPES/BIS experiments.

4. References

[1] Lander GH, Brooks MSS, Lebech B, Brown PJ, Vogt O and Mattenberger K 1990 Appl. Phys. Lett. **57**, 989-91.
[2] Palstra TTM, Menovsky AA, van den Berg J, Dirkmaat AJ, Kes PH, Nieuwenhuys GJ and Mydosh JA 1985 Phys. Rev. Lett. **55**, 2727-30.
[3] Werwinski M, Rusz J, Mydosh JA and Oppeneer PM 2014 Phys. Rev. B, **90**, 064430 (2014).
[4] Kutepov AL and Kutepova SG, 2003 J. Phys. Cond. Mat. **15**, 2607-24.
[5] Butterfield MT, Tobin JG, Teslich NE, Bliss RA, Wall MA, McMahan AK, Chung BW, Schwartz AJ and Kutepov AL 1996 Matl. Res. Soc. Symp. Proc. **893**, 95.
[6] Riseborough PS, Magalhaes SG, Calegari EJ and Cao G 2020 J. Phys., Cond. Mat. **32**, 4455601.
[7] Schoenes J, Frick B and Vogt O 1984 Phys. Rev. B **30**, 6578-85.
[8] Perkins NB, Nunez-Regueiro MD, Coqblin B and Iglesias JR 2007 Phys. Rev. B, **76**, 125101.
[9] Thomas C, Simoes ASD, Lacroix C, Iglesias JR and Coqblin B, 2014 J. Mag. Mag. Mat. **372**, 247-52.
[10] Yuan X, Riseborough PS, Calegari EJ and Magalhaes SG 2021 Electron. Struct. **3**, 024009.
[11] Dworin L and Narath A 1970 Phys. Rev. Lett. **25**, 1287-91.
[12] Poudel N, Jeffries J, Gofryk K 2021 Phys. Rev. B, **104** 014417.

Finite-temperature corrections to the Lorenz ratio at the $N = 3$ topological Kondo fixed point

Domenico Giuliano[(1,2)], **Andrea Nava**[(1,2)], **Reinhold Egger**[(3)], **Pasquale Sodano**[(4)], **and Francesco Buccheri**[(3)]

[(1)] Dipartimento di Fisica, Università della Calabria, Arcavacata di Rende I-87036, Italy
[(2)] I.N.F.N., Gruppo collegato di Cosenza, Arcavacata di Rende I-87036, Italy
[(3)] Institut für Theoretische Physik, Heinrich-Heine-Universität, D-40225 Düsseldorf, Germany
[(4)] I.N.F.N., Sezione di Perugia, Via A. Pascoli, I-06123, Perugia, Italy.

E-mail: domenico.giuliano@fis.unical.it

Abstract. We analyze the finite-temperature scaling of the Lorenz ratio at the topological Kondo fixed point realized at a junction of three interacting quantum wires connected to a floating superconducting island. Using the Tomonaga-Luttinger liquid approach to the quantum wires, we derive the full functional dependence of the finite-temperature correction on the Luttinger parameter g.

1. Introduction and conclusions

Whenever transport through an electronic system can be described in terms of quasiparticles, which carry both energy and charge, the ratio between thermal and electric conductance is linear in the temperature, which is known as the Wiedemann-Franz law (WFL). The proportionality constant takes an universal value L_0, the so called Lorenz number, given by $L_0 = \frac{\pi^2 k_B^2}{3e^2} \approx 2.44 \times 10^{-8} W\Omega K^{-2}$ [1].

While the WFL is generally well-verified in three-dimensional metals and semiconductors at room temperature, pairing correlations, impurities and electron-phonon scattering can lead to important deviations from the predicted Lorenz ratio [2, 3]. In mesoscopic systems, it has been reported that the interplay of dissipation and approximate conservation laws, as well as selective transport phenomena also lead to the breakdown of this relation [4, 5, 6]. Finally, a most notorious violation takes place in one-dimensional Luttinger Liquids, in which the collective excitations carry a noninteger electric charge and the Lorenz ratio is therefore renormalized by the inverse Luttinger parameter [7]. As argued above, in order for the WFL to apply, charge and heat must be carried by the same excitations, hence, it is eventually not surprising that the WFL still holds in systems such as the overscreened Kondo model [8], which are known to exhibit non-Fermi liquid correlations at low temperatures/energies [9].

Systems exhibiting phases in which charge and heat transport are "disentangled" from each other have been identified as strongly coupled boundary fixed points (SFPs) in junctions of interacting effectively one-dimensional systems, either fermionic [10, 11, 12, 13], or bosonic [14, 15, 16, 17]. Nevertheless, in order to make an SFP stable at low temperatures/energies, a strong attractive bulk interaction in each wire is required. It is worth noting that probing charge and heat transport through the junction typically requires connecting the wires to large, noninteracting Fermi liquid like reservoirs, which ultimately neutralizes the effects of bulk interactions in the Lorenz ratio [18, 19].

A striking possibility of recovering a stable SFP with an expected robust, and universal, violation of the WFL in a weakly interacting, or even noninteracting (in the bulk) junction of quantum wires, has been recently put forward in [20, 21]. Specifically, it has been shown that the WFL breaks down in a robust and predictable way in a junction of $N = 3$ spinless quantum wires connected to a superconducting island in its charging regime hosting a set of Majorana modes (MMs). At low temperatures, such a system enters a non-trivial phase, associated to the topological Kondo fixed point (TKFP) [22, 23]. The stability of the TKFP is indeed due to the localized MMs, which eventually turn the residual boundary interaction at the SFP irrelevant [24, 25]. As a result, the predicted violation of the WFL at the TKFP can be regarded as a mean to detect the so far pretty elusive emerging MMs. Yet, real measurements are performed at finite temperature. This requires the knowledge of the leading, finite-T corrections to the conductances near by the TKFP. In order to compute them, we employ the formalism developed in [20, 21] and as a result, beside recovering the main temperature dependence, we provide the explicit corrections of the conductances at the TKFP on the bulk interaction in the leads.

The paper is organized as follows:

- In Section 2 we present our model Hamiltonian and review the results for the charge and the heat transport at the TKFP.
- In Section 3 we compute the finite-T leading corrections to the conductances and to the Lorenz ratio close to the TKFP.

2. Model Hamiltonian and charge and heat transport at the strongly coupled fixed point

Our system consists of three interacting fermionic quantum wires connected to a floating superconducting island, hosting a set of MMs as the only subgap degrees of freedom. The island is characterized by the charging energy E_c and by the applied gate voltage V_g. To model the wires, we use the spinless Luttinger liquid Hamiltonian

$$H_0 = \sum_{a=1}^{3} \frac{u}{2} \int dx \, \left\{ g(\partial_x \phi_a(x))^2 + g^{-1}(\partial_x \theta_a(x))^2 \right\} \quad , \tag{1}$$

with $\{\phi_a\}$ being the bosonic Luttinger liquid plasmon fields and $\{\theta_a\}$ being their dual fields. The parameters u and g respectively denote the plasmon velocity and the Luttinger parameter of the leads, $g > 1 (< 1)$ corresponding to an attractive (repulsive) bulk interaction. The ends of the quantum wires are proximitized to the MMs with tunneling strength t. Calling ρ_0 the density of states around the Fermi energy, we focus on the limit $E_c \gg T, t^2\rho_0$ (Coulomb blockade) and tune V_g in such a way that the total charge at the island is integer. One can then sum over the leading cotunneling processes through the junction and recover an effective, purely bosonic tunneling Hamiltonian H_K

$$H_K = -2J_K \sum_{a=1}^{3} \cos\left[\sqrt{\pi}(\phi_a(0) - \phi_{a+1}(0))\right] \quad , (\phi_{a+3} \equiv \phi_a, J_K \sim t^2/E_c) \quad . \tag{2}$$

The total Hamiltonian $H = H_0 + H_K$ describes a purely bosonic junction [16, 17] in which, for $g \leq \frac{3}{4}$ the stable phase corresponds to the "disconnected" fixed point (DFP) while, for $g \geq 1$, it is described by a strongly coupled fixed point, the TKFP. Finally, for $\frac{3}{4} < g < 1$, both the DFP and the TKFP are stable, so that there is a finite-coupling, repulsive fixed point, encoding the quantum phase transition between the two stable phases [22, 23]. To compute the fixed point charge and the heat conductances, we resort to the splitting matrix approach put forward in [26]. The splitting matrix ρ is defined so that, denoting respectively with $\phi_{R,a}(x)$ and with $\phi_{L,a}(x)$ the right- and the left-handed components of the field $\phi_a(x)$, one has $\phi_{R,a}(0) = \sum_{b=1}^{3} \rho_{a,b}\phi_{L,b}(0)$. Given ρ and letting $G_{a,b}$ and $K_{a,b}$ respectively being the charge and the heat conductance tensor elements at the junction, one finds [20, 21]

$$G_{a,b} = \frac{e^2 g}{2\pi}\left\{\rho_{a,b} - \delta_{a,b}\right\}, \quad K_{a,b} = \frac{\pi k_B^2 T}{6}\{\rho_{a,b}^2 - \delta_{a,b}\} \quad . \tag{3}$$

At the DFP one has $\rho_{a,b} = \delta_{a,b}$, which implies $G_{a,b} = K_{a,b} = 0\ \forall a,b$. At variance, at the TKFP one finds $\rho_{a,b} = \frac{2}{3} - \delta_{a,b}$, which implies $G_{a,b} = \frac{e^2 g}{\pi}\left\{\frac{1}{3} - \delta_{a,b}\right\}$ and $K_{a,b} = \frac{2\pi k_B^2 T}{9}\left\{\frac{1}{3} - \delta_{a,b}\right\}$. As a result, taking the (Lorenz) ratio $\mathcal{L}$ between the heat and the charge conductance tensor matrix elements with the same indices, we find $\mathcal{L} = \frac{K_{a,b}}{TG_{a,b}} = \frac{2}{3g}L_0$. We therefore identify a "bulk" renormalization of L_0 by the nonuniversal factor g^{-1}, which is typical of a the WFL in a spinless Luttinger liquid [27, 28]. As stated above, such an effect disappears once the junction is connected to Fermi liquid reservoirs. In addition, we see the universal (that is, independent of the bulk interaction) factor $\frac{2}{3}$. This is purely determined by the junction dynamics and, in particular, by the onset of multi-particle scattering processes at the TKFP. Probing this renormalization should provide a remarkable piece of evidence of the presence of localized MMs at the junction, which are deemed to be the main mechanism stabilizing the TKFP [20, 21, 17].

Typical measurements are performed at low, yet finite, temperature. For this reason, it is important to derive how the TKFP result for $\mathcal{L}$ is modified at finite T, which is the topic we address next.

3. Finite-T corrections to the Lorenz ratio close to the strongly coupled fixed point

As a reference calculation, we briefly review the derivation of the finite-T corrections at the DFP in a perturbative expansion in J_K [20, 21]. At the DFP, the boundary interaction in Eq.(2) takes a scaling dimension g^{-1}. Therefore, the corrections will exhibit a nontrivial scaling behavior, as a function of the running scale T. Specifically, one finds

$$G_{a,b}(T) = (3\delta_{a,b} - 1)\frac{2\pi e^2 \Gamma^2(1/g)}{\Gamma(2/g)}\tilde{J}_K^2(T)\ , \ K_{a,b}(T) = T\frac{\Gamma(2/g)}{\pi g \Gamma^4(1/g)}\Phi(g^{-1}, g^{-1})L_0 G_{a,b} \quad , \tag{4}$$

with $\Gamma(z)$ being Euler Γ function and the running coupling $\tilde{J}_K(T) = J_K(2\pi k_B T)^{-1+g^{-1}}$, while the function $\Phi(x,y)$ is

$$\Phi(x,y) = \int dz dw\, \frac{w}{\sinh(\pi z)}|\Gamma(y + iw)|^2 |\Gamma(x + i(z-w))|^2 \quad . \tag{5}$$

In the non-interacting case $g = 1$, one finds no violation of the WFL even at finite-T.

The TKFP is accessed in the strongly coupled limit $J_K \to \infty$ in which, in order to minimize H_K, one has to "pin" the field operators entering H_K. To do so, we first note that, once resorting to the center-of-mass and to the relative-field basis, $X = \frac{1}{\sqrt{3}}\sum_{a=1}^{3}\phi_a$, $\varphi_1 = \frac{1}{\sqrt{2}}\{\phi_1 - \phi_2\}$, $\varphi_2 = \frac{1}{\sqrt{6}}\{\phi_1 + \phi_2 - 2\phi_3\}$, H_K only depends on $\varphi_1(0)$ and $\varphi_2(0)$. Thus, in order to minimize H_K, one has to pin $\varphi_1(0), \varphi_2(0)$. Accordingly, the leading boundary operator (LBO) at the TKFP must necessarely depend only on the relative dual fields, $\vartheta_1(0) = \frac{1}{\sqrt{2}}\{\theta_1(0) - \theta_2(0)\}$ and on $\vartheta_2(0) = \frac{1}{\sqrt{6}}\{\theta_1(0) + \theta_2(0) - 2\theta_3(0)\}$ [11, 16]. Specifically, the LBO $\tilde{H}_K$ is a linear combination of boundary operators describing instanton "jumps" between the minima of H_K [29, 30]

$$\tilde{H}_K = -2h\cos\left[\frac{4\sqrt{2\pi}}{3}\vartheta_2(0)\right] - 2h\sum_{\alpha=\pm 1}\cos\left[\frac{4\sqrt{2\pi}}{3}\left(-\frac{\vartheta_2(0)}{2} + \alpha\frac{\sqrt{3}\vartheta_1(0)}{2}\right)\right] \quad , \tag{6}$$

Journal of Physics: Conference Series **2164** (2022) 012060 doi:10.1088/1742-6596/2164/1/012060

h being the corresponding coupling strength. $\tilde{H}_K$ has scaling dimension $\tilde{d}_K = \frac{4g}{3}$, which implies the stability of the TKFP as soon as $g > \frac{3}{4}$. Applying the analysis of [21] to the perturbation Eq.(6), we obtain

$$G_{a,b}(T) = \frac{e^2 g}{\pi}\left\{1 - \mathcal{A}_G(g)\tilde{h}^2\right\}\left(\frac{1}{3} - \delta_{a,b}\right) \quad , \quad K_{a,b}(T) = \frac{2\pi k_B^2 T}{9}\left\{1 - \mathcal{A}_K(g)\tilde{h}^2\right\}\left(\frac{1}{3} - \delta_{a,b}\right) \quad , \tag{7}$$

with $\mathcal{A}_G(g) = 8\pi^2 g \Gamma^2\left[\frac{4g}{3}\right] / \left(3\Gamma\left[\frac{8g}{3}\right]\right)$, $\mathcal{A}_K(g) = 16\pi g\left\{5\frac{\Phi\left(\frac{2g}{9}, \frac{10g}{9}\right)}{\Gamma\left[\frac{4g}{9}\right]\Gamma\left[\frac{20g}{9}\right]} + \frac{\Phi\left(\frac{8g}{9}, \frac{4g}{9}\right)}{\Gamma\left[\frac{8g}{9}\right]\Gamma\left[\frac{4g}{9}\right]}\right\}$ and the running coupling $\tilde{h} = \tilde{h}(2\pi k_B T) = h(2\pi k_B T)^{-1+\frac{3g}{4}}$. Eq.(7) is the ultimate result of this paper. When considering the Lorenz ratio at finite T, $\mathcal{L}(T)$, from Eqs.(7) we obtain

$$\mathcal{L}(T) = \frac{2}{3g} L_0 \left\{1 - \mathcal{A}_{\mathcal{L}}(g)\tilde{h}^2(2\pi k_B T)\right\} \quad , \tag{8}$$

with $\mathcal{A}_{\mathcal{L}}(g) = \mathcal{A}_K(g) - \mathcal{A}_G(g)$. We conclude that, in proximity of the TKFP, $\mathcal{L}(T)$ takes a nontrivial dependence on T even in the absence of a bulk interaction in the leads.

References

[1] Franz R and Wiedemann G 1853 *Annalen der Physik* **165** 497–531
[2] Lavasani A, Bulmash D and Das Sarma S 2019 *Phys. Rev. B* **99** 085104
[3] Durst A C and Lee P A 2000 *Physical Review B* **62** 1270–1290
[4] Garg A, Rasch D, Shimshoni E and Rosch A 2009 *Phys. Rev. Lett.* **103**(9) 096402
[5] Kubala B, König J and Pekola J 2008 *Phys. Rev. Lett.* **100** 066801
[6] Majidi D, Josefsson M, Kumar M, Leijnse M, Samuelson L, Courtois H, Winkelmann C B and Maisi V F 2021 *Preprint arXiv:2106.06229*
[7] Kane C L and Fisher M P A 1992 *Phys. Rev. Lett.* **68**(8) 1220–1223
[8] Karki D B and Kiselev M N 2020 *Phys. Rev. B* **102** 241402(R)
[9] Affleck I and Ludwig A W 1991 *Nucl. Phys. B* **360** 641–696
[10] Chamon C, Oshikawa M and Affleck I 2003 *Phys. Rev. Lett.* **91**(20) 206403
[11] Oshikawa M, Chamon C and Affleck I 2006 *J. Stat. Mech. Theory Exp.* **2006** P02008
[12] Hou C Y, Rahmani A, Feiguin A E and Chamon C 2012 *Phys. Rev. B* **86**(7) 075451
[13] Giuliano D and Nava A 2015 *Phys. Rev. B* **92** 125138
[14] Tokuno A, Oshikawa M and Demler E 2008 *Phys. Rev. Lett.* **100** 140402
[15] Giuliano D and Sodano P 2008 *New Journal of Physics* **10** 093023
[16] Giuliano D and Sodano P 2009 *Nucl. Phys. B* **811** 395 – 419
[17] Mintchev M and Sorba P 2013 *Journal of Physics A: Mathematical and Theoretical* **46** 095006
[18] Safi I and Schulz H J 1995 *Phys. Rev. B* **52**(24) R17040–R17043
[19] Maslov D L and Stone M 1995 *Phys. Rev. B* **52**(8) R5539–R5542
[20] Buccheri F, Nava A, Sodano P, Egger R and Giuliano D 2021 *Preprint arXiv:2108.04156*
[21] Giuliano D, Nava A, Egger R, Sodano P and Buccheri F 2021 *Preprint arXiv:2108.04149*
[22] Beri B and Cooper N R 2012 *Phys. Rev. Lett.* **109** 156803
[23] Altland A and Egger R 2013 *Phys. Rev. Lett.* **110** 196401
[24] Beri B 2013 *Phys. Rev. Lett.* **110** 216803
[25] Giuliano D and Affleck I 2019 *Nucl. Phys. B* **944** 114645
[26] Eriksson E, Nava A, Mora C and Egger R 2014 *Phys. Rev. B* **90**(24) 245417
[27] Kane C L and Fisher M P A 1996 *Phys. Rev. Lett.* **76** 3192–3195
[28] Kane C L and Fisher M P A 1997 *Phys. Rev. B* **55** 15832–15837
[29] Yi H and Kane C L 1998 *Phys. Rev. B* **57**(10) R5579–R5582
[30] Herviou L, Le Hur K and Mora C 2016 *Phys. Rev. B* **94** 235102

Extremely High magnetoresistance and quantum Oscillation study of WTe_2 Weyl Semimetal

Abhishek Singh, Souvik Sasmal, Kartik K Iyer, A Thamizhavel and Kalobaran Maiti*

Tata Institute of Fundamental Research, Department of Condensed Matter Physics and Materials Science, Homi Bhabha Road, Colaba, Mumbai 400005, India

Email for corresponding author: kbmaiti@tifr.res.in

Abstract. Single crystals of WTe_2 have been prepared using the flux method. Analysis of the x-ray diffraction pattern shows the pure phase of the sample. Temperature dependent resistivity indicates the metallic nature of the sample. The residual resistance ratio of the sample has been found very high providing further evidence of the good quality of the crystals. Extremely high magnetoresistance has been observed in the sample. Quantum oscillations have been seen at high field. Fast Fourier transformation of the quantum oscillations shows nearly the same size of the electron and hole pocket in the sample.

1. Introduction

In recent years, topological materials have attracted considerable attention due to their novel properties [1-5]. Angle resolved photoemission spectroscopy (ARPES) experiment clearly identified single helical Dirac cone on the surfaces [6-10]. Properties such as topological superconductivity, the possibility of Majorana fermions, absence of backscattering from nonmagnetic impurities, etc. makes these materials unique and important for the emerging field of quantum computing [2-4]. Recently, a 3D topological semimetal WTe_2 has been identified as a type-II Weyl semimetal showing extremely large magnetoresistance (XMR), as high as 1.3×10^7 % with no sign of saturation up to 60T at 0.53K [1]. Materials having such a high value of magnetoresistance (MR) are very promising for advanced technologies involving magnetic sensors, magnetic switches, magnetic valves and hard drives in electronics and spintronics, *etc*. Usually, MR is small in simple metals and shows H^2 (H = magnetic field) dependence at low fields and saturates at high fields. Some materials show huge MR at low temperatures. Different materials specific mechanisms have been proposed to explain the origin of such large MR [1,11,12, 14,15]; the origin of MR is still not fully understood. In the case of XMR in WTe_2, a perfect electron-hole (e-h) compensation is proposed to be one explanation of the scenario [1]. One pair of electron and hole pockets of nearly the same size has been observed in a recent experiment which supports the idea of carrier compensation for XMR [11] whereas two pairs of electron and hole pockets have been observed by the another ARPES study [12]. Quantum oscillation study is also a powerful tool to probe the Fermi surface topology [13]. Based on these studies, Xiang *et al.* [14] have reported one electron and two-hole pockets, whereas two electron and two-hole pockets have been observed by Cai et al. [15]. Evidently, the origin of XMR in this material is still a matter of debate and further studies are required to understand the underlying physics. In this study, we have grown high quality single crystals of WTe_2 and studied their bulk properties. Our results provide an evidence of similar sized Fermi pockets in this system.

2. Sample growth and characterizations

Single crystals of WTe_2 were grown by the self-flux method. High purity powder of Tungsten (99.999%) and Tellurium (99.999%) have been mixed uniformly which was sealed in quartz tube after evacuating down to $\sim 10^{-6}$ torr. The sealed tube was heated up to 1050^0C, kept there for few hours and then cooled down slowly to 550^0C. The resulting crystals were needle-like in morphology with the crystallographic c axis perpendicular to the crystal surface. X-ray powder diffraction (XRD) measurement has been carried out with Cu Kα radiation using a Rigaku powder diffractometer to investigate the phase present in the sample. Electrical resistivity and magnetotransport measurements were performed in a Quantum Design Physical Property Measurement system (PPMS) using standard four-probe technique.

3. Results and Discussion

For the phase identification and purity of as prepared single crystals, we have arranged needle-shaped pieces of single crystals side by side on a glass sheet and performed powder X-ray diffraction (XRD) at room temperature. In Fig. 1, we show the XRD pattern. All the peaks in the XRD pattern have been indexed using the standard JCPDS file (JCPDS#04-003-2315) belonging to the space group $Pmn2_1$. The peak positions are consistent with the reported data [16]. All peaks match well with the standard Bragg's

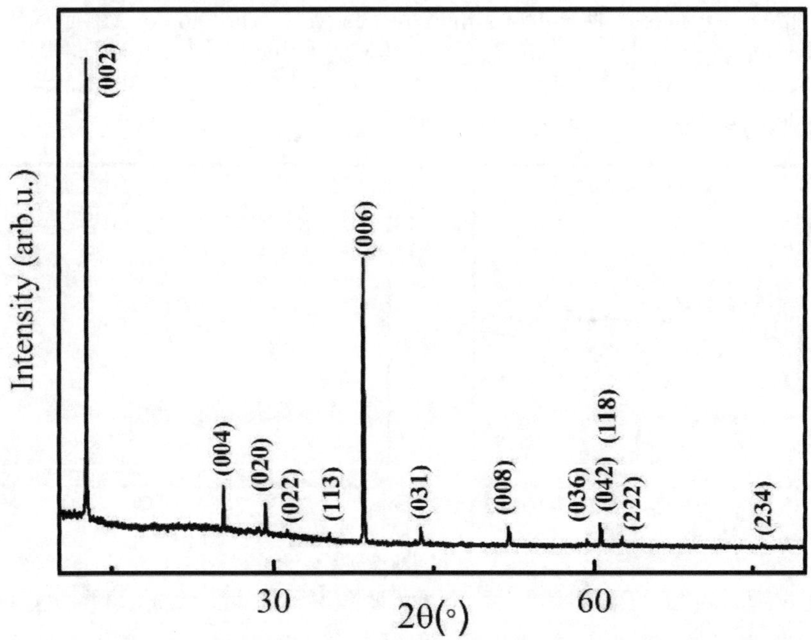

Figure 1. X-ray diffraction pattern of WTe_2 single crystal measured at room temperature.

positions of the orthorhombic structure and no extra peak has been observed. The absence of impurity peaks and sharp and intense XRD peaks indicate that the sample is pure in phase and is of excellent quality. We have estimated the lattice parameter using the Rietveld refinement method. The estimated value of lattice parameters was $a = 3.47$ Å, $b = 6.25$ Å and $c = 14.01$ Å.

Variation of electrical resistance as a function of temperature is shown in Fig. 2(a). The enhancement of the resistance with temperature indicates metallic nature of the sample. We have calculated the residual resistance ratio, RRR $\left(=\frac{R_{300K}}{R_{2K}}\right)$, where R_{300K} and R_{2K} are the resistances at 300K and 2K respectively. The RRR for this sample is found to be 551 which is larger than the reported values [12,15] and indicates very good quality of the crystals.

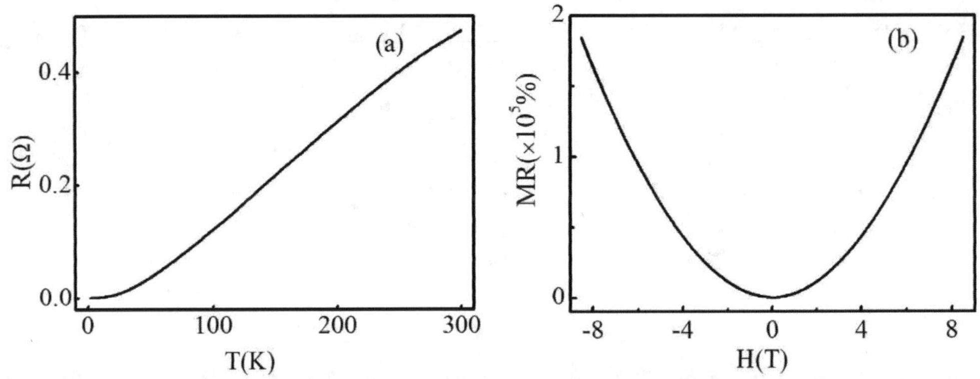

Figure 2. (a)Temperature dependent electrical resistance, (b) Magnetic field dependance of MR for WTe_2 single crystal at temperature 10K.

Here, the magnetoresistance has been defined as $MR = \frac{R(H)-R(0)}{R(0)} \times 100$; R(H) and R(0) are the resistances at applied magnetic field, H and zero field, respectively. Variation of MR as a function of applied magnetic field up to ± 8.5 T measured at 10 K is shown in Fig.2(b). Firstly, the MR is found to be very high (of the order of 10^5). MR gradually increases like a power law dependence with the magnetic field and has no sign of saturation up to the applied field of ± 8.5 T studied.

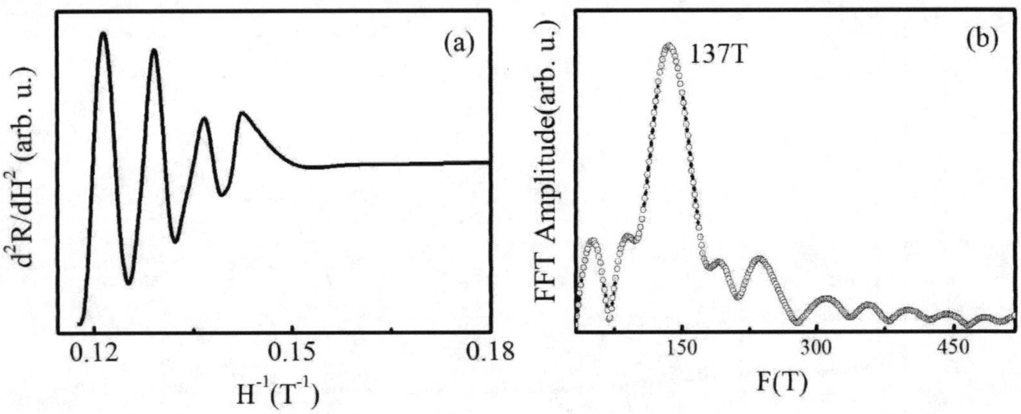

Figure 3. (a) Variation of d^2R/dH^2 as a function of H^{-1}, (b) FFT data obtained from d^2R/dH^2.

The obtained value of MR at 8.5 T is as high as 1.84×10^5 % which is also a signature of the excellent crystal quality.

Earlier studies proposed perfect electron-hole compensation and high mobility as the reason for such extremely high magnetoresistance. In order to probe the presence of such a scenario in this material, we have measured resistivity as a function of magnetic field at low temperature. We observe oscillations in the experimental data representing the presence of Shubnikov-de Hass oscillations (SdH) in this sample. To see the clear SdH oscillations, second derivative of resistance i.e., d^2R/dH^2 as a function of H^{-1} has been shown in Fig. 3(a) for the data at 10K. Fast Fourier Transform (FFT) of the SdH oscillations is shown in figure 3(b). Single peak corresponding to 137T has been observed in the FFT graph which indicates nearly same size of Fermi pockets involved in the SdH oscillations.

4. Conclusions

Single crystals of WTe_2 have been successfully prepared by the self-flux method. Analysis of the XRD pattern shows the pure phase of the samples. Electrical resistance measurement exhibits metallic nature of the sample along with a very high value of RRR that indicates excellent quality of the crystals. Extremely high value of MR up to 1.84×10^5 % has been found. SdH oscillations have been observed at 10 K as a function of magnetic field. FFT of the SdH oscillations shows single peak which indicates nearly same size of electron-hole pockets in this sample. These results support electron-hole compensation as a reason for XMR in this material.

5. Acknowledgement

A.S. acknowledges financial support from SERB, Government of India, under the project no. PDF/2018/003558. Financial support from DAE, Govt. of India (Project Identification No. RTI4003, DAE OM No. 1303/2/2019/R&D-II/DAE/2079 dated 11-02-2020). K.M. acknowledges financial assistance from DAE under the DAE-SRC-OI Award program.

References

[1] Ali M N, Xiong J, Flynn S, Tao J, Gibson Q D, Schoop L M, Liang T, Haldolaarachchige N, Hirschberger M, Ong N P and Cava R J 2014 *Nature* **514** 205
[2] Hasan M Z and Kane C L 2010 *Rev. Mod. Phys.* **2010** 823045
[3] Biswas D *et al.*, 2015 *Sci. Rep.* **5** 1
[4] Moore J E, 2010 *Nature* **464** 194
[5] Pramanik A *et al*, 2020 *Phys. Rev. B* **10** 035426
[6] Biswas D, and Maiti K 2016 *J. Electron Spectrosc. Relat. Phenom.* **208**, 90
[7] Arakane T, Sato T, Souma S, Kosaka K, Nakayama K, Komatsu M, Takahashi T, Ren Z, Segawa K and Ando Y 2012 *Nat. Commun* **24** 1
[8] Pramanik A *et al.* 2021 *Phys. Rev. B* **103** 155401
[9] Hao YJ, Liu P, Feng Y, Ma XM, Schwier EF, Arita M, Kumar S, Hu C, Zeng M, Wang Y, Hao Z 2019 *Phys. Rev. X* **9** 041038
[10] Biswas D and Maiti K 2015 *EPL* **110** 17001
[11] Pletikosić I, Ali M N, Fedorov AV, Cava RJ and Valla T 2014 *Phys. Rev. lett.* **113** 216601
[12] Jiang J *et al.* 2015 *Phys. Rev. lett.* **115** 166601
[13] Sakhya A P, Paulose P L, Thamizhavel A and Maiti K 2021 *Phys. Rev. Mater.* **5** 054201
[14] Xiang F X, Veldhorst M, Dou S X, Wang X L 2015 *EPL* **112** 37009
[15] Cai P L, Hu J, He L P, Pan J, Hong X C, Zhang Z, Zhang J, Wei J, Mao Z Q and Li SY 2015 *Phys. Rev. lett.* **115** 057202
[16] Mar A, Jobic S, Ibers J A 1992 *J. Am. Chem. Soc.* **114** 8963

Surface states in noncentrosymmetric superconductor BiPd

Arindam Pramanik[1], Ram Prakash Pandeya[1], Denis V Vyalikh[2,3], Alexander Generalov[4], Paolo Moras[5], Asish K Kundu[5], Polina M Sheverdyaeva[5], Carlo Carbone[5], Bhanu Joshi[1], A Thamizhavel[1], S Ramakrishnan[1], and Kalobaran Maiti[1*]

[1]Department of Condensed Matter Physics and Materials Science, Tata Institute of Fundamental Research, Homi Bhabha Road, Colaba, Mumbai - 400005, India
[2]Donostia International Physics Center (DIPC), 20018 Donostia San Sebastián, Basque Country, Spain
[3]IKERBASQUE, Basque Foundation for Science, 48013, Bilbao, Spain
[4]MAX IV Laboratory, Lund University, PO Box 118, 22100, Lund, Sweden
[5]Istituto di Struttura della Materia, Consiglio Nazionale delle Ricerche, I-34149 Trieste, Italy

*kbmaiti@tifr.res.in

Abstract. BiPd is a noncentrosymmetric superconductor with Dirac-like surface states on both (010) and $(0\bar{1}0)$ faces. The Dirac cone on (010) surface is intense and appears at 0.66 eV binding energy. These states have drawn much attention due to contradictory reports on dimensionality and the momentum of these Dirac fermions. We have studied the properties of these Dirac fermions using varied photon energies and different experimental conditions. The behavior of the Dirac cone is found to be two-dimensional. In addition, we found few more surface states appearing at higher binding energies compared to the Dirac cone.

1. Introduction

Topological insulators host symmetry protected states at the surface of band insulators. Bi_2Se_3 is one of the most studied topological insulators [1-4]. Efforts have been made to find such surface states in superconductors which have significant fundamental and technological interests. BiPd is proposed to be one such example which is a noncentrosymmetric superconductor with superconducting transition temperature of 3.8 K [5,6]. Extensive photoemission studies have shown that single crystalline BiPd harbors interesting surface and bulk electronic structure and the Dirac like surface states appear on both (010) and $(0\bar{1}0)$ surfaces [7-11]. Spin-resolved photoemission measurements [8] have confirmed the spin-polarization of the surface states appearing on (010) face. Scanning tunneling spectroscopy study [12] shows presence of surface states above the Fermi level also. It has been found [13] that the surface states appearing below the Fermi level are located at $\bar{S}$, while those appearing above the Fermi level are located at $\bar{\Gamma}$ (see figure 1(a)). The Dirac like surface states on (010) surface lie at 0.66 eV binding energy and are imaged clearly in ARPES experiments. But the states on $(0\bar{1}0)$ surface lie near the Fermi level in the vicinity of the bulk states and are not resolved properly in ARPES experiments. Because of twinning in the BiPd single crystals, simultaneous ARPES imaging of both (010) and

($0\bar{1}0$) surfaces is possible [10]. At the same time it becomes difficult to tell on which face a particular surface state resides.

The Dirac like states which appear at 0.66 eV binding energy on (010) face have drawn much attention due to conflicting reports about its properties. Both dimensionality and momentum of these bands are under scrutiny. While density functional theoretic calculations conclude it to be two-dimensional, photon energy dependent ARPES [9] shows its possible three-dimensional nature. Also, few studies report the Dirac point to be located at the centre of the surface Brilloiun zone $\bar{\Gamma}$ [8,9], while others report it to be at the edge of the surface Brillouin zone $\bar{S}$ (see figure1(a)) [10,13]. In this work [11], we have shown photon energy dependent results, which are consistent with the two-dimensional nature of the Dirac band. Apart from this Dirac state, we have discovered few more surface states.

2. Experimental Details

Single crystals of BiPd were prepared using modified Bridgman method by Joshi *et. al.* [5]. Good crystallinity was ensured by Laue diffraction method. Samples were cleaved inside the vacuum chamber along the *b* axis for ARPES measurements. ARPES were carried out in I05 beamline, Diamond light Source, UK and VUV beamline, Elettra, Italy. Total energy resolutions in the Diamond light source and Elettra were 5 meV and 10 meV, respectively.

3. Results and Discussions

To find the momentum and dimensionality of the Dirac bands, experiments were performed in two experimental geometries using various photon energies. Sample position was optimized at one photon energy to get a clear image of the Dirac point, i.e. without any energy gap at the Dirac point. But as the photon energy is varied, an energy gap at the Dirac point opens up in both cases. The energy

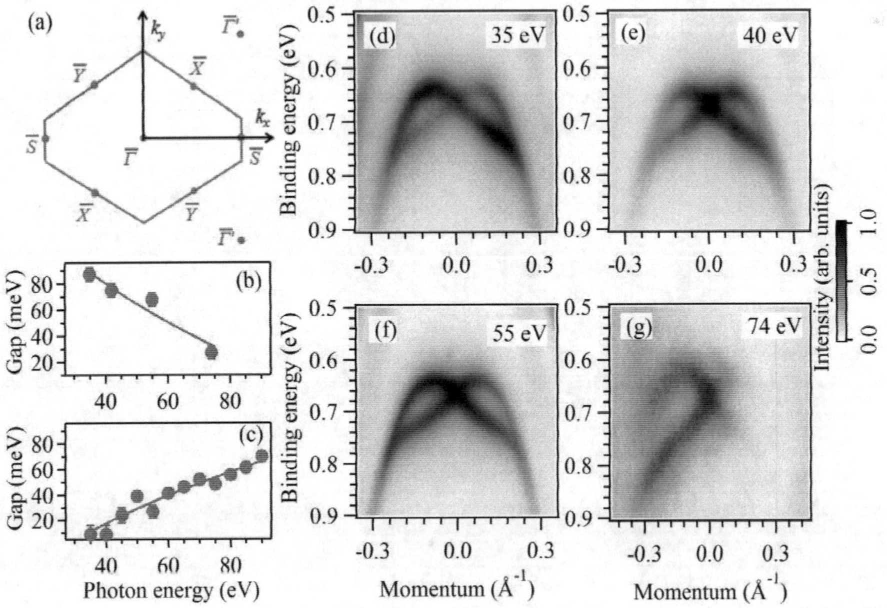

Figure 1. (a) Surface Brillouin zone of BiPd. Energy gap at the Dirac point is shown for two sets of data using the sample position optimized at (b) 74 eV and (c) 35 eV photon energies. Zero energy gap can be obtained by optimizing the sample at each photon energy. Spectra taken along $\bar{\Gamma}$-$\bar{S}$-$\bar{\Gamma}$ at (d) 35 eV, (e) 40 eV, (f) 55 eV, and (g) 74 eV photon energies are shown.

gaps at the Dirac point obtained at different photon energies for two sets of measurements are shown in figures 1(b) and (c). The gap shows opposite trends for the two sets. The gradual opening of energy gap with photon energy may indicate a possible three-dimensional character of the Dirac band. It is known that the properties of a three-dimensional band show repetition as the probing photon energy is varied. Because the photon energy is directly linked to the out-of-plane momentum k_z. But our study over a large photon energy range spanning several Brillouin zones shows no periodicity in the gap value which is clear from figure 1(c). Moreover, one can make the gap zero by reoptimizing the sample position at different non-equivalent photon energies.

Opening of energy gap at the Dirac point can also happen due to the sample geometry. Features appearing at $\bar{\Gamma}$ point will remain unchanged even if the photon energy is changed. Finite momentum of the Dirac point will correspond to different emission angle if the photon energy is changed. Thus, the above observations suggest that the position of the Dirac point maybe away from the $\bar{\Gamma}$-point. By analyzing the angular position of the Dirac point in the spectrum at various energies we found that the Dirac cone on (010) surface is located at $\bar{S}$ [11]. After accounting for the gap arising due to the sample geometry, we get zero-gap condition at each photon energy. Evidently, the bandgap at the Dirac point is not an inherent property of the sample. The sample position can be optimized at different photon energies to image the Dirac point without any energy gap. Figures 1(d)-(g) show the Dirac cone with zero-gap at few photon energies. This supports the two-dimensional nature of the Dirac band.

The dispersion of the Dirac band very quickly deviates from the linear dispersion as is clear from figures 1(d)-(g). Linear energy dispersion is maintained only very close to the Dirac point. It also looks very different from parabolic dispersion. Hence the consideration of the mass term, which is quadratic in momentum, is important; but it also cannot explain the dispersion well. Quadratic mass term and Rashba-like linear term together give rise to a parabolic dispersion. This suggests the importance of higher-order spin-orbit terms in the Hamiltonian of the Dirac states [14,15]. We find that the spin-orbit terms up to third order in momentum can explain all the experimental observations well [11].

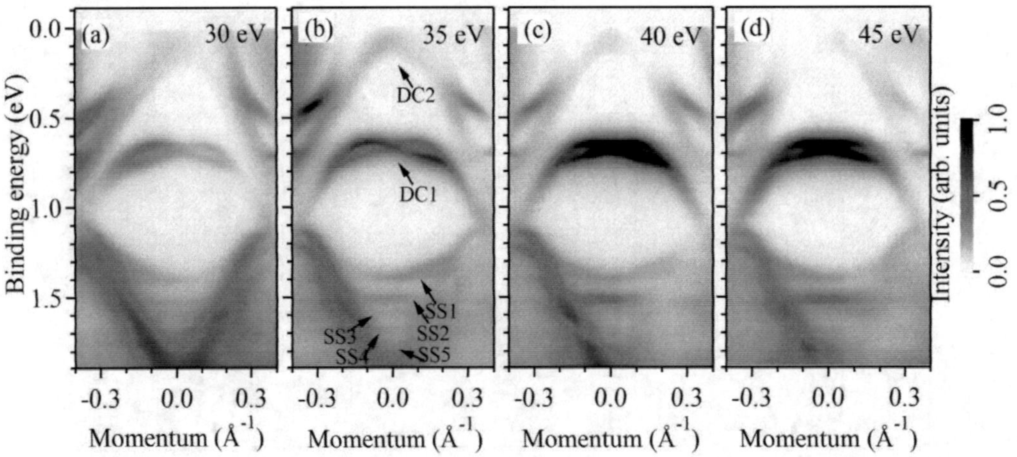

Figure 2. Spectra at (a) 30 eV, (b) 35 eV, (c) 40 eV, and (d) 45 eV. DC1 and DC2 in (b) denote Dirac cones on (010) and (0$\bar{1}$0) surface, respectively. SS1-SS5 denote additional surface states.

Next, we show that apart from the well-known Dirac cone there exist several other surface states in single-crystalline BiPd. Spectra collected at various photon energies are shown in figures 2(a)-(d). In figure 2(b), DC1 denotes the Dirac cone on (010) surface and DC2 is the Dirac cone on $(0\bar{1}0)$ surface. Apart from DC1 and DC2 we see that several other states labeled as SS1-SS5 are present in the spectrum. All of these states are present in the spectrum of every photon energy shown in figure 2. Binding energies of the bands SS1-SS5 do not vary as the photon energy is changed suggesting that they are two-dimensional in nature [11]. Also, these bands are not seen in the bulk band structure calculations. These results suggest that these bands represent the surface states of BiPd.

4. Conclusions

In summary, we have done photon energy-dependent ARPES on BiPd single crystals. The Dirac cone acquired using various photon energies and sample optimizations unambiguously suggest two dimensional behavior of the Dirac fermions and the Dirac point appears at $\bar{S}$. In addition, we discover several other surface bands at higher binding energies. These results suggest coexistence of the Dirac fermions and the normal surface states in this system. The binding energy of the Dirac Fermions is too high to have significant role in the ground state properties of this system.

Acknowledgements
The authors acknowledge financial support from the Department of Atomic Energy, Government of India, under Project No. 12-R&D-TFR-5.10-0100, Diamond Light Source for beamtime at beamline I05 under Proposal No. SI11512. D.V.V. acknowledges support from the Spanish Ministry of Economy (No. MAT-2017-88374-P). K.M. acknowledges financial support from DAE, Government of India, under the DAE-SRC-OI Award program.

References
[1] Zhang H, Liu C X, Qi X L, Dai X, Fang Z and Zhang S C 2009 *Nat. Phys.* **5** 438
[2] Chen Y L *et al.* 2010 *Science* **329** 659
[3] Kong D *et al.* 2011 *ACS Nano* **5** 4698
[4] Biswas D, Thakur S, Ali K, Balakrishnan G and Maiti K 2015 *Sci. Rep.* **5** 10260; Biswas D, Thakur S, Balakrishnan G and Maiti K 2015 *Sci. Rep.* **5** 17351; Biswas D and Maiti K 2015 *Europhys. Lett.* **110** 17001; Biswas D and Maiti K 2016 *J. Electron. Spectrosc. Relat. Phenom.* **208** 90
[5] Joshi B, Thamizhavel A and Ramakrishnan S 2011 *Phys. Rev.* B **84** 064518
[6] Mondal M, Joshi B, Kumar S, Kamlapure A, Ganguli S C, Thamizhavel A, Mandal S S, Ramakrishnan S and Raychaudhuri P 2012 *Phys. Rev.* B **86** 094520
[7] Pramanik A *et al.* 2020 *Phys. Rev.* B. **101** 035426
[8] Neupane M *et al.* 2016 *Nat. Commun.* **7** 13315
[9] Thirupathaiah S *et al.* 2016 *Phys. Rev. Lett.* **117** 177001
[10] Benia H M *et al.* 2016 *Phys. Rev.* B **94** 121407(R)
[11] Pramanik A *et al.* 2021 *Phys. Rev.* B. **103** 155401
[12] Sun Z, Enayat M, Maldonado A, Lithgow C, Yelland E, Peets D C, Yaresko A, Schnyder A P, and Wahl P 2015 *Nat. Commun.* **6** 6633
[13] Yaresko A, Schnyder A P, Benia H M, Yim C M, Levy G, Damascelli A, Ast C R, Peets D C, and Wahl P 2018 *Phys. Rev.* B **97** 075108
[14] Fu L 2009 *Phys. Rev. Lett.* **103** 266801
[15] Liu C X, Qi X L, Zhang H, Dai X, Fang Z and Zhang S C 2010 *Phys. Rev.* B **82** 045122

Quasi-Flat-Band in s1/s2 Pyrochlore Oxides and the Effect of Spin-Orbit Interaction

I Hase[1,*], Y Higashi[1] and T Yanagisawa[1]

[1] National Institute of Advanced Industrial Science and Technology (AIST), Tsukuba, 305-8568, Japan
*E-mail: i.hase@aist.go.jp

Abstract. An isotropic tight-binding model with the nearest-neighbour hopping on a pyrochlore lattice gives a rich variety of physical properties due to the emergence of the flat-band. Moreover, by introducing spin-orbit coupling into this model, the topological properties of the system changes significantly. This model is well applicable to some pyrochlore oxides called s1/s2 pyrochlores. In this paper we apply this model to an s2 pyrochlore oxide $Pb_2Ta_2O_7$ and found a characteristic quadratic touching of the quasi-flat band and dispersive band. Furthermore, when a ferromagnetic order appears due to this quasi-flat band, a pair of Weyl points appears in that direction.

1. Introduction

The energy of electrons in a crystal generally has dispersion. However, in a model such as a nearest-neighbor isotropic tight-binding (NNITB) model on pyrochlore lattice, flat bands without dispersion appear due to the quantum interference effect. In this flat band model, attractive physical properties such as perfect ferromagnetism, high-temperature superconductivity, and fractional quantum Hall effect are predicted [1,2]. However, in the actual pyrochlore compound, there has been no example so far in which an approximate flat band was observed near the Fermi level due to the influence of the anisotropy of the transfer integral. Based on first-principles calculations, we have found that such a system in which the uppermost part of the valence band is mainly the s-orbital of the A site, such as $Sn_2Nb_2O_7$ and $Tl_2Nb_2O_7$, is such an example [3-5]. Hereafter we call a system containing one s-electron in the outermost orbital as "s1 system" (for example $Tl_2Nb_2O_7$), and a system containing two s-electrons as "s2 system" (for example $Sn_2Nb_2O_7$). The bandwidth of the top of the valence band is very narrow, even though they are composed of extended s-orbitals. Hereafter we call this band as quasi-flat band. This quasi-flat band is sometimes found in two-dimensional materials, but extremely rare in three-dimensional materials. Recently, a calculation result has been reported that three-dimensional quasi-flat bands can be seen in a binary compound WN_2 [6].

NNITB model on the pyrochlore lattice is also drawing attention from the perspective of topological insulator [7]. Guo and Franz considered a model in which a spin-orbit coupling (SOC) term that does not break crystal translational symmetry is added to the NNITB model (GF model) [7]. At that time, the GF model could not be applied to specific material, but now it can be applied to the above s1/s2 pyrochlore oxides. In the GF model, the topological properties of the system are governed by the sign

of the SOC constant λ. Especially in the half-filled case, $\lambda < 0$ leads to a strong topological insulator (STI). On the other hand, $\lambda > 0$ leads to a semi-metallic state where the bands touch quadratically at the Γ point. Similar results have been obtained using other type of SOC by Kurita *et al.* [8]. These properties are expected to hold in real s1/s2 materials because the topological properties may not depend on the details of the band structure [9].
In this paper we try to reproduce the valence band of $Pb_2Ta_2O_7$ [9], which is one of the s2 compound, by using GF model with adjusting parameters. We chose this compound because $Pb_2Ta_2O_7$ has relatively narrow quasi-flat band and large SOC. We found that the magnitude of λ is about one order smaller than that shown in Ref. 7. Nevertheless, the band topology is unchanged and a characteristic quadratic touching of the bands at the Γ point. Furthermore, we considered a state peculiar to the flat-band system, i.e. a case where the system undergoes a phase transition to a ferromagnetic state due to doping. When $\lambda > 0$, the zero-gap semi-metallic state that was touched quadratically at the Γ point is almost gapped out by the Zeeman term, but a pair of Weyl points in the direction of the Zeeman field. This behavior is similar to the case where $\lambda < 0$, i.e. (topological) insulating phase transfers to the semi-metallic phase and the Weyl points appear by adding the Zeeman term [10].

This paper is organized as follows: The computational details are described in Section 2. The calculated results and the discussions are shown in Section 3. Finally, we give some conclusions in Section 4.

2. Computational Details

Guo-Franz (GF) model consists of two terms shown in the following Hamiltonian:

$$H = -t \sum_{<ij>\sigma} c^{\dagger}_{i\sigma} c_{j\sigma} + i\lambda \sum_{\ll ij \gg \alpha\beta} \left(\boldsymbol{d}^1_{ij} \times \boldsymbol{d}^2_{ij}\right) \cdot \boldsymbol{\sigma}_{\alpha\beta}\, c^{\dagger}_{i\alpha} c_{j\beta} \quad (1)$$

The first term denotes the non-interacting tight-binding Hamiltonian. t is the hopping parameter and $<ij>$ denotes nearest neighbors. The lattice sum is taken on the pyrochlore lattice. When the second term is zero (i.e. $\lambda = 0$), this Hamiltonian is exactly diagonalized and the eigenvalues are

$$E^{1,2}_k = -2t\left[1 \pm \sqrt{1 + A_k}\right],\ E^{3,4}_k = 2t \quad (2)$$

where $A_k = \cos(2k_x)\cos(2k_y) + \cos(2k_y)\cos(2k_z) + \cos(2k_z)\cos(2k_x)$. E^3 and E^4 do not depend on the wave vector $\boldsymbol{k}$, so they are called flat bands (FBs). Note that $A = 3$ at $\boldsymbol{k} = 0$ (Γ point), so $E^{1,2} = -6t$ and $2t$. Therefore, FBs touch one of the dispersive bands E^2 at the Γ point and forms a triply degenerated state (six-fold degenerated state if we include spin degree of freedom). Since we can see that $A = 3$ is the maximum value of A, the overall bandwidth of the entire system (E^1 to E^4) becomes $8t$.

Next, we include spin-orbit coupling (SOC) shown in the second term in Eqn. 1. This term is taken so as not to break crystal symmetry further, and to preserve time-reversal symmetry. Here λ is the SOC constant, $\boldsymbol{d}_{ij}^1$ and $\boldsymbol{d}_{ij}^2$ are the nearest-neighbor vectors traversed between second neighbors i and j. The summation $\ll ij \gg$ is taken on the pyrochlore lattice such that i and j are the (Euclid) second neighbor sites. $\boldsymbol{\sigma}$ denotes the vector of Pauli spin matrices. Due to this SOC term, we must diagonalize 8 x 8 matrix in $\boldsymbol{k}$-space. We have calculated the band structure of the Hamiltonian (1) by using the program package PythTB [11].

3. Results and Discussions

We calculate the band structure of the Hamiltonian (1) using the program package PythTB [11]. The results are shown in Figure 1(a). We take the parameters t = 0.3 eV and λ = 2.0 meV, which well reproduce the valence band structure of a s2 material $Pb_2Ta_2O_7$ [9]. Since the ratio $\lambda/t \sim 6.7\times10^{-3}$ which is much smaller than that shown in ref.7, the effect of SOC to the band dispersion is limited to the vicinity of the Γ point. Nevertheless, the topology of the band dispersion is the same with that in ref.7. Note that the quadratic touching of the bands at the Γ point is recognized as a quadratic contact point in s1-pyrochlore compound $Tl_2Nb_2O_7$ [12].

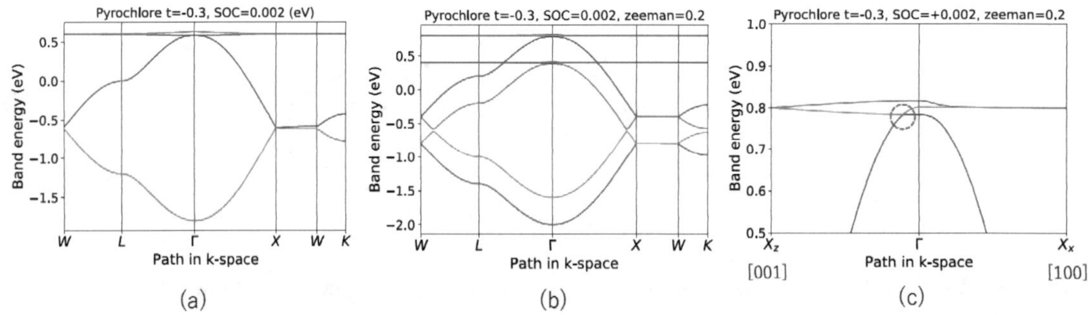

Figure 1. Energy dispersion curves of (a) GF model with t = 0.3 eV and λ = 2 meV, (b) GF model with t = 0.3 eV, λ = 2 meV with including Zeeman term λ_Z = 0.2 eV, and (c) an enlarged view of the panel (b) near the Γ point, while the $\boldsymbol{k}$-direction is [001] and [100]. A Weyl point is shown in dashed circle.

As shown in Introduction, (quasi-)flat band forms a very large density of states, and when the Fermi level (E_F) comes there, the nonmagnetic state becomes unstable, and the ferromagnetic state stabilizes [1-5]. We have found ferromagnetic ground state for the hole-doped s2 pyrochlore system using first-principles calculations [3-5]. Recently, Zhou *et al.* have shown that when SOC is introduced into this ferromagnetic state, a new type of Weyl points emerges [10]. $Sn_2Nb_2O_7$ has a topological gap at the Γ point, and the energy splitting Δ which is defined in the previous subsection is ~ -5 meV. They included the Zeeman term

$$H_Z=\lambda_Z \sum_{\alpha\beta} \sigma^z_{\alpha\beta} c^\dagger_{i\alpha} c_{i\beta} \tag{3}$$

into the GF-model with using the parameter λ_Z = 0.2 eV as half of the exchange splitting in hole-doped $Sn_2Nb_2O_7$.

Since most of the s1/s2 pyrochlore oxides have $\lambda > 0$, it is important to investigate the Zeeman effect to the GF model with $\lambda > 0$. We introduced the Zeeman term of λ_Z = 0.2 eV into the GF model with t = 0.3 eV and λ = 2 meV that modeled $Pb_2Ta_2O_7$. The direction of the Zeeman field was set along the z-axis [001]. The results are shown in Fig. 1(c).
As shown in Fig.1(a), since the GF-model has time-reversal symmetry and the pyrochlore lattice has also space inversion symmetry, all bands are doubly degenerated. The Zeeman term breaks the time-reversal symmetry, and all the bands are split as shown in Fig.1(b). Since $\lambda_z >> \lambda$, these split bands

almost have the character with spin-up or spin-down. The width of the topmost band ($E^{3,4}$ in Eqn.2) narrows significantly with the introduction of the Zeeman term. Figure 1(c) shows an enlarged view of the topmost band of Fig.1(b) near E_F, but the ***k***-direction is for [001] and [100]. When $\lambda_z = 0$, these two directions are equivalent due to the cubic lattice symmetry. However, the Zeeman term partially breaks this symmetry via SOC. Figure 1(c) apparently shows this symmetry breaking. As for the direction [001], which is parallel to the Zeeman field, the quasi-flat band and the dispersive band cross each other. This band crossing is also seen when $\lambda = -2$ meV (not shown), which has been shown to be a Weyl point [10]. This Weyl point at $(0, 0, k_z)$ in Fig.1(c) also appears at $(0, 0, -k_z)$ due to symmetry. As shown in the right half in Fig.3(c), Weyl points do not appear in any direction other than [001].
It is interesting to compare these results with the case of $\lambda < 0$, which seems to be realized in $Sn_2Nb_2O_7$ [10]. When $\lambda < 0$, a pair of Weyl points also appears when the Zeeman term is added. The difference from the present result is that when $\lambda < 0$, there is a parameter region in which four Weyl points appear, whereas when $\lambda > 0$, there are always two Weyl points. This is because the energy levels at the Γ point are reversed between these two cases. It is a future task to investigate the detailed properties of these Weyl points. We note that there are many s1/s2 pyrochlore oxides which have $\lambda > 0$ [9], and these compounds have ferromagnetic ground state when doped appropriately [5]. Therefore, our result could be applied for many realistic materials.

4. Conclusion

We have applied the Guo-Franz model to the valence band structure of the s2 pyrochlore oxide $Pb_2Ta_2O_7$. We determined the parameters of the Guo-Franz model and found that the effect of spin-orbit coupling is almost limited to the vicinity of the Γ point. We have found that the semi-metallic quadratic touching point when the spin-orbit coupling constant is positive. We also introduced the Zeeman term into this model, simulating the possible ferromagnetic ground state with doping. This Zeeman term creates a pair of Weyl points in direction of the Zeeman field.

Acknowledgments

This work was partially supported by KAKENHI (Grant No. JP19K03731) from Japan Society for the Promotion of Science (JSPS).

References

[1] Liu Z *et al.* 2014 *Chin. Phys. Lett.* B**23** 077308
[2] Derzhko O *et al.* 2015 *Int. J. Mod. Phys.* B**29** 1530007
[3] Hase I *et al.* 2018 *Phys. Rev. Lett.* **120** 196401
[4] Hase I *et al.* 2018 *Nanoscale. Res. Lett.* **13** 63
[5] Hase I *et al.* 2019 *Nanomaterials* **9** 876
[6] Ye X B, Tuo P and Pan B C, 2020 *J. Chem. Phys.* **152** 224503
[7] Guo H -M and Franz M 2009 *Phys. Rev.* Lett. **103** 206805
[8] Kurita M, Yamaji Y and Imada M 2011 J. Phys. Soc. Jpn. **80** 044708
[9] Hase I and Yanagisawa T 2020 *Symmetry* **12** 1076
[10] Zhou Y *et al.* 2019 *Phys. Rev.* B**99** 201105(R)
[11] Coh S and Vanderbilt D 2016 PythTB, *A Software Package Providing a Python Implementation of the Tight-Binding Approximation*: http://www.physics.rutgers.edu/pythtb/
[12] Zhang W *et al.* 2019 *npj Comput. Mater.* **5** 105

Journal of Physics: Conference Series **2164** (2022) 012064 doi:10.1088/1742-6596/2164/1/012064

Properties of the spin-momentum locked state

Mauro M. Doria
Instituto de Física, Universidade Federal do Rio de Janeiro, 21941-972 Rio de Janeiro, Brazil
E-mail: mauromdoria@gmail.com

Edinardo I. B. Rodrigues
Universidade Federal Rural de Pernambuco,54518-430,Cabo de Santo Agostinho, Pernambuco, Brazil

Abstract. Spin-momentum locking is becoming a cornerstone for the understanding and applications of novel condensed matter systems. Here we assume that it holds locally in position space and from it predict the existence of a local magnetic field. Although residual, this local magnetic field is important because it brings the topological stability that transforms particles into quasi-particles. The present approach shows that the Rashba term is already contained in the Schrödinger kinetic energy and a Dirac linear spectrum can be obtained without invoking a Dirac equation for the particles.

1. Introduction

Recently there has been growing evidence that spin is locked perpendicularly to the momentum (*spin momentum locking*) in topological insulators [1, 2], evanescent electromagnetic waves [3, 4] and high-temperature cuprate superconductors [5]. The development of angle-resolved photoemission spectroscopy able to determine also the spin (SARPES) has been instrumental to detect spin-momentum locking in several systems [5]. Here we assume that spin-momentum locking is locally satisfied by the physical state Ψ:

$$\vec{\sigma} \cdot \vec{D}\Psi = 0, \tag{1}$$

where $\vec{D} = \hbar\vec{\nabla}/\imath - e\vec{A}/c$ is the local gauge invariant momentum (e is the particle charge, $\vec{A}$ is the vector potential), and $\hbar\vec{\sigma}/2$ is the spin (σ_i, $i = 1, 2, 3$ are the Pauli matrices). We show here that a direct consequence of the spin-momentum locking condition is the presence of a local magnetic field $\vec{h}$, intimately connected to Eq.(1), due to the presence of a divergenceless field, namely, $\vec{\nabla} \cdot (\Psi^\dagger\vec{\sigma}\Psi) = 0$. This follows straightforwardly from the identity $\Psi^\dagger(\vec{\sigma} \cdot \vec{D}\Psi) - (\vec{\sigma} \cdot \vec{D}\Psi)^\dagger\Psi = 0$ which yields that $\Psi^\dagger(\vec{\sigma} \cdot \vec{\nabla}\Psi) + (\vec{\sigma} \cdot \vec{\nabla}\Psi)^\dagger\Psi = 0$. This could be interpreted as a signal of a conserved current but in reality, it is just one of Maxwell's equation being satisfied, $\vec{\nabla} \cdot \vec{h} = 0$, $\vec{h} \sim \Psi^\dagger\vec{\sigma}\Psi$. This assessment is a direct consequence of the *three-term decomposition* theorem of the kinetic energy.

2. The three-term decomposition theorem

The fermionic particles satisfy a standard parabolic dispersion relation, also referred as the Schrödinger kinetic energy:

$$K = \int d^3x \frac{1}{2m} |\vec{D}\Psi|^2, \quad \Psi = \begin{pmatrix} \psi_\uparrow \\ \psi_\downarrow \end{pmatrix}. \tag{2}$$

The three-term decomposition, also known as the Lichnerowicz-Weitzenböck decomposition [6], states that the Schrödinger kinetic energy is the sum of the square of the spin-momentum interaction, plus the curvature interaction, and the Rashba term or spin-orbit interaction. In summary, the kinetic energy of Eq.(2) admits a distinct but equivalent formulation given by,

$$K = \int d^3x \left\{ \frac{1}{2m} |\vec{\sigma} \cdot \vec{D}\Psi|^2 + \frac{\hbar e}{2mc} \vec{h} \cdot \left(\Psi^\dagger \vec{\sigma} \Psi\right) - \frac{\hbar}{4m} \vec{\nabla} \cdot \left[\Psi^\dagger \vec{\sigma} \times (\vec{D}\Psi) - (\vec{D}\Psi)^\dagger \times \vec{\sigma}\Psi\right] \right\}. \tag{3}$$

In case of a layer, the Rashba term becomes a two-dimensional integral along the plane $x_3 = 0$: $\int_{x_3>0} d^3x \{-\frac{\hbar}{4m}\vec{\nabla}\cdot[\Psi^\dagger\vec{\sigma}\times(\vec{D}\Psi)-(\vec{D}\Psi)^\dagger\times\vec{\sigma}\Psi]\} = \int d^2x \, \{\frac{\hbar}{4m}[\Psi^\dagger\hat{x}_3\cdot(\vec{\sigma}\times\vec{D})\Psi + c.c.]\}_{x_3=0}$. The three-dimensional space is sectioned by the layer in two parts, above ($x_3 > 0$) and below ($x_3 < 0$) in this formalism such that Maxwell's equations rejoin them through boundary conditions. Remarkably, according to this theorem, the Rashba term is contained in the Schrödinger kinetic energy. From the present point of view a strong spin-orbit interaction is just a way to trigger the breaking of the kinetic energy into its three parts. The spin-momentum locking can also be regarded as a helicity locking condition. The helicity is the total angular momentum projected along the momentum, $\vec{J} = \vec{x} \times \vec{D} + \hbar\vec{\sigma}/2$ and so, $\vec{J} \cdot \vec{D} = \hbar\vec{\sigma} \cdot \vec{D}/2$. Notice that the helicity, and consequently also the spin-momentum locking, are pseudo-scalar quantities, and so, not invariants under a reflection symmetry along the layer. In fact the arise of the local magnetic field is allowed by the breaking of the time reversal symmetry. Notice however that the kinetic energy remains symmetric because only the square of the spin-momentum locking equation enters on it. Thus we are in face of a typical example of a ground state with lower symmetry than the underlying energy.

Ampère's law determines the local magnetic field, as shown below. It is well known that the electromagnetic current can be obtained by varying the kinetic energy with respect to the vector potential, $\delta K = -\int d^2x \, \vec{J} \cdot \delta\vec{A}/c$. Starting with Eq.(2) the current is,

$$\vec{J} = \frac{e}{2m}\left(\Psi^\dagger \vec{D}\Psi + c.c.\right). \tag{4}$$

However the same procedure once applied to Eq.(3) yields a distinct but equivalent formulation of the current.

$$\vec{J} = \frac{e}{2m}\left[(\vec{\sigma}\Psi)^\dagger(\vec{\sigma}\cdot\vec{D}\Psi) + (\vec{\sigma}\cdot\vec{D}\Psi)^\dagger(\vec{\sigma}\Psi)\right] - \frac{\hbar e}{2m}\vec{\nabla}\times\left(\Psi^\dagger\vec{\sigma}\Psi\right). \tag{5}$$

Remarkably, in the latter case, Ampère's law admits an exact solution by the assumption of a spin-momentum locked state. To see this introduce Eq. (1) into the current expressed by Eq.(5), to obtain that,

$$\vec{J} = -\frac{\hbar e}{2m}\vec{\nabla}\times\left(\Psi^\dagger\vec{\sigma}\Psi\right). \tag{6}$$

Then one proceeds to solve Ampère's law, $\vec{\nabla} \times \vec{h} = 4\pi\vec{J}/c$, and obtain that the local magnetic field is,

$$\vec{h} = \vec{H} - 4\pi\mu_B\Psi^\dagger\vec{\sigma}\Psi, \tag{7}$$

where $\mu_B = \hbar e/2mc$ is Bohr's magnetron and $\vec{H}$ is an external applied field.

The first term of the three-term kinetic energy, Eq. (3), is intrinsically positive and vanishes by imposing Eq. (1). This suggests that the spin-momentum locking always lowers the kinetic energy. A scenario broader than the spin-momentum locking is to assume that the Weyl equation holds, [7, 8] namely, $\vec{\sigma} \cdot \vec{D}\Psi = \theta\Psi$, θ being a new parameter. The outcome is the onset of two Fermi surfaces, an inner one with low energy states as compared to the outer one.

The kinetic energy of the spin-momentum locked state is obtained by introducing Eqs.(1) and (7) into the kinetic energy, and this gives that,

$$K = 2\pi\mu_B^2 \int d^3x \left[-(\Psi^\dagger\vec{\sigma}\Psi)^2 + \frac{1}{2\pi r_e}\vec{\nabla}^2(\Psi^\dagger\Psi) \right], \tag{8}$$

where $r_e = e^2/mc^2$ is the electron's classical radius ($\hbar^2/4m = \mu_B^2/r_e$). Consider that $\Psi \sim 1/\sqrt{V}$, $V = L_3 A$, where L_3 is an arbitrary length and $A = L_1 L_2$ is the unit cell area in the layer. It follows that $\int d^3x\, (\Psi^\dagger\vec{\sigma}\Psi)^2 \sim 1/V$ and $\int d^3x \vec{\nabla}^2(\Psi^\dagger\Psi) \sim 1/A$. The laplacian acts in the plane and so, $\nabla^2 \sim 1/A$. Therefore these two terms contribute to the kinetic energy in the following fashion, $K \sim 2\pi\mu_B^2(-1/L_3 + 1/2\pi r_e)/A$. In summary, under the condition that $L_3 >> r_e$ the fourth order term is negligible in comparison with the second order one. L_3 is fixed by the velocity of the Dirac linear spectrum [7], $L_3 \sim 2.0 \times 10^{-10}$ nm which is much larger than $r_e = 2.8 \times 10^{-15}$ m. For such reason the fourth order term is abandoned and the leading kinetic energy is,

$$K \approx \frac{\hbar^2}{4m} \int d^3x\, \vec{\nabla}^2(\Psi^\dagger\Psi). \tag{9}$$

3. The Dirac linear spectrum

The above kinetic energy expression assumes that Ψ solves Eq. (1), and this reduces it to a surface integral in the layer, located at $x_3 = 0$. To show that it corresponds to a Dirac spectrum one must take into account the solution of Eq.(1) around a layer at $x_3 = 0$.

$$\Psi(\vec{x}) = \frac{1}{\sqrt{V}} \sum_{\vec{k},\, k>0} c_{\vec{k}}\, e^{\imath\vec{k}\cdot\vec{x}} e^{-k|x_3|} \begin{pmatrix} 1 \\ -\imath\frac{x_3}{|x_3|}\frac{k_+}{k} \end{pmatrix}, \tag{10}$$

where $\vec{k} = k_1\hat{x}_1 + k_2\hat{x}_2$, $k = |\vec{k}|$, and $k_\pm = k_1 \pm \imath k_2$. Then one obtains that in momentum space the above kinetic energy becomes,

$$K = \sum_{\vec{k},\, k\geq 0} \frac{2\hbar^2}{mL_3} |\vec{k}| c_{\vec{k}}{}^\dagger c_{\vec{k}}. \tag{11}$$

The in-plane velocity v can be the independent parameter instead of L_3: $v = 2\hbar/mL_3$, and so, the Dirac spectrum is retrieved, $\hbar v|\vec{k}|$. The free coefficients $c_{\vec{k}}$ of Eq. (10) are interpreted as second quantized operators that annihilate spin-momentum locked particles. Consider a single particle state given by $|\chi\rangle = \sum_{\vec{k}} f(\vec{k})\, c^\dagger_{\vec{k}}|0\rangle$, such that $\sum_{\vec{k}} |f(\vec{k})|^2 = 1$. Then the kinetic energy expectation value is straightforwardly obtained, $\langle\chi|K|\chi\rangle = \hbar v \sum_{\vec{k},k\neq 0} |\vec{k}|\, |f(\vec{k})|^2$.

The magnetic field although small is important because it is responsible for the topological stability that turns particles into quasi-particles. From the above expression for Ψ the general features of the local field can be deduced. The field components h_1 and h_2 flip sign from one side of the layer to the other while h_3 does not. Eqs.(7) and (10) give that $\Psi^\dagger\sigma_1\Psi \sim x_3/|x_3|$, $\Psi^\dagger\sigma_2\Psi \sim x_3/|x_3|$ whereas $\Psi^\dagger\sigma_3\Psi$ does not depend on $x_3/|x_3|$. For a given state $|\chi\rangle$ one is interested to calculate $\langle\vec{h}\rangle = \langle\chi|\vec{h}|\chi\rangle = -4\pi\mu_B\langle\chi|\Psi^\dagger\vec{\sigma}\Psi|\chi\rangle$. It holds that $\langle\vec{\nabla}\cdot\vec{h}\rangle = 0$, and so, a streamline crosses the layer twice because the in-plane component, $\langle h_1\hat{x}_1 + h_2\hat{x}_2\rangle$, flips sign from above to below the layer, but not the h_3 component. The topological stability stems from the Chern-Simons index, which is calculated by taking $\hat{h} = \langle\vec{h}\rangle/|\langle\vec{h}\rangle|$.

$$Q = \frac{1}{4\pi}\int_{x_3=0^+}\left(\frac{\partial\hat{h}}{\partial x_1}\times\frac{\partial\hat{h}}{\partial x_2}\right)\cdot\hat{h}\,d^2x. \quad (12)$$

In previous publications [7, 8] $\langle\vec{h}\rangle$ was explicitly calculated and solutions such that $Q \neq 0$ found. Fig. 1 displays a pictorial view of the found local magnetic field and current of these previously calculated states [7, 8].

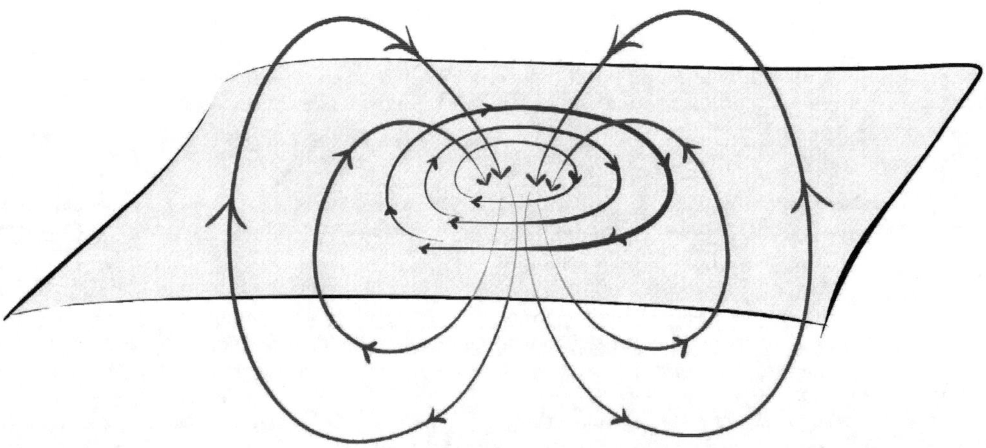

Figure 1. Pictorial view of a quasi-particle which is a spin-momentum locked state dressed by its own magnetic field and currents.

Acknowledgements
M. M. Doria thanks the Brazilian Agency Fundação de Amparo Pesquisa do Estado do Rio de Janeiro (FAPERJ) for financial support under process No. E-26/010.001497/2019.

References

[1] Kohda M, Okayasu T, and Nitta J, *Scientific Reports* 2019 **9** 1909.
[2] He P, Zhang S S-L, Zhu D, Shi S,Heinonen O G, Vignale G, and Yang H, *Phys. Rev. Lett.* 2019 **123** 016801.
[3] Mechelen T V and Jacob Z, *Optica* 2016 **3** 118.
[4] Triolo C, Cacciola A, Patanè S, Saija R, Savasta S, and Nori F, *ACS Photonics* 2017 **4** 2242.
[5] Gotlieb K, Lin C-Y, Serbyn M, Zhang W, Smallwood C L, Jozwiak C, Eisaki H, Hussain Z, Vishwanath A, and Lanzara A, *Science* 2018 **362** 1271.
[6] Vargas-Paredes A A, Doria M M, and Neto J A H, *J. of Math. Phys.* 2013 **54** 013101.
[7] Doria M M and Perali A, 2017 *Europhys. Lett.* **119** 21001.
[8] Doria M M, 2019 *Euro. Phys. J. B* **92** 64.

Anomalous transport properties of the antiferromagnetic Weyl semimetals Mn_3X (X = Sn, Ge)

Takahiro Tomita[1,2], Susumu Minami[3], Muhammad Ikhlas[1,2], Hiroto Nakamura[1], Ryotaro Arita[2,4], and Satoru Nakatsuji[1,2,3,5,6]

[1]Institute for Solid State Physics, University of Tokyo, Kashiwa, Chiba 277-8581, Japan
[2]CREST, Japan Science and Technology Agency (JST), Kawaguchi, Saitama 332-0012, Japan
[3]Department of Physics, University of Tokyo, Tokyo 113-0033, Japan
[4]Department of Applied Physics, University of Tokyo, Tokyo 113-8656, Japan
[5] Trans-scale Quantum Science Institute, University of Tokyo, Bunkyo, Tokyo 113-0033, Japan
[6] Institute for Quantum Matter and Department of Physics and Astronomy, Johns Hopkins University, Baltimore, Maryland 21218, USA

E-mail: ttomita@issp.u-tokyo.ac.jp or ttomita.takahiro@gmail.com

Abstract. Noncollinear antiferromagnets Mn_3X (X = Sn, Ge) are characterized by a large anomalous Hall effect originating from a large Berry curvature despite a vanishingly small magnetization. From recent first-principle theories, the large Berry curvature is predicted to be induced by a existence of Weyl nodes broken time-reversal symmetry. The large anomalous Nernst effect is also contributed by the magnetic Weyl state around the Fermi level E_F, and likely shares its origin with the anomalous Hall effect. The thermoelectric transport $S(T)$ and thermomagnetic transport $S_{ji}(T)$ are thus investigated in single crystals of Mn_3X. Here, Mn_3X exhibits a large anomalous Nernst effect; in particular, the signal magnitude of Mn_3Ge exceeds $1\mu V/K$, which is 1.5 times that of Mn_3Sn. The Weyl properties are discussed by analyzing the thermal conductivity, specific heat, and Seebeck and Nernst effects. We also evaluate the zero-field Nernst-driven thermoelectric figure of merit for device applications in the antiferromagnets Mn_3X.

1. Introduction

The large anomalous Hall effect (AHE) observed in the noncollinear antiferromagnets (AFMs) Mn_3X (X = Sn and Ge) is known to be given by the intrinsic contribution due to large Berry curvature generated by time-reversal-breaking Weyl metals [1, 2]. The existence of the magnetic Weyl nodes at Fermi level E_F are suggested by recent first-principles calculations of the electronic band structure [3, 4, 5, 6]. Additionally, the planar Hall effect accompanied by positive longitudinal magnetoconductance has been observed recently in magnetotransport measurements. Such a phenomenon is considered to be induced by the chiral anomaly corroborated with the magnetic Weyl semimetal owing to the clear evidence of Weyl nodes, which is provided by angle-resolved photoemission spectroscopy and the corresponding theoretical band structure.

The sum of the Berry curvature over the occupied states below the E_F theoretically constitutes the intrinsic contribution of AHE. When the E_F are tuned just at the Weyl nodes, an ideally maximized AHE appears. The theoretically calculated value of AHE agrees reasonably well with the experimental value [7]. The Berry curvature is also detected using the thermoelectric counterpart of the AHE, the so-called anomalous Nernst effect (ANE). In fact, the observed anomalous Nernst effect is stronger than the effect reported for ferromagnets in the antiferromagnets Mn_3X [7, 8]. When the finite Berry curvature at Weyl nodes appears only in the k_BT energy region around E_F, the ANE is provided by the contribution of the Weyl nodes in the k-space because the ANE intrinsically determines the spectrum of the Berry curvature at E_F [7]. The ANE thus provides us important properties on the Berry curvature and its relation to the Weyl nodes near E_F [6, 7].

Here, a comprehensive study of the large spontaneous ANE are performed in the chiral antiferromagnets Mn_3X. In antiferromagnet Mn_3Sn, the inverse-triangular spin structure is only magnetically stable from the spin-glass transition temperature T_s ($\cong$ 50 K) to the antiferromagnetic Néel temperature T_N ($\cong$ 430 K). Although Mn_3Ge has the same type of spin structure, it is stable above 300 mK and has no additional transition below $\cong$ 380 K. The low-temperature Nernst effect S_{ji} is affected by the spectrum of the Berry curvature at the effective E_F and reveals a unique property of Weyl metals. It is thus essential to investigate the low-temperature ANE of Mn_3Ge in elucidating the magnetic Weyl phenomenon for future application to evaluate the thermodynamic properties of the Nernst-type modules as well as our previous studies of Mn_3Sn [9, 10].

2. Experimental and Results

Figures 1a and 1b present the specific heats $C_{ab}(T)$ and $C_{ab}(T)/T$ of Mn_3Ge (Mn_3Sn). The peak is observed at 65 K (70 K) for Mn_3Ge (Mn_3Sn). According to the Einstein–Debye equation of $C_{ab}/T = \gamma + \beta T^2$, where γ and β are respectively parameters of the electronic and lattice contributions to the specific heat, the best parameter fittings to the Mn_3Ge (Mn_3Sn) data are obtained in the low-temperature regime of 2–15 K as γ = 22.5 mJ/mol K^2 (γ =32.5 mJ/mol K^2) and β = 156 μJ/mol (β = 300 μJ/mol). Here, the Debye temperature is evaluated as T_θ = 291 K (T_θ = 234 K) using $C_{ab}/T = (12\pi^4 N_k)/(5T_\theta^3)T^2$ and Boltzmann's constant N_k [5]. In Figure 1c, the positive Seebeck coefficient S_{ii} is shown at 300 K and indicates that hole carriers are dominated by electron carriers [5, 10]. $S_{ii}(T)$ changes sign at $\sim$ 100 K and peaks at $\sim$ 50 K in both the $Q||[2\bar{1}\bar{1}0]$ and $Q||[01\bar{1}0]$ of Mn_3Ge. Additionally, we obtain from the estimated Debye temperatures a crossover temperature of $T_\theta/5$ = 58.6 and 46.8 K for the minimum Seebeck coefficient of Mn_3Ge and Mn_3Sn respectively. Figure 1d displays the zero-field thermal conductivity $\kappa(T)$ in the warming process after annealing at a temperature above T_N, which aligns the domain along B. The value κ(300 K) $\sim$ 8–12 W/Km of Mn_3Sn is slightly more conductive than that of Mn_3Ge. The spontaneous Nernst voltage can be applied to thermoelectric modules that convert only a temperature gradient into electrical energy and ordinary Seebeck-type at room temperature [11]. As shown in Fig. 2a, the temperature dependence of S_{ji} is maximized at 100 K (200 K) for Mn_3Ge (Mn_3Sn). We find that the Nernst coefficient is largest in Mn_3Ge as the antiferromagnet. The Seebeck-driven thermoelectric figure of merit $ZT = (S^2/\rho\kappa)T$ is a quantity used to characterize the performance of a device and defined for a maximally efficient thermoelectric generator. Similarly, the Nernst-driven $Z_NT = (S_{ji}^2/\rho\kappa)T$ is defined in Mn_3Ge and is only $\sim 9 \times 10^{-6}$ at 300 K, which indicates higher value compared with the value for Mn_3Sn ($\sim 2 \times 10^{-6}$ at 300 K) reported in our previous study [10] and is still quite low for fabricating a Nernst-driven thermoelectric device.

3. Conclusion

A sizable Nernst coefficient S_{ji} is obtained for antiferromagnetic Mn_3Ge, which exceeds 1 μV/K at around 100 K. This value is comparable to the Nernst coefficients of ordinal ferromagnets and twice that of Mn_3Sn. The ANE in Mn_3Ge is thus assumed to be induced by large

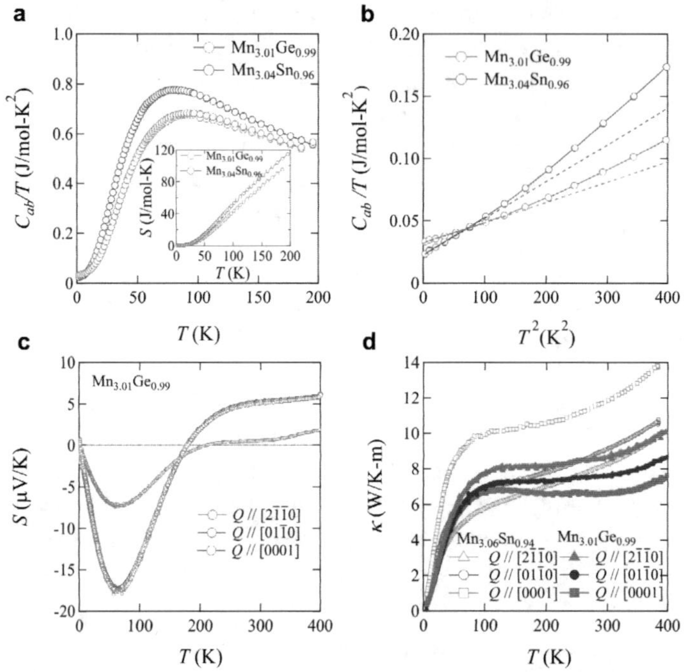

Figure 1. (a) Zero-field specific heat divided by temperature $C_{ab}(T)/T$ of Mn_3Ge (red) and Mn_3Sn (blue) in (a) the temperature range of 2–200 K and (b) the low-temperature range below 20 K. The calculated entropy $S(T)$ is plotted in the inset in (a). The dashed lines show the linear fitting of $C_{ab}/T = \gamma + \beta T^2$ to the low-temperature data in C_{ab}/T vs T^2 of panel (b). (c) Zero-field Seebeck coefficients $S(T)$ of Mn_3Ge in three different configurations: $Q||[2\bar{1}\bar{1}0]$ (green), $Q||[01\bar{1}0]$ (blue), and $Q||[0001]$ (red). (d) Zero-field thermal conductivity $\kappa(T)$ for the three different configurations of Mn_3X (X = Sn, Ge): $Q||[2\bar{1}\bar{1}0]$ (green), $Q||[01\bar{1}0]$ (blue), and $Q||[0001]$ (red).

Berry curvature, indicating the existence of paired Weyl points. In both Mn_3Sn and Mn_3Ge, the AHE and ANE are considered to share the same origin. The large anomalous Nernst coefficient S_{ji} for Mn_3Ge beyond the magnetization scaling relation are shown in Fig. 2b [8]. Rrom a viewpoint of the ANE, however, the zero-field Nernst-driven thermopower of antiferromagnetic Weyl metals still does not have sufficient ZT for practical application, in contrast with Seebeck-driven thermopower in semiconductors. In future practical applications, antiferromagnetic Weyl semimetals are expected to be available for topological memory besides Nernst-driven thermoelectric modules as discussed here. These Weyl semimetals are suitable for microfabricated devices owing to the weak magnetic leakage due to integration. We need to investigate further their possible use in spintronic applications [12, 13] and continue searching for the antiferromagnetic type of such novel materials.

acknowledgments
This work is partially supported by CREST (JPMJCR18T3), Japan Science and Technology Agency, by Grants-in-Aids for Scientific Research on Innovative Areas (15H05882 and 15H05883) from the Ministry of Education, Culture, Sports, Science, and Technology of Japan, and by Grants-in-Aid for Scientific Research (19H00650) from the Japanese Society for the Promotion

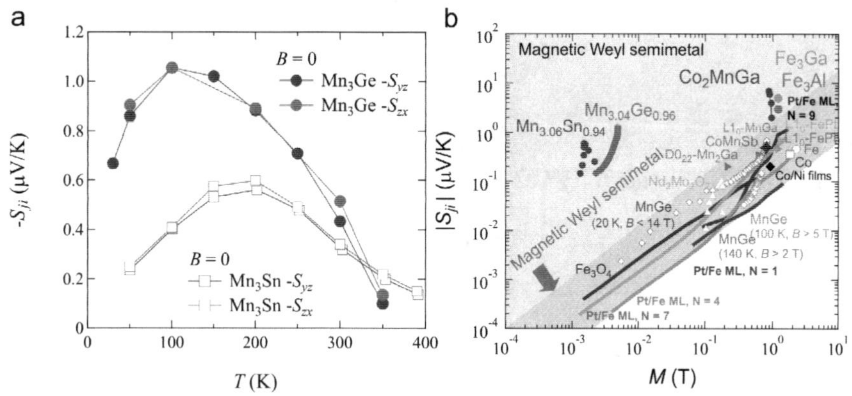

Figure 2. (a) Zero-field Nernst signals $S_{ji}(T)$ of Mn_3X (with open and closed symbols indicating data for Mn_3Sn and Mn_3Ge respectively). (b) Nernst coefficient S_{ji} as a function of magnetization M in Mn_3X and various ferromagnetic materials [7, 8]. The red-shaded area indicates the regions in which $S_{ij}(M)$ follows the linear scaling relation for conventional ferromagnets [7, 8].

of Science (JSPS). The work at IQM was supported by the US Department of Energy, Office of Basic Energy Sciences, Division of Materials Sciences and Engineering under grant DE-FG02-08ER46544. The work on the first-principles calculation was supported in part by JSPS Grant-in-Aid for Scientific Research on Innovative Areas (18H04481 and 19H05825) and by MEXT as a social and scientific priority issue (Creation of new functional devices and high-performance materials to support next-generation industries) to be tackled using the Post-K computer (hp180206 and hp190169). The use of the facilities of the Materials Design and Characterization Laboratory at the Institute for Solid State Physics, The University of Tokyo, is gratefully acknowledged.

References

[1] Nakatsuji S, Kiyohara N and Higo T 2015 *Nature* **527** 212–215
[2] Kiyohara N, Tomita T and Nakatsuji S 2016 *Phys. Rev. Applied* **5**(6) 064009
[3] Yang H, Sun Y, Zhang Y, Shi W J, Parkin S S P and Yan B 2017 *New J. Phys.* **19** 015008
[4] Suzuki M T, Koretsune T, Ochi M and Arita R 2017 *Phys. Rev. B* **95**(9) 094406
[5] Ziman J M 1960 *Electrons and phonons: the theory of transport phenomena in solids* (Oxford University Press)
[6] Kuroda K, Tomita T, Suzuki M T, Bareille C, Nugroho A A, Goswami P, Ochi M, Ikhlas M, Nakayama M, Akebi S, Noguchi R, Ishii R, Inami N, Ono K, Kumigashira H, Varykhalov A, Muro T, Koretsune T, Arita R, Shin S, Kondo T and Nakatsuji S 2017 *Nature Materials* **16** 1090–1095
[7] Ikhlas M, Tomita T, Koretsune T, Suzuki M T, Nishio-Hamane D, Arita R, Otani Y and Nakatsuji S 2017 *Nature Phys* **13** 1085–1090
[8] Chen T, Tomita T, Minami S, Fu M, Koretsune T, Kitatani M, Muhammad I, Nishio-Hamane D, Ishii R, Ishii F, Arita R and Nakatsuji S 2020 *Nature Communications* **12** 572
[9] Sugii K, Imai Y, Shimozawa M, Ikhlas M, Kiyohara N, Tomita T, Suzuki M T, Koretsune T, Arita R, Nakatsuji S and Yamashita M *arXiv preprint arXiv:1902.06601*
[10] Tomita T, Ikhlas M and Nakatsuji S 2020 *JPS Conference Proceedings* **30** 011009
[11] Snyder G J and Toberer E S 2008 *Nature Materials* **7** 105–114
[12] Sugimoto S, Nakatani Y, Yamane Y, Ikhlas M, Kondou K, Kimata M, Tomita T, Nakatsuji S and Otani Y 2020 *Communications Physics* **3** 111
[13] Miwa S, Iihama S, Nomoto T, Tomita T, Higo T, Ikhlas M, Sakamoto S, Otani Y, Mizukami S, Arita R and Nakatsuji S 2021 *Small Science* **1** 2000062

The correlation between magnetic interactions and lattice distortions in Pr and Eu substituted $NdCrO_3$

Pragya Gupta and D. Pal

Department of Physics, Indian Institute of Technology Guwahati, Guwahati, Assam 781039, India

E-mail: dpal@iitg.ac.in

Abstract. The study reveals the comparison of $Nd_{0.9}Pr_{0.1}CrO_3$ and $Nd_{0.9}Eu_{0.1}CrO_3$ samples where the A-site is doped by Pr^{3+} having the larger ionic radius (1.126 A°) corresponding to less chemical pressure and Eu^{3+} (1.066 A°) having the smaller ionic radius corresponding to enhanced chemical pressure exerted on the $NdCrO_3$, with the help of X-ray diffraction, Raman spetroscopy, UV- visible spectroscopy, and dc magnetization measurements. The different spin configuration Pr^{3+} (J = 4) and Eu^{3+} (J = 0) altering the strong coupling between Nd^{3+} (J = 9/2) and Cr^{3+} (S = 3/2) spins result in the variation of the optical band gap, fermi energy and exchange bias.

1. Introduction

The rare earth chromites RCrO$_3$ (R = rare earth) consist of octahedrally coordinated B-sites occupied by Cr ions and A-sites occupied by R-ions. The complex interaction between magnetic moments of R and Cr ions gives rise to the canted antiferromagnetism. The $NdCrO_3$ is a multifunctional material with antiferromagnetic (AFM) ordering of Cr spins at T_N = 219 K followed by a spin-reorientation transition at T_{SR} = 35 K. In this compound, the origin of exchange bias (EB) is mostly spin-phonon induced [1]. The RCrO$_3$ shows semiconductor behavior in the visible frequency range, which is related to structural distortion. In few chromites, the effect of the external pressure leads to structural distortions, which modifies band structure and resulting interesting magnetic properties [2]. In this paper, we explored the effect of chemical pressure by substituting the A-site with Pr^{3+} and Eu^{3+} ions.

2. Experimental details

Polycrystalline samples of $Nd_{0.9}Pr_{0.1}CrO_3$ and $Nd_{0.9}Eu_{0.1}CrO_3$ were prepared by standard solid-state reaction route. The structural characterization of the samples was performed using the X-ray diffractometer (Rigaku, Smart Lab)and Raman spectrometer (Horiba Jobin Yvon, LabRam HR). The absorption spectra have been investigated by using a UV-Vis spectrometer (Perkin Elmer). Magnetization measurements were performed by PPMS (Quantum Design).

3. Results and discussions

3.1. Structural properties

The structural characterization of $Nd_{0.9}Pr_{0.1}CrO_3$ and $Nd_{0.9}Eu_{0.1}CrO_3$ samples done by XRD analysis (figure 1) confirms the orthorhombic structure ($Pnma$ space group) similar to $NdCrO_3$.

The substitution of Pr^{3+} ions gives the increment in the cell parameters, whereas the decrement in the overall cell parameters is obtained due to Eu^{3+} ions with the changes in CrO_6 octahedral bond lengths and angles for the $Nd_{0.9}R_{0.1}CrO_3$ sample, listed in table 1. The influence of the

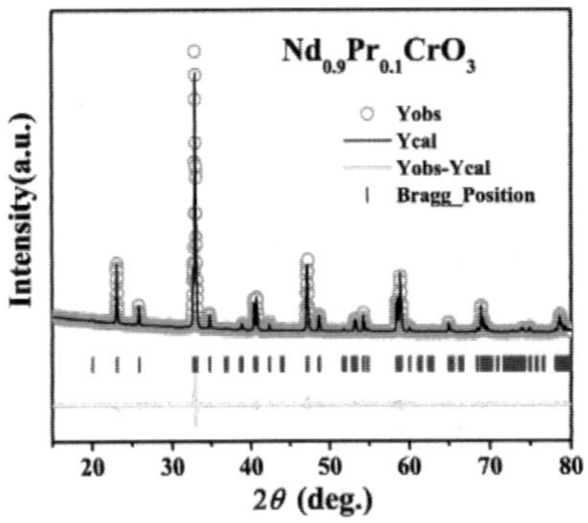

Figure 1. Rietveld refinement of XRD pattern of $Nd_{0.9}Pr_{0.1}CrO_3$ sample using FULLPROF program.

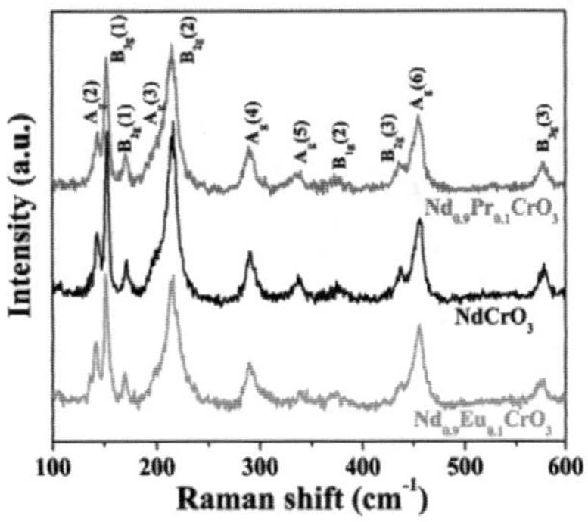

Figure 2. Raman spectra of $Nd_{0.9}Pr_{0.1}CrO_3$, $NdCrO_3$, and $Nd_{0.9}Eu_{0.1}CrO_3$ samples from wavenumber 100 to 600 cm^{-1}.

dopant on the average A-site ionic radius, r_{avg} accounts for this variation. The distortion of cell edges from the ideal cubic perovskite is calculated using Goldschmidt's tolerance factor, $t = (r_R + r_O)/\sqrt{2}(r_{Cr} + r_O)$ and the orthorhombic strain, $s = 2(a - c)/(a + c)$. The decrease in t and increase in s values are observed for decreasing r_{avg} (table 1). The microstructural analysis revealed the good quality of the homogenous bulk samples. The Raman spectra (figure 2) shows that the modes $B_{2g}(2)$, $A_g(4)$ belonging to R-O1 vibration, $B_{1g}(2)$, $B_{2g}(3)$ related to CrO_6 octahedral rotations and $A_g(6)$ corresponds to CrO_6 octahedral bending, exhibit noticeable

Table 1. Average A-site ionic radius r_{avg}, tolerance factor (t), orthorhombic strain (s), cell parameters and Cr-O2 bond lengths, and Cr-O2-Cr bond angles obtained from the Rietveld refinement of $Nd_{0.9}R_{0.1}CrO_3$ (R = Pr, Eu) with parent sample.

	$Nd_{0.9}Pr_{0.1}CrO_3$	$NdCrO_3$	$Nd_{0.9}Eu_{0.1}CrO_3$
r_{avg} (A°)	1.111	1.109	1.105
a (A°)	5.481	5.482	5.484
b (A°)	7.690	7.689	7.688
c (A°)	5.419	5.418	5.415
V ($A^{\circ 3}$)	228.45	228.37	228.30
t	0.884	0.883	0.882
s	0.0114	0.0117	0.0127
Cr-O2 (Å)	1.963	1.939	1.901
Cr-O2-Cr (deg.)	156.19	156.3	161.6

red shift (Pr doping) and blue shift (Eu doping), possibly due to changes in average ionic radii and atomic mass. Thus, structural analysis suggests that lattice distortions are less for $Nd_{0.9}Pr_{0.1}CrO_3$ and more for $Nd_{0.9}Eu_{0.1}CrO_3$ system compared to $NdCrO_3$.

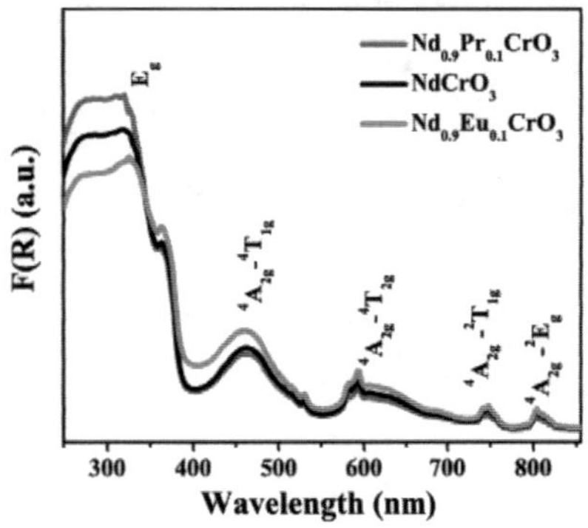

Figure 3. UV-Vis DRS spectrum of $Nd_{0.9}Pr_{0.1}CrO_3$, $NdCrO_3$, and $Nd_{0.9}Eu_{0.1}CrO_3$ samples in the range of 250 to 850 nm.

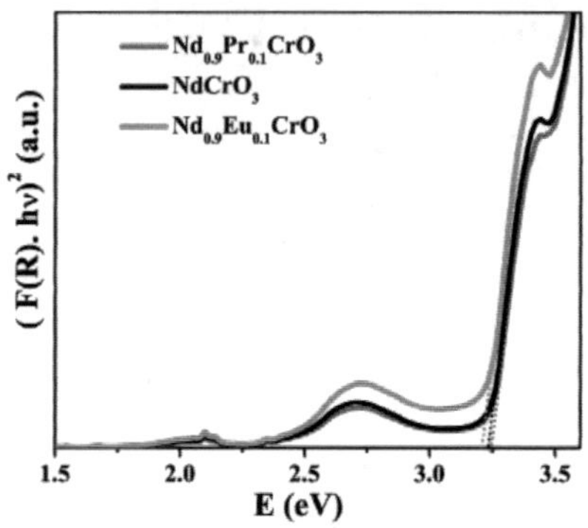

Figure 4. Estimation of bandgap E_g from Tauc plots for $Nd_{0.9}Pr_{0.1}CrO_3$, and $NdCrO_3$ $Nd_{0.9}Eu_{0.1}CrO_3$ samples.

3.2. Optical properties

The optical behavior of the samples is investigated by the UV–Vis diffuse reflectance spectroscopy technique. The UV–Vis spectrum of the samples are analyzed by optical absorption, i.e., Kubelka-Munk (KM) function F(R) [3], which can be expressed as, $F(R) = (1\text{-}R)^2/2R$, where R is the diffuse reflectance of the sample. From F(R) versus wavelength plot in figure 3, we see the $Nd_{0.9}R_{0.1}CrO_3$ system has a multiband electronic structure. The electronic structure of $NdCrO_3$ consists of a filled O $2p$ band and partially filled Cr $3d^3$ and Nd $4f^4$ orbitals. The broad absorption band located around 250-390 nm is attributed to bandgap E_g of the material due to O $2p \rightarrow$ Cr $3d$ transitions. The other two intense absorption bands at about 463 nm and 593 nm corresponds to $^4A_{2g}$–$^4T_{1g}$ and $^4A_{2g}$–$^4T_{2g}$ transition. The two weak transitions at 748 nm and 804 nm corresponds to $^4A_{2g}$–$^2T_{1g}$ and $^4A_{2g}$–^2E_g transition. These transitions are related to the octahedral crystal field splitting of Cr^{3+} d-orbitals that allow the corresponding $d(Cr) \rightarrow d(Cr)$ transitions. The optical energy band gap can be calculated from the Tauc relation $F(R)h\nu = B(h\nu - E_g)^n$. The value of n depends on the direct (n = 1/2) or indirect (n = 2) band gap of the material. The Tauc plot of $(F(R).h\nu)^2$ versus $(h\nu)$ is used to determine the direct band gap (E_g) as shown in figure 4, which shows the samples have the bandgap E_g = 3.24 eV (Pr doping) and 3.20 eV (Eu doping) lies in the visible region. We have also utilized the Density functional theory (DFT) with the quantum ESPRESSO suite to calculate the fermi energy of the $Nd_{0.9}R_{0.1}CrO_3$ system. The obtained fermi energy values are E_F = 14.48 eV (Pr doping) and E_F = 14.57 eV (Eu doping). From the magneto-optical study of $NdCrO_3$, we know that the coupling between Nd^{3+} and Cr^{3+} spins are extremely strong [4], hence the substitution of different R on Nd-site can influence Cr-O overlap integral, which can also modify the structural distortion via the distortion of CrO_6 octahedra, resulting in the change in the optical bandgap and fermi energy.

3.3. Magnetic properties

The magnetization measurements (figure 5) confirm the AFM ordering of Cr^{3+} spins occurring at T_N = 226.7 K, 224.1 K, and the spin reorientation transition due to reorientation of Cr^{3+} spins at T_{SR} = 41.8 K, 39.2 K for $Nd_{0.9}Pr_{0.1}CrO_3$ and $Nd_{0.9}Eu_{0.1}CrO_3$ samples respectively. The magnetic behavior below T_N varies significantly due to different alignment of Cr sublattice, $|Nd+Pr|$ and $|Nd+Eu|$ sublattice that weakens the reorientation of Cr^{3+} spins in $Nd_{0.9}Pr_{0.1}CrO_3$ but leaves the $Nd_{0.9}Eu_{0.1}CrO_3$ system slightly affected. The negative exchange bias (figure 6)

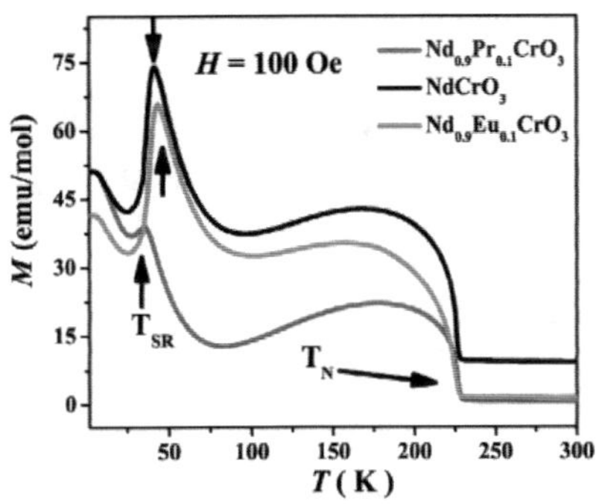

Figure 5. Temperature dependence of magnetization measured in FC mode for the applied field of 100 Oe for $Nd_{0.9}Pr_{0.1}CrO_3$, $NdCrO_3$, and $Nd_{0.9}Eu_{0.1}CrO_3$ samples demonstrating the transitions.

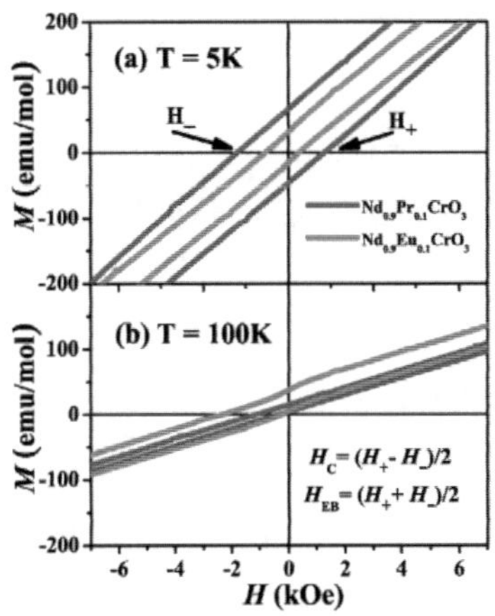

Figure 6. Enlarged portions of field cooled M-H loops at (a) 5 K and (b) 100 K of $Nd_{0.9}Pr_{0.1}CrO_3$ and $Nd_{0.9}Eu_{0.1}CrO_3$ samples.

having the exchange bias field of H_{EB} = -258 Oe and -209 Oe with the coercive field H_C =1528 Oe and 583 Oe at 5 K ($T < T_{SR}$) is obtained for $Nd_{0.9}Pr_{0.1}CrO_3$ and $Nd_{0.9}Eu_{0.1}CrO_3$ samples. The value of (EB) field increases at 100 K ($T < T_N$) such that H_{EB} = -748 Oe and -1280 Oe with H_C = 405 Oe and 1049 Oe is observed for $Nd_{0.9}Pr_{0.1}CrO_3$ and $Nd_{0.9}Eu_{0.1}CrO_3$ samples, respectively. The high EB field at 100 K for $Nd_{0.9}Eu_{0.1}CrO_3$ is obtained due to high chemical pressure acting on competing AFM interaction between the weak ferromagnetic component of Cr^{3+} moments and the paramagnetic moment of Nd^{3+}/R^{3+} ions. However, the value of EB field for the parent compound [5] is significantly higher than both these substituted compounds. It is possible that coupling between the A-site and the B-site has been sufficiantly modified, leading to the decrease in EB effect.

4. Conclusions

The single-phase $Nd_{0.9}R_{0.1}CrO_3$ (R = Pr, Eu) samples were formed by the solid-state reaction method. The structural analysis confirmed the increase in structural distortion with the variation of R-ion. The change in optical band gap and fermi energy via the Cr-O overlap integral is observed. The coupling between Nd^{3+}/R^{3+} and Cr^{3+} spins was altered, resulting in the change in T_N and T_{SR} values and the negative exchange bias effect. Our results clearly demonstrate that the magnetic interaction between Nd/R and Cr moments varies with the average A-site ionic radius due to different chemical pressure in $Nd_{0.9}R_{0.1}CrO_3$ (R = Pr, Eu) system.

5. Acknowledgments

The authors would like to acknowledge the FIST program of Department of Science and Technology, India, for partial support of this work [Ref. No. SR/FST/PSII-037/2016] and CIF IIT Guwahati for providing its facilities.

References

[1] Gupta P and Pal D 2021 *Journal of Physics: Condensed Matter* **33** 135806
[2] Yin S, Seehra M S, Guild C J, Suib S L, Poudel N, Lorenz B and Jain M 2017 *Physical Review B* **95** 184421
[3] López R and Gómez R 2012 *Journal of sol-gel science and technology* **61** 1–7
[4] Hornreich R, Komet Y, Nolan R, Wanklyn B and Yaeger I 1975 *Physical Review B* **12** 5094
[5] Indra A, Dey K, Midya A, Mandal P, Gutowski O, Rütt U, Majumdar S and Giri S 2016 *Journal of Physics: Condensed Matter* **28** 166005

Mössbauer effect study of a polycrystalline $Fe_{1+x}Cr_{2-x}O_4$ spinel grown by solid-state synthesis

M A Cherosov[1], A L Zinnatullin[1], R G Batulin[1], A G Kiiamov[1], R V Yusupov[1] and D A Taurskii[1]

[1] Institute of Physics, Kazan Federal University, Kazan 420008, Russia

mcherosov@gmail.com

Abstract. Polycrystalline $FeCr_2O_4$ samples have been synthesized by the solid-state reaction method and studied with X-ray diffraction and Mössbauer spectroscopy. Rietveld refinement of X-ray diffraction pattern has been performed; obtained unit cell parameters are in a good agreement with ones in the literature. Mössbauer spectra were measured at room temperature and reveal by two components originating from the A and B sites of the spinel structure occupied by iron ions.

1. Introduction

A number of studies have been performed on the magnetic spinel compounds with a general formula AB_2O_4 (where A and B are the metal ions or a combination of them) due to their interesting magnetic, dielectric, and electrical properties [1-7]. Among them, a spinel-structure oxide $FeCr_2O_4$ (or ferrous chromite spinel) is a well-known compound with a cubic structure at room temperature (space group Fd3m) and the lattice constant a = 8.378 Å. The A-site with the tetrahedral coordination is occupied by the Fe^{2+} ions ($3d^6$ electronic configuration) and the octahedrally-coordinated B-site is occupied by the Cr^{3+} ions ($3d^3$ electronic configuration). A different cation distribution takes place in the so-called inverse spinels: here, one-half of the cations B are in A positions and another one-half are share in a random way the B sites with the A ions. For the intermediate, or a mixed, case the general formula can be written as $A_\delta B_{1-\delta}[A_{1-\delta}B_{1+\delta}]_2O_4$, where the cations in the B site are shown in brackets, and δ=1 corresponds to the normal spinel and δ=0 to the inverse one. Recently, they have attracted an attention due to the peculiar magnetic and electronic properties [8,9].

It is well known that if the ferrous oxide is used as one of the starting materials for a solid state synthesis (even in an appropriate inert atmosphere or vacuum) the grown crystal has a modified spinel formula $Fe^{2+}Fe^{3+}{}_xCr_{2-x}O_4$ due to the partial iron Fe^{2+} ions oxidation to the Fe^{3+} state. However, the XRD analysis cannot distinguish this cation distribution and there is lack of data in the literature where the Mössbauer spectroscopy is used for investigations of the samples grown by this method.

We present here the polycrystalline samples grown by a solid-state synthesis and Mössbauer spectroscopy studies of as grown and annealed crystals.

The article is organized as follows. Next section describes the growth procedure and XRD analysis studies. In section 3, the Mössbauer spectroscopy results are presented.

2. Sample preparation and powder X-ray diffraction

Polycrystalline $FeCr_2O_4$ samples were synthesized by the solid-state synthesis using Fe_2O_3 (Alfa Aesar, 99%), metallic Fe (99.97%) and Cr_2O_3 (Alfa Aesar, 99%) as starting reagents. The oxides were dried at a temperature 200°C for 16 hours. The stoichiometric mixture was thoroughly ground in an agate mortar for 3 hours in a glovebox with an inert atmosphere to avoid oxidation. The mixture in an alumina crucible was placed into the a vertical tube furnace. The chamber was evacuated down to 10^{-2} mbar for several times using a vacuum pump and purged with a pure Ar (99.995%). The synthesis was carried out with a continuous evacuation of the chamber at a temperature of 1400°C for 12 hours. The resulting reaction product was checked for the formation of the desired phase and composition by the powder X-ray diffraction analysis (Bruker D8 ADVANCE, Cu Kα). Three synthesis attempts have been made. The reaction product of the first synthesis attempt (Sample 1) was visually homogeneous. The reaction products of two next syntheses (Sample 2 and 3) consisted of two components – it had the small black-colored inclusions in the total volume of a brown-colored sample. The powder X-ray diffraction pattern shows the spinel phase $FeCr_2O_4$ and a small amount of the residual Cr_2O_3. X-ray diffraction patterns of all the samples are very similar. The X-ray diffraction pattern of Sample 1 and of the brown-colored part of Sample 2 and their Rietveld refinement are shown in figure 1. The refined unit cell parameter values are 8.3754 and 8.3763 Å, respectively, and are in good agreement with the data obtained by Francombe [10]. Rietveld refinement was carried out in the FULLPROF software [11].

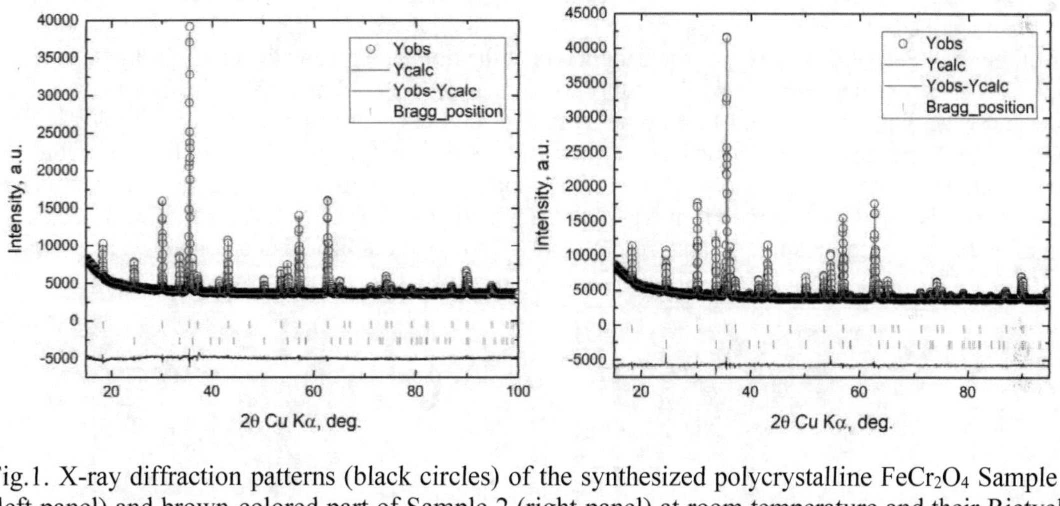

Fig.1. X-ray diffraction patterns (black circles) of the synthesized polycrystalline $FeCr_2O_4$ Sample 1 (left panel) and brown-colored part of Sample 2 (right panel) at room temperature and their Rietveld refinement (red line)

3. Mössbauer spectroscopy studies

^{57}Fe Mössbauer spectroscopy was conducted on synthesized samples in order to examine the valence and spin states of the iron ions in transmission geometry on a conventional spectrometer (*WissEl Gmbh*), operating in a constant acceleration mode. Measurements were carried out at room temperature. ^{57}Co(Rh) (*Ritverc Gmbh*) with the activity of about 40 mCi was used as the source of a resonance radiation. The spectrometer velocity scale was calibrated using a thin metallic iron foil at room temperature (RT). The spectra were fitted using the *SpectrRelax* software [12]. Values of the isomer shifts are reported versus the center of gravity of α-Fe spectrum at RT. For the measurements, polycrystalline samples were carefully ground in an agate mortar to the fine powder state. The powder samples were used to prepare thin Mössbauer absorbers.

The Mössbauer spectrum of the Sample 1 recorded at room temperature is shown in Figure 2a. Two components are found in this spectrum, namely, a major doublet (spectral area is 76(1) %) and a minor doublet (spectral area is 24(1) %). The major doublet had the following hyperfine parameters: isomer shift of 0.94(1) mm/sec, quadrupole splitting of 0.42(1) mm/sec, and the full-width at half-maximum (FWHM) of 0.31(1) mm/sec. The minor doublet had the following hyperfine parameters: isomer shift of 0.36(1) mm/sec, quadrupole splitting of 0.42(1) mm/sec, and FWHM of 0.30(1) mm/sec. The isomer shift value of the major doublet is characteristic for the high-spin ferrous (Fe^{2+}) ions in an oxygen tetrahedral coordination, and the isomer shift value of the minor doublet is characteristic for the high-spin ferric (Fe^{3+}) ions in an oxygen octahedral environment. We suppose that the major and the minor doublets are associated with iron ions in the A and B sites of the spinel structure, respectively, and therefore Fe^{3+} ions substitute for the Cr^{3+} ions. Quadrupole splitting is determined by quantization of quadrupole moment of nuclei in the electric field gradient (EFG). Consequently, the departure of observed ferrous doublet from expected ferrous singlet for ideal $FeCr_2O_4$ [13] should be assigned to the emergence of EFG due to aforementioned Fe^{3+} substitution in the B sites.

Mössbauer spectra of brown-colored parts of Samples 2 and 3 are almost identical to each other. The Mössbauer spectrum of brown-colored part of Sample 2 at room temperature is shown in Figure 2b. Two components manifest itself in this spectrum, namely, a singlet (spectral area 87(1) %) and a doublet (spectral area 13(1) %). The isomer shift for the singlet is 0.92(1) mm/sec with FWHM of 0.51(1) mm/sec. Another component of the spectrum, the doublet, has the following hyperfine parameters: isomer shift 0.36(1) mm/sec, quadrupole splitting 0.41(1) mm/sec, and FWHM of 0.32(1) mm/sec. The isomer shift value of the singlet is characteristic for the high-spin ferrous ions in an oxygen tetrahedral environment and matches to the value reported earlier for the ferrous ions in A sites of the ideal $FeCr_2O_4$ [13]. Hyperfine parameters of the doublet agree with values for the ferric ion doublet observed in the spectrum of the Sample 1. However, spectral area of this component in the spectrum of the Sample 2 is almost two times less than in the spectrum of the Sample 1. The lesser content of Fe^{3+} ions in the B sites determines low EFG and, consequently, leads to a small value or even to total vanishing of quadrupole splitting for component related to ferrous ions in the A sites. The appearance of the singlet in the spectrum is associated with this fact. The Mössbauer spectrum of black-colored inclusions in the Sample 2 (not presented here) is very complicated and contains a number of spectral components associated to a various valence and spin states of iron.

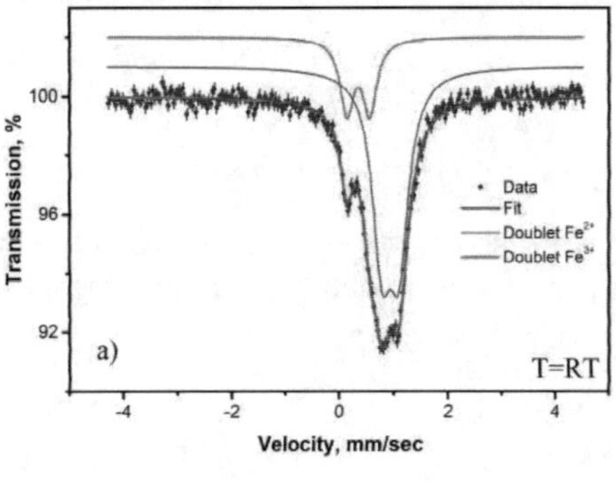

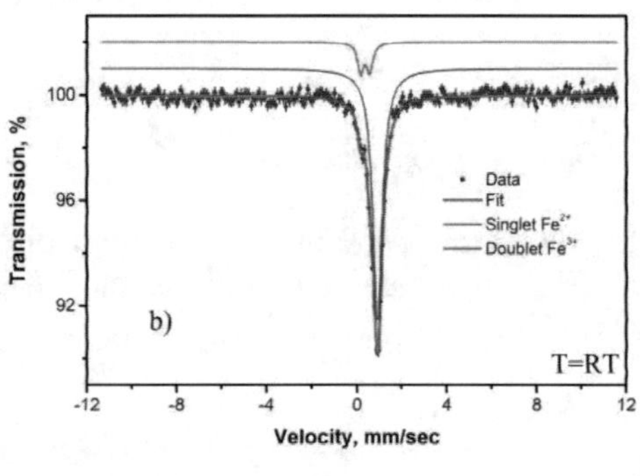

Fig. 2. Room temperature Mössbauer spectra of: a) Sample 1 and b) brown-colored part of Sample 2 (the synthesis product without black inclusions)

4. Conclusion

Polycrystalline $FeCr_2O_4$ samples were synthesized by the solid-state synthesis. To avoid ferrous ions oxidation the following precautions were taken: the reagents were ground in a glovebox with an inert atmosphere of Ar, the furnace chamber was evacuated and purged with pure Ar several times, and a pump was used to continuously evacuate the chamber during sample synthesis procedure. Powder X-ray diffraction pattern shows two structural phases – $FeCr_2O_4$ and a small amount of unreacted Cr_2O_3. Crystal structure parameters were calculated using the Rietveld refinement technique. Information about the iron ions environments and its electronic configurations was obtained by Mössbauer spectroscopy.

As the $FeCr_2O_4$ compound is susceptible to iron oxidation from the Fe^{2+} to Fe^{3+} state and chromium and iron ions have almost identical ionic radii and atomic scattering factors, X-ray diffraction alone is not enough to characterize the synthesized $FeCr_2O_4$ spinel. Analysis of the Mössbauer spectroscopy data is capable to provide an information on the iron ions environment and electronic configurations.

To avoid intermediate oxidation of iron ions during the solid state reaction we propose to use ferrous oxalate as an initial reagent since the whole number of iron ions in pre-synthesized mixture will be in the divalent state. $FeCr_2O_4$ synthesis results with the ferrous oxalate and chromium oxide used as starting materials will be published elsewhere.

5. Acknowledgment

This work was supported by the subsidy allocated to Kazan Federal University for the state assignment in the sphere of scientific activities (Project No. 0671-2020-0050).

References

[1] Krupicka S and Novak P 1982 *Handb. Magn. Mater.* **3** 189
[2] Lee S-H, Broholm C, Ratcliff W, Gasparovic G, Huang Q, Kim T H and Cheong S-W 2002 *Nature* **418** 856
[3] Singh K, Maignan A, Simon C and Martin C 2011 *Appl. Phys. Lett.* **99** 172903
[4] Tomiyasu K, Fukunaga J and Suzuki H 2004 *Phys. Rev.* B **70** 214434
[5] Yamasaki Y, Miyasaka S, Kaneko Y, He J P, Arima T and Tokura Y 2006 *Phys. Rev. Lett.* **96** 207204.
[6] Radaelli P. G. 2005 *New J. Phys.* **7** 53
[7] Kocsis V, Bordács S, Deisenhofer J, Kiss L F, Ohgushi K, Kaneko Y, Tokura Y and Kézsmárki I 2018 *Phys. Rev.* B **97** 125140
[8] Benhalima C, Amari S, Beldi L and Bouhafs B 2019 *SPIN* **9** 1950014
[9] Santos-Carballal D, Roldan A, Grau-Crespo R and de Leeuw N H 2015 *Phys. Rev.* B **91** 195106
[10] Francombe M H J. 1957 *Phys. Chem. Solids* **3** 37
[11] Rodriguez-Carvajal 1993 *Physica* B **192** 55
[12] Matsnev M E and Rusakov V S 2012 *AIP Conf. Proc.* **1489** 178
[13] Tanaka M Tokoro T and Aiyama Y *J.* 1966 *Phys. Soc. Japan* **21** 262

Electrical conductivity in helical and conical magnetic states

Shun Okumura[1], Takahiro Morimoto[2], Yasuyuki Kato[2], and Yukitoshi Motome[2]

[1]The Institute for Solid State Physics, The University of Tokyo, Kashiwa 277-8581, Japan
[2]Department of Applied Physics, The University of Tokyo, Tokyo 113-8656, Japan

E-mail: okumura@issp.u-tokyo.ac.jp

Abstract. We theoretically study the electrical conductivity in a one-dimensional helimagnet whose spin texture changes from helimagnetic to conical magnetic, and to forced ferromagnetic state while increasing the magnetic field along the helical axis. We find that the conductivity in the helimagnetic state at zero field depends on the electron filling and the coefficient of the spin-charge coupling. We also find that the conductivity in the conical magnetic state changes nonlinearly to the applied field, and the magnetoresistance becomes negative and positive depending on the model parameters.

1. Introduction

A helical magnetic state (HM) is an archetype of the noncollinear magnetic texture, which has a periodically twisted spin structure, as shown in figure 1(a). The HM has attracted considerable attention in condensed matter physics since they give rise to peculiar physical phenomena in multiferroics [1] and spintronics [2] due to lack of the spatial-inversion symmetry. In an external magnetic field parallel to the helical axis, the HM turns into a noncoplanar magnetic texture, which is called the conical magnetic state (CM) [figure 1(b)], and finally relaxes to a forced ferromagnetic state (FFM). Recently, the CM in the magnetic conductors has attracted many interests due to the peculiar electrical transport, for instance, the magnetoresistance effect [3] and the electric magnetochiral effect [4, 5]. While the spin transport was studied in both HM and CM [6, 7, 8], the charge transport has not been fully clarified yet.

In this paper, we report our theoretical study on the electronic transport properties in a one-dimensional helimagnet. We calculate the electrical conductivity by the linear response theory varying the spin-charge coupling constant, the electron filling, and the magnetization. We find that the conductivity in the HM at zero field is maximized around quarter filling similar to the FFM case, whereas the peak shifts to higher filling as the spin-charge coupling decreases. We also find that the CM in the magnetic field exhibits not only negative but also positive magnetoresistance depending on the model parameters.

2. Model

We consider a one-dimensional spin-charge coupled model, whose Hamiltonian is given by

$$H = -t\sum_{l,\mu}(c^{\dagger}_{l\mu}c_{l+1\mu} + \text{h.c.}) - J\sum_{l,\mu,\nu} c^{\dagger}_{l\mu}\boldsymbol{\sigma}_{\mu\nu}c_{l\nu}\cdot\mathbf{S}_l - h^z\sum_{l} S^z_l, \tag{1}$$

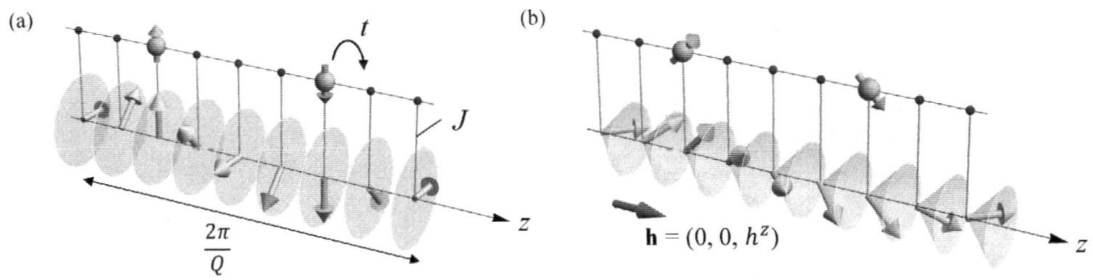

Figure 1. Schematic pictures of the model (1) in the (a) helical and (b) conical magnetic states with the helical pitch $\frac{2\pi}{Q}$. The localized spins (yellow) are coupled to spins of itinerant electrons (cyan) by the coupling constant J. The gray arrow in (b) represents the magnetic field parallel to the helical axis, which is taken as the z axis.

where $c_{l\mu}$ ($c^\dagger_{l\mu}$) is an annihilation (creation) operator for an itinerant electron with spin $\mu =\uparrow,\downarrow$ at site l, and $\boldsymbol{\sigma} = (\sigma^x, \sigma^y, \sigma^z)$ is the vector of Pauli matrices. The first term describes the kinetic energy of itinerant electrons with the nearest-neighbor hopping t and the second term represents the onsite coupling between the itinerant electrons and the localized classical spins $\mathbf{S}_l$ with the coupling constant J. The last term is the Zeeman coupling to the external magnetic field along the chain direction (the z axis) which is taken into account only for the localized spins for simplicity. We show schematic pictures of the model (1) in figure 1.

We study the ground state of the model (1) by assuming a swirling magnetic texture for the localized spins, $\mathbf{S}_l = (\sqrt{1-m^2}\cos(Ql), \sqrt{1-m^2}\sin(Ql), m)$, which describes the HM at $m = 0$ [figure 1(a)], the CM for $0 < m < 1$ [figure 1(b)], and the FFM at $m = 1$; m represents the magnetization per spin induced by the external magnetic field, and Q is the wavenumber specifying the helical pitch, $2\pi/Q$. By the Fourier transformation of (1), we obtain two energy bands split by $\sim 2J$ [9], whose energy dispersions are given by

$$\varepsilon_\lambda(k) = -2t\cos\frac{Q}{2}\cos k + \lambda\sqrt{4t^2\sin^2\frac{Q}{2}\sin^2 k + 4tJm\sin\frac{Q}{2}\sin k + J^2}, \tag{2}$$

where $\lambda = +$ and $-$ represent the higher- and lower-energy bands, respectively, and k is the shifted wavenumber defined as $k = \tilde{k} + \frac{Q}{2}\lambda$, where $\tilde{k}$ is the original wavenumber. Here and hereafter, we take the lattice constant unity.

In the strong coupling limit $J \to \infty$, the spins of itinerant electrons are parallel to the localized spins at each site, and hence, the effective hopping amplitude depends on the relative angle between the neighboring localized spins [10]. For the swirling spin configuration, the energy dispersions are asymptotically described by the simple cosine form as

$$\varepsilon_\lambda(k) = -2t\sqrt{1-(1-m^2)\sin^2\frac{Q}{2}}\cos(k-\varphi) + \lambda J, \tag{3}$$

where φ satisfies $\tan\varphi = m\lambda\tan\frac{Q}{2}$.

3. Linear response theory

We calculate the electrical conductivity for the model (1) by using the linear response theory as

$$\sigma = \frac{e^2}{\hbar}\int\frac{dk}{2\pi}\left[\tau\sum_\lambda\left(\frac{\partial f_\lambda}{\partial\varepsilon_\lambda}\right)\langle\lambda|\partial_k H|\lambda\rangle^2 + i\sum_{\lambda\neq\lambda'}\frac{f_\lambda - f_{\lambda'}}{\varepsilon_{\lambda'}-\varepsilon_\lambda}\frac{\langle\lambda|\partial_k H|\lambda'\rangle\langle\lambda'|\partial_k H|\lambda\rangle}{\varepsilon_\lambda - \varepsilon_{\lambda'} + i/\tau}\right], \tag{4}$$

where $|\lambda\rangle$ and $f_\lambda = f(\varepsilon_\lambda)$ are the eigenvector and the Fermi distribution function, respectively; τ represents the relaxation time, which is assumed to be a positive constant for simplicity. The first and second terms in (4) represent the Drude and interband contributions, respectively. In the ground state, we can ignore the second term for $J \geq 2t$ because the two energy bands in (2) do not overlap with each other. Then, (4) is reduced to

$$\sigma = \frac{e^2\tau}{\pi\hbar}\sum_{\lambda,j}\left| t\cos\frac{Q}{2}\sin k_{\mathrm{F}j}^{\lambda} + \lambda\frac{t^2\sin^2\frac{Q}{2}\sin(2k_{\mathrm{F}j}^{\lambda}) + tJm\sin\frac{Q}{2}\cos k_{\mathrm{F}j}^{\lambda}}{\sqrt{4t^2\sin^2 k_{\mathrm{F}j}^{\lambda}\sin^2\frac{Q}{2} + 4tJm\sin\frac{Q}{2}\sin k_{\mathrm{F}j}^{\lambda} + J^2}}\right|, \tag{5}$$

where $k_{\mathrm{F}j}^{\lambda}$ represents the jth Fermi wavenumber of $\varepsilon_\lambda(k)$. Furthermore, in the limit of $J \to \infty$, using (3), we obtain

$$\sigma = \sqrt{1 - (1 - m^2)\sin^2\frac{Q}{2}}\,\sigma_{\mathrm{FFM}}, \tag{6}$$

where $\sigma_{\mathrm{FFM}} = \frac{e^2\tau}{\pi\hbar}2t|\sin(\pi n)|$ represents the electrical conductivity in the FFM at $m = 1$; n is the electron filling (the average number of electrons per site). In the following calculation, we focus on the situation when the Fermi energy is in the lower-energy band and $J \geq 2t$, i.e., $0 < n < 1$, and we set $t = 1$ as the energy unit.

4. Calculation results

First, we show the results for the HM at $m = 0$ in figure 2. As shown in figure 2(a), the energy dispersion $\varepsilon_-(k)$ deviates from the cosine curve in (3) as J decreases. Since $\varepsilon_-(k)$ is symmetric with respect to k at $m = 0$, the Fermi wavenumber is given by n as $k_{\mathrm{F}1}^- = -k_{\mathrm{F}2}^- = \pi n$ for $\frac{2\pi}{Q} \geq 4$. We note that $\varepsilon_-(k)$ shows a double-well structure when the helical pitch becomes shorter and satisfies $2\pi/Q < 4$, while it is not shown in figure 2(a).

In figure 2(b), we plot the electrical conductivity in the HM, σ_{HM}, calculated by (5) at $Q = \frac{2\pi}{6}$ as functions of n for several values of J. At $m = 0$, the first term in (5) corresponds to the strong coupling limit in (6), and the second term represents the deviation from the strong coupling limit. We find that σ_{HM} is a symmetric dome-shaped function of n in the limit of $J \to \infty$ similar to σ_{FFM}, while the peak is shifted from $n = 0.5$ to higher n as J decreases due to the modulation of $\varepsilon_-(k)$ in figure 2(a). Note that the second term in (5) vanishes irrespective of J at $n = 0.5$, and hence the electrical conductivity is constant as $\sigma_{\mathrm{HM}} = \frac{2te^2\tau}{\pi\hbar}\cos\frac{Q}{2}$.

We also plot the ratio $\sigma_{\mathrm{HM}}/\sigma_{\mathrm{FFM}}$ in figure 2(c). In the limit of $J \to \infty$, $\sigma_{\mathrm{HM}}/\sigma_{\mathrm{FFM}} = \cos\frac{Q}{2}$ irrespective of n, while $\sigma_{\mathrm{HM}}/\sigma_{\mathrm{FFM}}$ becomes smaller (larger) as decreasing J for $n < 0.5$ ($n > 0.5$).

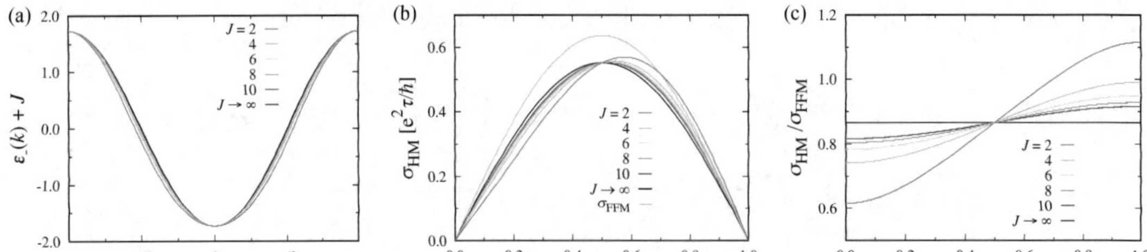

Figure 2. (a) Energy dispersion for the lower-energy band, $\varepsilon_-(k)$, plotted with the offset of J, and (b) n dependence of the electrical conductivity in the helical magnetic state at $m = 0$, σ_{HM}, for several J. We take $Q = \frac{2\pi}{6}$. The gray curve shows the data in the forced ferromagnetic state at $m = 1$, σ_{FFM}. (c) n dependence of $\sigma_{\mathrm{HM}}/\sigma_{\mathrm{FFM}}$.

Interestingly, the ratio can be larger than 1 in the high n and small J region, which indicates unusual positive magnetoresistance.

In figure 3, we show the magnetization dependence of the electrical conductivity, whose inverse gives the magnetoresistance. The energy dispersion is asymmetrically distorted in the CM for $0 < m < 1$ [9], and the Fermi wavenumber is no longer symmetric with respect to k, $k^-_{\mathrm{F1}} \neq -k^-_{\mathrm{F2}}$. This distortion gives rise to the further modulation of σ as given by the second term of (5). Figures 3(a) and 3(b) show the data for $n = 0.2$ and $n = 0.5$, respectively, indicating nonlinear negative magnetoresistance. On the other hand, the results for $n = 0.8$ in figure 3(c) show that the magnetoresistance changes from negative to positive while decreasing J, as suggested in figure 2(c).

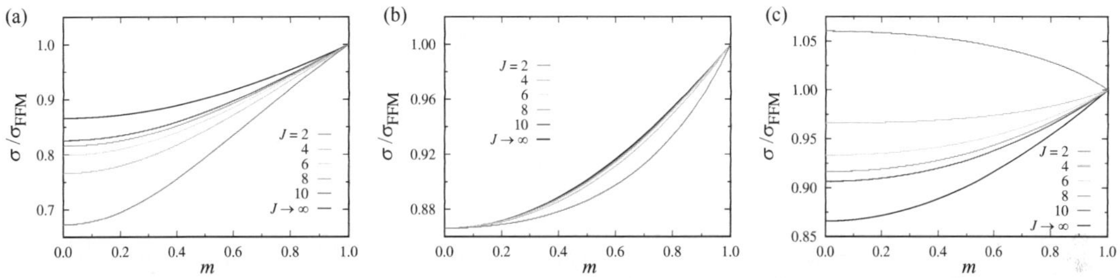

Figure 3. m dependence of the electrical conductivity normalized by the value in the FFM, $\sigma/\sigma_{\mathrm{FFM}}$, at (a) $n = 0.2$, (b) 0.5, and (c) 0.8 for $Q = \frac{2\pi}{6}$ and several J.

5. Summary

In summary, we have studied the electrical conductivity in the HM and CM by using the linear response theory for the one-dimensional spin-charge coupled model. We found that the conductivity in the HM at zero field becomes the dome-like shape as a function of the electron filling, while the functional form depends on the spin-charge coupling reflecting the modulation of the energy dispersion by the spiral magnetic texture. We also found that the conductivity in the CM exhibits the nonlinear magnetic field dependence, including unusual positive magnetoresistance in the hole doped region from half filling. Our results would be experimentally relevant to the quasi-one-dimensional helimagnets, e.g., $CrNb_3S_6$ [3]. Our findings give a starting point for further study of peculiar transport phenomena in magnets showing noncollinear and noncoplanar magnetic textures.

Acknowledgments

This research was supported by JST CREST (Nos. JPMJCR18T2 and JPMJCR19T3), JST PRESTO (No. JPMJPR19L9), the JSPS KAKENHI (Nos. JP19H05822 and JP19H05825).

References

[1] Tokura Y, Seki S and Nagaosa N 2014 *Rep. Prog. Phys.* **77** 076501
[2] Yang S H, Naaman R, Paltiel Y and Parkin S 2021 *Nat. Rev. Phys.* **3** 328
[3] Yonemura J *et al.* 2017 *Phys. Rev. B* **96** 184423
[4] Aoki R, Kousaka Y and Togawa Y 2019 *Phys. Rev. Lett.* **122** 057206
[5] Jiang N, Nii Y, Arisawa H, Saitoh E and Onose Y 2020 *Nat. Commun.* **11** 1601
[6] Watanabe H, Hoshi K and Ohe J 2016 *Phys. Rev. B* **94** 125143
[7] Okumura S, Ishizuka H, Kato Y, Ohe J and Motome Y 2019 *Appl. Phys. Lett.* **115** 012401
[8] Ustinov V and Yasyulevich I 2020 *Phys. Met. Metallogr.* **121** 223
[9] Okumura S, Morimoto T, Kato Y and Motome Y 2021 Quadratic Optical Responses in a Chiral Magnet (*Preprint* `arXiv:2108.00674`)
[10] Anderson P W and Hasegawa H 1955 *Phys. Rev.* **100** 675

Phase shift, ellipticity, angle, and topological number in skyrmion lattices

Kotaro Shimizu[1], **Shun Okumura**[2], **Yasuyuki Kato**[1], **and Yukitoshi Motome**[1]

[1] Department of Applied Physics, The University of Tokyo, Tokyo 113-8656, Japan
[2] Institute for Solid State Physics, University of Tokyo, Kashiwa 277-8581, Japan

E-mail: k.shimizu@aion.t.u-tokyo.ac.jp

Abstract. We theoretically study skyrmion lattices realized in a Kondo lattice model on a triangular lattice, focusing on the phase, ellipticity, and angle of the constituent multiple-Q waves. Analyzing the numerical data obtained in the previous study [Ozawa R, Hayami S and Motome Y 2017 *Phys. Rev. Lett.* **118** 147205], we extract these parameters for the two types of skyrmion lattices with the skyrmion number of 1 and 2. We show that the topological transition between the two skyrmion lattices driven by an external magnetic field is accompanied by significant modulations of all three parameters.

1. Introduction

The magnetic skyrmion lattice (SkL) is a two-dimensional topological spin texture [1, 2, 3, 4, 5]. It is described by a superposition of multiple spin density waves, and hence, called a multiple-Q spin state. In such a state, the topological property can be changed by modulating the constituent waves. For instance, the symmetry of the spin textures and the skyrmion number $N_{\rm sk}$, which characterizes the topology of the SkL, are changed by phase shift in the constituent waves [6, 7, 8]. Since there are many other parameters in the superpositions, further studies are desired to establish the way of controlling the topological property of multiple-Q states.

In this report, we clarify the evolution of the superposed waves in the SkLs discovered in the previous study [9]. Analyzing the numerical data of the real-space spin configurations, we extract the phases of the constituent waves, and the ellipticity and the angle of the propagating planes for two types of SkLs with $|N_{\rm sk}| = 1$ and 2. We show that all these three parameters change abruptly at the topological phase transition from $|N_{\rm sk}| = 2$ to 1 while increasing the magnetic field, indicating that the constituent waves are modulated through the topological change.

2. Model and method

We analyze the spin configurations numerically obtained in the ground state of the Kondo lattice model on a triangular lattice [9]. The Hamiltonian is given by

$$\mathcal{H} = -\sum_{\mathbf{r},\mathbf{r}',\sigma} t_{\mathbf{r}\mathbf{r}'} \hat{c}^{\dagger}_{\mathbf{r}\sigma}\hat{c}_{\mathbf{r}'\sigma} - J\sum_{\mathbf{r},\sigma,\sigma'} \mathbf{S}_{\mathbf{r}} \cdot \hat{c}^{\dagger}_{\mathbf{r}\sigma}\boldsymbol{\sigma}_{\sigma\sigma'}\hat{c}_{\mathbf{r}\sigma'} - h\sum_{\mathbf{r}} S^z_{\mathbf{r}}, \tag{1}$$

where the operator $\hat{c}^{\dagger}_{\mathbf{r}\sigma}$ ($\hat{c}_{\mathbf{r}\sigma}$) creates (annihilates) an electron with spin index $\sigma = \pm$ at site $\mathbf{r}$ and $\boldsymbol{\sigma}$ is the vector of Pauli matrices; the localized spins $\mathbf{S}_{\mathbf{r}}$ are treated as classical vectors with $|\mathbf{S}_{\mathbf{r}}| = 1$. See [9] for the details of the parameters.

In the previous study, two types of SkLs were found in the model (1): SkL with an unusual skyrmion number of $N_{\rm sk} = 2$ in the low field region including at zero field, and SkL with $N_{\rm sk} = 1$ in an intermediate field region. Since both of them are regarded as superpositions of three spin density waves (i.e., $3Q$ states) characterized by the wave vectors $\mathbf{q}_\eta = R_{\triangle}^{\eta-1}(\pi/3, 0)^{\mathsf{T}}$ ($\eta = 1, 2, 3$), where $R_\triangle$ is a two-dimensional $\frac{2\pi}{3}$-rotation operator and T denotes the transpose of the vector, we assume the following functional form to approximately describe them:

$$\mathbf{S}^{3Q}_{\mathbf{r}}(\{\varphi_\eta\}, m, \theta, \epsilon, \Gamma; \hat{\mathbf{n}}, \xi) \propto R(\hat{\mathbf{n}}, \xi)\left[\sum_{\eta=1}^{3} \frac{1}{\sqrt{3}}\left(\mathbf{e}^1_\eta(\theta,\Gamma)\cos\mathcal{Q}_\eta + \epsilon\mathbf{e}^2_\eta(\theta,\Gamma)\sin\mathcal{Q}_\eta\right) + m\hat{\mathbf{z}}\right], \tag{2}$$

where $R(\hat{\mathbf{n}}, \xi)$ denotes the three-dimensional rotation matrix about the unit vector $\hat{\mathbf{n}}$ with the angle ξ, $\mathbf{e}^1_\eta(\theta,\Gamma) = \left[R\left(\hat{\mathbf{z}}, (-1)^\Gamma \frac{2\pi}{3}\right)\right]^{\eta-1}(0, (-1)^\Gamma \sin\theta, \cos\theta)^{\mathsf{T}}$, $\mathbf{e}^2_\eta(\theta,\Gamma) = \left[R\left(\hat{\mathbf{z}}, (-1)^\Gamma \frac{2\pi}{3}\right)\right]^{\eta-1}(0, -(-1)^\Gamma \cos\theta, \sin\theta)^{\mathsf{T}}$, $\mathcal{Q}_\eta = \mathbf{q}_\eta \cdot \mathbf{r} + \varphi_\eta$, Γ takes 0 or 1, and $\hat{\mathbf{z}} = (0, 0, 1)^{\mathsf{T}}$. In (2), φ_η, ϵ, and θ denote the phase, the ellipticity, and the angle between the z axis and the major axis of the ellipse, respectively, which are relevant parameters to the energy of (2) for the model (1). We note that the model in (1) does not change the energy by flipping all the y components of the localized spins, which corresponds to the change of Γ between 0 and 1 in (2). This operation reverses the sign of $N_{\rm sk}$.

For the spin configurations obtained numerically, $\{\mathbf{S}^{\rm GS}_{\mathbf{r}}\}$, we optimize the parameters in (2) to minimize the cost function defined by

$$U(\{\varphi_\eta\}, m, \theta, \epsilon, \Gamma; \hat{\mathbf{n}}, \xi) = \frac{1}{N}\sum_{\mathbf{r}}\left(1 - \mathbf{S}^{\rm GS}_{\mathbf{r}} \cdot \mathbf{S}^{3Q}_{\mathbf{r}}(\{\varphi_\eta\}, m, \theta, \epsilon, \Gamma; \hat{\mathbf{n}}, \xi)\right), \tag{3}$$

where N is the number of sites. In the following, we focus on the optimal values of the phases φ^*_η, the ellipticity ϵ^*, and the angle θ^*. For the phases, paying attention to the sixfold rotational symmetry of the model, the symmetry by the transformation from φ_η to $-\varphi_\eta$, and the 2π periodicity of $\sum_\eta \varphi^*_\eta$, we define the sum of φ^*_η in the form of

$$\tilde{\varphi} = \pi - \left|\pi - \mathrm{Mod}\left[\sum_{\eta=1}^{3}\varphi^*_\eta, 2\pi\right]\right|. \tag{4}$$

3. Results

Figure 1 shows h dependences of the cost function in (3) and the difference of the grand potentials between the optimal state $\{\mathbf{S}^{3Q}_{\mathbf{r}}\}^*$ and the numerically obtained ground state $\{\mathbf{S}^{\rm GS}_{\mathbf{r}}\}$, $\Delta\Omega = \Omega^{3Q} - \Omega^{\rm GS}$. Here, the grand potential is defined by $\Omega = E - \mu n$, where $E = \langle\mathcal{H}\rangle/N$ and $n = \sum_{\mathbf{r},\sigma}\langle c^{\dagger}_{\mathbf{r}\sigma}c_{\mathbf{r}\sigma}\rangle/N$. In the entire region of h calculated, we obtain sufficiently small values of $U < 0.006$, indicating that $\{\mathbf{S}^{3Q}_{\mathbf{r}}\}^*$ gives a good approximation to $\{\mathbf{S}^{\rm GS}_{\mathbf{r}}\}$, as exemplified in the left and right panels for $h = 0.000$ and 0.005, respectively. $\Delta\Omega$ is also small enough, whose relative error is less than $\simeq 5 \times 10^{-5}$ for all h.

Figure 2(a) shows the skyrmion number per magnetic unit cell, $|N_{\rm sk}|$, calculated for the optimal spin configuration $\{\mathbf{S}^{3Q}_{\mathbf{r}}\}^*$; there is a phase transition at $h \simeq 0.00325$ accompanied by the topological change from $|N_{\rm sk}| = 2$ to 1. The results well reproduce those for $\{\mathbf{S}^{\rm GS}_{\mathbf{r}}\}$ [9].

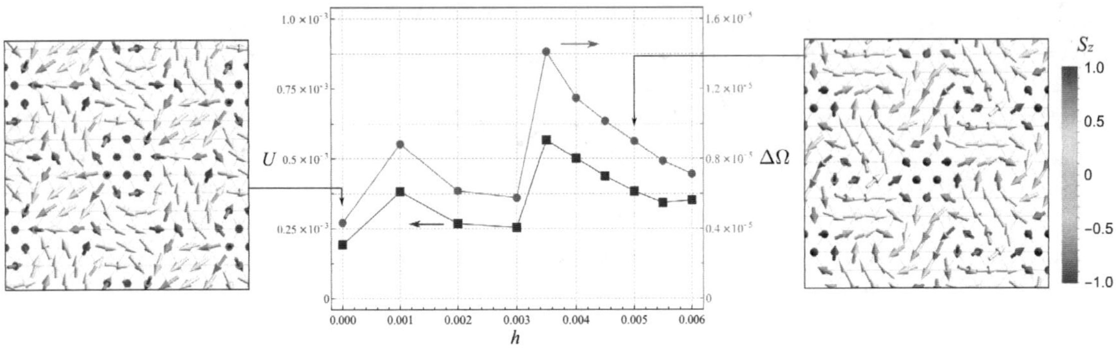

Figure 1. Magnetic field dependences of U in (3) and the difference of the grand potentials between the optimal state and the numerically-obtained ground state, $\Delta\Omega = \Omega^{3Q} - \Omega^{\mathrm{GS}}$. The real-space spin configurations obtained by the optimization, $\{\mathbf{S}_{\mathbf{r}}^{3Q}\}^*$, for $h = 0.000$ and 0.005 are plotted in the left and right panels, respectively; the arrows and color denote the xy and z components of $\{\mathbf{S}_{\mathbf{r}}^{3Q}\}^*$, respectively.

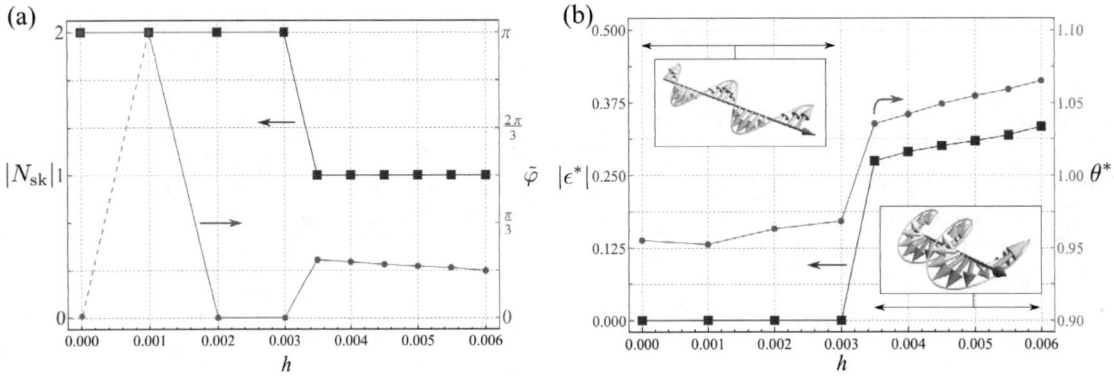

Figure 2. (a) Magnetic field dependences of the topological number $|N_{\mathrm{sk}}|$ and the sum of phases $\tilde{\varphi}$ in (4). (b) Magnetic field dependences of the ellipticity ϵ^* and the angle θ^*. The insets in (b) represent the schematic illustrations of one of three constituent waves.

The sum of phases $\tilde{\varphi}$ in (4) is also shown in figure 2(a). For the $|N_{\mathrm{sk}}| = 2$ state at $h = 0$, we obtain $\tilde{\varphi} \simeq \pi$, where $\tilde{\varphi}$ is equivalent to $\tilde{\varphi} - \pi$ because of time-reversal symmetry; to show this explicitly, we plot both points. While increasing h, we find that $\tilde{\varphi}$ remains $\simeq 0$ in the $|N_{\mathrm{sk}}| = 2$ region for $h \lesssim 0.00325$, except for $\tilde{\varphi} \simeq \pi$ at $h = 0.001$. We speculate that this exception is a numerical artifact; the simulation in the previous study presumably failed to find the true ground state with $\tilde{\varphi} \simeq 0$ at $h = 0.001$, as the energy difference between the states with $\tilde{\varphi} \simeq 0$ and π becomes smaller for lower h. When entering into the $|N_{\mathrm{sk}}| = 1$ region for $h \gtrsim 0.00325$, $\tilde{\varphi}$ is suddenly shifted from 0 to $\sim \frac{\pi}{6}$ and gradually decreases while increasing h. This clearly shows that the phase shift takes place simultaneously with the topological phase transition from $|N_{\mathrm{sk}}| = 2$ to 1.

Figure 2(b) shows the ellipticity ϵ^* and the angle θ^* in (2) as functions of h. In the $|N_{\mathrm{sk}}| = 2$ state for $0 \leq h \lesssim 0.00325$, we obtain $\epsilon^* \simeq 0$ and $\theta^* \sim \arccos(1/\sqrt{3}) \simeq 0.955$ which indicates that the constituent waves are sinusoidal waves as schematically depicted in the upper-left inset and the vectors $\mathbf{e}_{\eta}^{1}(\theta^*, \Gamma^*)$ in (2) are almost orthogonal to each other. On the other hand, both ϵ^* and θ^* change suddenly at the topological transition at $h \simeq 0.00325$, and gradually increase while increasing h. The results indicate that the constituent waves in this region are elliptically modulated as exemplified in the lower-right inset and the angle between the major axes of the ellipses, $\mathbf{e}^1_\eta(\theta^*, \Gamma^*)$, is larger than $\pi/2$. Thus, we find that the topological transition from $|N_{\rm sk}| = 2$ to 1 is accompanied by not only the phase shift but also the changes in the ellipticity and the angle of the propagating planes of the constituent waves.

4. Concluding remarks

To summarize, we analyzed the real-space spin configurations of two types of triple-Q SkLs previously found in the Kondo lattice model, and clarified the phase, the ellipticity, and the angle of the superposed waves. We elucidated that these parameters abruptly change at the topological phase transition from $|N_{\rm sk}| = 2$ to 1 while increasing the magnetic field. We note that our results are qualitatively consistent with those obtained by the recent variational study in [8], where the relative phase between the constituent sinusoidal and cosinusoidal waves, which partly corresponds to $\tilde{\varphi}$, ϵ, and θ in our analysis, is shifted at the topological phase transitions induced by the magnetic anisotropy and the magnetic field. Our results indicate that the parameters to characterize the constituent waves underlie the change of the topological properties in the triple-Q SkLs.

Our finding provides a first step toward the systematic investigation on other topological phase transitions in terms of the detailed parameters in the constituent waves, not only for the SkLs but also for other topological spin textures, such as the three-dimensional hedgehog lattices [10, 11, 12, 13]. It is desired to establish the mechanism to control the constituent waves, e.g., by the external magnetic field and pressure, in the future studies. It is also intriguing to study the relations between the nature of the constituent waves and the electronic, transport, and dynamical properties in the topological multiple-Q states.

Acknowledgments

We thank R. Ozawa for providing us the numerical data and S. Hayami for fruitful discussions. This research was supported by Grant-in-Aid for Scientific Research Grants (Nos. JP18K03447, JP19H05822, JP19H05825, and JP21J20812), JST CREST (No. JP-MJCR18T2), and the Chirality Research Center in Hiroshima University and JSPS Core-to-Core Program, Advanced Research Networks.

References

[1] Bogdanov A N and Yablonskii D 1989 *Zh. Eksp. Teor. Fiz* **95** 178
[2] Bogdanov A and Hubert A 1994 *J. Magn. Magn. Mater.* **138** 255
[3] Roessler U K, Bogdanov A and Pfleiderer C 2006 *Nature* **442** 797
[4] Mühlbauer S, Binz B, Jonietz F, Pfleiderer C, Rosch A, Neubauer A, Georgii R and Böni P 2009 *Science* **323** 915
[5] Yu X Z, Onose Y, Kanazawa N, Park J H, Han J H, Matsui Y, Nagaosa N and Tokura Y 2010 *Nature* **465** 901
[6] Kurumaji T *et al.* 2019 *Science* **365** 914
[7] Hayami S, Okubo T and Motome Y 2020 Phase Shift in Skyrmion Crystals *Preprint* 2005.03168
[8] Hayami S and Yambe R 2021 Locking of skyrmion cores on a centrosymmetric discrete lattice: onsite versus offsite *Preprint* 2107.03500
[9] Ozawa R, Hayami S and Motome Y 2017 *Phys. Rev. Lett.* **118** 147205
[10] Binz B and Vishwanath A 2006 *Phys. Rev. B* **74** 214408
[11] Park J H and Han J H 2011 *Phys. Rev. B* **83** 184406
[12] Fujishiro Y, Kanazawa N and Tokura Y 2020 *Appl. Phys. Lett.* **116** 090501
[13] Okumura S, Hayami S, Kato Y and Motome Y 2020 *Phys. Rev. B* **101** 144416

Journal of Physics: Conference Series **2164** (2022) 012070 doi:10.1088/1742-6596/2164/1/012070

The surface tension at the metal-insulator phase transition in the magnetic field.

L.B. Dubovskii

NRC "Kurchatov Institute", Moscow 123182, Russia,

Abstract. Any metal-insulator transition in a crystalline matter must be the transition from a situation in which the electronic bands overlap to that when they do not [1]. On the basis of the phenomenological theory [2], various singularities are considered in the magnetic field for the surface tension at the metal-insulator contact. The surface tension displays also the quantum magnetic oscillations at low temperatures [3]. The consideration is applied to the Mott insulator in the magnetic field.

1. Introduction.

It is well known [4, 5] that the characteristic magnitudes for the surface tension in insulators are one order of magnitude lower than those in metals. The physical mechanism underlying this difference is related to the conduction electrons penetrating outside from the interface to vacuum at a distance of order of about atomic layer spacing in the metal bulk [6, 7]. At the same time, the surface tension of semimetals with the characteristic electron density differing from that of usual metals by the factor $10^{-6} - 10^{-5}$ is about the same order of that of usual metals [5].

The metal-insulator transitions (MIT) in crystals are (at least at zero temperature) transitions from the state where the electron bands overlap to the state with no overlapping [1]. The weak overlapping is responsible for the transition to the metallic semimetal-type state with the number of charge carriers per unit cell being far below unity and typical for the case of semimetals.

We consider the self-consistent equations of MIT in the external magnetic field. The MIT belongs to the first-order phase transition with the order parameter corresponding to the transition of matter density $\rho(\vec{r})$ for two different phases, metallic and isolating. The MIT is accompanied by appearing the second-order parameter, namely, electron density vanishing in the insulator phase.

The new order parameter is the electron and holes densities in two bands. The density of electrons and holes can be described by a vector in the Hilbert space with electronic and hole components $\Upsilon(\vec{r})$. Here $\Upsilon(\vec{r})$ is a vertical column and $\Upsilon^+(\vec{r})$ is a horizontal row:

$$\Upsilon(\vec{r}) = (\,u\,)_\uparrow(\vec{r})u_\downarrow(\vec{r})v_\uparrow(\vec{r})v_\downarrow(\vec{r})\,,\ \Upsilon^+(\vec{r}) = (u^*_\uparrow(\vec{r}), u^*_\downarrow(\vec{r}), v^*_\uparrow(\vec{r}), v^*_\downarrow(\vec{r})). \tag{1}$$

The components of vector $\Upsilon(\vec{r})$ are the functions of electrons and holes for the up and down spins. In the absence of magnetic field all these components of the vector $\Upsilon(\vec{r})$ are real functions.

2. The Phenomenological Theory of Metal-Insulator phase transition.

The role of these two order parameters $\rho(\vec{r})$ and $\Upsilon(\vec{r})$ is quite different. The density $\rho(\vec{r})$ governs a position of the MIT as a function of pressure and temperature (p, T). The order parameter $\Upsilon(\vec{r})$ becomes nonzero just in the point $\rho = \rho_c(p, T)$. However, the physical properties of the system depend weakly on ρ alone. On the other hand, the value of $\Upsilon(\vec{r})$ can drastically change the physical properties of the system. In particular, the surface tension Σ_ρ of the MI interface depends only weakly on ρ and is determined to a great extent by the variation of $\Upsilon(\vec{r})$. The physical reason of the phenomenon is an outflow of conduction electrons to the MI interface [6, 7]. This property of metal boundaries is analogous to the well-known property of the phonon spectrum of metals, which is in the great extent determined by the conduction electrons [8].

We can put the following invariant scalar quantity

$$n(\vec{r}) = n_e(\vec{r}) + n_h(\vec{r}) \equiv (\Upsilon^+\Upsilon) \ , \tag{2}$$

$n_e(\vec{r}) = |u_\uparrow(\vec{r})|^2 + |u_\downarrow(\vec{r})|^2$, $n_h(\vec{r}) = |v_\uparrow(\vec{r})|^2 + |v_\downarrow(\vec{r})|^2$. This quantity determines the densities of electrons and holes in the metal phase and vanishes in the insulating phase at $T = 0$. We take the identity $n_h(\vec{r}) = n_e(\vec{r})$ due to the local electrical neutrality within the metal phase.

In the MIT vicinity, the Ginzburg-Landau (GL) functional can be assumed as a sum of contributions from the order parameters $\rho(\vec{r})$ and $\Upsilon(\vec{r})$:

$$\Phi = \Phi[\rho, \Upsilon, \Upsilon^+] = \int d\vec{r} F(\rho, \Upsilon, \Upsilon^+) ,$$
$$F = F_\rho(\rho(\vec{r})) + F_\Upsilon(\Upsilon(\vec{r})) + F_{int}(\rho(\vec{r}), \Upsilon(\vec{r})). \tag{3}$$

The parameter $\rho(\vec{r})$ describes the first-order phase transition between the insulating and metal phases whereas the vector function $\Upsilon(\vec{r})$ describes the second-order phase transition. We expand the functional $F_\rho(\rho(\vec{r}))$ according to the Cahn – Hillard approximation [9], see also [10]:

$$F_\rho(\rho(\vec{r})) = \varphi(\rho) + \frac{1}{2}\lambda(\rho)(\nabla\rho)^2. \tag{4}$$

We use the Ginzburg-Landau expansion [11] for $\Upsilon(\vec{r})$:

$$F_\Upsilon(\Upsilon(\vec{r})) = \alpha\,(\Upsilon^+\Upsilon) + \frac{1}{2}\beta(\Upsilon^+\Upsilon)^2 + \frac{\hbar}{2m}\left(\nabla\Upsilon^+(\vec{r})\nabla\Upsilon(\vec{r})\right);$$
$$\alpha = \alpha_0(\rho - \rho_c). \tag{5}$$

Here we have expanded coefficient α linearly in ρ in the vicinity of ρ_c. It is analogous to the well-known expansion of the appropriate parameter in temperature T [11] for the Landau theory of the second-order phase transition. The expansion proposed is convenient for investigating the MIT at a fixed temperature, say, at $T = 0$. The interaction term $F_{int}(\rho(\vec{r}), \Upsilon(\vec{r}))$ is taken as a linear function of $n(\vec{r})$ (cf. [2]):

$$F_{int}(\rho(\vec{r}), \Upsilon(\vec{r})) = g(\rho)\,(\Upsilon^+\Upsilon)\,. \tag{6}$$

The variation of $\Phi = \Phi[\rho, \Upsilon, \Upsilon^+]$ in $\rho(\vec{r})$ and $\Upsilon^+(\vec{r})$ gives the self-consistent equations of the Ginzburg-Landau type for the MI phase transition:

$$\lambda^{1/2}(\rho)\nabla\left(\lambda^{1/2}(\rho)\nabla\rho(\vec{r})\right) = \frac{d\varphi}{d\rho}(\rho) + \frac{dg}{d\rho}\,(\Upsilon^+\Upsilon)\,,$$
$$-\frac{\hbar}{2m}\nabla^2\Upsilon + (\alpha + g)\,\Upsilon + \beta\,(\Upsilon^+\Upsilon)\,\Upsilon = 0. \tag{7}$$

In the case $g \equiv 0$, the parameters ρ and Υ are independent and the surface tension Σ at the interface between the metallic and the insulating phases can be calculated separately for ρ and Υ, correspondingly Σ_ρ and Σ_Υ. The surface tension Σ_ρ can readily be calculated as [2, 10]:

$$\Sigma_\rho = \int_{\rho_{ins}}^{\rho_{met}} d\rho \sqrt{\varphi(\rho)\lambda(\rho)/2}\,. \tag{8}$$

3. The surface tension at the MI interface in the magnetic field

We consider the surface tension in the external magnetic field. In the functional (5) we should include the energy of the magnetic field, namely $H^2/8\pi$. Also we should involve $[\nabla - e/\hbar c\vec{A}]$ instead of nabla ∇ and also $[\mu(\Upsilon^+\vec{H}\vec{\sigma}\Upsilon)]$ [2]. Here $\vec{\sigma}$ is the Pauli matrix. So, we arrive to the following equation, where x is the coordinate perpendicular to the interface:

$$-\frac{\bar{\hbar}}{2m}\left(\frac{d^2\Upsilon}{dx^2}\right) + (\alpha + g\rho)\,\Upsilon + \beta\left(\Upsilon^+\Upsilon\right)\Upsilon + \hat{\Lambda}(\vec{H})\Upsilon = 0\,. \tag{9}$$

Here the operator $\hat{\Lambda}(\vec{H})$ is the four-row matrix [2] determined by the magnetic field $\vec{H} = \vec{i}H_x + \vec{k}H_z$. The vector $\vec{k}$ is directed along the metal-insulator boundary and the vector $\vec{i}$ is parallel to the boundary. The Equation (9) can be reduced to the following one:

$$-\xi_{\vec{H}}^2(T)\frac{d^2 f}{dx^2} - f + f^3 = 0\,,\ \ \xi_{\vec{H}}(T) = \sqrt{\frac{\bar{\hbar}}{2m|\alpha|}}\,,\ \ g = 0\,. \tag{10}$$

Here the dimensionless function $f(x)$ has the following solution $f(x) = \tanh\left[\frac{x}{\sqrt{2}\xi_{\vec{H}}(T)}\right]$. Here $\xi_{\vec{H}}(T)$ and $f(x)$ are the functions of the magnetic field $\vec{H}$. The calculation of the surface energy Σ_Υ is straightforward and gives:

$$\Sigma_\Upsilon = \frac{2\sqrt{2}}{3}\xi_{\vec{H}}(T)\beta n_0^2. \tag{11}$$

Here n_0 is the bulk density of electrons in the metal. If we put n_0 to be of the order of unity per crystal cell, for Σ_Υ (11) we get the typical magnitude for metals.

In addition, we consider the surface tension oscillations in the external magnetic field. In this case we should replace $\vec{H}$ for $\vec{H} + 4\pi\hat{\vec{M}}$ in (9)[2, 3]. Here we have [2, 3] to substitute

$$\hat{\vec{M}} = \vec{H} - (\Upsilon^+\Upsilon)\frac{\partial\Omega}{(\Upsilon^+\Upsilon)}\,.$$

The quantity Ω is the thermodynamical potential [11]. The oscillating magnetic moment has the following structure [3]:

$$\hat{\vec{M}} = \Im\vec{A}\exp[iN(\epsilon_F)]\exp\left(-\frac{2\pi^2 T}{\bar{h}\omega}\right)\,,$$

$$N(\epsilon_F) = \left(\frac{\epsilon_F}{\bar{h}\omega}\right)\,,\ \ \bar{h}\omega = \frac{eH}{mc}\,.$$

The quantity $N(\epsilon_F)$ is the number of Landau levels under the Fermi surface, $N(\epsilon_F) \gg 1$. And $|\vec{A}| \cong |\sqrt{N(\epsilon_F)} * M_{mon}$ [3, 11].

4. Conclusions.

Based on the Ginzburg-Landau expansion, the phenomenological theory of metal-insulator phase transition in the magnetic field is considered.

The surface tension Σ_Υ (7) is investigated for the various situations. It is shown that the typical magnitude of the metal surface tension is much larger than that for insulators. The singularities and the surface tension oscillations in the external magnetic field are considered.

So far, we have considered the system of non-interacting electrons and holes. If we replace the constant β (5) with an operator $\hat{\beta}$ that takes the electron-hole interaction into account, we will come to an insulator even for the overlapping bands, namely, to the Mott insulator. However, this approach leads only to an effective description of the phenomenon. The self-consistent approach is to take into account the Coulomb interaction based on the Hubbard interaction [12, 13, 14]. This requires the splitting of the electronic band into two subbands (Hubbard bands [14] based on the consideration in Ref. [12]) and leads to a significant modification of the electronic order parameter $\Upsilon(\vec{r})$.

At last, it should be emphasized that the Mott insulator under the influence of magnetic field can turn into a metallic, or more precisely, semi-metallic state. In this case, we are talking about a sufficiently weak Hubbard interaction, so that the magnetic fields of $\sim 10^5$ kOe are sufficiently strong for the transition to the semi-metallic state. This can be compared with the transition of semimetals Bi and Sn to the insulator state (see [15]). In the Mott insulator the surface tension is much less as compared with the case when the system becomes semi-metallic one.

5. References

[1] *N.F. Mott. Metal-Insulator, 2nd ed.* (Taylor@Francis, London, 1990).
[2] *L.B. Dubovskii, Physica B* **536**, 469, 2018; *JETP Letters* **99**, *No.1*, 22-26, 2014.
[3] *L.B. Dubovskii, J. Low. Temp. Phys. 182 182*, 192-205, 2016.
[4] *W. Monch, Rep. Prog. Phys. 53*, 221, 1990.
[5] *Tables of Physical Values. The Manuel, Ed. by I.K. Kikoin (Atomizdat, Moscow, 1976)* [in Russian].
[6] N.D. Lang and W. Kohn Phys. Rev. B, 28, 258, 1970.
[7] *M.B. Partenskii, Sov. Phys. Usp. 22*, 330, 1979.
[8] *E. G. Brovman and Yu. M. Kagan, Phonons in nontransition metals,Physics-Uspekhi*, **17**, 125-152, 1974.
[9] J. W. Cahn and J. E. Hillard, J. Chem. Phys. **28** ,258 (1958); ibid. **42** , 93, 1964.
[10] I. M. Lifshits and Yu. Kagan. Sov. Phys. JETP 35, 206, 1972.
[11] *L.D.Landau and E.M.Lifshits, Statistical Physics. Part 1: Course of Theoretical Physics. Vol. 5 (3rd ed.)* Publisher; Butterworth-Heineman: 3 edition, 1980.
[12] *J. Hubbard, Proc. Roy. Soc. A 276*, 238, 1963.
[13] *J. Hubbard, Proc. Roy. Soc. A 277*, 237, 1964.
[14] *J. Hubbard, Proc. Roy. Soc. A 281*, 41, 1963.
[15] *M. Lifshitz, M. Ya. Azbel' and M. I. Kaganov. Electron Theory of Metals* Nauka, M. 1971 [in Russian].

Field-angle dependence of thermal transport in Kitaev-Γ model

Shinnosuke Koyama and Joji Nasu

Department of Physics, Yokohama National University, Hodogaya, Yokohama 240-8501, Japan

E-mail: koyama-shinnsuke-kz@ynu.jp

Abstract. We investigate the magnetic field effect on the Kitaev model with the Γ-type interaction, which plays an essential role in understanding the magnetic properties of real materials. We examine the mean-field phase diagram and thermal Hall conductivity using the linear spin-wave theory. We find that the Γ interaction substantially changes the field-angle dependence of the thermal Hall conductivity and suppresses its magnitude. The suppression is caused by the enhancement of the low-energy magnon gap while increasing the Γ interaction.

1. Introduction

The quantum spin model proposed by A. Kitaev has attracted considerable attention in strongly correlated electron systems [1, 2]. This is called the Kitaev model and realizes a canonical quantum spin liquid (QSL) as a ground state due to strong quantum fluctuations and correlations. The Kitaev model is believed to dominate magnetism in compounds including 4*d* or 5*d* transition metal ions with strong spin-orbit coupling [3]. Among them, A_2IrO_3 (A = Li, Na) and α-$RuCl_3$ whose magnetism are governed by $j_{\rm eff} = 1/2$ pseudospins, have been intensively studied as candidate materials of the Kitaev QSL [4]. However, other interactions, such as the Heisenberg and off-diagonal Γ terms, are present in real materials. These contributions lead to the appearance of a magnetic order [5]. While the additional interactions destabilize the Kitaev QSL state, they cause a wide variety of magnetically ordered states [6, 7]. In particular, a zigzag-type magnetic order, which appears in A_2IrO_3 (A = Li, Na) and α-$RuCl_3$ at low temperatures, is stabilized by the competition between the Kitaev and Heisenberg/Γ interactions.

Recent experiments clarified that the Kitaev candidate material α-$RuCl_3$ exhibits the half-quantization of the thermal Hall conductivity in the region where the zigzag-type magnetic order disappears by applying the magnetic field [8], and the thermal Hall conductivity is strongly depends on the magnetic-field direction [9]. These observations are considered to be convincing evidence for a Majorana chiral edge mode. Stimulated by these results and related experiments, magnetic field effects on the Kitaev-related systems have been examined theoretically [10]. In addition to the Majorana picture consistent with the result in the Kitaev model [11], theoretical calculations based on the magnon picture have been made in the extended Kitaev models with the Heisenberg and Γ interactions [6, 7]. Although the thermal Hall effect has been examined in the Kitaev-Heisenberg-Γ model [7, 12, 13] for specific magnetic-field directions, the field-angle dependence of the thermal Hall effect ascribed to the magnons remains unclear in the Kitaev-Γ model.

In this study, we investigate the thermal transport in the Kitaev-Γ model on a honeycomb lattice under magnetic fields. By applying mean-field approximation and linear spin-wave theory, we examine the thermal Hall conductivity by changing the direction systematically. We find that the direction of the magnetization deviates from that of the magnetic field by introducing the Γ interaction and a peculiar magnetic order appears around the perpendicular direction to the honeycomb plane. Moreover, the Γ interaction suppresses the thermal Hall conductivity by enhancing the low-energy magnon gap. We also clarify that the field direction vanishing the thermal Hall conductivity depends on the magnitude of the Γ interaction, which is in contrast to the effect of the Heisenberg interaction.

2. Model and Method

We consider the Kitaev-Γ model with $S = 1/2$ on the honeycomb lattice, which is given by

$$\mathcal{H} = \sum_{\gamma \in xyz} \sum_{\langle ij \rangle_\gamma} \left[K S_i^\gamma S_j^\gamma + \Gamma (S_i^\alpha S_j^\beta + S_i^\beta S_j^\alpha) \right] - \boldsymbol{h} \cdot \sum_i \boldsymbol{S}_i, \tag{1}$$

where K and Γ is the Kitaev and off-diagonal interaction, respectively, and $\boldsymbol{h}$ is the magnetic field. $\langle ij \rangle_\gamma$ stands for the nearest neighbor sites on the γ bond ($\gamma = x, y, z$) presented in Fig. 1(a). The indexes α and β are given such that (α, β, γ) is a cyclic permutation of (x, y, z). In this study, we focus on the ferromagnetic Kitaev coupling, namely, $K = -1$, and the magnetic field is fixed to be $h = 0.05$.

We apply the mean-field approximation to the Kitaev-Γ model given in Eq. (1) and determine the stable magnetic-order pattern at zero temperature. Under the spin configuration, we calculate the magnon dispersions and thermal Hall conductivity using the linear spin-wave theory. In this theory, the Hamiltonian is approximately written as a bilinear form of the bosonic operators describing the magnon excitations as follows:

$$\mathcal{H}_{\rm sw} = \frac{1}{2} \sum_{\boldsymbol{k}} \mathcal{A}_{\boldsymbol{k}}^\dagger \mathcal{M}_{\boldsymbol{k}} \mathcal{A}_{\boldsymbol{k}}, \tag{2}$$

where

$$\mathcal{A}_{\boldsymbol{k}}^\dagger = \left(a_{1,\boldsymbol{k}}^\dagger \; a_{2,\boldsymbol{k}}^\dagger \; \cdots \; a_{M,\boldsymbol{k}}^\dagger \; a_{1,-\boldsymbol{k}} \; a_{2,-\boldsymbol{k}} \; \cdots \; a_{M,-\boldsymbol{k}} \right). \tag{3}$$

Here, $a_{l,\boldsymbol{k}}$ is the annihilation operator of a boson with momentum $\boldsymbol{k}$ on sublattice l with M being the number of sublattices. In Eq. (2), $\mathcal{M}_{\boldsymbol{k}}$ is a $2M \times 2M$ Hermitian matrix, which is obtained by the Holstein-Primakoff transformation [14]. By diagonalizing $\mathcal{M}_{\boldsymbol{k}}$ using the Bogoliubov transformation with the transform matrix $\mathcal{J}_{\boldsymbol{k}}$, we evaluate the magnon energy $\varepsilon_{n,\boldsymbol{k}}$ for the nth branch. We also calculate the thermal Hall conductivity κ^{ab}, which is defined by $\langle J_Q^a \rangle_{\nabla T} = \kappa^{ab}(-\nabla_b T)$, where $\langle J_Q^a \rangle_{\nabla T}$ is the expectation value of the thermal current along the a axis in the presence of the thermal gradient $\nabla_b T$ along the b axis [see Fig. 1(a)]. This quantity is evaluated using the linear spin-wave theory, which is written as [15]

$$\kappa^{ab} = -\frac{T}{V} \sum_{\boldsymbol{k}} \sum_{n=1}^{M} c_2(f_{\rm BE}(\varepsilon_{n,\boldsymbol{k}})) \Omega_{n,\boldsymbol{k}}, \tag{4}$$

where V is the volume, $c_2(x) = \int_0^x dt \, [\ln(1+t) - \ln t]^2$, and $\Omega_{n,\boldsymbol{k}}$ is the Berry curvature given as

$$\Omega_{n\boldsymbol{k}} = i \left[\sigma_3 \frac{\partial \mathcal{J}_{\boldsymbol{k}}^\dagger}{\partial k_a} \sigma_3 \frac{\partial \mathcal{J}_{\boldsymbol{k}}}{\partial k_b} \right]_{nn} + \text{H.c.}, \tag{5}$$

with $(\sigma_3)_n = +1$ for $n \leq M$ and $(\sigma_3)_n = -1$ for $n > M$.

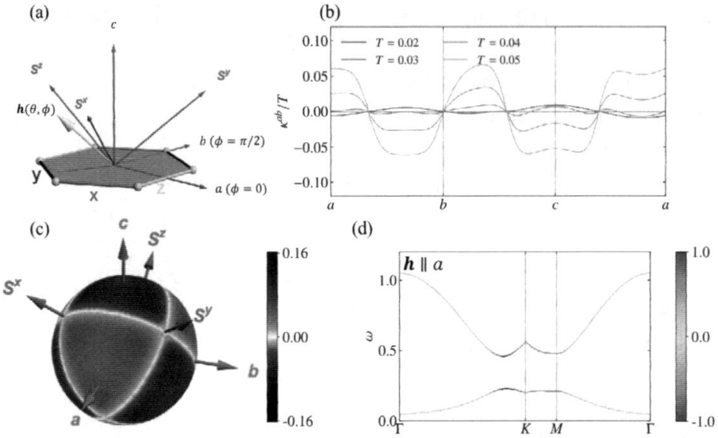

Figure 1. (a) Schematic picture of the honeycomb lattice on which the Kitaev-Γ model is defined. The lattice axes, a, b, and c, and the spin axes, S^x, S^y, and S^z, are also presented. (b) Thermal Hall conductivity κ^{ab}/T as a function of the field direction in the Kitaev model with $\Gamma = 0$ for $h = 0.05$ at several temperatures, and (c) its spherical plot at $T = 0.05$. (d) Spin-wave dispersions under the magnetic fields applied along the a direction at $T = 0.05$. The line color indicates the value of $\Omega_{n,\boldsymbol{k}}$ associated with the magnon branch $\varepsilon_{n,\boldsymbol{k}}$.

3. Result and Discussion

3.1. Magnetic field effect on pure Kitaev model

First, we focus on the case without the Γ interaction. In this case, a spin-polarized state appears regardless of the field direction. Figure 1(b) shows the field-angle dependence of κ^{ab}/T at several temperatures in the ferromagnetic Kitaev model under the magnetic field. With increasing temperature, κ^{ab}/T is enhanced, but there are several nodal points where it remains zero for the temperature evolution. In particular, for the magnetic field parallel to the b axis, κ^{ab} vanishes because of the symmetry [14]. The other nodal points are located on the planes spanned by two of three spin axes S^x, S^y, and S^z. This property is clearly seen in Fig. 1(c), and it is maintained even in the presence of the Heisenberg interaction but is not for the Γ one [14].

The sign of κ^{ab} depends on the direction of the applied magnetic field. The thermal Hall conductivity is positive for $\boldsymbol{h} \parallel a$ while it is negative for $\boldsymbol{h} \parallel c$ [Figs. 1(b) and 1(c)]. The nonzero thermal Hall conductivity comes from the finite Berry curvature $\Omega_{n,\boldsymbol{k}}$. Figure 1(d) shows the Berry curvature for the magnon excitations in the magnetic field applied to the a direction. In this case, the low-energy magnon branch possesses the negative Berry curvature around the K and M points, which leads to the positive thermal Hall conductivity.

3.2. Effect of Γ interaction

Next, we show the results in the presence of the nonzero Γ interaction. Figures 2(a) and 2(b) present the mean-field phase diagram on the plane of the field direction parametrized by (θ, ϕ) [see Fig. 1(a)] at $h = 0.05$ for $\Gamma = 0.1$ and 0.2, respectively. As shown in these figures, the 6-site spin order presented in Fig. 2(g) appears in the magnetic field applied around the c direction. On the other hand, when the magnetic field is away from the c axis, the magnetic order disappears, and the spin-polarized state is stabilized. In this polarized state, the total magnetic moment is not parallel to the field direction, unlike the case without the Γ interaction.

The field-angle dependence of κ^{xy}/T for $\Gamma = 0.1$ and 0.2 is shown on the sphere in Figs. 2(c) and 2(d), respectively. The feature seen in the pure Kitaev model is not observed in the presence of the Γ

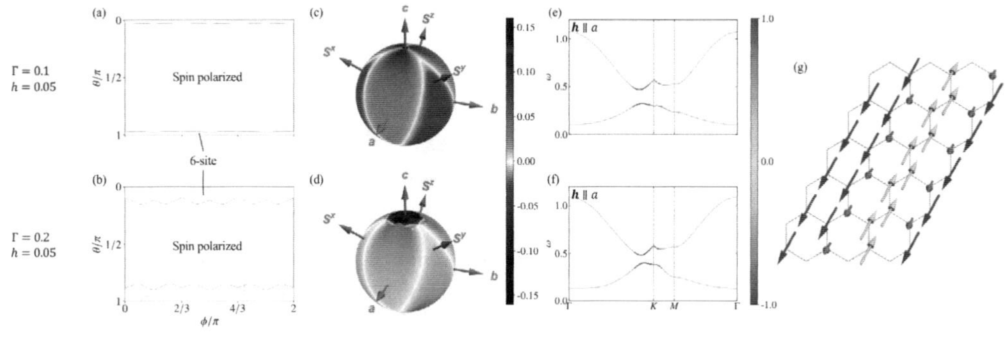

Figure 2. (a),(b) Mean-field phase diagram of the Kitaev-Γ model under the magnetic field whose direction is parametrized by (θ, ϕ) with (a) $\Gamma = 0.1$ and (b) $\Gamma = 0.2$. (c),(d) Spherical plot of κ^{ab}/T with (c) $\Gamma = 0.1$ and (d) $\Gamma = 0.2$ at $T = 0.05$. (e),(f) Spin-wave dispersions and Berry curvature for (e) $\Gamma = 0.1$ and (f) $\Gamma = 0.2$. The magnitude of the magnetic field is fixed to $h = 0.05$. (g) Spin configuration of the 6-site ordered state.

interaction; the nodal lines are not on the planes spanned by the spin axes. We also find a large negative thermal Hall conductivity in the 6-site ordered phase and the suppression of κ^{xy}/T in the spin-polarized phase by the introduction of the Γ interaction. To clarify the origin of this suppression effect, we examine the Berry curvature for $\boldsymbol{h} \parallel a$. As shown in Figs. 2(e) and 2(f), the momentum dependence of $\Omega_{n,\boldsymbol{k}}$ is largely intact despite the increase of the Γ interaction, but the excitation gap increases with increasing Γ. The enhancement of the gap leads to the suppression of the thermal Hall conductivity.

4. Conclusion

We have revealed the magnetic-field angle dependence of the spin configuration and thermal transport in the Kitaev-Γ model using the mean-field approximation and the linear spin-wave theory. We have demonstrated that the Γ interaction changes the nodal-line structure in the field-angle dependence of the thermal Hall conductivity. We also find that the 6-site magnetic order appears when the magnetic field is perpendicular to the honeycomb plane. We also find that the Γ interaction suppresses the thermal Hall conductivity because of enhancing the low-energy magnon gap.

References

[1] Kitaev A 2006 *Ann. Phys.* **321** 2–111 ISSN 0003-4916
[2] Motome Y and Nasu J 2020 *J. Phys. Soc. Jpn.* **89** 012002
[3] Jackeli G and Khaliullin G 2009 *Phys. Rev. Lett.* **102**(1) 017205
[4] Yamaji Y, Nomura Y, Kurita M, Arita R and Imada M 2014 *Phys. Rev. Lett.* **113**(10) 107201
[5] Rau J G, Lee E K H and Kee H Y 2014 *Phys. Rev. Lett* **112** 077204
[6] Janssen L, Andrade E C and Vojta M 2016 *Phys. Rev. Lett* **117** 277202
[7] Chern L E, Kaneko R, Lee H Y and Kim Y B 2020 *Phys. Rev. Research* **2**(1) 013014
[8] Kasahara Y, Ohnishi T, Mizukami Y, Tanaka O, Ma S, Sugii K, Kurita N, Tanaka H, Nasu J, Motome Y, Shibauchi T and Matsuda Y 2018 *Nature* **559** 227–231
[9] Yokoi T, Ma S, Kasahara Y, Kasahara S, Shibauchi T, Kurita N, Tanaka H, Nasu J, Motome Y, Hickey C, Trebst S and Matsuda Y 2021 *Science* **373** 568–572
[10] Gordon J S, Catuneanu A, Sørensen E S and Kee H Y 2019 *Nat. Commun.* **10** 2470
[11] Nasu J, Yoshitake J and Motome Y 2017 *Phys. Rev. Lett.* **119**(12) 127204
[12] Zhang E Z, Chern L E and Kim Y B 2021 *Phys. Rev. B* **103**(17) 174402
[13] McClarty P A, Dong X Y, Gohlke M, Rau J G, Pollmann F, Moessner R and Penc K 2018 *Phys. Rev. B* **98**(6) 060404
[14] Koyama S and Nasu J 2021 *Phys. Rev. B* **104**(7) 075121
[15] Matsumoto R, Sindou R and Murakami S 2014 *Phys. Rev. B* **89** 054420

Angular-field magnetic phase diagram of b-plane at 4 K of YAlGe-type TbAlGe with zigzag-chain

Kaiki Gouda, Takashi Nishioka

Dep. of Phys., Fac. of Sci. and Tech., Kochi Univ., Kochi 780-8520, Japan

E-mail: b20m6g03@s.kochi-u.ac.jp

Abstract. Orthorhombic YAlGe-type TbAlGe is expected to have an interesting magnetic anisotropy due to zigzag chains of the Tb ions. We have grown the single crystal for the first time and measured the AC magnetic susceptibility and specific heat from 1.3 K to 60 K, and the vector magnetization for the *b*-plane up to 7 T at 4 K. The specific heat and AC magnetic susceptibility indicate that there are two antiferromagnetic transitions at T_{N1} = 38 K and T_{N2} = 7.6 K, where the transition at T_{N2} is first-order like. The magnetization curve at 4 K for the *a*-axis shows a large hysteresis, and metamagnetic transition appears at H_1 = 1.6 T in the field increasing process, and another metamagnetic transition at H_2 = 3.5 T in addition to H_1 in the decreasing field process. The magnetization curves of the *b*- and *c*-axis are linear up to 7 T. The measurement of vector magnetization at 4 K reflects the hysteresis of the magnetization curve, and there is a large hysteresis. From this vector magnetization measurement, we have made the angular magnetic field phase diagram at 4 K for the *b*-plane. In this phase diagram, there are phase lines that cannot be obtained by ordinary magnetization measurement.

1. Introduction

*R*AlGe (*R* = rare earth) crystallizes in the tetragonal α-$ThSi_2$ type in light rare earths and the orthorhombic YAlGe type in heavy rare earths. The boundary is R = Gd, where the high temperature phase is α-$ThSi_2$type, and the low temperature phase is YAlGe type [?]. These two structures are related each other that they have a triangular prism with Ge in the body. The α-$ThSi_2$ type is a triangular prism with [R6] as the vertices, and the YAlGe type is a triangular prism with [R2Al4] as the vertices. One feature of the YAlGe type structure is that the *R* ions have a zigzag-chain in the *a*-plane and extend in the *c*-axis. Zigzag-chain of *R* ions is also found in YFb_2Al_{10} type structures and is studied by Ohno *et al.* [?]. They found that $TbRu_2Al_{10}$ has a very peculiar magnetic phase diagram that depends on the angle, and therefore TbAlGe is also expected to be interesting. However, the existence of YAlGe-type RAlGe has only been clarified only as a polycrystal, and its physical properties have not been measured. The purpose of this study is to grow a single crystal of TbAlGe, and measure the vector magnetization.

2. Experiment procedure

Single crystal growth of TbAlGe was carried out by the Al self-flux method. The growth was carried out in a Tanman crucible enclosed in a quartz tube with an argon atmosphere. The starting composition ratio was Tb:Al:Ge = 1:1:10. The quartz tube was heated to 1050°C, maintained for 8 hours, and then cooled to 700 °C at rate of 3.5 °C/h. The Al fulx was removed by centrifugation at the temperature, the remaining flux was removed by etching with 3M NaOH aqueous solution. The resulting single crystal axes were determined by an XRD.

The AC magnetic susceptibility was measured by the usual Hartshorn bridge method, and the specific heat was measured by the adiabatic method using a GM refrigerator [?], both at 1.4K to 60K. The vector magnetization was measured at 4K for the b-plane using our hand-made syste [?].

3. Results and Discussion

Figure 1 shows the temperature dependence of the AC magnetic susceptibility χ. Two cusps corresponding to the antiferromagnetic transition are found at $T_{N1}(\chi) = 41.5$ K and $T_{N2}(\chi) = 8.2$ K. Furthermore, χ increases below 4 K, suggesting that there is a further phase transition at low temperatures. Figure 2 shows the temperature dependence of the specific heat C divided by temperature T. Anomalies are seen corresponding to the two anomalies in χ, but the temperatures are slightly different, $T_{N1}(C) = 39$ K and $T_{N2}(C) = 7.6$ K. The phase transition at $T_{N2}(C)$ seems to be divergent, suggesting a first-order transition. In addition, although there is a slight upturn below 1.4 K that seems to correspond to the increase in χ. Lower temperature measurements are needed.

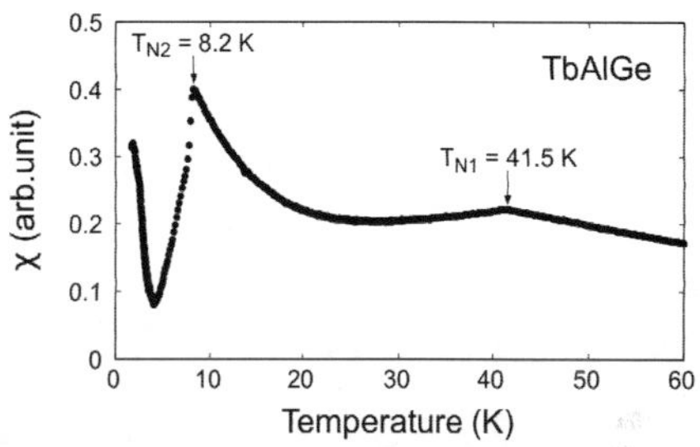

Figure 1. The AC magnetic susceptibility of TbAlGe along the a-axis.

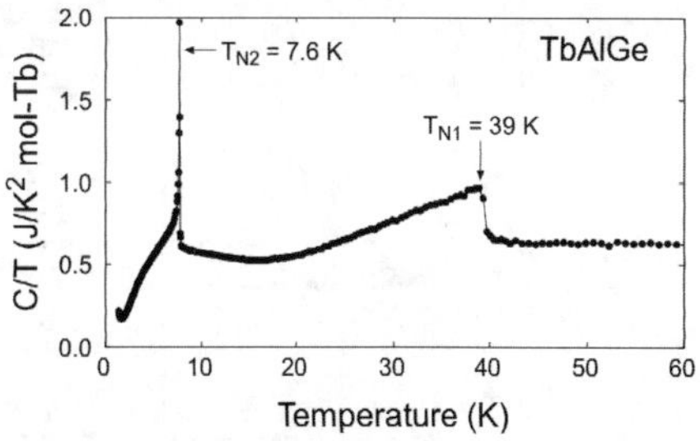

Figure 2. The C/T vs T plot of TbAlGe.

Figure 3 shows the magnetic field (H)-dependence of the magnetization parallel to H direction (M_x) at 4 K along the a- and c-axis. On the a-axis, there are metamagnetic transitions (MT) with large hysteresis. The MT is observed at $H_1 = 1.7$ T in the H-up process, and at $H_2 = 3.5$ T in addition to H_1 in the H down process. At $H > 6.0$ T, the magnetization appears to be saturated, but the magnetic moment is smaller than the theoretical value of Tb (9.0 μ_B). Therefore, there may be another MT at higher fields. The c-axis magnetization curve is linear without hysteresis. Figures 3 (b) and 3 (c) show the magnetization curves when H is applied at an angle formed by the H-direction with the a-axis, divided into H-up process and H-down process, respectively. In Fig. 3 (c), there is a bend near 4 T at 50°, while it disappears at 60°. Comparing the H-up process and the H-down process, the hysteresis disappears from 60°. The MT at $H = 1.7$ T becomes broad in the H-down process.

We have measured the angle dependence of M_x and M_y (the perpendicular magnetization to H-direction) of clockwise (cw) and counterclockwise (ccw) rotation at intervals of 0.5 T up to 6.5 T at 4 K. Figure 4 (a) shows the results at 6.5 T, where black and red points are cw and ccw, respectively. $\theta = 0°$ and 90° correspond to a- and c-axis, respectively. Since the peculiar behavior of cw and ccw is not clear in M_x, we pay attention to M_y. We can see that there is a large hysteresis of angular dependence. It can also be seen that the period of magnetization is 180°. There is also an interesting magnetization behavior at 0° and 360°. These angles are on the same a-axis, but the magnetization is not the same. This is due to the rotation hysteresis, and depends on the start angle. In Fig. 4 (a),There is no peak at 10° that appears at 190°, because the measurement started at 0°. However, when we start the measurement from −20°, we can see this peak at 10°. Figures 4 (b) and (c) show M_x and M_y in cw each 0.5 T from 0° to

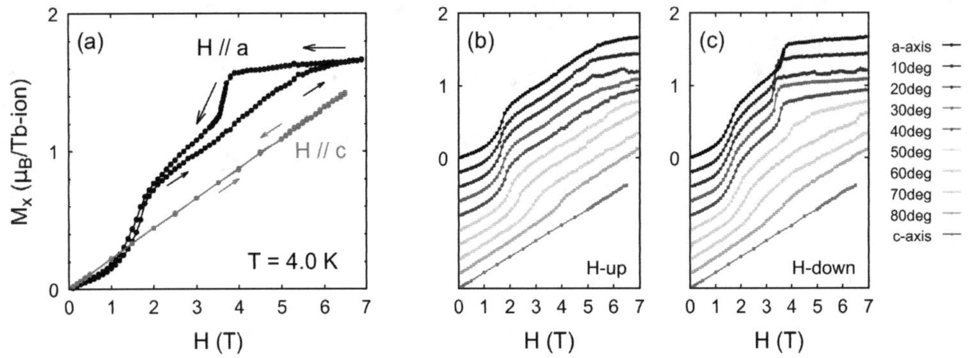

Figure 3. Magnetic field (H) dependence of M_x of TbAlGe at $T = 4$ K (a) H up and H down process of a-axis and c-axis (b) H up process every 10° (c) H down process each 10°

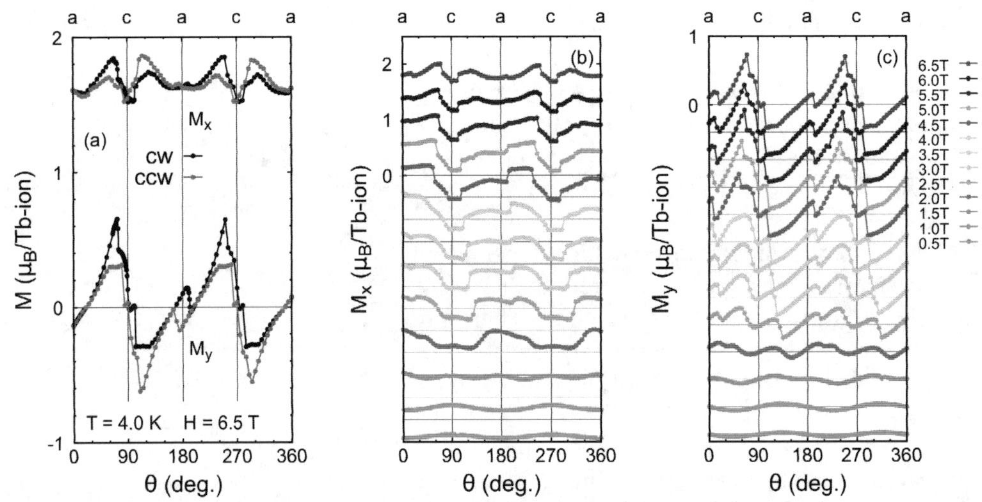

Figure 4. Angular dependence of magnetization for TbAlGe at $T = 4$ K . (a) b-plane, H = 6.5 T. The abbreviation of cw and ccw indicate clockwise and counterclockwise rotation, respectively. Angular dependence of (b) M_x and (c) M_y for several fields. .

360°, the measurements are shown in the range from −20° to 380°. At high fields, it is difficult to see the change in M_x, but if we focus on M_y, we can see the angle at which the M_y changes abruptly (10°, 70°, 100°, 190°, 250°, 280° for cw).

Figure 5 shows a magnetic phase diagram obtained in this study for the angle and field at 4 K for b-plane in polar coordinate, where the angular rotation is the cw rotation. The blue and red circles are the anomalies that represent the obtained from the magnetization curve (Fig. 3) and the angular rotation (Fig. 4), respectively. For convenience of explanation, the phase lines are shown in the alphabet in the figure.

First, we consider the phase diagram from 0° to 90°. Since this corresponds to the H-down process of the magnetization curve, two MT appear on the a-axis (0°). The phase line (A) from H_1 = 2.0 T and the phase line (B) from H_2 = 3.5 T extend along the c-axis. The phase line from A extends in the circumferential direction and disappears at 60°. The anomaly is also seen in angular rotation, where the phase line of the magnetization curve extends further along the c-axis from where it disappears (D), but disappears at 4.0 T. Although the phase line B extends along the circumference and disappears at 40°, the anomaly of angular rotation connects from the phase line of B and extends along the c-axis (E). The phase line of F is found only in angular rotation and appears in a higher magnetic field than the phase line of A.

Next, we consider the phase diagram from 90° to 180°. This rotation corresponds to the H-up process and there is only one MT. The phase line (C) from the MT at $H_1 = 1.7$ T on the a-axis (180°) disappears at 50°, but continues to the anormalies in the angular rotation, and the phase line is along the c-axis (G).

Finally, let us consider the relationship between the magnetic moment $\boldsymbol{\mu}$ as vector and the phases. Since we measured both M_x and M_y, the direction and magnitude of $\boldsymbol{\mu}$ at each point can be evaluated. The $\boldsymbol{\mu}(\boldsymbol{H}, \theta)$ is represented by an arrow, and the starting point is the coordinates $(\boldsymbol{H}, \theta)$. The arrow length is drawn exactly in proportion to the magnitude of $\boldsymbol{\mu}$. The μ increases strongly across the phase line A obtained by the magnetization curve. This reflects the MT. When the phase line of angular rotation is straddled in the angular direction, the $\boldsymbol{\mu}$ direction changes. The phase line of E appears to continue from the position where the phase line disappeared on the magnetization curve. In the result of the magnetization curve, the hysteresis disappears at 60°, but with angular rotation, there is a hysteresis up to 70°. This result is a hysteresis that appears depending on the angle of rotation and is not expected from the hysteresis seen in the magnetization curve. Similarly, Similarly, it is difficult to infer the phase line F independent on fields from the magnetization curve.

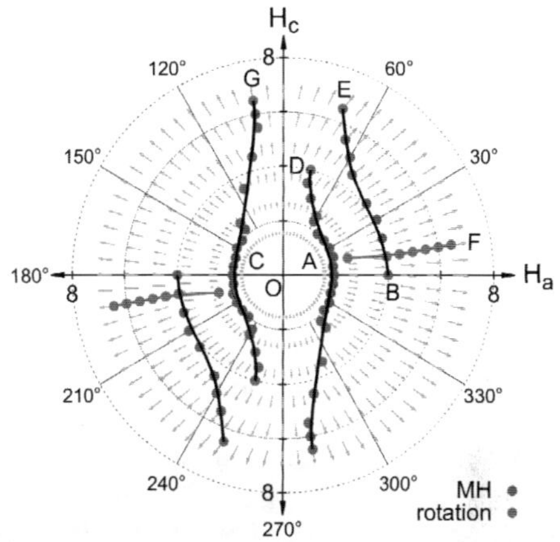

Figure 5. Magnetic phase diagram of the magnetic field and angle of TbAlGe for the b-plane at 4 K in polar coordinates.

4. Summary

We have measured the field and angular dependence of parallel and perpendicular magnetization for b-plane of TbAlGe at 4 K under fields up to 7 T, and obtained a field and angular magnetic phase diagram. As a result, the magnetic anomaly such as MT and hysteresis was observed in the a-axis direction, and the behavior of magnetization changed depending on the direction of rotation, and a hysteresis loop with a period of 180° was observed in M_y. The a-axis, in which MTs are clearly seen, is the direction perpendicular to the zigzag chain, which is very similar to $TbFe_2Al_{10}$, though there is no hysteresis in $TbFe_2Al_{10}$ [?]. Therefore, from an experimental point of view, this strange magnetic phase diagram and the zigzag structure are considered to be related.

Acknowledgments

This work was supported by the Grant-in-Aid for Scientific Research(C) (KAKENHI No. 18K03520).

References

[1] S. Pukas, Y. Lutsyshyn, M. Manyako, E. Gladyshevskii 2004 *J. Alloys and Compd.* **367**, 162.
[2] K. Ohno, F. Tajima, T. Nishioka 2020 *JPS Conf. Proc.,* **30** 011163.
[3] T. Nishioka, K. Moriyama, T. Okazaki, S. Miyamoto, T. Kado 2021 *SCES2020.*
[4] K. Ohno, 2020 *M.A. thesis, Kochi Univ.*

Adiabatic specific heat measurement at very low temperatures using GM refrigerator

Takashi Nishioka[1], **Kazuaki Moriyama**[1], **Takuto Okazaki**[1], **Satoru Miyamoto**[2], **Tsuneo Kado**[2]

[1]Dep. of Phys., Fac. of Sci. and Tech., Kochi Univ., Kochi 780-8520, Japan
[2]Rockgate Co., Tokyo 113-0034, Japan

E-mail: nisioka@kochi-u.ac.jp

Abstract. We have developed an adiabatic specific heat measurementt using a 4 K GM refrigerator system that can measure down to 1.3 K with exactly the same accuracy as liquid helium environmen. The cryogenic environment can be obtained by liquefying about 20 L of helium gas at room temperature into a 1 K pot by a cascade method and evacuating it with a rotary pump. The 1 K pot temperature reaches 1.2 K. The specific heat measurement probe is mounted on the 1 K pot and is covered with three radiation shields, and the helium liquefied in the 1 K pot is maintained for about 70 minutes after the refrigerator is stopped, and the 1 K pot temperature maintains below 1.5 K. In this state, the specific heat from 1.3 K to 10 K can be measured. By tightly fixing the adenda to the support (base), temperature fluctuations due to mechanical vibration of the refrigerator can be greatly suppressed, and above 2.5 K, specific heat measurement with almost the same accuracy as when the refrigerator is stopped can be performed. By combining these two types of measurements, we can measure the specific heat with exactly the same accuracy as in a liquid helium environment.

1. Introduction

Specific heat measurements at very low temperatures down to below 1 K are very important in magnetic studies of such as heavy fermion system[1]. In order to measure the specific heat accurately, it is important that the temperature fluctuation under adiabatic state is small. In liquid helium experiments, the t emperature fluctuation is not a problem at all. However, recently, the amount of helium available to researchers has decreased rapidly, and it is expected that it will be difficult for researchers to experiment with liquid helium in the near future. Therefore, it is necessary to develop a method for doing the cryogenic experiments without wasting helium. One way to obtain the cryogenic temperatures without using liquid helium is to use a commercially available 4 K GM refrigerator.

The 4 K GM refrigerator has the advantage that the temperature can be decreased to about 3 K just by turning on the power without using liquid helium, but it cannot be lowered to a temperature below 3 K, and the temperature oscillation is larger. In the case of the refrigerator we are using, the minimum temperature is 3.7 K, and the temperature oscillation at the temperature is about 300 mK. To resolve these problems, we attached a 4 K plate and a 1 K pot on the 2nd stage of the refrigerator and covered them with a 50 K shield made of Al wrapped with super insulation that is in thermal contact with the 1st stage. As a result, when 23.5 L of helium gas at room temperature is liquefied in the 4 K plate, about 30 cc of liquid helium can be stored in the 1 K pot. By evacuating this with a rotary pump, 1.2 K can be maintained for up to 12 hours[2]. By attaching the ^{3}He pot, it can also maintain 0.4 K for more than a day with 3 L ^{3}He gas[3]. Since the heat capacity of liquid helium is very large, the temperature fluctuation in 1 K pot is less than 0.1 mK, and the measurement of electrical resistivity and AC magnetic susceptibility that can be measured by immersing in the pot is completely the same as the measurement using liquid helium. However, measurements that need to be performed under adiabatic state outside the pot, such as specific heat, are not easy because the effects of mechanical vibration of the refrigerator cannot be avoided.

The purpose of this study is to measure the specific heat using a 4 K GM refrigerator with the same accuracy as the liquid helium environment. The only way to make addenda temperature fluctuations the same as in a liquid helium environment is to stop the refrigerator. We stopped the refrigerator and tried to measure the specific heat.

2. Experimental procedure

Figure 1 shows our design drawing of the specific heat measurement cryostat made this time. The refrigerator we used was RDK-205D manufactured by Sumitomo Heavy Industries, Ltd., whose cooling power at 4.2 K is 0.5 W[4]. The hatch part is the body part of the GM refrigerator. A 4 K plate and a 1 K pot are attached on the 2nd stage of the refrigerator, and a specific heat measurement probe is installed under the 1 K pot. These are covered with 1 K shield, 4 K shield, and 50 K shield. 23.5 L of helium gas at room temperature is liquefied in the 4 K plate and stored in the 1 K pot. When this is evacuated with a rotary pump, the temperature inside the 1 K pot reaches 1.2 K, and the temperature fluctuation becomes about 0.1 mK. The temperature reaches within 3 minutes with only the 50 K shield, but since addenda is in adiabatic state and has a small cooling power, the temperature decreases only to about 12 K for only the 50 K shield. By attaching a 4 K shield, the temperature decreases to the same temperature as the base one, but as we will see later, a 1 K shield is required for specific heat measurement. We call the support part of addenda base and corresponds to the heat bath. Although there is mechanical vibration of the refrigerator, the temperature and temperature fluctuation of the base are almost the same as those of the 1 K pot. Therefore, it is considered that the cause of the temperature fluctuation of addenda is the vibration deviating from the mechanical vibration of the refrigerator[5]. Therefore, as shown in the enlarged view, addenda is tightly fixed to the base, which is hardened by impregnating the medicine wrapping paper with varnish.

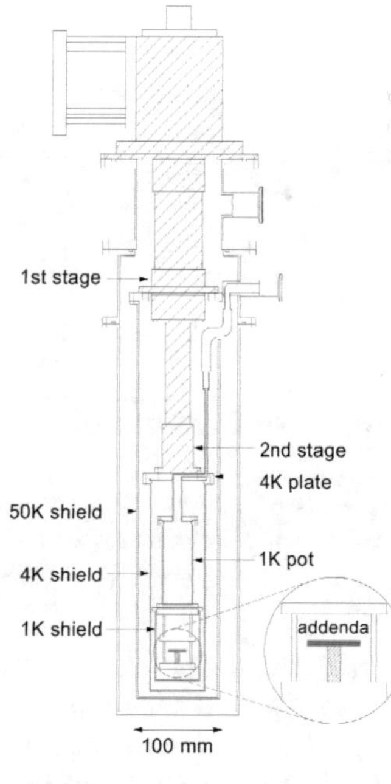

Figure 1. Design drawing of specific heat measurement system. The hatch part is the main body of the GM refrigerator. The support that fixes the addenda is made of medicine wrapping paper soaked with varnish and hardened.

3. Results and Discussion

Figure 2 shows the temperature dependence of addenda, which turned off the GM about 1000 seconds after the 1 K pot was evacuated, with and without the 1 K shield. From Fig. 2 (b), it can be seen that when the GM is on and there is no 1 K shield, the temperature reached by addenda is 1.45 K, and there is a temperature fluctuation of about 50 mK. By attaching the

1 K shield, the temperature reached by the adenda will be reduced by about 100 mK, and the temperature fluctuation will be reduced to about 5 mK. The temperature of the base is 1.2 K regardless of the presence of the 1 K shield, and the temperature fluctuation is about 0.1 mK. In the adiabatic specific heat measurement, the temperature rise due to the heat pulse is usually about 1∼2%, and at 1.4 K it is about 20 mK. In order to measure the specific heat accurately, the temperature fluctuation of addend needs to be sufficiently smaller than this temperature rise, so it can be seen that accurate specific heat measurement at this temperature is difficult even with a 1 K shield. However, by stopping the refrigerator, this fluctuation disappears completely and becomes about 0.1 mK, which enables accurate specific heat measurement. However, since the amount of liquid helium we store in a 1 K pot is only about 30 cc, it is important how much liquid helium has after the refrigerator is stopped. From Fig. 2 (a), this duration is about 70 minutes regardless of the presence or absence of the 1 K shield, which is sufficient for specific heat measurement as described later. Within 10 minutes after the refrigerator is stopped, the temperature is maintained at about 1.3 K regardless of the presence or absence of the 1 K shield, but without the 1 K shield, the base temperature is maintained below 1.5 K. However, the temperature of the adenda increases monotonically up to about 5 K.

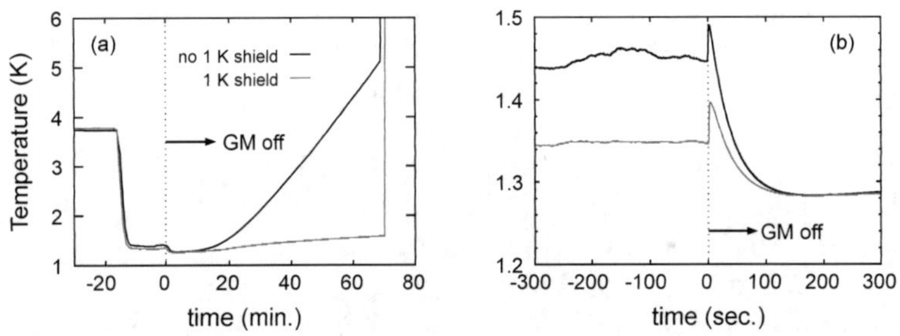

Figure 2. (a) Temperature dependence of addenda when pumping 1 K pot from the minimum temperature of the refrigerator and GM off after about 1000 seconds with and without 1 K shield. The time when GM is off is set to 0. (b) Enlarged view before and after GM off.

Figure 3 shows the effect of GM on/off of the heating curve during measurement near 3.7 K and near the lowest temperature (below 1.5 K). In this experiment, the temperature rise is about 1.5 % of the temperature. As the temperature decreases, the temperature rise becomes smaller and the temperature fluctuation becomes larger. Therefore, the fluctuation with respect to the heating curve becomes more remarkable as the temperature becomes lower. At 3.8 K, the effect of GM on/off of the heating curve is hardly seen, but at GM on below 1.5 K, the effect of fluctuation is quite large, and it is difficult to measure the specific heat accurately. You can see that a clear heating curve can be obtained by stopping the refrigerator.

Figure 4 shows the the specific heat of addenda in the GM on state and the GM off state. The red points are the result fo GM off, and the solid line is the result for GM off. As expected from the heating curve, the lower the temperature, the greater the error of the specific heat. However, above 2.5 K, the effect of GM on/off is hardly seen, as shown in the figure. Therefore, in order to measure the specific heat with GM off, a maximum of about 4 K is sufficient. In the actual experiment, a measurement above 4 K is performed before pumping the 1 K pot, and after that, after the sample temperature decreases sufficiently, the 1 K pot was pumped and the GM was off, and measure the specific heat from 1.3 K to 4 K.

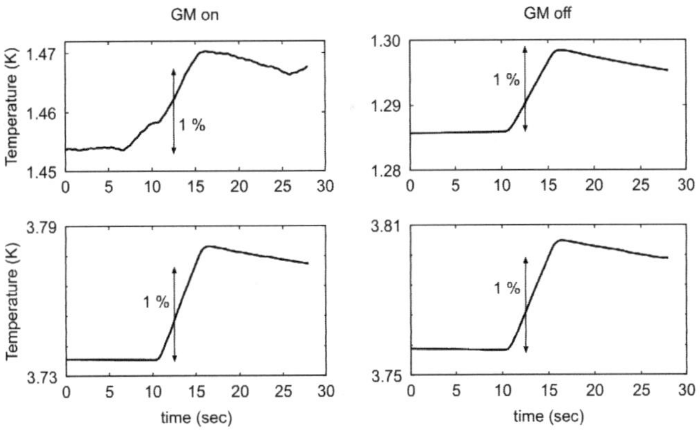

Figure 3. Heating curves around 1.5 K and around 4 K for GM on / off. The size of the double-headed arrow indicates 1% of the temperature.

In order to investigate whether our specific heat measurement system gives the correct specific heat, we measured Ag as a standard sample and compared it with NBS data (not shown). The difference is less than 1%, which shows that our system is reliable.

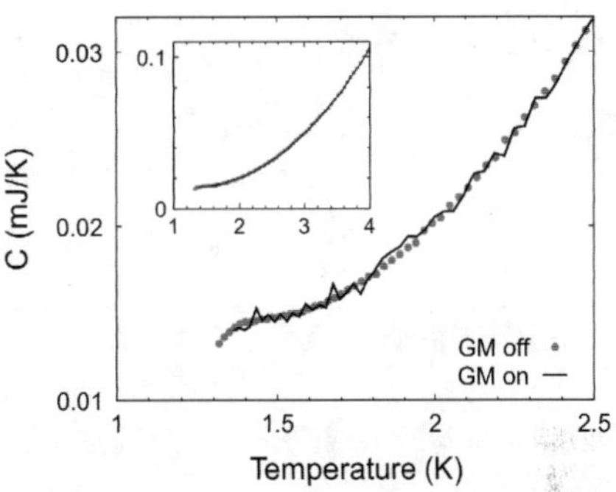

Figure 4. Temperature dependence of heat capacity of addenda: GM on (closed circles), GM off (solid line)

4. Summary

We have succeeded in measuring the adiabatic specific heat using a GM refrigerator with the same accuracy as the environment of liquid helium. To achieve this, it is necessary to tightly fix the addenda to the support and to cover the addenda with three shields, 1 K shield, 4 K shield, and 50 K shield. In this way, specific heat measurements above 2.5 K can be performed with the same accuracy as in a liquid helium environment even in the GM on state. Measurements below 2.5 K can be performed by pumping a 1 K pot and stopping the refrigerator after it has reached a steady state. The duration of liquid helium in a 1 K pot is about 70 min., which is sufficient for measuring specific heat from 1.3 K to 4 K.

Acknowledgments

This work was supported by the Grant-in-Aid for Scientific Research(C) (KAKENHI No. 18K03520).

References

[1] See for example, Hewson 1993 *The Kondo Problem to Heavy Fermions* (Cambridge Studies in Magnetism).
[2] H. Koyama, K. Nakanishi, T. Nishioka 2020 *JPS Conf. Proc.* **30** 011195.
[3] T. Nishioka, T. Sumida, T. Takesaka, Y. Kawamura, H. Kato and M. Matsumura 2009 *J. Phys: Conf. Ser.* **150** 012030.
[4] Sumitomo Heavy Industries, Ltd, 2019 *Cryocooler product catalogue.*
[5] S. Nishioka, T. Nishioka, Y. Yagi, H. Mukuta 2020 *Osaka Univ. Low Temp. News (in Japanese)* **170** 8

AUTHOR INDEX